Crystallo-
graphic
Computing

Crystallographic Computing

Proceedings of an International Summer School organized by

THE COMMISSION ON CRYSTALLOGRAPHIC COMPUTING OF THE INTERNATIONAL UNION OF CRYSTALLOGRAPHY

and held in Ottawa, 4 to 11 August 1969

Edited by
F. R. AHMED

with Co-Editors
S. R. HALL and C. P. HUBER

SCANDINAVIAN·UNIVERSITY·BOOKS

MUNKSGAARD
INTERNATIONAL BOOKSELLERS and PUBLISHERS LTD., COPENHAGEN.

Cover design: C. K. Johnson and P. Breuning

Printed in Denmark
by P. J. Schmidt, Vojens

ISBN 87 16 00179 6

PREFACE

This book contains the proceedings of an *International Summer School on Crystallographic Computing,* organized by the *Commission on Crystallographic Computing* of the *International Union of Crystallography*, and held in *Ottawa* at *Carleton University*, 4–11 August 1969. The school was attended by 170 crystallographers and research students from 18 different countries, including 38 invited lecturers who spoke on 14 topics covering the methods of crystal structure analysis and the techniques of their effective implementation on present-day computers. The topics for discussion were selected to cover the aspects of crystallographic computing which are expected to have wide applications. Thus, some of the topics cover well-established methods which will continue to be employed extensively, while other topics describe fairly new procedures with expected high potential.

The school's Organizing Committee consisted of F. R. Ahmed (Chairman of the Commission), G. Bassi, G. Y. Chao, D. W. J. Cruickshank, S. R. Hall, C. K. Johnson, K. Sasvari, and J. M. Stewart. This committee selected and outlined the topics, allotted the hours of lectures on each, and then invited 11 Topic Chairmen to take charge of arranging the details and inviting the speakers for their topics. The careful work of these Topic Chairmen and the invited lecturers contributed largely to the success of the school.

It is a great pleasure to acknowledge the generous support given to the school by the following sponsors: *Carleton University* (host to the School); *Department of Energy, Mines and Resources of Canada; Digital Equipment of Canada, Ltd.; International Union of Crystallography; National Research Council of Canada;* and *UNESCO*. The book of proceedings has been prepared with the financial assistance of UNESCO as part of the UNESCO Pilot Project on the Teaching of Crystallography in relation to the Physics and Chemistry of Solids.

The early publication of this book and its effective presentation and distribution should be credited partly to the co-operation of the authors, and partly to MUNKSGAARD, International Booksellers and Publishers, Ltd., Copenhagen, who through their faith in the usefulness of the book have undertaken the financial responsibility.

The plans for holding the school were initiated while D. W. J. Cruickshank was Chairman of the Commission on Crystallographic Computing. He contributed a great deal to the school as a member of the Organizing Committee, as Topic Chairman, and as Treasurer of the IUCr. The school owes a great deal to his efforts, and to the efforts of G. Boom, the IUCr General Secretary, for valuable administrative advice. Many people, particularly L. D. Calvert, E. J. Gabe, M. E. Pippy, and five NRC postdoctoral fellows assisted with the local arrangements and contributed greatly to the smooth functioning of the school.

Ottawa, Canada
September, 1969

F. R. Ahmed
S. R. Hall
C. P. Huber

CONTENTS

TOPIC A - Chairman: J. KARLE

Direct Methods of Phase Determination

TOPIC B - Chairman: C.E. NORDMAN

Automatic Interpretation of Vector Maps

TOPIC C - Chairman: G. KARTHA

Utilisation of Anamalous Dispersion and Isomorphous Replacement for Elucidation of Large Molecules

TOPIC D - Chairman: J. S. ROLLETT

Least-Squares Procedures in Crystal Structure Analysis

TOPIC E - Chairman: D. W. J. CRUICKSHANK

Least-Squares Refinement of Atomic Parameters

TOPIC F - Chairman: C. K. JOHNSON

Analysis of Atomic Thermal Vibration

TOPIC G - Chairman: P. COPPENS

Absorbtion and Extinction Corrections

TOPIC H - Chairman: J.M. STEWART

Organization of Crystallographic Computer Systems

TOPIC I - Chairman: W.R. BUSING

Programs for Computer-Controlled Diffractometers

TOPIC J - Chairman: G. KARTHA

Protein Programs

TOPIC K - Chairman: F.R. AHMED

Fourier Techniques

TOPIC L - Chairman: J.M. STEWART

Data Files

TOPIC M - Chairman: G. C. BASSI

Algol

TOPIC N - Chairman: J.M. STEWART

Standard Tests

TOPIC A

DIRECT METHODS OF PHASE DETERMINATION

CHAIRMAN: J. KARLE

Selection of Origin Defining Reflections

J. KARLE

U.S. Naval Research Laboratory, Washington, D.C., U.S.A.

The information required for specifying the origin for the various space groups by assigning the values of appropriate phases is summarized in Tables I–IV (Hauptman & Karle, 1953, 1956, 1959; Karle & Hauptman, 1961). The formal aspects of this subject will first be briefly discussed and then the practical aspects of relating this formalism to phase determination by the symbolic addition procedure will be illustrated by several examples. Tables II and IV concerning space groups with conventionally centered cells are based on transformations to coordinates which represent the space group relative to a primitive unit cell. The coordinates are obtained from those corresponding to a non-primitive unit cell (*International Tables*, 1952) by means of the following matrices:

$$C \to P, \quad \begin{pmatrix} 1 & 1 & 0 \\ 1 & -1 & 0 \\ 0 & 0 & -1 \end{pmatrix}$$

$$A \to P, \quad \begin{pmatrix} -1 & 0 & 0 \\ 0 & -1 & 1 \\ 0 & 1 & 1 \end{pmatrix}$$

$$I \to P, \quad \begin{pmatrix} 0 & 1 & 1 \\ 1 & 0 & 1 \\ 1 & 1 & 0 \end{pmatrix}$$

$$F \to P, \quad \begin{pmatrix} -1 & 1 & 1 \\ 1 & -1 & 1 \\ 1 & 1 & -1 \end{pmatrix}$$

I. Definitions and procedures for centrosymmetric space groups

Seminvariants are linear combinations of phases whose values are determined by the observed intensities, once a functional form for the structure factor has been chosen. For the centrosymmetric space groups, the permissible origins are the eight centers of symmetry in the primitive unit cell. Often, these eight centers are not all equivalent, since different centers may be situated differently with respect to the symmetry elements. Non-equivalent origins are associated with different functional forms for the structure factors. Therefore a choice of the functional form, when non-equivalent origins exist, limits the number of permitted origins. There may be one, two, four or eight classes of equivalent origins. It is then necessary to assign the values of three, two, one, or zero appropriate phases, respectively, in order to specify the origin.

The seminvariant vector and seminvariant modulus characterize the nature of the subscripts of the linear combinations of phases forming the seminvariants. The components of the seminvariant vector for the centered cells may be obtained from those for the corresponding primitive cell by performing the required transformations utilizing the above matrices. The components of the vector indicate which combinations of subscripts need to be considered in each case and the corresponding moduli describe the mathematical relationship which must be fulfilled by the subscripts. This relationship states that the given combination of subscripts, h_i, must be divisible by the modulus, ω_i or $h_i \equiv 0 \pmod{\omega_i}$. Individual phases may be seminvariants and the characteristics of their subscripts are readily derivable from knowledge of the seminvariant vector and seminvariant modulus. Seminvariant phases are listed in Tables I–IV.

If we consider, for example, type $3P_2$, the seminvariant vector components h_i are all the sums, $h+k+l$, which are divisible by 2. As indicated in Table I, the sum of the subscripts for each of the seminvariant phases in type $3P_2$ is even (g and u denote even and odd, respectively). Seminvariants may consist of the sum of more than one phase. Although any individual member of the four parity groups not listed for type $3P_2$ can not be a seminvariant phase, the sum of any two of them can be. This is confirmed by summing the corresponding parities of the two phases in question and noting that the resultant parity is one of the four listed in Table II. If we denote the seminvariant by $\varphi_{\mathbf{h}_1}+\varphi_{\mathbf{h}_2}$, the condition which is being fulfilled is that $h_1+k_1+l_1+h_2+k_2+l_2 \equiv 0 \pmod 2$.

In centrosymmetric crystals, insight into origin specification is readily facilitated by the fact that the X-ray intensities determine the value of the seminvariants and that a seminvariant can be constructed from the sum of two properly chosen phases. Consider the case when $\varphi_{\mathbf{h}_1}+\varphi_{\mathbf{h}_2}$ form a seminvariant, although neither phase is a seminvariant phase. Given the value of one of the phases, the value of the second is obtainable from the value of the seminvariant as computed from the X-ray intensities. In the example of type $3P_2$, the value of one phase which is not a seminvariant may evidently be specified. In principle all remaining phases are then determined by forming appropriate seminvariants based on this initial specification. In practice the choice of reflections for which the phases are specified is based not only on the necessary parity criteria but also on additional considerations such as being associated with the larger intensities.

Table I. Seminvariant vector, modulus, and phases for the conventionally primitive centrosymmetric space groups. (*International Tables for X-ray Crystallography*)

Type	2P	2P₁	3P₂	3P₃	4P
Space Groups	C2/m C2/c Cmcm Cmca Cmmm Cccm Cmma Ccca	Immm Ibam Ibca Imma	Fmmm Fddd Fm3 Fd3 Fm3m Fm3c Fd3m Fd3c	I4/m $I4_1/a$ I4/mmm I4/mcm $I4_1/amd$ $I4_1/acd$	Im3 Ia3 Im3m Ia3d
Vector $\vec{h}$ seminvariantly associated with $\varphi_{\vec{h}}$, referred to primitive cell	(h + k, ℓ)	(h + k, k + ℓ, ℓ + h)	(h + k + ℓ)	(h + k)	(h, k, ℓ)
Seminvariant modulus	(2, 2).	(2, 2, 2)	(2)	(2)	(1, 1, 1)
Seminvariant phases	φ_{ggg} φ_{uug}	φ_{ggg} φ_{uuu}	φ_{ggg} φ_{uug} φ_{ugu} φ_{guu}	φ_{ggg} φ_{ggu} φ_{uug} φ_{uuu}	any $\varphi_{hk\ell}$
For fixed form of structure factor, number of phases linearly semi-independent to be specified	2	2	1	1	0

Table II. Seminvariant vector, modulus, and phases for the conventionally centered centrosymmetric space groups, referred to a primitive unit cell. (*International Tables for X-ray Crystallography*)

Type	1P		2P		3P₁	3P₂
Space Groups	P$\bar{1}$ P2/m $P2_1/m$ P2/c $P2_1/c$ Pmmm Pnnn Pccm Pban Pmma Pnna	Pmna Pcca Pbam Pccn Pbcm Pnnm Pmmn Pbcn Pbca Pnma	P4/m $P4_2/m$ P4/n $P4_2/n$ P4/mmm P4/mcc P4/nbm P4/nnc P4/mbm P4/mnc	P4/nmm P4/ncc $P4_2/mmc$ $P4_2/mcm$ $P4_2/nbc$ $P4_2/nnm$ $P4_2/mbc$ $P4_2/mnm$ $P4_2/nmc$ $P4_2/ncm$	P$\bar{3}$ P$\bar{3}$1m P$\bar{3}$1c P$\bar{3}$m1 P$\bar{3}$c1 P6/m $P6_3/m$ P6/mmm P6/mcc $P6_3/mcm$ $P6_3/mmc$	R$\bar{3}$ R$\bar{3}$m R$\bar{3}$c Pm3 Pn3 Pa3 Pm3m Pn3n Pm3n Pn3m
Vector $\vec{h}$ seminvariantly associated with $\varphi_{\vec{h}}$	(h, k, ℓ)		(h + k, ℓ)		(ℓ)	(h + k + ℓ)
Seminvariant modulus	(2, 2, 2)		(2, 2)		(2)	(2)
Seminvariant phases	φ_{ggg}		φ_{ggg} φ_{uug}		φ_{ggg} φ_{ugg} φ_{gug} φ_{uug}	φ_{ggg} φ_{uug} φ_{ugu} φ_{guu}
For fixed form of structure factor, number of phases linearly semi-independent to be specified	3		2		1	1

Origin specification then consists of assigning the values of an independent set of phases so that the values of the remaining phases can be obtained from the values of the seminvariants defined by the X-ray intensities. In order to identify the independent set, it is only necessary to know what the seminvariant vector and seminvariant modulus are and exclude seminvariants from this set, recalling that the values of seminvariants are origin independent, being determined only by the intensities and the functional form for the structure factor. The seminvariant vectors and moduli are given for centrosymmetric crystals in Tables I and II.

II. Definitions and procedures for noncentrosymmetric space groups

The specification of the origin in noncentrosymmetric space groups proceeds in a similar fashion to that for centrosymmetric space groups. However, it is necessary to introduce some additional concepts in order to adequately define the procedures.

In addition to assuming that a sufficient number of X-ray intensities have been measured in order to determine the magnitudes of the seminvariants, it is also assumed that the sign of one such magnitude has been specified. Evidently the magnitude chosen should differ in value from 0 or π. The arbitrary assignment of the sign to the magnitude of a seminvariant distinguishes the two enantiomorphous structures when they are distinct, or the permissible reference frames when they are distinct, or both. For the four enantiomorphous pairs of space groups ($P4_122$, $P4_322$), ($P4_12_12$, $P4_32_12$) ($P3_1$, $P3_2$), and ($P3_112$, $P3_212$), the values of the seminvariants are determined so that no specification of sign may be made. The signs of all the seminvariants are also uniquely determined for space group $P(I4_122)$, *i.e.*, space group $I4_122$ expressed in terms of a primitive unit cell. Values of the phases are considered to be in the interval $-\pi < \varphi < \pi$.

As in the case of centrosymmetric crystals, origin specification for noncentrosymmetric crystals results in the possibility of obtaining the values for all the phases from the values of the seminvariants. In order to attain this, it is necessary to put certain restrictions on the values of the subscripts of the independent phases whose values are specified. These restrictions have been described in terms of the concept of primitivity which requires that certain functions of the subscripts, depending upon the particular space group, be equal to ± 1. The permitted values for the independent phases are listed in Tables III and IV. When the values must differ by $\pi/2$ or π, it is generally possible to choose indices for which the phases may have the convenient values of 0 or $\pi/2$. The appropriate indices can be determined by reference to the *International Tables* Vol. I and Karle & Hauptman (1961). The relation $h_i \equiv 0 \pmod 0$ implies that $h_i = 0$.

III. Practical aspects

The point of departure between the foregoing formalisms and the symbolic addition procedure is that the assumption has been made that the values of the seminvariants or, in the case of most noncentrosymmetric crystals, their magnitudes have been accurately calculated from the measured X-ray intensities. In applying the main formulas of the symbolic addition procedure, it is assumed, as the probability formulas indicate, that the values of the particular seminvariants $\varphi_{\mathbf{h}} - \varphi_{\mathbf{k}} - \varphi_{\mathbf{h}-\mathbf{k}}$ associated with the larger $|E|$'s are close to zero. The success of the method depends in part on the existence of the means for recognizing and taking into account those exceptional seminvariants which have values significantly different from zero. Some of the most evident manifestations are conflicting phase indications and on occasion the necessity for assigning a new symbol in order to proceed with the phase determination. The question of significance here concerns the extent to which the specification of origin, enantiomorph and/or reference frame is affected by primarily utilizing seminvariants in a phase determining procedure whose values are assumed to be probably near zero. So far as specifying the origin is concerned, there is no effect for centrosymmetric crystals and only a small one for noncentrosymmetric ones which may even be regarded as an advantage. It derives from the fact that the primitivity concept, involving the restriction that certain functions of the subscripts be equal to ± 1, can be ignored in the assignment of phases. The primitivity concept in the seminvariant theory assures that all phase values will be determinate from the initial specifications and the known magnitudes of the seminvariants. In the symbolic addition procedure, by contrast, any phases associated with the larger $|E|$'s which are not accessible from the initial specifications as the phase determination proceeds can be made so by the assignment of additional unknown symbols. It is also a fairly simple matter to specify the enantiomorph and/or reference frame. Once an origin is specified, this is accomplished in practice by assigning the sign of the imaginary part of a phase whose magnitude is known to be near $\pi/2$. In a successful phase determination in space group $P2_1$, for example, at least one of the symbols should have a magnitude significantly different from 0 or π. It is readily recognizable at times because there may be some indications that its value is zero and other indications are that its value is π, leading to an indeterminacy. Actually it has a value in between the two. In the investigation of the structure of panamine (Karle & Karle, 1966) which crystallized in space group $P2_1$, the assignment of the enantiomorph was very simple since there was only one unknown symbol. The phase represented by this symbol had to have a significant imaginary part for the phase

Table III. Seminvariant vector, modulus, and phases for the conventionally primitive noncentrosymmetric space groups. Possible values for the independent phases are given, where ||2|| means two possible values differing by π, ||3|| means three possible values differing by $2\pi/3$, ||4|| means four possible values differing by $\pi/2$, and ||∞|| means arbitrary values in the range $-\pi < \varphi \leq \pi$. (*International Tables for X-ray Crystallography*)

Type	1P000	1P202	1P020	1P222	1P220	2P20	2P22	3P30	3P32	$3P_10$	$3P_12$	$3P_20$	$3P_22$
Space Groups	P1	P2 $P2_1$	Pm Pc	P222 $P222_1$ $P2_12_12$ $P2_12_12_1$	Pmm2 $Pmc2_1$ Pcc2 Pma2 $Pca2_1$ Pnc2 $Pmn2_1$ Pba2 $Pna2_1$ Pnn2	P4 $P4_1$ $P4_2$ $P4_3$ P4mm P4bm $P4_2cm$ $P4_2nm$ P4cc P4nc $P4_2mc$ $P4_2bc$	$P\bar{4}$ P422 $P42_12$ $P4_122$ $P4_12_12$ $P4_222$ $P4_22_12$ $P4_322$ $P4_32_12$ $P\bar{4}2m$ $P\bar{4}2c$ $P\bar{4}2_1m$ $P\bar{4}2_1c$ $P\bar{4}m2$ $P\bar{4}c2$ $P\bar{4}b2$ $P\bar{4}n2$	P3 $P3_1$ $P3_2$ P3m1 P3c1	P312 $P3_112$ $P3_212$ $P\bar{6}$ $P\bar{6}m2$ $P\bar{6}c2$	P31m P31c P6 $P6_1$ $P6_5$ $P6_4$ $P6_3$ P6mm P6cc $P6_3cm$ $P6_3mc$	P321 $P3_121$ $P3_221$ P622 $P6_122$ $P6_522$ $P6_222$ $P6_422$ $P6_322$ $P\bar{6}2m$ $P\bar{6}2c$	R3 R3m R3c	R32 P23 $P2_13$ P432 $P4_232$ $P4_332$ $P4_132$ $P\bar{4}3m$ $P\bar{4}3n$
Vector $\vec{h}$ seminvariantly associated with $\varphi_{\vec{h}}$	(h, k, ℓ)					(h + k, ℓ)		(h - k, ℓ)		(ℓ)		(h + k + ℓ)	
Seminvariant modulus	(0, 0, 0)	(2, 0, 2)	(0, 2, 0)	(2, 2, 2)	(2, 2, 0)	(2, 0)	(2, 2)	(3, 0)	(3, 2)	(0)	(2)	(0)	(2)
Seminvariant phases	φ_{000}	φ_{g0g}	φ_{0g0}	φ_{ggg}	φ_{gg0}	φ_{gg0} φ_{uu0}	φ_{ggg} φ_{uug}	φ_{hk0}, $h + k \equiv 0$ (mod 3)	φ_{hkg} $h + k \equiv 0$ (mod 3)	φ_{hk0}	φ_{hkg}	$\varphi_{h,k,\bar{h}+\bar{k}}$	φ_{ggg} φ_{uug} φ_{ugu} φ_{uug}
Permitted values for independent phases	‖∞‖,	‖∞‖, except ‖2‖ if k = 0	‖∞‖, except ‖2‖ if h = ℓ = 0	‖2‖	‖∞‖, except ‖2‖ if ℓ = 0	‖∞‖, except ‖2‖ if ℓ = 0	‖2‖	‖∞‖, except ‖3‖ if ℓ = 0	‖2‖ if $h \equiv k$ (mod 3) ‖3‖ if $\ell \equiv 0$ (mod 2)	‖∞‖	‖2‖	‖∞‖	‖2‖
For fixed form of structure factor, number of phases linearly semi-independent to be specified	3					2				1			

determination to lead to a sensible structure.

A photolysis product of a mescaline derivative (Karle & Karle, 1969) crystallized in the space group *Aba*2. In the present context, this space group is of interest not only because it is non-primitive but also because it is an example where the enantiomorphs are not distinct. In this case, in addition to specifying the origin, an additional specification is required in order to fix the reference frame. As seen in Table IV, the various quantities listed under space group *Aba*2 imply that an origin can be specified in the primitive cell by assigning two phases whose subscripts could be, for example, $uk\bar{k}$ and $gk(\bar{k} \pm u)$. It can be verified from the transformation matrices that these subscripts transform to ones of the general type $ug0$ and guu, respectively. The type $ug0$ can be associated only with phase values of 0 or π while type guu can be associated with arbitrary phase values. In the study of the mescaline photoproduct, it was convenient to assign a phase of 0 to a $ug0$ and to a $gu1$. Among the additional assignments involving symbols were two which were assigned to general phases. Either or both had to have significant imaginary parts, or the phase determination would have led to all real phases. The specification of the reference frame was accomplished by assigning the sign of the imaginary part of one of the two symbols. This is quite comparable to the manner of specification of the enantiomorph in space group $P2_1$, as described above.

References

Hauptman, H. & Karle, J. (1953). *Solution of the Phase Problem.* I. *The Centrosymmetric Crystal.* A.C.A. Monograph No. **3**. Pittsburgh: Polycrystal Book Service.

Hauptman, H. & Karle, J. (1956). *Acta Cryst.* **9**, 45.

Hauptman, H. & Karle, J. (1959). *Acta Cryst.* **12**, 93.

Karle, J. & Hauptman, H. (1961). *Acta Cryst.* **14**, 217.

Table IV. Seminvariant vector, modulus, and phases for the conventionally centered noncentrosymmetric space groups, referred to a primitive unit cell. Possible values for the independent phases are given, where the symbols are defined in Table III. (*International Tables for X-ray Crystallography*)

Type	$2P_002$	2P00	2P20	2P22	$2P_120$	$2P_1022$	$2P_1222$	$3P_22$	$3P_24$	$3P_30$	$3P_32$	$3P_34$	$3P_40$	4P111
Space Groups	C2	Cm Cc	Cmm2 $Cmc2_1$ Ccc2	C222 $C222_1$	Amm2 Abm2 Ama2 Aba2	Imm2 Iba2 Ima2	I222 $I2_12_12_1$	F432 $F4_132$	F222 F23 $F\bar{4}3m$ $F\bar{4}3c$	I4 $I4_1$ I4mm I4cm $I4_1md$ $I4_1cd$	I422 $I4_122$	$I\bar{4}$ $I\bar{4}m2$ $I\bar{4}c2$ $I\bar{4}2m$ $I\bar{4}2d$	Fmm2 Fdd2	I23 $I2_13$ I432 $I4_132$ $I\bar{4}3m$ $I\bar{4}3d$
Vector $\vec{h}$ seminvariantly associated with $\varphi_{\vec{h}}$, referred to primitive cell	$(h - k, \ell)$	$(h + k, \ell)$			$(h, k + \ell)$	$(h + k, k + \ell, \ell + h)$		$(h + k + \ell)$		$(h + k)$		$(h - k + 2\ell)$	$(h + k - \ell)$	(h, k, ℓ)
Seminvariant modulus	(0, 2)	(0, 0)	(2, 0)	(2, 2)	(2, 0)	(0, 2, 2)	(2, 2, 2)	(2)	(4)	(0)	(2)	(4)	(0)	(1, 1, 1)
Seminvariant phases	φ_{hhg}	$\varphi_{h\bar{h}0}$	φ_{gg0}, φ_{uu0}	φ_{ggg}, φ_{uug}	$\varphi_{gk\bar{k}}$	$\varphi_{h\bar{h}\ell}$, $h + \ell \equiv 0$ (mod 2)	φ_{ggg}, φ_{uuu}	φ_{ggg}, φ_{uug}, φ_{ugu}, φ_{guu}	$\varphi_{hk\ell}$, $h + k + \ell \equiv 0$ (mod 4)	$\varphi_{h\bar{h}\ell}$	$\varphi_{gg\ell}$, $\varphi_{uu\ell}$	$\varphi_{hk\ell}$, $h - k + 2\ell \equiv 0$ (mod 4)	$\varphi_{h,k,h+k}$	any $\varphi_{hk\ell}$
Permitted values for independent phases	$\Vert\infty\Vert$, except $\Vert 2\Vert$ if $h - k = 0$	$\Vert\infty\Vert$	$\Vert\infty\Vert$, except $\Vert 2\Vert$ if $\ell = 0$	$\Vert 2\Vert$	$\Vert\infty\Vert$, except $\Vert 2\Vert$ if $k + \ell = 0$	$\Vert\infty\Vert$, except $\Vert 2\Vert$ if $h + k = 0$	$\Vert 2\Vert$	$\Vert 2\Vert$	$\Vert 2\Vert$ if $h + k + \ell \equiv 0$ (mod 2) and $\Vert 4\Vert$ if $h + k + \ell \equiv 1$ (mod 2)	$\Vert\infty\Vert$	$\Vert 2\Vert$	$\Vert 2\Vert$ if $h - k + 2\ell \equiv 0$ (mod 2) and $\Vert 4\Vert$ if $h - k + 2\ell \equiv 1$ (mod 2)	$\Vert\infty\Vert$	every phase is a seminvariant
For fixed form of structure factor, number of phases linearly semi-independent to be specified	2							1						0

Karle, I. L. & Karle, J. (1966). *Acta Cryst.* **21**, 860.
Karle, I. L. & Karle, J. *Acta Cryst.* In press.

Discussion

SOLTZBERG: All other factors being equal, is there any reason for choosing three-dimensional over two-dimensional reflections in origin specification? I have noted in earlier work that there was some reluctance to using 2-D reflections.

J. KARLE: No, we use 2-D reflections for origin determining purposes when we possibly can. The only constraints are the magnitudes of the E-values involved and the extent to which these reflections interact with other reflections as the phasing proceeds. The point to be careful about in using 2-D reflections is that these phases are used directly to provide the phases of 3-D reflections and that immediate indications for 2-D reflections are avoided.

SOLTZBERG: Does this mean that there is not very much to be gained by carefully selecting the origin defining sets but rather simply choosing ones which obey the parity rules?

J. KARLE: Not quite; one must be very careful about the conditions I have just described. These will enable a number of alternate origin defining sets to be selected.

HARDGROVE: The tables described in your talk use primitive cells in place of multiple (centred) cells. Do the indices in your tables refer to the primitive cells?

J. KARLE: All these tables involve the primitive cell but you may use the inverse of the matrices supplied to obtain the subscripts corresponding to the non-primitive cell and do your determination this way.

DE VRIES: Referring to your example where $b=2c$, you made the choice that $c=\pm\pi/2$. Would it have been equally reasonable to choose $b=\pm\pi/2$?

J. KARLE: This choice probably would have been the next one if the first one had not worked satisfactorily.

DEWAR: I refer to your statement that reflection phases were not chosen as zero since this would lead to a situation where the entire phase set was real. Presumably one could be unlucky and choose phases with just this property. Have you ever run into this problem?

J. KARLE: No, if your phase determination is progressing properly it should force you to assign a symbol in such cases. This is best exemplified in the centrosymmetric space group $P\bar{1}$ where just origin specification cannot lead to a negative sign. However, a proper phase determination, except where there is a heavy atom at the origin, will force you to choose at least one symbol that is a minus. Another feature of this procedure is if you specify the origin correctly, but only partially, you will not be able to proceed through the phase determination.

General Procedure for Phase Determination

ISABELLA L. KARLE

Laboratory for the Structure of Matter, U.S. Naval Research Laboratory, Washington, D.C., U.S.A.

The determination of the crystal structure of an unknown material using only the magnitudes of the X-ray reflections and no other knowledge, except, perhaps, the empirical formula, has become quite feasible. The phases of only a relatively few properly chosen reflections are required to be known in order to proceed with the phase determination by the symbolic addition procedure. In general, there are many sets of phases which are essentially internally consistent with respect to the phase determining relations. The introduction of probability measures facilitates choosing the physically correct set.

The formulas which have proved to be most useful in determining phases are for centrosymmetric space groups:

$$E_{\mathbf{h}} \approx \sum_{\mathbf{k}} E_{\mathbf{k}} E_{\mathbf{h}-\mathbf{k}} \tag{1}$$

(Karle & Hauptman, 1950; Hauptman & Karle, 1953) and the associated probability expression

$$P_{+}(E_{\mathbf{h}}) = \tfrac{1}{2} + \tfrac{1}{2} \tanh \frac{\sigma_3 |E_{\mathbf{h}}| \sum_{\mathbf{k}} E_{\mathbf{k}} E_{\mathbf{h}-\mathbf{k}}}{\sigma_2^{3/2}} \tag{2}$$

(Woolfson, 1954; Hauptman & Karle, 1953) where

$$\sigma_n = \sum_{j=1}^{N} Z_j^n;$$

and for noncentrosymmetric space groups:

$$\varphi_{\mathbf{h}} \approx \langle \varphi_{\mathbf{k}} + \varphi_{\mathbf{h}-\mathbf{k}} \rangle_{\mathbf{k}_r} \tag{3}$$

(Karle & Hauptman, 1950; Karle & Karle, 1964a, 1966), and

$$\tan \varphi_{\mathbf{h}} = \frac{\sum_{\mathbf{k}} |E_{\mathbf{k}} E_{\mathbf{h}-\mathbf{k}}| \sin(\varphi_{\mathbf{k}} + \varphi_{\mathbf{h}-\mathbf{k}})}{\sum_{\mathbf{k}} |E_{\mathbf{k}} E_{\mathbf{h}-\mathbf{k}}| \cos(\varphi_{\mathbf{k}} + \varphi_{\mathbf{h}-\mathbf{k}})} \tag{4}$$

(Karle & Hauptman, 1956). The variance associated with determination of $\varphi_{\mathbf{h}}$ can be derived from a probability formula of Cochran (1955) and is given by

$$V = \frac{\pi^2}{3} + [I_0(\alpha)]^{-1} \sum_{n=1}^{\infty} \frac{I_{2n}(\alpha)}{n^2} - 4[I_0(\alpha)]^{-1} \sum_{n=0}^{\infty} \frac{I_{2n+1}(\alpha)}{(2n+1)^2} \tag{5}$$

where

$$\alpha = \{[\sum_{\mathbf{k}_r} 2\sigma_3\sigma_2^{-3/2} |E_{\mathbf{h}} E_{\mathbf{k}} E_{\mathbf{h}-\mathbf{k}}| \times \cos(\varphi_{\mathbf{k}} + \varphi_{\mathbf{h}-\mathbf{k}})]^2 + [\sum_{\mathbf{k}_r} 2\sigma_3\sigma_2^{-3/2} \times |E_{\mathbf{h}} E_{\mathbf{k}} E_{\mathbf{h}-\mathbf{k}}| \sin(\varphi_{\mathbf{k}} + \varphi_{\mathbf{h}-\mathbf{k}})]^2\}^{1/2} \tag{6}$$

(Karle & Karle, 1966). The subscripts **h** and **k** are Miller indices and the normalized structure factors $E_{\mathbf{h}}$ are defined by

$$E_{\mathbf{h}}^2 = F_{\mathbf{h}}^2 / \varepsilon \sum_{j=1}^{N} f_{j\mathbf{h}}^2 \tag{7}$$

(Hauptman & Karle, 1953) where f_j is the scattering factor for the jth atom in a unit cell containing N atoms and ε is a small integer which is space group dependent. It is

$$\varepsilon = \frac{m_2^0 + m_0^2}{n}$$

where n is the symmetry number of the

space group,

$$m_2^0 = \iiint \xi^2(x,y,z;\mathbf{h})\,dx\,dy\,dz$$

and

$$m_0^2 = \iiint \eta^2(x,y,z;\mathbf{h})\,dx\,dy\,dz$$

where ξ and η are the trigonometric functions for the real and imaginary parts of the structure factor as defined in the *International Tables for X-ray Crystallography*. For reflections which are pure real $m_0^2=0$, for reflections which are pure imaginary $m_2^0=0$, and for general reflections $m_2^0=m_0^2$. It is interesting to note that both Eqs. (1) and (3) are contained in the first non-trivial inequality (34) in the complete set of inequalities derived from the non-negativity principle (Karle & Hauptman, 1950).

As much scattering data as possible are desirable in applying these phase determining formulas. For a successful determination, in general, the minimum number of data should be more than that contained in one-half the copper sphere.

Noncentrosymmetric crystals will be emphasized in the description of the application of these formulas since they present, in general, considerably more difficulty than centrosymmetric structures. No computer programs will be discussed since detailed information about computer programs and the logical procedures and criteria used in them will be presented by others. The structure determination of two crystals in space group $P2_12_12_1$ will be described in detail.

The first structure determination, that of 3,4-dihydroxy-L-proline (Karle, 1969), is an example that proceeds with textbook simplicity. The second example, that of an unknown product from a photo-rearrangement reaction (Karle, Gibson & Karle, 1969), presents some problems.

Table I. Reflection data with $|E| > 1{\cdot}5$ for $C_5H_9NO_4$

g	*g*	*g*	$\|E\|$	*u*	*g*	*g*	$\|E\|$	*g*	*u*	*g*	$\|E\|$	*u*	*u*	*g*	$\|E\|$
0	4	10	2·69	3	0	2	2·52	6	5	0	2·24	7	3	0	2·19
2	0	8	2·34	1	0	8	1·98	6	3	0	2·01	1	1	8	2·03
0	0	10	2·32	5	0	2	1·97	2	1	10	1·88				
4	2	2	2·18	3	2	4	1·91	8	1	2	1·73				
6	4	4	1·92	3	4	8	1·82	4	3	0	1·71				
8	0	0	1·74	9	2	4	1·78	2	3	2	1·52				
4	0	6	1·71	5	2	0	1·74	2	5	4	1·52				
6	2	4	1·68	1	2	4	1·73								
2	4	2	1·61	1	0	6	1·72								
2	2	2	1·61	3	6	0	1·67								
0	2	6	1·59	5	4	8	1·54								
2	6	2	1·50												

g	*g*	*u*	$\|E\|$	*u*	*g*	*u*	$\|E\|$	*g*	*u*	*u*	$\|E\|$	*u*	*u*	*u*	$\|E\|$
0	2	3	1·75	5	4	3	2·61	4	3	5	2·27	1	3	7	2·42
				3	6	5	2·25	0	5	3	2·12	3	3	1	2·16
				5	0	9	2·10	6	1	3	2·06	9	1	5	2·03
				7	0	7	1·64	2	1	1	2·01	1	3	5	1·93
				1	0	7	1·63	4	5	5	2·00	7	1	3	1·79
				9	0	1	1·59	4	5	3	1·72	3	1	5	1·70
				3	2	5	1·57	2	3	1	1·65	3	3	7	1·65
				5	6	3	1·56	8	1	1	1·64	1	1	5	1·60
				5	0	7	1·53	2	3	7	1·51	1	1	9	1·60
				7	4	1	1·52	2	5	5	1·50	1	1	1	1·55
				3	2	7	1·51	10	1	1	1·50	5	5	3	1·53
				3	2	9	1·50					7	3	7	1·52
												3	1	9	1·51

The data are normalized, separated into eight subgroups and listed in order of decreasing $|E|$. Table I has such a listing for the amino acid. It terminates at $|E| = 1{\cdot}5$ since phases for weaker reflections usually are not determined in the initial stages. It is important to use the reflections with the largest $|E|$ magnitudes at the beginning of the phase determination to assure the highest probability of a correct determination. In the case of centrosymmetric crystals, an indication for a phase is not accepted unless the probability (Eq. 2) is greater than 0·98. This rule has to be relaxed somewhat in some triclinic structures. For noncentrosymmetric space groups it is desirable that the variance (Eq. 5) be less than about 0·5 before accepting a phase indication. Figure 2 of Karle & Karle (1966) shows a graph of the variance as a function of α (Eq. 6). As the phase determination procedure progresses, phases for reflections with somewhat lower $|E|$ magnitudes may be determined.

The first step is to assign an origin, as described in the preceding paper, and an enantiomorph, if possible. Sometimes it is not convenient to designate the enantiomorph at the beginning of the determination, especially if two-dimensional reflections in $P2_12_12_1$, for example, have small $|E|$ magnitudes. In such cases, one of the general reflections whose phase has been assigned by a symbol can ultimately be used to designate the enantiomorph. Table II shows the assignment of the origin and enantiomorph. In addition, three other reflections were assigned symbols to facilitate the application of Eq. 3. The symbol a is either 0 or π, as required by the space group restrictions, whereas the symbols p and q can have values between $-\pi$ and $+\pi$. The phase relationships in space group $P2_12_12_1$ are shown in Table III.

Table II

h	$\|E\|$	φ	
3 0 2	2·52	0	origin specification
7 3 0	2·19	$+\pi/2$	origin specification
0 5 3	2·12	$+\pi/2$	origin specification
7 0 7	1·64	$+\pi/2$	enantiomorph
2 0 8	2·34	a	
1 3 7	2·42	p	
5 4 3	2·61	q	

The successive application of Eq. 3 is illustrated in Table IV. In the eighth step, the phase for 124 can be obtained in three different ways with the same indication in each case. In the eleventh step, the phase for 650 appears to be p. Since a $gu0$ in this space group has a 0 or π phase, it would appear that p must be near 0 or π. Actually p is near $+\pi/2$ as will be seen from many other indications and Eq. 3 does not

Table III. Phase relations in $P2_12_12_1$

	hkl	$\bar{h}kl$	$h\bar{k}l$	$hk\bar{l}$
ggg uuu	α	$-\alpha$	$-\alpha$	$-\alpha$
ggu uug	α	$-\alpha$	$\pi-\alpha$	$\pi-\alpha$
gug ugu	α	$\pi-\alpha$	$\pi-\alpha$	$-\alpha$
guu ugg	α	$\pi-\alpha$	$-\alpha$	$\pi-\alpha$

$0gg$	$0, \pi$	$g0g$	$0, \pi$	$gg0$	$0, \pi$
$0uu$	$\pm\pi/2$	$u0g$	$0, \pi$	$ug0$	$\pm\pi/2$
$0ug$	$\pm\pi/2$	$g0u$	$\pm\pi/2$	$gu0$	$0, \pi$
$0gu$	$0, \pi$	$u0u$	$\pm\pi/2$	$uu0$	$\pm\pi/2$

Table IV. The application of Eq. 3.

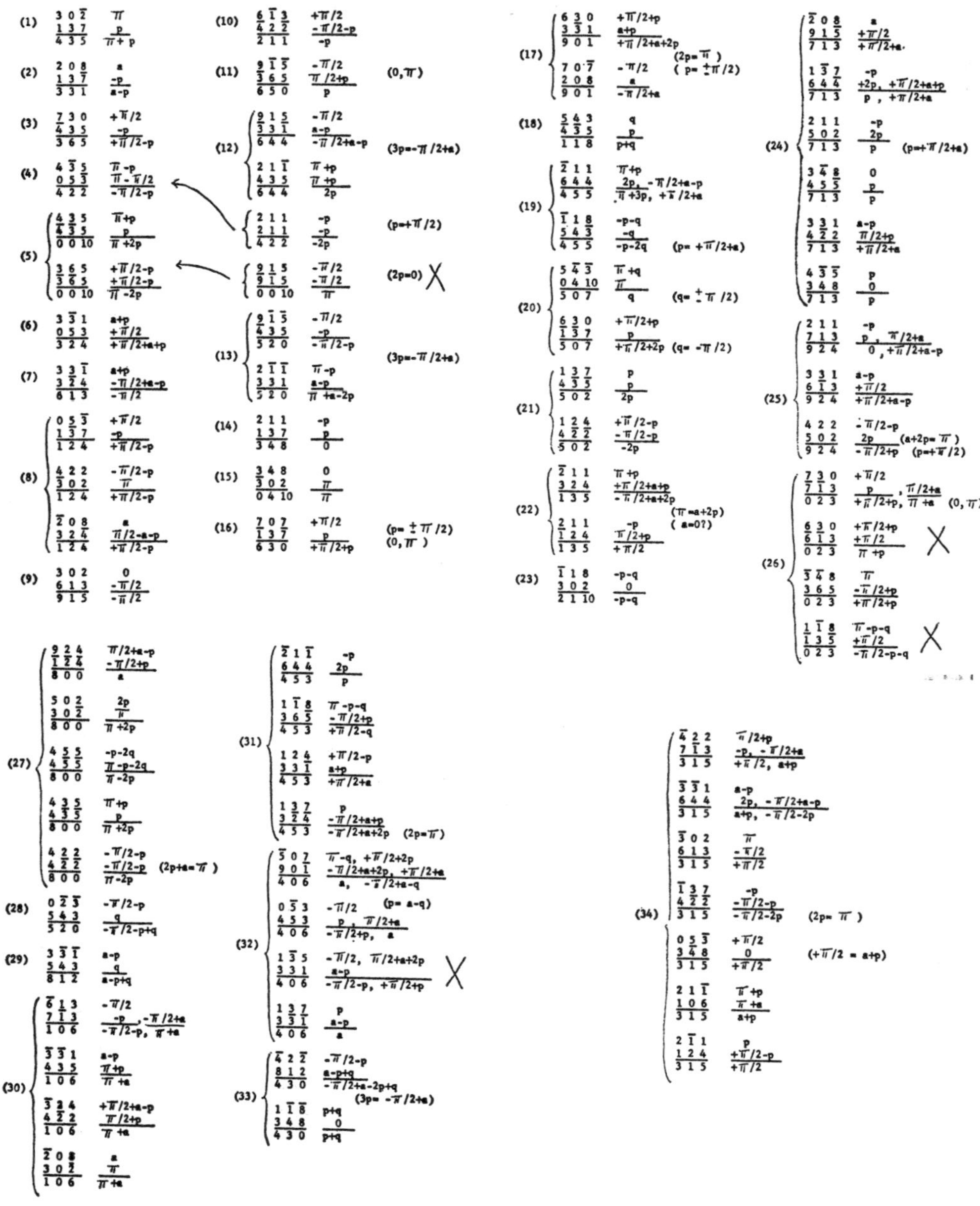

hold particularly well for this set of three reflections. Care must be taken not to determine the value of a symbol from single indications. In the twelfth and thirteenth steps as well as in later steps, the following relationships recur:

$$\begin{cases} 3p = -\pi/2 + a \\ p = +\pi/2 + a \\ 2p = \pi \\ 2p = \pi + a \end{cases}$$

All these relationships are satisfied by $a=0$ and $p=+\pi/2$. The very few indications that $p=0$ or π can be ignored. There are only two relationships for q,

$$\begin{cases} q=+\pi/2+2p \\ q=a-p \end{cases}$$

which are satisfied by $q=-\pi/2$. Although two indications are not considered definitive, they suggest that the assignment $q=-\pi/2$ be given first choice. These assignments for a, p, and q proved to be correct. It is important to have phases for some reflections in each of the eight subgroups in Table I. It may be necessary to assign a symbol to an additional reflection in order to have some phases determined in each subgroup.

Three two-dimensional reflections, such as the second set in the 17th step in Table IV, should not be used as the sole phase indication since these types of combinations very often violate Eq. 3 (or in centrosymmetric space groups, Eq. 1) and lead to wrong answers. These combinations should be avoided except in cases where they corroborate other indications as in step 27 or 30.

The 39 reflections for phases determined by Eq. 3 are used as input in the tangent formula, Eq. 4. Usually they are recycled in the tangent formula to effect a refinement on the phases. Additional reflections whose phases are not known added in order of decreasing $|E|$. The number added in each cycle should not exceed greatly the number of reflections with known phases. The known phases are usually kept constant in the succeeding cycles as new reflections are added. Criteria for accepting phases in each cycle are: (1) that the number of contributors in the summations in Eq. 4 should be greater than some minimum, *e.g.* >3; (2) that the value of the phase does not change greatly upon successive reiteration; (3) that $|E_{\mathbf{h}}|_{calc}$ is greater than some minimum, *e.g.* $>0{\cdot}4$. The values of $|E_{\mathbf{h}}|^2_{calc}$ are obtained by taking the sum of the squares of the numerator and denominator of Eq. 4 and

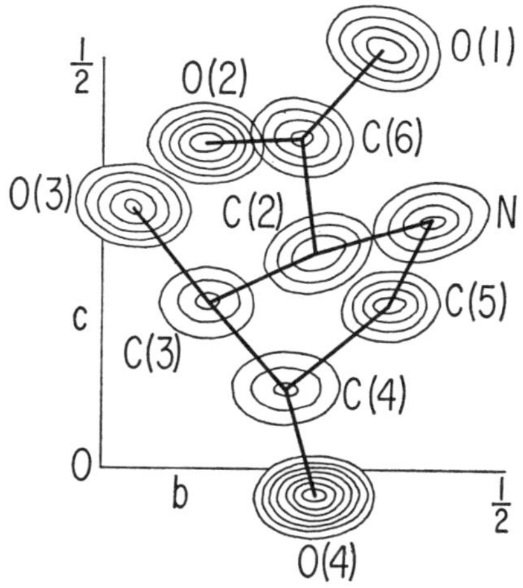

Fig. 1. Sections from a three-dimensional E-map for 3,4-dihydroxy-L-proline computed with 214 reflections with $|E|>1{\cdot}1$ whose phases were determined directly from the structure factor magnitudes.

rescaling them by matching their average to the average of the corresponding $|E_{\mathbf{h}}|^2_{obs}$ (Karle & Karle, 1966). Usually phases are obtained for almost the all reflections with $|E|>1{\cdot}0-1{\cdot}1$.

For 3,4 dihydroxy L-proline, 214 phases for reflections with $|E|>1{\cdot}1$ were obtained by using the sum of angles formula and the tangent formula. Sections from an E-map (Karle, Hauptman, Karle & Wing, 1958) computed with these reflections are shown in Fig. 1. The maxima for the ten peaks ranged in relative value from 151 for C(4) to 356 for O(4). There were no other peaks above the background ripple where sharp oscillations were at most ± 80 and usually less than ± 50 on the same relative scale.

The phase determination for the crystal of

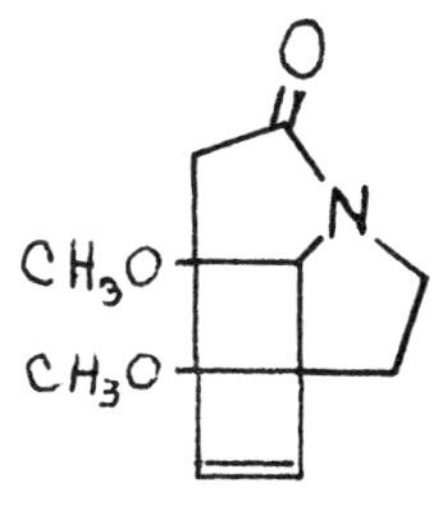

whose molecular formula was not known prior to the solution of the structure was not as direct as for the amino acid. Several different attempts were made with different sets of starting phases for origin assignment, etc., before the correct *E*-map was obtained. The difficulty lay in the fact that Eq. 3 was not sufficiently well satisfied by some of the contributors. If such a phase indication which is in error by $\pi/2$ to π is used early in the phase determination, then phases derived from it are likely to be in error. Since, *a priori*, it is not known which phase is causing the difficulty, a different route through the phase determination is required to circumvent the difficulty. For this determination the serious trouble-makers were:

$\|E\|$		φ (in radians)
1·99	9 0 6	0
2·38	1 1 $\bar{1}$	−1·48
2·16	10 1 5	−1·48 (should be +2·50)
1·99	9 0 6	0
2·38	$\bar{1}$ 1 $\bar{1}$	+1·48
2·15	8 1 5	+1·48 (should be −0·79)

Choosing 13 0 8 ($E=1{\cdot}98$) instead of 9 0 6 ($E=1{\cdot}99$) to specify the origin avoided these particular phase relations at the beginning of the process and led directly to an *E*-map showing 14 of the 16 atoms in this molecule. (Karle, Gibson & Karle, 1969).

There are many different paths that can be taken in a phase determination. Furthermore the procedure described in this paper, as well as variations thereof in which symbols are not used but trial values assigned to some of the phases, can be, in many cases, executed automatically on a computer as will be illustrated in the succeeding papers. Some of the virtues in starting the phase determination by hand are that more options are available for varying the unknowns and that, in stubborn cases, there are often clues as to which reflections are misbehaving.

The phase determination for centrosymmetric crystals proceeds in the same fashion, except, of course, all the phases are 0 or π. For centrosymmetric space groups, the correct *E*-map almost always reveals the complete molecule in equal atom cases. In space group $P\bar{1}$ it is often the case that the most consistent set of phases (other than all positive phases) is not the correct physical answer (*e.g.* Karle & Karle, 1964b). Computer programs should have the option of testing several different sets of phases obeying the criterion of high probability.

In noncentrosymmetric space groups, sometimes the complete molecule is visible in the *E*-map and sometimes only a few atoms appear. When only a part of the molecule appears, it is usually that part with the lowest thermal motion. A procedure for obtaining the structure from knowledge of only part of it (Karle, 1968) will also be discussed in a succeeding paper.

References

Cochran, W. (1955). *Acta Cryst.* **8**, 473.

Hauptman, H. & Karle, J. (1953). *Solution of the Phase Problem.* I. *The Centrosymmetric Crystal.* A.C.A. Monograph No. **3**. Pittsburgh: Polycrystal Book Service.

Karle, I. L. (1969). *Acta Cryst.* In press.

Karle, I. L., Gibson, J. W. & Karle, J. (1969). *Acta Cryst.* **B25**, 2034.

Karle, I. L., Hauptman, H., Karle, J. & Wing, A. B. (1958). *Acta Cryst.* **11**, 257.

Karle, I. L. & Karle, J. (1964a). *Acta Cryst.* **17**, 835.

Karle, I. L. & Karle, J. (1964b). *Acta Cryst.* **17**, 1356.

Karle, J. (1968). *Acta Cryst.* B**24**, 182.

Karle, J. & Hauptman, H. (1950). *Acta Cryst.* **3**, 181.

Karle, J. & Hauptman, H. (1956). *Acta Cryst.* **9**, 635.

Karle, J. & Karle, I. L. (1966). *Acta Cryst.* **21**, 849.

Woolfson, M. M. (1954). *Acta Cryst.* **7**, 61.

Discussion

BEURSKENS: Is the Σ_1 relationship, based on the two-fold screw axis, still of use in the phasing process?

I. KARLE: Yes, it does not always generate phases we can use but we use them whenever possible.

BEURSKENS: In some cases the Σ_1 relationships have been considered unreliable. Are there any criteria on which to judge this?

J. KARLE: The use of the Σ_1 relationship has changed in an important respect since its first derivation. It now tends to be used at the end of a phase determination in order to decide the possible alternatives you may have for a symbol. In general, our point of view, as far as all auxiliary phase determining formulae are concerned, is to use them at the end of a phasing procedure to reduce the number of alternate E-maps to be calculated. In this way we do not depend as much on them being correct.

BEURSKENS: In non-centrosymmetric structures one can readily find starting phases of centrosymmetric projection reflections by using the $B_{3\cdot0}$ formula of Karle & Hauptman (1958), with the complete set of three-dimensional data.

I. KARLE: Yes, that is an important point.

DE VRIES: Is there a theoretical explanation for the curved Wilson plot?

J. KARLE: I think so; if atoms in a structure have quite different temperature factors then there is no reason that some average representation of the structure factor will have a purely Gaussian exponent.

COPPENS: I would expect, however, that the Wilson plot should curve the other way due to the contributions of atoms with lower temperature factors dominating in the high order reflections.

MASLEN: I would be grateful for a little more detail on your reason for preferring the K-curve to Wilson statistics in calculating the E-value. We know from Rogers' work that departures from 'ideal' Wilson statistics may result from the presence of symmetry elements or frequent occurrence of a particular interatomic spacing in a structure. One might therefore be concerned that the use of the K-curve might dampen out real structural information in the data.

J. KARLE: It is clear that there may be genuine structural reasons why $\langle E^2 \rangle$ may be more or less than one in particular regions of reciprocal space. With the use of a slowly varying form for the K-curve such structural information should be preserved.

HOPPE: In the original paper on inequalities was the relationship $F_{\mathbf{h}} \propto \langle F_{\mathbf{k}} \cdot F_{\mathbf{h-k}} \rangle_{\mathbf{k}}$ explicitly stated?

J. KARLE: No, this formula was stated as a single term and the statement was also made that there were many indications of a given phase. What was left out was that when you have many indications they are averaged.

ALCOCK: You indicated that phase relationships involving 2-D relationships are unreliable. Firstly, can you offer an explanation of this and, secondly, how does this interfere with phase determination in projection?

I. KARLE: In projection, relationships of the sort $(6, 5, 0) + (-3, -2, 0) \rightarrow (3, 3, 0)$ have often been found to disagree. Sometimes the signs from these interactions are correct but often they are not. This is probably due to overlap of the atom sites in projection.

BEURSKENS: One must also be careful of reflections like *hhl* in some cases.

I. KARLE: Yes, this is probably the case in some of the more symmetric space groups.

COPPENS: Do you remove the factor ε from the E-values before the E-map is calculated in order to make it more comparable with the F_o-Fourier?

J. KARLE: No, I think this is a personal decision, as is the case for various weighted Fouriers, since one already has significant series termination errors due to the small sample of data being used.

Some Automatic Procedures for the Solution of Crystal Structures with Direct Methods and Phase Correction

W. HOPPE, J. GASSMANN & K. ZECHMEISTER

Abteilung für Röntgenstrukturforschung am Max-Planck-Institut für Eiweiss- und Lederforschung, München und Physikalisch-Chemisches Institut der Technischen Hochschule München, Abteilung für Strukturforschung, München, West Germany.

One of the aims in crystal structure determination is to proceed to automatic methods. In this paper we wish to explain some of the structure determination methods developed and used in our laboratory. The programs for these methods have been built into a new version of X-ray–67 and can be combined into a fully automatic procedure. Fig. 6 shows the general flow chart. After data reduction has been done using the X-ray–67 routines (DATRDN, DATFIX), three possible ways to determine initial phases have been built into our system. This talk will not be concerned with the heavy atom technique [Programs for the automatic heavy atom search have been developed (Zechmeister *et al.*, 1967) and included in our version of X-ray–67], but with our experience with direct methods (*A* in Fig. 6) in acentric cases. In another paper, Dr. Huber will discuss the convolution molecule method (*B* in Fig. 6). The second part of this paper will be devoted to the development of the final structure from the crude initial phases (phase correction, *C* in Fig. 6).

Direct methods

The relationship derived by Cochran (1955) can be used to determine the phases of acentric structures from the structure factor amplitudes. This gives the most probable value of the phase $\varphi_{\mathbf{h}}$ for a limited number (m) of $\mathbf{h}'$ terms as

$$\langle\varphi_{\mathbf{h}}\rangle = \text{phase of} \sum_{1}^{m} |U_{\mathbf{h}'}|\,|U_{\mathbf{h}-\mathbf{h}'}| \times \exp i(\varphi_{\mathbf{h}'} + \varphi_{\mathbf{h}-\mathbf{h}'}) \quad (1)$$

where $U_{\mathbf{h}}$ is the unitary structure factor.

An analogous relation known as the tangent formula was given by Karle & Hauptman (1956). For $m \to \infty$ Eq. (1) becomes the Sayre equation (1952) applied to unitary structure factors. For $m=1$ it becomes the triple product relation.

The statistical error $\Delta\varphi_{\mathbf{h}} = |\varphi_{\mathbf{h}} - \langle\varphi_{\mathbf{h}}\rangle|$ associated with each term in Eq. (1) was given by Cochran (1955) for equal atoms as

$$\Delta\varphi_{\mathbf{h}} = \frac{1}{\pi \,.\, I_0(X)} \int_0^{\pi} \Delta\varphi_{\mathbf{h}} \exp(2X \,.\, \times \cos\Delta\varphi_{\mathbf{h}}) \,.\, d\Delta\varphi_{\mathbf{h}} \quad (2)$$

where

$$X = 2N \,.\, |U_{\mathbf{h}} \,.\, U_{\mathbf{h}'} \,.\, U_{\mathbf{h}-\mathbf{h}'}| \quad (3)$$

$I_0(X)$ = modified Bessel function of the second kind.

N = number of atoms in the whole unit cell.

We have developed a computer program (written in FORTRAN IV) which determines and refines further phases by applying Eq. (1) in a cyclic procedure to a given set of initial phases. Eq. (2) is used to limit the procedure. The first program link (PREPAR) prepares data for the cyclic procedure. First the largest asymmetric set of $|U_{\mathbf{h}}|$ (generally about 300 to 500) are arranged in descending order of magnitude. Next the largest products $|U_{\mathbf{h}'} \,.\, U_{\mathbf{h}-\mathbf{h}'}|$ (about 50 to 100) are determined for each $|U_{\mathbf{h}}|$. All this information, as well as the tabulated $\Delta\varphi_{\mathbf{h}}(X)$, is stored on an external device in such a way that it can be read into the core memory at any time. Of prime importance in the success of a phase determination procedure using Eq. (1) is the choice of the initial phase set (or sets). It is well known that the phases of three linearly independent reflections may be chosen arbitrarily in the general asymmetric case ($P1$) as the origin of the unit cell is arbitrary. One further phase can take one of two possible values, depending on the relevant enantiomorph. In acentric space groups with centrosymmetric projections, however, the origin may only take on certain special positions (Hauptman & Karle, 1956). It is, therefore, common to fix the origin by assigning initial phases to two-dimensional reflections.

One special feature of our method is that the analysis can start from an inappropriate origin. This allows the use of the strongest reflections (and not only the strongest projection reflections) in the starting set. In this new procedure (Hoppe, 1969; Zechmeister, 1969), the origin of the cell is temporarily fixed by assigning arbitrary phases to the three-dimensional starting reflections. The important point is that these phases will not be definitely fixed but continually redetermined during the cyclic procedure. It will be shown later that this procedure is organized in such a way that the origin moves to one of the positions which are allowed by the centrosymmetric projection reflections (whose phases may only assume the values 0 or π and $\pi/2$ or $3\pi/2$ in some space groups). This procedure proved extremely useful in the structure determination of Phorbol (Hoppe *et al.*, 1969). Here, only the minimum starting set of 3 three-dimensional reflections were required for the success of the method.

The initial set of the three (or four) phases can be extended in our program. The phases of these additional reflections, however, have to be permutated (in increments) in the usual way (see Germain & Woolfson, 1968), and therefore the number of possible initial sets increases rapidly with the number of additional reflections. The advantages of such a multisolution process have been discussed by Germain & Woolfson. Another way to extend the starting set has been developed using the symbolic addition procedure (Karle & Karle, 1963); it has been used in our corresponding program for centric space groups (which will not be discussed in this paper).

Let us now describe the simple organization of our cyclic program for the acentric case: for each initial phase set, new phases are calculated and the others refined by the iterative application of Eq. (1) (using the program link SAYRE). The summation runs in a loop over all terms for which $\varphi_{\mathbf{h}'}$ and $\varphi_{\mathbf{h}-\mathbf{h}'}$ have already been determined. A new $\langle\varphi_{\mathbf{h}}\rangle$ is calculated only if the statistical error [Eq. (2)] is below a given limit for at least one term.* During this procedure also phase angles of projection reflections will be calculated. The essential point is that the program replaces these calculated phase angles by the nearest 'projection' phase angles 0, π (or $\pi/2$, $3\pi/2$). These projection reflections now contribute to Eq. (1) and cause phase shifts of the general reflections in the starting set, thereby moving the origin. When no further appreciable phase changes occur the probability limit [Eq. (2)] is lowered to allow the determination of further phases in the next cycle.

If several initial phase sets are used, a limited number of cycles are calculated until about 100 to 150 phases have been found. The 'best' solution (or solutions) are then recycled until all phases have been determined. We defined the 'best' solution by Eq. (4) below; it is the one for which the phases of the individual $\mathbf{h}'$ terms have

* In this case Eq. (1) will be reduced to the triple product relationship (acentric case). It is obvious that in the beginning, mainly this relationship will occur.

the highest correlation [q in Eq. (4) will then have the minimal value].

$$q = 1 - \frac{\sum^{\mathbf{h}} |U_{\mathbf{h}}| \left\{ |\Sigma^{\mathbf{h}'}| U_{\mathbf{h}'} U_{\mathbf{h}-\mathbf{h}'}| \exp i(\varphi_{\mathbf{h}'} + \varphi_{\mathbf{h}-\mathbf{h}'})| / \Sigma^{\mathbf{h}'} |U_{\mathbf{h}'} U_{\mathbf{h}-\mathbf{h}'}| \right\}}{\Sigma_{\mathbf{h}} |U_{\mathbf{h}}|} \quad (4)$$

Two structures that have been determined by this direct procedure are Phorbol and Ingenol. Both were isolated by Hecker and co-workers at the Institute of Cancer Research, Heidelberg. In the case of Ingenol, the chemical constitution was unknown.

Crystallographic data of Phorbol ($C_{20}H_{28}O_6 + C_2H_5OH$)

Space group $P2_12_12$, $Z=4$; $a=18{\cdot}575$ Å; $b=12{\cdot}777$ Å; $c=9{\cdot}611$ Å; $D_{calc}=1{\cdot}20$ g/cm³; $D_{meas}=1.21$g/cm³. 1916 reflections were measured with Cu $K\alpha$ radiation using a Siemens single crystal diffractometer ($\theta \leq 60°$, $\theta/2\theta$ scan, 5-point measuring procedure). Direct method with the 334 largest unitary structure factors.

Starting sets

A. conventional

h	U	φ
10 7 0	0·32	0°
0 1 5	0·31	90
13 6 0	0·29	0
0 8 7	0·25	90

$q=0{\cdot}56$
U-Fourier failed.

B. three-dimensional data

h	U	φ
4 5 7	0·34	150°
5 10 8	0·31	45
7 7 6	0·29	225

$q=0{\cdot}41$
U-Fourier contains the whole structure; average phase error 12°.

During the cyclic procedure, the origin moves to one of the eight allowed positions. Spectacular phase shifts occur for the starting set: +40°, +82°, −53°. A U-Fourier synthesis using the 334 phases determined yielded all 26 atoms of the Phorbol skeleton. The height of the atomic peaks in this Fourier synthesis varied between 150 and 37 (arbitrary scale); the highest spurious peak was 45; among the 32 highest peaks were all 26 atoms. The final R-factor (including all 1916 structure factors) was 5·4%. The ethanol molecule occurs in two statistical positions for which the oxygen atoms coincide. The positions of 27 (out of 28) hydrogen atoms could be found.

Crystallographic data of Ingenol $C_{26}H_{34}O_8$

Space group $P2_12_12_1$; $Z=4$; $a=17{\cdot}58$ Å $b=21{\cdot}55_2$ Å; $c=6{\cdot}75_3$ Å. $D_{calc}=1{\cdot}23$ g/cm³; $D_{meas}=1{\cdot}22$ g/cm³; 2265 reflections measured with Cu $K\alpha$ radiation on a Siemens single crystal diffractometer. Direct method with the 457 largest unitary structure factors.

Four starting sets:

h	U	φ	
5 1 0	0·26	90	origin determining and enantiomorph selecting
1 18 0	0·26	90	
5 18 1	0·22	0	
2 13 5	0·22	0	
10 8 0	0·24	0/180°	
11 11 0	0·24	90/270	

10 cycles
113 phases determined
$q=0{\cdot}28/0{\cdot}43/0{\cdot}38/0{\cdot}41$.

The structure determination with four starting reflections alone did not converge ($q=0{\cdot}65$). Only after having added two more projection reflections and then further recycling the best set [according to the q-criterion Eq. (4)] did the correct structure appear ($q = 0{\cdot}50$ for 457 phases; average phase error 12°). The phase shifts of the two three-dimensional reflections during the cyclic procedure were −32° and −5°. A subsequent *U*-Fourier synthesis immediately showed 31 out of 34 atoms of the molecule. Peak heights for atoms in this Fourier synthesis varied from 110 to 30 (arbitrary scale); the highest spurious peak had 37; among 35 highest peaks were 31 atoms. The next Fourier synthesis using all 2265 structure factors, based on the model of 31 atoms, showed the whole structure. After refinement ($R = 6{\cdot}2\%$) 33 out of 34 hydrogen atoms could be located.

In the next section of this paper, we come to the question of the determination of the final model using the not very precise initial phases. The usual Fourier refinement process is not very suitable for automation; it needs careful handling and intuitivity of the crystallographer. We have replaced it by the methods of phase correction.

Phase correction

I. *Theoretical aspects*

Phase correction is a method which starts from approximate phases and reaches better phases through modification of electron density (Hoppe & Gassmann, 1968). It is not comparable to the least squares refinement procedures since it does not require a nearly correct model which includes all atomic sites. Phase correction, therefore, is most appropriate after the initial step of structure determination. The basic equation for phase correction is

$$\varrho^* = g(\varrho) \cdot \varrho \qquad (5)$$

The initial electron density $\varrho(\mathbf{r})$ is weighted by a suitably chosen function $g(\varrho)$ to give a 'better' electron density $\varrho^*(\mathbf{r})$. The calculated corrected phases $\varphi_{\mathbf{h}}$ of $\varrho^*(\mathbf{r})$ are then combined with the correct experimental structure factor amplitudes $|F_{\mathbf{h}}|$ to give a new electron density. This new electron density again is subjected to the modification of Eq. (5) and as many times as necessary cyclically repeated. It is easy to see that the procedure can only work successfully if the modification $\varrho(\mathbf{r})$ to $\varrho^*(\mathbf{r})$ leads to a structure with better phases but worse amplitudes. The last fact is obvious; $\varrho(\mathbf{r})$ has been calculated with correct amplitudes; its distortion to $\varrho^*(\mathbf{r})$ cannot lead to 'better' amplitudes.

The essential features of this cyclic procedure are:

A. Reduction of the enhanced maxima at known atomic sites.

B. Enhancement of the reduced maxima at unknown atomic sites.

C. Reduction of the reduced maxima at wrongly placed atomic sites.

D. Reduction of background and negative regions.

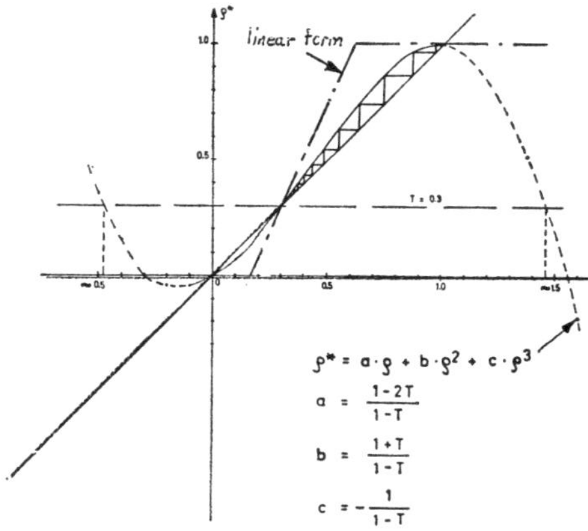

Fig. 1. *Different forms of weighting in phase correction. The polynomial form can be used in reciprocal space, the linear form may be applied in direct space.*

E. Improvement of site positions for correct atoms.

F. No human intervention or interpretation until the final Fourier synthesis.

In the first structures studied by

phase correction (Phyllochlorine ester $C_{33}N_4O_2H_{38}$ and Sulfolipid $RbSO_{10}C_9H_{17}$), the weighting function $g(\varrho)$ was taken as a Taylor series expansion $g(\varrho)=a+b\varrho+c\varrho^2$, constrained by the conditions A–D. This weighting function could then be applied through convolutions in reciprocal space without calculating any intermediate Fourier synthesis. Though it proved very effective also in following structures, the rate of change in the electron density was small, as the weighting was 'too carefully' done. Other weighting schemes were used to reduce the number of phase correction cycles. A very convenient weighting scheme that we are now using is the 'linear' form, which will be described later. As the changes induced in the electron density are much greater, fewer cycles are required than for the Taylor expansion. Because of the analytical form, modification using linear weight could be done in direct space through Fourier synthesis and reciprocal Fourier synthesis.

Table I. Comparison between the Fourier syntheses of Ecdyson before and after phase correction. Initial phases result from the incorrect cholestane model (wrong atoms are primed, spurious peaks are marked P), the line in the O. cycle indicates the end of the partial structure used for phasing

Ecdyson initial structure: Cholesterine-model

O. cycle			25. cycle		
Value	Name		Value	Name	
134·59	C 20		131·24	O 4	
132·14	C 16		131·07	O 1	
129·65	C 18		128·77	O 3	
129·65	C 18		128·77	O 3	
128·85	C 19		128·66	O 2	
122·89	C 15		125·32	C 8	
120·45		C 3′	123·02	O 5	
119·95	C 6		121·35	O 6	
117·79		C 7′	116·71	C 17	
113·30	C 5		116·17	C 15	
111·08	C 13		116·00	C 5	
110·05	C 1		115·18	C 19	
109·16	C 14		114·54	C 14	
105·62	C 8		114·52	C 7	
102·78	C 17		113·49	C 1	
94·81	C 9		112·98	C 2	
93·98	C 11		111·99	C 13	
92·61	C 12		111·92	C 4	
92·29	C 10		110·19	C 6	
90·33		C 2′	108·58	C 12	
84·80		C 4′	106·02	C 20	
48·91	C 4		103·33	C 16	
39·32	O 4		96·21	C 11	
36·78	C 2		95·71	C 18	
32·03		P	88·17	C 23	
30·33	O 3		87·70	C 22	
29·15	O 1		83·45	C 27	
27·98		P	79·70	C 3	
27·49		P	68·42	C 10	
27·48		P	66·12	C 9	
26·37	C 3		61·26	C 25	
26·05		P	44·00	C 21	
25·66		P	35·42	C 24	
24·92	O 5		24·40	C 26	
24·21	C 23		18·86		C 2′

Phase correction can be applied equally well both to heavy atom and to light atom structures. The heavy atom contribution must be subtracted before application of phase correction.

Phase correction using non-local weighting schemes

In Eq. (5) the electron density ϱ is a function of the space point $\varrho = \varrho(\mathbf{r})$ so one could also write:

$$\varrho^*(\mathbf{r}) = g[\varrho(\mathbf{r})] \cdot \varrho(\mathbf{r}) \qquad (5a)$$

indicating that the weight $g[\varrho(\mathbf{r})]$ is a function of the electron density present at the point $\mathbf{r}$. This weighting is not affected by the electron density at other spatial points $\mathbf{r}'$. The physical reasons for taking into account other regions of space for weighting at the point $\mathbf{r}$ are obvious. Molecular structures consists of atoms which are

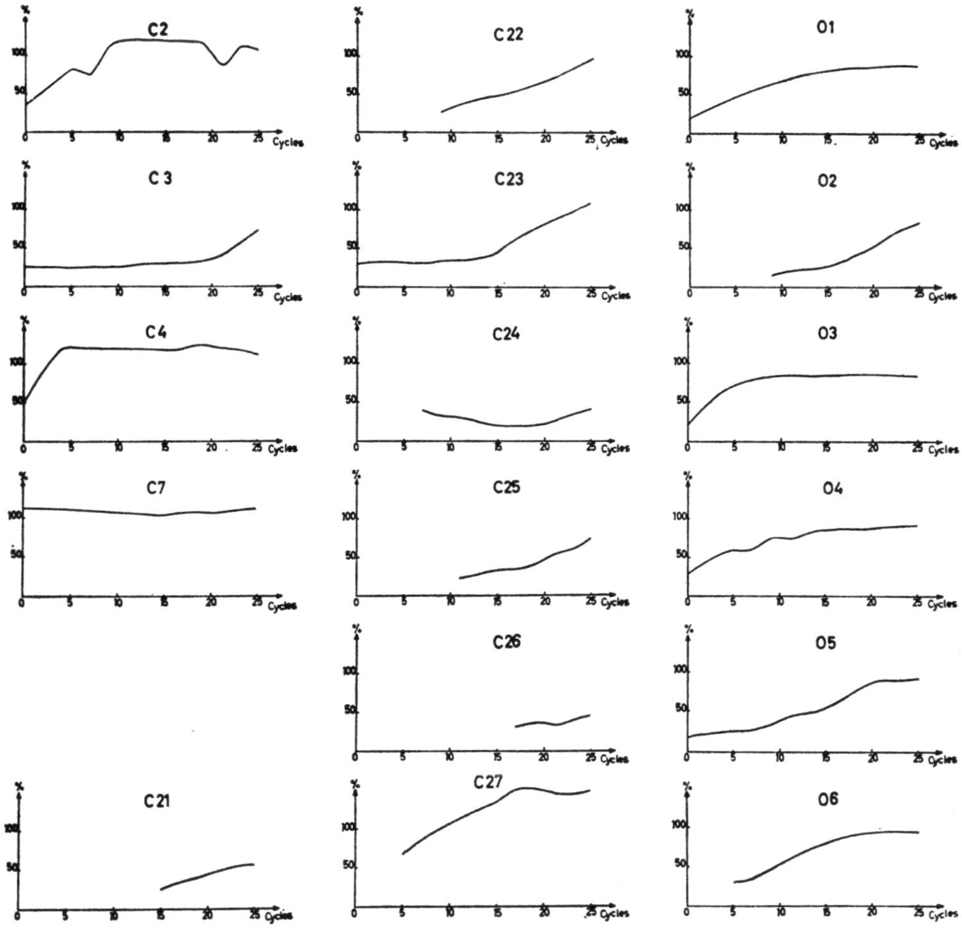

Fig. 2. Enhancement of unknown atomic peaks of ecdyson during phase correction Wrong positions for initial phasing were assumed for C 2, C 3, C 4, *and* C 7. *These wrong atomic positions (primed in Table I) disappeared during phase correction. The correct positions had been indicated already in the first Fourier synthesis by smaller peaks. Since the initial wrong position of* C 7 *was only* 1·2 Å *from the correct site, this peak which shifted slowly to the correct place was always taken as atom* C 7, *thus always having full height.*

restricted in spatial arrangement to certain distances and angles. Two atoms have a minimal distance; the triangle formed by three atoms may not have an angle below a minimum.

Non-local weighting takes into account these constraints. The value of the weight function $g(\mathbf{r})$ will be influenced mostly by the surrounding radial arrangement of electron density, thus taking into account bond distances. Secondly the angular density distribution restricts weighting to permissible bond angles.

More constraints for the weighting function can be found from chemical reasoning (*e.g.* planarity, packing of molecule) and the crystallographic properties (*e.g.* dimensions of elementary cell, space group). Restrictions due to symmetry elements can be incorporated very easily into phase correction as a general feature, whereas chemical information has to be patched for each structure. For non-local weighting, Eq. (5) will be generalized to

$$\varrho^*(\mathbf{r}) = \int G(\varrho', \varrho) \, . \, d\tau' \, . \, \varrho(\mathbf{r}) \qquad (5b)$$

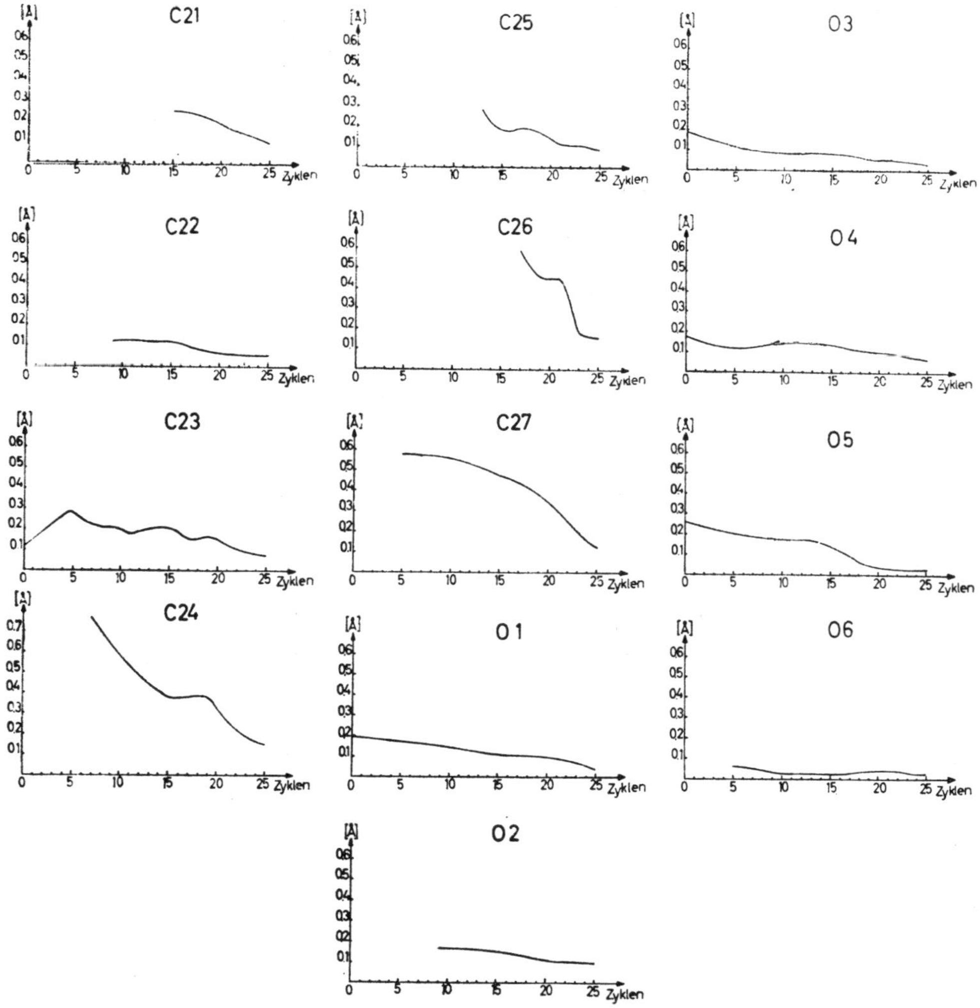

Fig. 3. Improvement of atomic positions during phase correction for the skeleton atoms of ecdyson.

where $G(\varrho',\varrho)$ is the weight which results through correspondence of the electron density $\varrho'(\mathbf{r}')$ at a point $\mathbf{r}'$ of direct space. In this non-local weighting is it no longer necessary to separate the different atom types (*e.g.* to subtract heavy atoms before applying local weighted phase correction).

Parts of this non-local weighting, such as suppression of spurious peaks near known atoms and near symmetry elements (axes, planes), are incorporated in the programs for locally weighted phase correction. Locally weighted phase correction works only with data allowing atomic resolution. Appropriate non-local weighting schemes are at present being investigated for use in cases where atomic resolution cannot be achieved.

II. *Programming of phase correction*

There are two different forms of the program PHASCØR depending on the kind of weighting in reciprocal or direct space.

A. *Reciprocal space*: transforming the weight scheme $\varrho^* = (a + b\varrho + c\varrho^2).\varrho$ into reciprocal space, one gets the corresponding convolutions

$$F_{\mathbf{h}}^{\text{mod}} = aF_{\mathbf{h}} + b\langle F_{\mathbf{h}'} F_{\mathbf{h}-\mathbf{h}'}\rangle_{\mathbf{h}'} + c\langle\!\langle F_{\mathbf{h}''} F_{\mathbf{h}'-\mathbf{h}''} F_{\mathbf{h}-\mathbf{h}'}\rangle\!\rangle_{\mathbf{h}'\mathbf{h}''}\,. \quad (6)$$

These complex numbers $F_{\mathbf{h}}^{\text{mod}}$ are calculated first. A new structure factor

$$F_{\mathbf{h}}^{\text{new}} = |F_{\mathbf{h}}^{\text{ex}}|\,.\,\exp{(i\,.\,\varphi_{\mathbf{h}}^{\text{mod}})} \quad (6a)$$

is constructed from the experimental structure factor amplitude and the phase $\varphi_{\mathbf{h}}^{\text{mod}}$ of the modified structure factor $F_{\mathbf{h}}^{\text{new}}$. The value of $F_{\mathbf{h}}^{\text{new}}$ is then stored to be used as the starting value for the new cycle. During the whole process $|F_{\mathbf{h}}^{\text{new}}| = |F_{\mathbf{h}}^{\text{ex}}|$, the amplitudes are not changed.

B. *Direct space*: from initial phases and experimental structure factor amplitudes a Fourier synthesis is calculated and

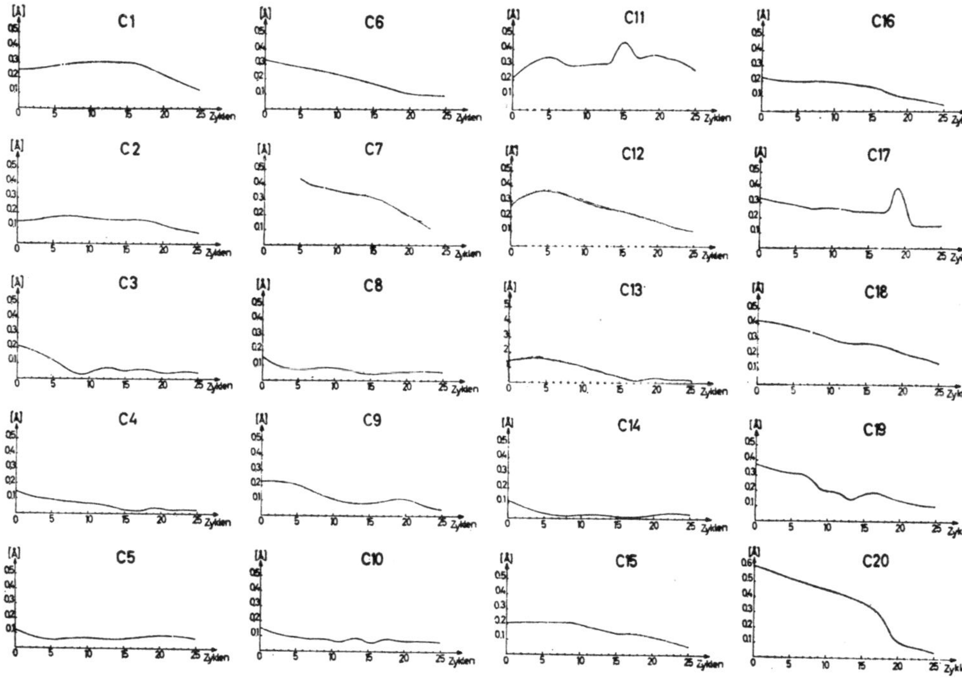

Fig. 4. Improvement of atomic positions during phase correction for the rest of the molecule of ecdyson.

stored. Modifications are then applied to the electron density. A reciprocal Fourier synthesis gives new phases and amplitudes for each reflection. The amplitudes are deleted, the phases combined with the experimental amplitudes to calculate the next Fourier synthesis.* The cycle is repeated. A reliability factor indicates the rate of progress.

III. *Experiments on phase correction*

1. *Ecdyson* $C_{27}O_6H_{28}$, space group $P2_12_12_1$: the structure of this insect hormone has been solved by Huber & Hoppe (1965) using the convolution molecule technique with the steroid structure as the known part of the molecule. The Fourier synthesis using the cholestane folded sterine for initial phasing could not be interpreted; it turned out that the molecule had koprostane folding.

This uninterpretable cholesterine-model Fourier synthesis was processed by phase correction. Of the initial 20 atoms used

0.cycle

Value	Name	
20.38		P
19.58	O 4	
18.75	C 4	
16.00	C 11	
15.88		P
15.41	C 18	
15.39	O 8	
15.34		P
14.99	C 12	
14.76	O 7	
14.62	C 6	
13.65	C 8	
13.14		P
12.90	C 2	
12.89	O 9	
12.75	C 19	
12.65	C 15	
12.64	O 3	
12.57	O 6	
12.55		P
12.43		P
12.42	C 3	
12.40		P
12.37		P
12.32	C 13	
12.29	C 23	
12.27	C 28	
11.97		P
11.95	O 5	
11.90	C 5	
11.58	C 30	
11.42		P
11.35		P
11.27	C 22	
11.26		P
11.25	C 1	
11.24		P
11.22	O 2	
11.20		P
11.17	C 20	
11.08		P
10.96		P
10.94	C 7	
10.94		P
10.81		P
10.67	C 9	
10.46		P
10.29	C 14	
10.27		P
10.21		P
10.04	C 31	
9.98		P
9.94		P
9.88		P
9.85	C 10	
9.77		P
9.75		P
9.74		P
9.66		P
9.63		P
9.47	O 1	
9.33		P
9.32		P
9.32		P

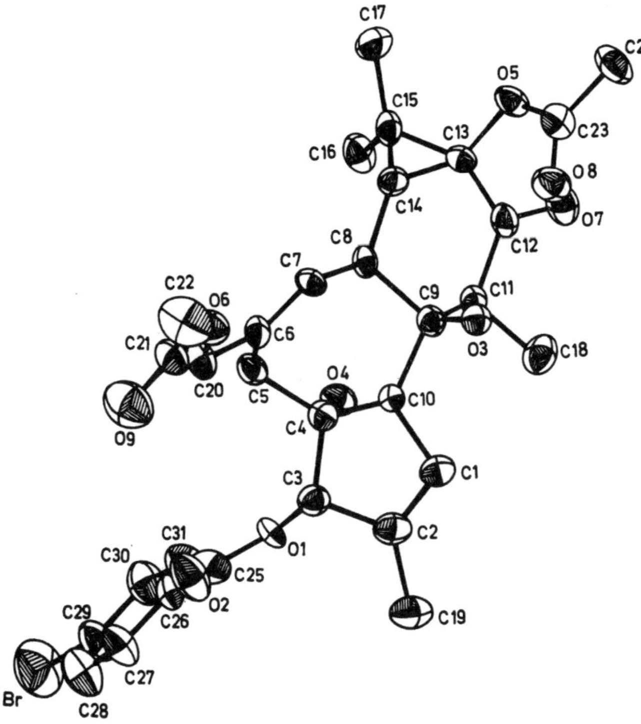

Fig. 5. Neophorbol 13, 20-*diacetate*-3-*p*-*bromo*-*benzoate and the new peaks in a Fourier synthesis phased with the bromobenzoyl part of the molecule.*

* Input cards, program description, and listing are available upon request.

for phasing, four were in wrong positions. After 25 cycles of phase correction wrong atoms and spurious peaks had been suppressed, and the 33 highest peaks proved to be the correct atoms; even distinction between oxygen and carbon was possible. Not only had the height of the peaks at atomic sites been increased, but also their positions had considerably improved.

2. *Lumiphorbol* $C_{33}O_{10}H_nBr$: this structure (space group $P2_1$) has been solved fully automatically (Hoppe, Gassmann, Röhrl & Abel, 1968). The position of the bromine atom was determined from a Patterson synthesis. From the Fourier synthesis phased with the bromine atom, one further peak was selected to destroy the pseudosymmetry. After 18 cycles phase correction was stopped. Among the 48 highest peaks were all 44 atoms.

In principle, one could find not only the bromine atom from a Patterson synthesis but also the attached benzene ring for use

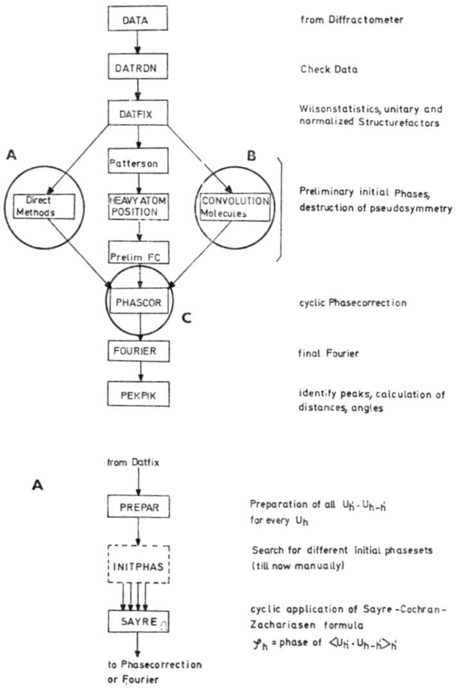

Fig. 6. Flow chart of programs in our version of crystallographic system X-Ray–67.

in initial phasing. However, the determination of the translation parameter for the benzene ring from convolution molecule technique was ambiguous and two solutions were possible. After having determined the structure, we tested these two solutions of phase correction, starting with the correct translation. After 12 cycles the molecule was complete. The wrong translation parameter, as expected, gave no complete molecule; too many mirror images of atoms appeared in the Fourier synthesis.

3. *Neophorboldiacetate* $C_{31}O_9H_{35}Br$: this structure was solved by successive Fourier syntheses and phase correction. The bromine-benzene part of the molecule was used for initial phasing (Hoppe *et al.*, 1969; Hecker, 1968).

4. *Cu-complex of phyllochlorine ester* $C_{33}N_4O_2H_{38}Cu$: this structure is nearly isomorphous with phyllochlorine ester and was solved with phase correction. The position of the copper atom was determined from a Patterson synthesis; one additional peak was used for initial phasing to destroy pseudosymmetry ($P2_1$).

5. *Lucidusculine Hydriodide* $C_{24}H_{35}O_4N$ .HI: this structure was determined by successive Fourier syntheses with some problems (split iodide atoms). A test with phase correction proceeded quite normally and showed the structure clearly after 10 cycles.

6. *Vitamin* B_{12} (in collaboration with H. G. Abel): this structure was used as a test for phase correction since it represents one of the largest molecules determined by successive Fourier syntheses (108 atoms/asymmetric unit).

For initial phasing we used the cobalt and phosphorus atom which constitute 6% of all diffracting electrons. The 2640 observed structure factors with a final *R*-factor above 20% and measured to a resolution of $\sin\theta/\lambda = 0{\cdot}45$ were subjected to phase correction. Since it was known from test calculations that normal phase correction does not work below atomic resolution, it was not surprising to find only limited improvements in the electron density. After 20 cycles only 58 atoms could be identified among 73 peaks with a height greater than $\frac{1}{2}$ C-atom.

To avoid the restrictions resulting from low accuracy of $|F|$-values and resolution, we calculated $|F|$ values from the atomic positions out to 0·8 Å resolution and used only the normalized structure factors above $E \geq 1{\cdot}2$. The initial Fourier synthesis revealed 77 atoms with 23 interspersed spurious peaks. After six cycles of phase correction 97 atoms could be found with only 8 interspersed spurious peaks.

This test shows very clearly that the phase correction method may be used successfully with large molecules if the experimental data is of sufficient accuracy and resolution. These two conditions are no restriction for a modern diffractometer, provided the crystal permits data to be collected at high resolution. Therefore, in the case of proteins where the resolution is generally limited to 1·5–2 Å, phase correction in the present form cannot be applied.

References

Cochran, W. (1955). *Acta Cryst.* **8**, 473.

Germain, G. & Woolfson, M. M. (1968). *Acta Cryst.* B**24**, 91.

Hauptman, H. & Karle, J. (1956). *Acta Cryst.* **9**, 45.

Hecker, E. (1968). Private communication.

Hoppe, W. (1969). *J. P. Appl. Chem.* In press.

Hoppe, W., Brandl, F., Strell, I., Röhrl, M. & Gassmann, J. (1967). *Angew. Chem.* **79**, 824.

Hoppe, W. & Gassmann, J. (1968). *Acta Cryst.* B**24**, 97.

Hoppe, W., Gassmann, J., Röhrl, M. & Abel, H. (1968). *Angew. Chem.* **7**, 913.

Hoppe, W., Zechmeister, K., Röhrl, M., Brandl, F., Hecker, E., Kreibich, G. & Bartsch H. (1969). *Tetrahedron Letters* **8**, 667.

Huber, R. & Hoppe, W. (1965). *Chem. Ber.* **98**, 2403.

Karle, J. & Hauptman, H. (1956). *Acta Cryst.* **9**, 635.

Karle, J. & Karle, I. (1963). *Acta Cryst.* **16**, 969.

Sayre, D. (1952). *Acta Cryst.* **5**, 60.

Yoshino, A. & Jitaka, Y. (1966). *Acta Cryst.* **21**, 57.

Zechmeister, K. (1969). Thesis TH München.

Zechmeister, K., Brandl, F., Hoppe, W., Hecker, E. & Opferkuch, H. J. In preparation.

Zechmeister, K., Hechtfischer, S. & Hoppe, W. (1967). 9. *Diskussionstagung der Sektion für Kristallkunde in der Deutschen Mineralogischen Gesellschaft in Bonn.*

Discussion

CHRISTOPH: How much computer time do these calculations take?

GASSMANN: A recent example, using an IBM 360/50, took twelve minutes for 8 to 10 cycles, and 40 atoms were located from the 44 highest peaks in the map.*

ROLLETT: Is the resolution of 1·1 Å, an experimental observation or have you an analytical way of determining it?

HOPPE: No, it is an experimental value.

KARTHA: It seems the minimum resolution of 1·1 Å is unduly restrictive. I have been using a similar procedure for refining protein phases at lower resolutions. Some test computations suggest that with good data one can start this type of refinement cycle at resolutions as low as 2·5 Å. What is needed here, presumably, is a sufficient number of good initial phases to obtain additional phase indications.

* Using a Fourier transform after the Cooley-Tookey algorithm cut computer time by a factor of 20-40.

Partial Structures and the Tangent Formula

J. KARLE

U.S. Naval Research Laboratory, Washington, D.C.

A partial structure rather than a complete one is often obtained for non-centrosymmetric crystals from a phase determination employing the symbolic addition procedure. It is therefore quite desirable to have a method for generating the complete structure once a structural fragment has been located in a unit cell. The tangent formula can readily serve this purpose (Karle, 1968). Its application in this regard is further extended owing to the fact that the Patterson function can also be employed to locate structural fragments. This applies not only to heavy atom partial structures but also to light atom fragments (Hoppe & Will, 1960; Nordman & Nakatsu, 1963).

The tangent formula (Karle & Hauptman, 1956) is

$$\tan \varphi_{\mathbf{h}} \approx \frac{\sum_{\mathbf{k}} |E_{\mathbf{k}} E_{\mathbf{h}-\mathbf{k}}| \sin (\varphi_{\mathbf{k}} + \varphi_{\mathbf{h}-\mathbf{k}})}{\sum_{\mathbf{k}} |E_{\mathbf{k}} E_{\mathbf{h}-\mathbf{k}}| \cos (\varphi_{\mathbf{k}} + \varphi_{\mathbf{h}-\mathbf{k}})}, \quad (1)$$

where $\varphi_{\mathbf{h}}$ is the phase associated with the normalized structure factor $E_{\mathbf{h}}$. This formula can be evidently implemented when a certain number of phases $\varphi_{\mathbf{k}}$ and $\varphi_{\mathbf{h}-\mathbf{k}}$ are available. When a partial structure has been located in a unit cell, it may be used to generate phases for use with Eq. (1). It is remarkable that a relatively small structural fragment can produce phases through structure factor calculations which are sufficiently accurate to be employed in Eq. (1).

Only certain of the phases computed from the partial structure are suitable for use with Eq. (1). These phases are chosen by means of an acceptance criterion which is based on the scattering power of the corresponding reflections as obtained from experiment and the relative scattering power of the structural fragment. Specifically, a phase as computed from a partial structure is accepted if $|F_{\mathbf{h}}|_{\text{calc}} \geq p|F_{\mathbf{h}}|_{\text{obs}}$, where p is the fraction of the total scattering power contained in the fragment and where $|F_{\mathbf{h}}|_{\text{obs}}$ is associated with an $|E_{\mathbf{h}}|_{\text{obs}} \geq 1{\cdot}5$. The quantity $|F_{\mathbf{h}}|_{\text{calc}}$ is the value of the structure factor computed from the partial structure and $|F_{\mathbf{h}}|_{\text{obs}}$ is the value for the complete structure observed experimentally. As a working rule, if $p < 0{\cdot}25$, it is replaced by $0{\cdot}25$, or, at least, a number somewhat larger than p. Also if $p > 0{\cdot}6$, it is replaced by $0{\cdot}6$.

It is found that the phases obtained by means of the acceptance criterion do not change greatly throughout the recycling of the procedure as more and more atoms are added to the initial partial structure. In several studies starting with about 20 per cent of the structure, it was found that the average change was on the order of 0·7 radians. It has also been found that the placement of the atoms in the initial partial structure changes very little, generally much less than 0·02 of a cell edge, in the course of the development of the complete structure. If an incorrect peak is identified as an atom, it will disappear in the recycling procedure. For these many reasons, it becomes somewhat more understandable why a relatively small fraction of a structure can be developed into a complete structure by means of the tangent formula.

Two examples of the development of a partial structure into a complete one by means of the tangent formula are the investigations of reserpine (Karle & Karle, 1968) and 6-hydroxycrinamine (Karle, Estlin & Karle, 1967). In the sequence shown in Fig. 1, sixteen peaks denoted by the solid circles, which were found from the application of the symbolic addition procedure, were used to start the recycling procedure. The peaks were all weighted as carbon atoms and corresponding phases were accepted if both $|F_{\mathbf{h}}|_{\text{calc}} > 0{\cdot}33\, |F_{\mathbf{h}}|_{\text{obs}}$ and $|E_{\mathbf{h}}| > 1{\cdot}5$. By use of the tangent formula, the values of these phases were refined and additional phases were obtained for those reflections with $|E| > 1{\cdot}1$. Of the 39 strongest peaks obtained in the *E*-map from the new phases, 16 corresponded to the initial set, 21 additional ones made good chemical sense, and 2 of the weaker ones were clearly spurious. The 21 new atoms are represented by shaded circles in Fig. 1. The procedure was repeated with the 37 atoms considered as carbon using the acceptance criterion for new phases, $|F_{\mathbf{h}}|_{\text{calc}} > 0{\cdot}6\, |F_{\mathbf{h}}|_{\text{obs}}$ and $|E_{\mathbf{h}}| > 1{\cdot}5$. The additional atoms obtained in this cycle are shown as open circles in Fig. 1. In addition, the spurious peak included in the initial set of atoms, denoted by the solid circle with a cross, became weak and its position began to approach the location of C(29), not yet found from the *E*-maps. In the last cycle, the remaining atoms, denoted by × in Fig. 1, were found. They consisted of several of the carbon atoms of peripheral methyl groups which have relatively large thermal factors.

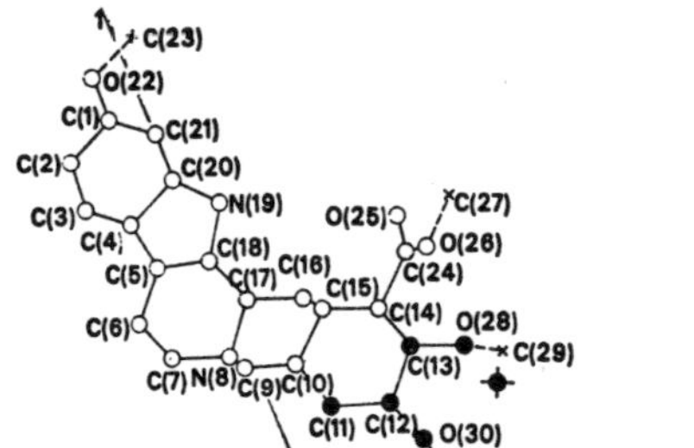

Fig. 1. *The order of appearance of atoms in successive E-maps. The solid circles mark atoms which appeared in the initial E-map based on phases from the symbolic addition procedure. The shaded circles, open circles, and x's mark the atoms as they appeared in three successive E-maps based on phases obtained from the partially known structure and refined and expanded by recycling with the tangent formula. The* –●– *mark indicates a spurious peak which disappeared in the later E-maps* (*Karle & Karle,* 1968).

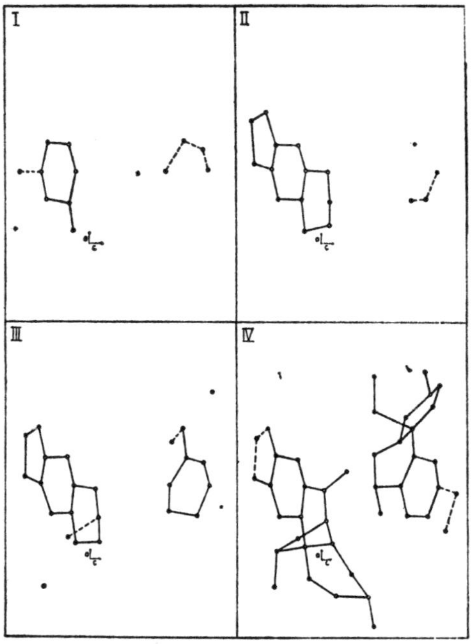

Fig. 2. *The open circles in part I represent the seven largest peaks in the E-map computed from the phases obtained from the symbolic addition procedure. Of the seven next largest peaks, the closed circles represent meaningful peaks whereas the x marks are spurious. Only the seven atoms represented by the open circles were used to obtain initial phases for recycling the tangent formula to obtain phases for the next E-map whose* 17 *largest peaks are shown in part II. Three repetitions of the procedure yielded the atoms shown in parts III and IV and finally all* 46 *atoms* (*not illustrated*) (*Karle, Estlin & Karle,* 1967).

The successive development of partial structures for 6-hydroxycrinamine is illustrated in Fig. 2. The seven largest peaks obtained from the symbolic addition procedure, which were used as the basis for proceeding with the recycling with the tangent formula, are shown as open circles in Fig. 2, Part I. Among the next seven largest peaks were five meaningful peaks indicated by the solid circles and two indicated by a cross which were spurious. The initial partial structure was limited to the seven largest peaks not only because they were the largest peaks, but also because they clearly formed a meaningful chemical unit. The acceptance criterion used with this partial structure was $|F_{\mathbf{h}}|_{calc} > 0{\cdot}20\ |F_{\mathbf{h}}|_{obs}$ and $|E_{\mathbf{h}}| > 1{\cdot}5$. Phases were obtained for reflections with $|E| > 1{\cdot}2$ using the tangent formula and the new E-map gave a meaningful three-ring system shown in Fig. 2, Part II. Successive cycles are shown in Parts III and IV. In each cycle the coefficient of $|F_{\mathbf{h}}|_{obs}$ in the acceptance criterion was increased since the number of atoms in the partial structure increased. The complete structure was found in cycle V.

Some additional almost equal atom noncentrosymmetric structures which have been solved from a partial structure are a photolysis product of a derivative of tyrosine (Karle, Karle & Estlin, 1967) digitoxigenin (Karle & Karle, 1969 a), anhydrodigitoxigenin (Gilardi & Karle, 1969), a mescaline photoproduct (Karle & Karle, 1969 b), and a cyclotetradepsipeptide (Konnert & Karle, 1969). They range in complexity from 17 to 46 atoms, aside from hydrogen, in the asymmetric unit and from 68 to 184 in the unit cell. In choosing the initial partial structures consisting of about a half dozen atoms or more, only those were selected, as noted above from the initial E-map which could form a good bonded unit making chemical sense. It was found that generally the determination was completed in about four cycles. Occasionally it was convenient to locate the last few atoms associated with large thermal factors by means of a difference map.

The tangent formula is also effective for developing structures containing heavy atoms after the heavy atoms have been located by means of a Patterson map. The heavy atoms act as the initial partial structure for generating phases and additional phases are obtained from the tangent formula. Some applications have been made to two anti-radiation agents containing sulfur (Karle, Flippen & Karle, 1967; Karle, Karle & Mitchell, 1969) and to the anti-heroin agent cyclazocine (Karle, Gilardi, Fratini & Karle, 1969) obtained as the hydrobromide. The latter substance crystallized in space group $P2_1$. In order to obtain initial phases other than zero or pi, it was necessary to add two light atoms to the bromine atom in the initial partial structure. It was also necessary in this case to hold the phases obtained from the structural fragment constant during the course of the development of new phases by means of the tangent formula. An attempt to refine the initial phases would cause a deterioration of the set to the centrosymmetric phases corresponding to the bromine atom position. With the heavy atom partial structures, a complete structure was obtained in one or two cycles of application of the tangent formula.

As a matter of interest, a test was made to determine whether a pair of atoms would suffice as a starting partial structure with anhydrodigitoxigenin (Gilardi & Karle, 1969), a cyclotetradepsipeptide (Konnert & Karle, 1969), and a photorearrangement product from a tyrosine derivative (Karle, Karle & Estlin, 1967), having 26, 32, and 17 non-hydrogen atoms in the asymmetric unit, respectively. In each case the two atoms could be readily developed into a complete structure implying that two atoms properly placed in the unit cell could also suffice as an initial fragment for still larger structures.

References

Gilardi, R. D. & Karle, I. L. (1969). Abstracts, Meeting of the International Union of Crystallography, Stony Brook, N.Y.

Hoppe, W. & Will, G. (1960). *Z. Kristallogr.* **113**, 104.

Karle, I. L., Gilardi, R. D., Fratini, A. V. & Karle, J. (1969). *Acta Cryst.* B**25**, 1469.

Karle, I. L. & Karle, J. (1968). *Acta Cryst.* B**24**, 81.

Karle, I. L. & Karle, J. (1969 a). *Acta Cryst.* B**25**, 434.

Karle, I. L. & Karle, J. (1969 b). *Acta Cryst.* In press.

Karle, I. L., Karle, J. & Estlin, J. A. (1967). *Acta Cryst.* **23**, 494.

Karle, J. (1968). *Acta Cryst.* **9**, 635.

Karle, J., Estlin, J. A. & Karle, I. L. (1967). *J. Amer. Chem. Soc.* **89**, 6510.

Karle, J., Flippen, J. & Karle, I. L. (1967). *Z. Kristallogr.* **125**, 201.

Karle, J. & Hauptman, H. (1956). *Acta Cryst.* **9**, 635.

Karle, J., Karle, I. L. & Mitchell, D. (1969). *Acta Cryst.* B**25**, 866

Konnert, J. & Karle, I. L. (1969). *J. Amer. Chem. Soc.* **91**, 4888.

Nordman, C. E. & Nakatsu, K. (1963). *J. Amer. Chem. Soc.* **85**, 353.

Discussion

TSOUCARIS: I would like to point out that the criteria for accepting or rejecting reflections in the phasing process could be incorporated into a weighting scheme and used directly.

WEBER: What do you mean by slow refinement of the phases in the last $P2_1$ structure you discussed which contained a bromine atom?

J. KARLE: You should fix the phases accepted by the $|F_{calc}| > p|F_{obs}|$ criterion and refine only the new values. Otherwise the phases will return to the heavy atom values.

ROLLETT: What is the resolution required in the data?

J. KARLE: The lowest resolution data we have used is half the Cu sphere; however we always collect as much data as possible. Resolution does not seem to be an important limitation for the type of organic molecules we study; half to a full Cu sphere seems to be suitable.

KARTHA: It seems to me that the tangent formula and the Fourier transform procedures are in principle equivalent, and the choice of the method depends on personal preferences and computing convenience. Dealing with small molecules and few reflections, the tangent formula has the advantage of computing efficiency. However, when we are dealing with larger molecules the Fourier method has a computational edge. This presumably is due to the fact that the computing volume increases with the size of the molecule much more rapidly in application of tangent formula than Fourier refinement.

J. KARLE: This well may be; I really would not like to make a value judgment at this time.

The Calculation of Structure Invariants

HERBERT HAUPTMAN

Naval Research Laboratory, Washington, D.C., U.S.A.

§ 1. Background

For simplicity it will be assumed that the unit cell of a crystal consists of N identical atoms, the position vector of the jth one of which is denoted by $\mathbf{r}_j$. Then the (complex) normalized structure factors $E_{\mathbf{h}}$ are defined by means of

$$E_{\mathbf{h}} = |E_{\mathbf{h}}| \exp(i\varphi_{\mathbf{h}}) = \frac{1}{N^{1/2}} \sum_{j=1}^{N} \exp(2\pi i \mathbf{h} \cdot \mathbf{r}_j) . \quad (1.1)$$

Evidently, then

$$\langle E_{\mathbf{h}} \exp(-2\pi i \mathbf{h} \cdot \mathbf{r}) \rangle_{\mathbf{h}} = \frac{1}{N^{1/2}} \left\langle \sum_{j=1}^{N} \exp 2\pi i h \cdot (\mathbf{r}_j - \mathbf{r}) \right\rangle_{\mathbf{h}} \begin{aligned} &= \frac{1}{N^{1/2}} \text{ if } \mathbf{r} = \mathbf{r}_j \\ &= 0 \quad \text{if } \mathbf{r} \neq \mathbf{r}_j . \end{aligned} \quad (1.2)$$

Eq. 1.2 implies that the normalized structure factors $E_{\mathbf{h}}$ determine the crystal structure. However, Eq. 1.1 does not imply that, conversely, the crystal structure determines the values of the normalized structure factors, $E_{\mathbf{h}}$, since the position vectors $\mathbf{r}_j$ depend not only on the crystal structure but on the choice of origin as well. It turns out nevertheless that the magnitudes, $|E_{\mathbf{h}}|$, of the normalized structure factors are in fact uniquely determined by the crystal structure but that the values of the phases, $\varphi_{\mathbf{h}}$, depend also on the choice of origin (Hauptman & Karle, 1956; Hauptman, 1963).

A finite number of magnitudes, $|E_{\mathbf{h}}|$, are obtainable from experiment whereas the phases, $\varphi_{\mathbf{h}}$, cannot be measured. Nevertheless, Eq. 1.2 can still be used to determine crystal structures since the prior knowledge that the crystal consists of N identical point atoms severely restricts the values of the phases, $\varphi_{\mathbf{h}}$. In fact, once the origin has been specified (as well as the enantiomorph when appropriate), the magnitudes $|E_{\mathbf{h}}|$ uniquely determine, in general, those phases $\varphi_{\mathbf{h}}$ having the property that the left side of Eq. 1.2, as a function of $\mathbf{r}$, is then zero everywhere except at N points where it has the value $1/N^{1/2}$. (Naturally, owing to the finite number of data obtainable from experiment, Eq. 1.2 is non-vanishing in N discrete regions, not points, and is only approximately zero elsewhere. In practice therefore it is the maxima of Eq. 1.2 which yield the atomic positions.)

Although the values of the individual phases depend on the crystal structure and the choice of origin, it turns out that there exist certain linear combinations of the phases, the so-called structure invariants, whose values are determined by the structure alone and are independent of the choice of origin. The most important set of structure invariants are the linear combinations of the phases

$$\varphi_{\mathbf{h}_1} + \varphi_{\mathbf{h}_2} + \varphi_{\mathbf{h}_3} \quad (1.3)$$

where

$$\mathbf{h}_1 + \mathbf{h}_2 + \mathbf{h}_3 = 0 . \quad (1.4)$$

It is not difficult to show also that the magnitudes of the normalized structure factors, $|E_{\mathbf{h}}|$, determine the magnitudes of the structure invariants or, equivalently, the values of the cosines,

$$\cos(\varphi_{\mathbf{h}_1}+\varphi_{\mathbf{h}_2}+\varphi_{\mathbf{h}_3}), \qquad (1.5)$$

where Eq. 1.4 holds. In view of these facts the problem of determining the values of the phases of the structure factors, given only their magnitudes, may be replaced by two others. The first is to calculate the structure invariants 1.5 from the magnitudes of the structure factors. The second is to determine the values of the individual phases assuming that the values of the structure invariants, 1.5, have been found. The present article is devoted to a description of a program for phase determination along these lines.

The calculation of the structure invariants is strongly dependent on a recent study of the different interactions which may occur among the Patterson peaks (Hauptman, 1964). Not only does this work show that an earlier formula, leading from the magnitudes of the structure factors to the magnitudes of the structure invariants, is not valid if the Patterson function contains overlapping peaks, but it leads also to an improved formula in the case that interatomic vectors do, in fact, coincide. However, a further modification turns out to be necessary when the overlap in the Patterson peaks is only approximate, a frequent occurrence. The method also depends in an essential way on the conditional probability distribution of the cosines of the structure invariants as well as on a novel least-squares technique leading from the approximate values of the structure invariants, 1.5, to initial values of the individual phases.

§ 2. Overview of the method

For definiteness it will be assumed that the space group is $P2_1$. The same method, with obvious modifications, is valid for the other non-centrosymmetric space groups.

As many phases $\varphi_{2h\,0\,2l}=0$ or π as possible are first determined by means of Σ_1 (Hauptman & Karle, 1953):

$$E_{2h\,0\,2l} \approx N^{1/2}\langle(-1)^k(|E_{hkl}|^2-1)\rangle_k . \qquad (2.1)$$

Next, the values of a linearly independent pair of phases $\varphi_{\mathbf{h}_1}=\varphi_{h_1\,0\,l_1}$, $\varphi_{\mathbf{h}_2}=\varphi_{h_2\,0\,l_2}$ are arbitrarily specified (*i.e.*, either 0 or π) and the value of a third phase $\varphi_{\mathbf{h}_3}=\varphi_{h_3\,1\,l_3}$ is also arbitrarily specified (between 0 and 2π, *e.g.* 0), thus uniquely fixing the origin (Hauptman & Karle, 1956). The vectors $\mathbf{h}_1$, $\mathbf{h}_2$, $\mathbf{h}_3$ are so chosen that $|E_{\mathbf{h}_1}|$, $|E_{\mathbf{h}_2}|$, and $|E_{\mathbf{h}_3}|$ are large and in such a way that they 'interact' with each other (*i.e.*, many of the $|E_{\mathbf{h}_i\pm\mathbf{h}_j}|$, $i, j=1, 2, 3$, are large) and with those two-dimensional vectors $\mathbf{h}$ such that the values of the $\varphi_{\mathbf{h}}$ have been previously determined by Σ_1.

Next, owing to the space group symmetry, knowledge of the phase φ_{hkl} implies knowledge of the three additional phases $\varphi_{\bar{h}\bar{k}l}$, $\varphi_{\bar{h}k\bar{l}}$, $\varphi_{\bar{h}\bar{k}\bar{l}}$ if $k\neq 0$ while, if $k=0$, only one additional phase, $\varphi_{\bar{h}\,0\,\bar{l}}$, is known.

The whole phase determination rests on the basic set of phases consisting of those determined by means of Σ_1 and the three used to fix the origin. Ordinarily one attempts to find the values of only those phases, $\varphi_{\mathbf{h}}$, whose corresponding magnitudes, $|E_{\mathbf{h}}|$, are large, say, greater than unity. Such phases will be said to be permissible. Preliminary values of a limited set of permissible phases are then determined iteratively as follows.

Corresponding to each pair $\varphi_{\mathbf{k}}$, $\varphi_{-\mathbf{h}-\mathbf{k}}$ of the basic set having the property that the phase $\varphi_{\mathbf{h}}$ is permissible, one constructs the associated structure invariant $\varphi_{\mathbf{h}}+\varphi_{\mathbf{k}}+\varphi_{-\mathbf{h}-\mathbf{k}}$ and the product

$$\frac{2}{N^{1/2}}|E_{\mathbf{h}}E_{\mathbf{k}}E_{\mathbf{h}+\mathbf{k}}|=A_{\mathbf{k}} . \qquad (2.2)$$

For each such vector $\mathbf{h}$, the sum $\sum_{\mathbf{k}} A_{\mathbf{k}}$ (which may consist of only a single term), taken over all allowed vectors $\mathbf{k}$, is computed. These sums (one for each **h**) are arranged in decreasing order and the largest one selected, thus defining a unique vector **h** and several (perhaps only one) structure invariants $\cos(\varphi_{\mathbf{h}}+\varphi_{\mathbf{k}}+\varphi_{-\mathbf{h}-\mathbf{k}})$ the values of which will be determined in the manner described in § 3. Since the values of the $\varphi_{\mathbf{k}}$, $\varphi_{-\mathbf{h}-\mathbf{k}}$ and the several $\cos(\varphi_{\mathbf{h}}+\varphi_{\mathbf{k}}+\varphi_{-\mathbf{h}-\mathbf{k}})$ are thus presumed to be known, the value of $\varphi_{\mathbf{h}}$ is actually overdetermined (in general) and its evaluation by least-squares will be described in § 4. The new phase, $\varphi_{\mathbf{h}}$, and its three (or possibly only one) symmetry-related phases are added to the basic set of known phases and the process repeated. During the second cycle, however, the three largest sums $\sum_{\mathbf{k}} A_{\mathbf{k}}$ are selected, so that three new phases are determined, rather than only one as in the first cycle. During the third, fourth, fifth, ..., cycles, the numbers of new phases determined are five, seven, nine, ... Thus, the order in which the values of the phases are to be obtained is determined and the identity of the structure invariants, $\cos(\varphi_{\mathbf{h}}+\varphi_{\mathbf{k}}+\varphi_{-\mathbf{h}-\mathbf{k}})$, whose values are required, is also found. Clearly, the rate at which new phases are picked up may be varied if desired.

The first structure invariant, $\cos(\varphi_{\mathbf{h}}+\varphi_{\mathbf{k}}+\varphi_{-\mathbf{h}-\mathbf{k}})$, whose value is different from ± 1 leads to two possible values for $\varphi_{\mathbf{h}}+\varphi_{\mathbf{k}}+\varphi_{-\mathbf{h}-\mathbf{k}}$ differing only in sign. The sign of this invariant, $\varphi_{\mathbf{h}}+\varphi_{\mathbf{k}}+\varphi_{-\mathbf{h}-\mathbf{k}}$, may be arbitrarily specified, thus fixing the enantiomorph and leading to a unique value for $\varphi_{\mathbf{h}}$ (Hauptman & Karle, 1956). Equivalently, there are two values for $\varphi_{\mathbf{h}}$ yielding the same minimum sum of squares. One value is chosen arbitrarily and the enantiomorph is thus fixed. All subsequent phases are then uniquely determined.

§ 3. Calculation of the structure invariants

In order to carry out several cycles of the process described in § 2, it is necessary to compute the values of some one or two thousand structure invariants $\cos(\varphi_{\mathbf{h}}+\varphi_{\mathbf{h}}+\varphi_{-\mathbf{h}-\mathbf{k}})$. To this end a variant of Eq. 5.12 (Hauptman, 1964),

$$\begin{aligned}|E_1E_2E_3|\cos(\varphi_1+\varphi_2+\varphi_3) &\approx \frac{\sigma_2^3-3\sigma_1\sigma_2+2\sigma_3}{\sigma_2^{3/2}\langle(|E_{\mathbf{l}}|^2-1)^3\rangle_{\mathbf{l}}} \\ &\times \langle(|E_{\mathbf{l}}|^2-1)(|E_{\mathbf{h}_1+\mathbf{l}}|^2-1) \\ &\times (|E_{-\mathbf{h}_3+\mathbf{l}}|^2-1)\rangle_{\mathbf{l}} \\ &+\frac{\sigma_3}{\sigma_2^{3/2}}(|E_1|^2+|E_2|^2 \\ &+|E_3|^2-2)\,, \end{aligned} \tag{3.1}$$

where **l** ranges over all possible vectors corresponding to observed intensities, has proved to be useful. It was shown in the latter reference that earlier versions of this formula (*e.g.* Eq. 2.2 of Karle & Hauptman, 1957; Eq. 2.1.3 of Karle & Hauptman, 1958; or Eq. 16 of Vaughan, 1958) are not valid if the structure contains a considerable number of induced or chance interactions (*e.g.* if Patterson peaks overlap). Since substantial overlap in the Patterson function is a frequent occurrence in real crystals, the earlier formulas are not useful and Eq. 3.1 represents a significant improvement over them. However, owing to the fact that overlap among Patterson peaks is usually only approximate, even the improved formula 3.1 is still inadequate. Instead the following variant (Hauptman *et al.*, 1969) turns out to be more useful:

$$\begin{aligned}|E_1E_2E_3|\cos(\varphi_1+\varphi_2+\varphi_3) &\approx K\langle(|E_{\mathbf{l}}|^{1/2}-\overline{|E|^{1/2}})(|E_{\mathbf{h}_1+\mathbf{l}}|^{1/2}-\overline{|E|^{1/2}}) \\ &\times(|E_{-\mathbf{h}_3+\mathbf{l}}|^{1/2}-\overline{|E|^{1/2}})\rangle_{\mathbf{l}}+R_3\,, \end{aligned} \tag{3.2}$$

where

$$\begin{aligned}R_3=\frac{1}{4N^{1/2}}&(\tfrac{3}{2}(|E_1E_2|^2+|E_2E_3|^2+|E_3E_1|^2) \\ &+|E_1|^2+|E_2|^2+|E_3|^2-\tfrac{7}{2})\,, \end{aligned} \tag{3.3}$$

$$\overline{|E|^{1/2}}=\langle|E_{\mathbf{l}}|^{1/2}\rangle_{\mathbf{l}}\,, \tag{3.4}$$

K is a sliding scale factor whose value depends on $|E_1E_2E_3|$ in a manner to be described, and the abbreviations

$$|E_1|=|E_{\mathbf{h}_1}|=|E_{\mathbf{h}}|,\ |E_2|=|E_{\mathbf{h}_2}|=|E_{\mathbf{k}}|,\ |E_3|=|E_{\mathbf{h}_3}|=|E_{-\mathbf{h}-\mathbf{k}}|\,, \quad (3.5)$$

$$\varphi_1=\varphi_{\mathbf{h}_1}=\varphi_{\mathbf{h}},\ \varphi_2=\varphi_{\mathbf{h}_2}=\varphi_{\mathbf{k}},\ \varphi_3=\varphi_{\mathbf{h}_3}=\varphi_{-\mathbf{h}-\mathbf{k}} \quad (3.6)$$

have been used, so that Eq. 1.4 is satisfied.

In order to determine the sliding scale factor K which appears in Eq. 3.2, the conditional probability distribution of $\cos(\varphi_1+\varphi_2+\varphi_3)$, given $|E_1E_2E_3|$, which is described in the Appendix is employed. It turns out that K is a function of $|E_1E_2E_3|$ and the value of K is chosen so that the resulting frequency distribution of the $\cos(\varphi_1+\varphi_2+\varphi_3)$ agrees, for fixed $|E_1E_2E_3|$, with the known theoretical conditional probability distribution of $\cos(\varphi_1+\varphi_2+\varphi_3)$ as given in the Appendix. One way of achieving this is to arrange the computed averages in Eq. 3.2 in decreasing order of $|E_1E_2E_3|$ and group them into sets having 200 to 300 members in each set. Thus, in each set the values of $|E_1E_2E_3|$ are essentially constant. The value of K in each such set is determined in such a way as to force the frequency and theoretical distributions of the $\cos(\varphi_1+\varphi_2+\varphi_3)$ to coincide.

An alternative scheme is to rank, within each set of 200 or 300, the computed averages of Eq. 3.2 in increasing order and to assign those values to the corresponding $\cos(\varphi_1+\varphi_2+\varphi_3)$ which will satisfy the theoretical probability distribution and, at the same time, preserve the order of the computed averages. Both this method and the one described in the previous paragraph have been successfully applied.

§ 4. Approximate evaluation of initial phases by means of least squares

Suppose now that the structure invariants $\cos(\varphi_{\mathbf{h}}+\varphi_{\mathbf{k}}+\varphi_{-\mathbf{h}-\mathbf{k}})$ called for by the iterative process described in § 2 have been computed by the method of § 3. Thus, for each fixed vector $\mathbf{h}$ whose corresponding phase $\varphi_{\mathbf{h}}$ is to be determined the values of several structure invariants

$$\cos(\varphi_{\mathbf{h}}+\varphi_{\mathbf{k}}+\varphi_{-\mathbf{h}-\mathbf{k}})=c_{\mathbf{k}} \quad (4.1)$$

are presumed to be known. The values of the several phases $\varphi_{\mathbf{h}}$, $\varphi_{-\mathbf{h}-\mathbf{k}}$ are also assumed to be known in accordance with § 2. It is natural to suppose further that the larger the number n of contributors to the average in Eq. 3.2 from which $c_{\mathbf{k}}$ is computed and the larger the value of $|E_{\mathbf{h}}E_{\mathbf{k}}E_{\mathbf{h}+\mathbf{k}}|$, the more accurate the computed value of $c_{\mathbf{k}}$ will be. Hence a weight $w_{\mathbf{k}}$ is defined by means of

$$w_{\mathbf{k}}=|E_{\mathbf{h}}E_{\mathbf{k}}E_{\mathbf{h}+\mathbf{k}}|\sqrt{n}\,. \quad (4.2)$$

The phase $\varphi_{\mathbf{h}}$ is then determined by minimizing, in accordance with the principle of least-squares,

$$\Phi=\frac{\sum_{\mathbf{k}} w_{\mathbf{k}}(\cos(\varphi_{\mathbf{h}}+\varphi_{\mathbf{k}}+\varphi_{-\mathbf{h}-\mathbf{k}})-c_{\mathbf{k}})^2}{\sum_{\mathbf{k}} w_{\mathbf{k}}} \quad (4.3)$$

which, after a straightforward but lengthy calculation, finally reduces to

$$\Phi=\tfrac{1}{2}C_2\cos 2\varphi_{\mathbf{h}}-\tfrac{1}{2}S_2\sin 2\varphi_{\mathbf{h}} -2C_1\cos\varphi_{\mathbf{h}}+2S_1\sin\varphi_{\mathbf{h}}+c\,, \quad (4.4)$$

where

$$C_2=\frac{\sum_{\mathbf{k}} w_{\mathbf{k}}\cos 2(\varphi_{\mathbf{k}}+\varphi_{-\mathbf{h}-\mathbf{k}})}{\sum_{\mathbf{k}} w_{\mathbf{k}}} =\langle\cos 2(\varphi_{\mathbf{k}}+\varphi_{-\mathbf{h}-\mathbf{h}})\rangle_{\mathbf{k}} \quad (4.5)$$

$$S_2=\frac{\sum_{\mathbf{k}} w_{\mathbf{k}}\sin 2(\varphi_{\mathbf{k}}+\varphi_{-\mathbf{h}-\mathbf{k}})}{\sum_{\mathbf{k}} w_{\mathbf{k}}} =\langle\sin 2(\varphi_{\mathbf{k}}+\varphi_{-\mathbf{h}-\mathbf{k}})\rangle_{\mathbf{k}} \quad (4.6)$$

$$C_1 = \frac{\sum\limits_{\mathbf{k}} w_{\mathbf{k}} c_{\mathbf{k}} \cos(\varphi_{\mathbf{k}} + \varphi_{-\mathbf{h}-\mathbf{k}})}{\sum\limits_{\mathbf{k}} w_{\mathbf{k}}} = \langle c_{\mathbf{k}} \cos(\varphi_{\mathbf{k}} + \varphi_{-\mathbf{h}-\mathbf{k}})\rangle_{\mathbf{k}} \quad (4.7)$$

$$S_1 = \frac{\sum\limits_{\mathbf{k}} w_{\mathbf{k}} c_{\mathbf{k}} \sin(\varphi_{\mathbf{k}} + \varphi_{-\mathbf{h}-\mathbf{k}})}{\sum\limits_{\mathbf{k}} w_{\mathbf{k}}} = \langle c_{\mathbf{k}} \sin(\varphi_{\mathbf{k}} + \varphi_{-\mathbf{h}-\mathbf{k}})\rangle_{\mathbf{k}} \quad (4.8)$$

$$c = \frac{\sum\limits_{\mathbf{k}} w_{\mathbf{k}}(\tfrac{1}{2} + c_{\mathbf{k}}^2)}{\sum\limits_{\mathbf{k}} w_{\mathbf{k}}} = \langle \tfrac{1}{2} + c_{\mathbf{k}}^2 \rangle_{\mathbf{k}}, \quad (4.9)$$

so that the quantities C_2, S_2, C_1, S_1, and c are all known. Then $\varphi_{\mathbf{h}}$ is uniquely determined by minimizing Φ as given by (4.4). While the problem of minimizing Φ may be solved by standard techniques and leads ultimately to the solution of a quartic equation, it turns out that, with the aid of modern automatic computers, Φ is more readily minimized by computing Φ as a function of $\varphi_{\mathbf{h}}$ (using Eq. 4.4–4.9) and then choosing the (in general unique) value of $\varphi_{\mathbf{h}}$ which minimizes Φ. It may happen occasionally, particularly in one of the early cycles of the iteration, that two distinct values of $\varphi_{\mathbf{h}}$ lead to essentially the same minimum value for Φ. In this case one is led to two or more solutions and the incorrect ones will have to be rejected by inspection of the corresponding Fourier series. Finally, the first structure invariant, $\cos(\varphi_{\mathbf{h}} + \varphi_{\mathbf{k}} + \varphi_{-\mathbf{h}-\mathbf{k}})$, whose value differs significantly from ± 1 yields two possible values for $\varphi_{\mathbf{h}} + \varphi_{\mathbf{k}} + \varphi_{-\mathbf{h}-\mathbf{k}}$, differing only in sign, and therefore to two possible values for $\varphi_{\mathbf{h}}$. Choosing either of these two values for $\varphi_{\mathbf{h}}$ is equivalent to choosing one of the two possible signs for $\varphi_{\mathbf{h}} + \varphi_{\mathbf{h}} + \varphi_{-\mathbf{h}-\mathbf{k}}$ and therefore to selecting one of the two enantiomorphs. Once this is done, the remaining phases are then uniquely determined. It has been found in practice that it is feasible to determine the values of several dozen phases by this process. The more efficient tangent formula (Karle & Hauptman, 1956, Eq. 5.62),

$$\tan \varphi_{\mathbf{h}} = \frac{\sum\limits_{\mathbf{k}} |E_{\mathbf{k}} E_{\mathbf{h}-\mathbf{k}}| \sin(\varphi_{\mathbf{k}} + \varphi_{\mathbf{h}-\mathbf{k}})}{\sum\limits_{\mathbf{k}} |E_{\mathbf{k}} E_{\mathbf{h}-\mathbf{k}}| \cos(\varphi_{\mathbf{k}} + \varphi_{\mathbf{h}-\mathbf{k}})}, \quad (4.10)$$

may then be used in order to evaluate the remaining permissible phases.

§ 5. Applications

5.1. *The iterative process*

It will be instructive to describe in some detailed the application of the present method to the phase determination for estriol, $C_{18}H_{24}O_3$ (Hauptman *et al.*, 1969). The space group is $P2_1$ and there are four molecules in the unit cell. The phase $\varphi_{400} = 0$ was determined by means of Σ_1 and the phase $\varphi_{200} = 0$ was found by a special argument which is not relevant here. The three phases $\varphi_{30\bar{9}}$, $\varphi_{70\bar{4}}$, and $\varphi_{21\bar{1}}$ were arbitrarily specified in order to fix the origin. Thus the phase determination commenced with a basic set of five independent phases or, taking into account the space group symmetry, twelve phases in all.

Table I shows the computer output of the first three cycles of the iterative process described in § 2. Although φ_{200} is in the

Table I. Three cycles of the iterative process.

$\vec{h}$	$\lvert E_{\vec{h}} \rvert$	Cycle	$\langle A_{\vec{-k}} \rangle$	No. of Contrib.	$\sum_{\vec{k}} A_{\vec{-k}}$	
Cycle 1						
2 0 0	5.3961	0	12.7680	1	12.7680	
4 0 0	2.0433	0	12.7680	1	12.7600	
4 1 $\bar{1}$	2.6041	1	10.4616	1	10.4616	***
Cycle 2						
0 1 1	2.0013	2	5.1626	2	10.3252	***
1 0 $\bar{9}$	2.8268	2	9.5935	1	9.5935	***
2 0 0	5.3961	0	11.2304	3	33.6911	
2 1 $\bar{1}$	3.4692	0	10.4616	1	10.4616	
4 0 0	2.0433	0	12.7680	1	12.7680	
4 1 $\bar{1}$	2.6041	1	10.4616	1	10.4616	
6 1 $\bar{1}$	1.6771	2	3.8043	2	7.6086	***
Cycle 3						
0 1 1	2.0013	2	5.1626	2	10.3252	
1 0 $\bar{9}$	2.8268	2	9.5935	1	9.5935	
1 1 8	2.7441	3	5.1346	2	10.2693	***
2 0 0	5.3961	0	8.6849	8	69.4792	
2 1 $\bar{1}$	3.4692	0	7.0176	3	21.0527	
3 0 $\bar{9}$	2.9307	0	9.5935	1	9.5935	
4 0 $\bar{2}$	1.4864	3	2.3880	3	7.1639	***
4 0 0	2.0433	0	4.4882	5	22.4409	
4 1 $\bar{1}$	2.6041	1	5.9347	3	17.8042	
4 2 $\bar{2}$	1.5521	3	2.8723	2	5.7446	***
5 0 $\bar{9}$	1.7643	3	4.0873	2	8.1745	***
5 0 $\bar{4}$	1.7579	3	4.2050	1	4.2050	***
6 1 $\bar{1}$	1.6771	2	3.8043	2	7.6086	

basic set, this phase still appears in Cycle 1 in view of the identification

$$\varphi_{200}+\varphi_{200}+\varphi_{\bar{4}00}=\varphi_{\mathbf{h}}+\varphi_{\mathbf{k}}+\varphi_{-\mathbf{h}-\mathbf{k}}$$

where both φ_{200} and $\varphi_{\bar{4}00}$ are in the basic set. Similarly, the identity

$$\varphi_{400}+\varphi_{\bar{2}00}+\varphi_{\bar{2}00}=\varphi_{\mathbf{h}}+\varphi_{\mathbf{k}}+\varphi_{-\mathbf{h}-\mathbf{k}}$$

where $\varphi_{\bar{2}00}$ is in the basic set explains the appearance of φ_{400} in this cycle. The phase $\varphi_{41\bar{1}}$ occurs in this cycle because of

$$\varphi_{41\bar{1}}+\varphi_{\bar{2}\bar{1}1}+\varphi_{\bar{2}00}=\varphi_{\mathbf{h}}+\varphi_{\mathbf{k}}+\varphi_{-\mathbf{h}-\mathbf{k}}$$

and is the only new phase to appear. As it happened, the computed value of $\cos(\varphi_{41\bar{1}}+\varphi_{\bar{2}1\bar{1}}+\varphi_{\bar{2}00})$ turned out to be $+1{\cdot}0000$, so that the enantiomorph could not be chosen at this point. Hence the phase $\varphi_{41\bar{1}}$ was computed in the first cycle and it is expected that its value will be 0. (See Table III.) Column 2 is self-explanatory. Column 3 gives the first cycle in which the corresponding phase appears. Thus the entry 0 means that the phase is one of five in the basic set; the entry 1 signifies that the corresponding phase appears, for the first time, in the first cycle; etc. For each fixed **h**, $A_{\mathbf{k}}$ is defined by Eq. 2.2 so that Column 6 is self-explanatory. Column 5 gives the number of contributors to the sum $\sum_{\mathbf{k}} A_{\mathbf{k}}$ so that Column 4 is also self-evident. The asterisks in the last column indicate the new phases in each cycle. In view of

$$\varphi_{011}+\varphi_{2\bar{1}\bar{1}}+\varphi_{\bar{2}00}=\varphi_{\mathbf{h}}+\varphi_{\mathbf{k}}+\varphi_{-\mathbf{h}-\mathbf{k}},$$

$$\varphi_{011}+\varphi_{4\bar{1}\bar{1}}+\varphi_{\bar{4}00}=\varphi_{\mathbf{h}}+\varphi_{\mathbf{k}}+\varphi_{-\mathbf{h}-\mathbf{k}},$$

where $\varphi_{2\bar{1}\bar{1}}$, $\varphi_{\bar{2}00,\bar{4}00}$ φ are in the basic set and $\varphi_{4\bar{1}\bar{1}}$ appears in the earlier Cycle 1, the new phase φ_{011} occurs in Cycle 2 with the entry 2 in Column 5. The remaining entries in Table I are self-explanatory. Twelve cycles of this iterative procedure, of which only the first three are shown in Table I, yielded 144 new phases which, together with the 5 in the basic set, led to a total of 149 phases. However, the values of 1805 structure invariants, $\cos(\varphi_1+\varphi_2+\varphi_3)$, identified by the iterative process and exhibited, in part, in Table II, were first required before individual phases could be computed. How these were found is described in the next section.

5.2. *The calculation of the structure invariants,* $\cos(\varphi_1+\varphi_2+\varphi_3)$

Twelve cycles of the iterative process described in § 5.1 led to 1805 structure in-

Table II. Values of $\cos(\varphi_1+\varphi\ +\varphi\ )$ arranged in decreasing order of **A**.

$\vec{h}_1$	$\vec{h}_2$	$\vec{h}_3$	$\lvert E_1\rvert$	$\lvert E_2\rvert$	$\lvert E_3\rvert$	A	n	$\langle\ \rangle$	computed $\cos(\varphi_1+\varphi_2+\varphi_3)$	true $\cos(\varphi_1+\varphi_2+\varphi_3)$
$2\,0\,0$	$0\,\bar{4}\,\bar{6}$	$\bar{2}\,4\,6$	5.3961	3.4964	3.3412	13.5281	5871	0.003162	0.9458	0.9888
$2\,0\,0$	$2\,1\,\bar{1}$	$\bar{4}\,\bar{1}\,1$	5.3961	3.4692	2.6041	10.4616	9149	0.003717	1.0000	0.9988
$2\,0\,0$	$\bar{2}\,\bar{5}\,6$	$0\,5\,\bar{6}$	5.3961	3.0813	2.8552	10.1878	5809	0.003176	1.0000	0.9991
$2\,0\,0$	$\bar{3}\,0\,9$	$1\,0\,\bar{9}$	5.3961	2.9307	2.8268	9.5935	3717	0.001957	0.8424	1.0000
$2\,0\,0$	$2\,6\,\bar{7}$	$\bar{4}\,\bar{6}\,7$	5.3961	2.8029	2.6917	8.7366	5120	0.004017	1.0000	0.9820
$2\,0\,0$	$7\,3\,\bar{8}$	$\bar{9}\,\bar{3}\,8$	5.3961	2.6929	2.6385	8.2279	3856	0.000774	0.6485	0.9553
$2\,0\,0$	$\bar{4}\,\bar{2}\,1$	$2\,2\,\bar{1}$	5.3961	2.6415	2.6311	8.0482	9093	0.002909	1.0000	0.9838
$2\,0\,0$	$\bar{2}\,\bar{1}\,1$	$0\,1\,\bar{1}$	5.3961	3.4692	2.0013	8.0399	10195	0.002024	0.9629	0.9926
$2\,0\,0$	$4\,2\,\bar{1}$	$\bar{6}\,\bar{2}\,1$	5.3961	2.6415	2.1928	6.7075	7587	0.003007	1.0000	0.9976
$2\,4\,\bar{6}$	$\bar{4}\,\bar{4}\,6$	$2\,0\,0$	3.3412	1.7292	5.3961	6.6905	6014	0.003013	1.0000	0.9737
$2\,0\,0$	$\bar{2}\,\bar{6}\,7$	$0\,6\,\bar{7}$	5.3961	2.8029	1.9896	6.4578	4935	0.002794	1.0000	0.9976
$2\,0\,0$	$0\,12\,0$	$\bar{2}\,\overline{12}\,0$	5.3961	2.5173	2.1986	6.4090	7738	0.001875	0.9293	0.9888
$2\,0\,0$	$\bar{2}\,\bar{4}\,\bar{5}$	$0\,4\,5$	5.3961	2.3732	2.3034	6.3301	6201	0.001738	0.8970	0.9759
$2\,1\,\bar{1}$	$2\,5\,\bar{6}$	$\bar{4}\,\bar{6}\,7$	3.4692	3.0813	2.6917	6.1748	5489	0.002609	1.0000	0.9960
$4\,5\,\bar{6}$	$\bar{2}\,\bar{5}\,6$	$\bar{2}\,0\,0$	1.7033	3.0813	5.3961	6.0776	5949	0.002589	1.0000	0.9992
$0\,5\,6$	$\bar{2}\,\bar{5}\,\bar{6}$	$2\,0\,0$	2.8552	1.8213	5.3961	6.0218	5306	0.001786	0.9358	0.9998
$3\,0\,\bar{9}$	$\bar{5}\,0\,9$	$2\,0\,0$	2.9307	1.7643	5.3961	5.9876	3850	0.001874	0.9975	1.0000
$2\,1\,\bar{1}$	$0\,5\,\bar{6}$	$\bar{2}\,\bar{6}\,7$	3.4692	2.8552	2.8029	5.9580	5364	0.002656	1.0000	0.9717
$2\,1\,\bar{1}$	$\bar{1}\,0\,9$	$\bar{1}\,\bar{1}\,\bar{8}$	3.4692	2.8268	2.7441	5.7750	3755	0.001446	0.7416	0.9664
$2\,0\,0$	$4\,4\,\bar{7}$	$\bar{6}\,\bar{4}\,7$	5.3961	2.4821	1.9951	5.7345	5092	0.002082	1.0000	0.9689
$2\,1\,\bar{1}$	$2\,\bar{5}\,\bar{6}$	$\bar{4}\,4\,7$	3.4692	3.0813	2.4821	5.6939	5569	0.001094	0.6617	0.9606
$4\,6\,\bar{7}$	$\bar{6}\,\bar{6}\,7$	$2\,0\,0$	2.6917	1.7792	5.3961	5.5458	4977	0.002073	1.0000	0.9928
$0\,4\,6$	$2\,\bar{6}\,\bar{7}$	$\bar{2}\,2\,1$	3.4964	2.8029	2.6311	5.5334	5362	0.001202	0.6876	0.7452
$0\,4\,6$	$2\,\bar{1}\,\bar{1}$	$\bar{2}\,\bar{3}\,\bar{5}$	3.4964	3.4692	2.1218	5.5231	5840	0.001701	0.8523	0.9988
$2\,4\,\bar{7}$	$\bar{4}\,\bar{4}\,7$	$2\,0\,0$	1.8856	2.4821	5.3961	5.4197	5236	0.001753	0.8866	0.9916

variants, $\cos(\varphi_1+\varphi_2+\varphi_3)$, whose values were needed in order to determine initial values for 144 new phases. The first 25 of these invariants, arranged in decreasing order of $A=A_{\mathbf{k}}$, are listed in Column 10 of Table II. The first seven columns of Table II are self-explanatory. Column 8 gives the number, n, of contributors to the average on the right side of Eq. 3.2 and the entries in Column 9 are the values of these averages. The values of R_3 needed in Eq. 3.2 are easily computed from Eq. 3.3 but are not listed in Table II. Instead the values of the invariants, $\cos(\varphi_1+\varphi_2+\varphi_3)$, as computed from Eq. 3.2, are listed in Column 10 whereas the true values of these invariants, as determined from the refined structure, are shown in Column 11 for comparison with the computed values.

The dependence of K on $A=2|E_1E_2E_3|/N^{1/2}$, obtained as described in § 3 and required by Eq. 3.2 in order to compute the invariants, $\cos(\varphi_1+\varphi_2+\varphi_3)$, is shown in Fig. 1. It is noteworthy that the computed values of K (indicated by dots) fall almost precisely on a smooth, monotonically increasing curve.

Table III. Five cycles of least-squares determination of the phases.

$\vec{h}$	$\|E_{\vec{h}}\|$	No. of Contrib.	$\sum_{\vec{k}} A_{\vec{k}}$	$\langle A_{\vec{k}}\rangle$	$\varphi_{\vec{h}}$	Residual	$\varphi_{\vec{h}}$	Residual
Cycle 1								
** 4 1 $\bar{1}$	2.6041	1	10.46	10.462	0.0012	0.000000		
Cycle Fig. of Merit		0.000 000.						
0 HOL Fig. of Merit		0.000 000.						
Cycle 2								
2 1 $\bar{1}$	3.4692	1	10.46	10.462	0.0017	0.000009		
4 1 $\bar{1}$	2.6041	1	10.46	10.462	0.0012	0.000000		
2 0 0	5.3961	2	20.92	10.462	0.0011	0.000009		
Cycle Fig. of Merit		0.000 000.						
1 HOL Fig. of Merit		0.001 125.						
Cycle 3								
** 1 0 $\bar{9}$	2.8268	1	9.59	9.593	-0.5690	0.000021	0.5690	0.000040
2 1 $\bar{1}$	3.4692	1	10.46	10.462	0.0017	0.000009		
4 1 $\bar{1}$	2.6041	1	10.46	10.462	0.0012	0.000000		
** 6 1 $\bar{1}$	1.6771	2	7.61	3.804	0.0015	0.000008		
2 0 0	5.3961	2	20.92	10.462	0.0011	0.000009		
** 0 1 1	2.0013	2	10.33	5.163	-2.8982	0.014934	2.8991	0.015162
Cycle Fig. of Merit		0.005761.						
2 HOL Fig. of Merit		0.319052.						
Cycle 4								
1 0 $\bar{9}$	2.8268	1	9.59	9.593	-0.5690	0.000021	0.5690	0.000040
3 0 $\bar{9}$	2.9307	1	9.59	9.593	-0.5690	0.000021	0.5690	0.000040
2 1 $\bar{1}$	3.4692	3	21.05	7.018	-0.0294	0.000372		
4 1 $\bar{1}$	2.6041	3	17.80	5.935	0.0843	0.005590		
6 1 $\bar{1}$	1.6771	2	7.61	3.804	0.0015	0.000008		
2 0 0	5.3961	7	56.71	8.102	0.0630	0.053548	-0.0628	0.053548
4 0 0	2.0433	4	9.67	2.418	0.0000	0.020416		
0 1 1	2.0013	2	10.33	5.163	-2.8982	0.014934	2.8991	0.015162
Cycle Fig. of Merit		0.034506.						
4 HOL Fig. of Merit		0.274289.						
Cycle 5								
1 0 $\bar{9}$	2.8268	1	9.59	9.593	-0.5690	0.000021	0.5690	0.000040
3 0 $\bar{9}$	2.9307	1	9.59	9.593	-0.5690	0.000021	0.5690	0.000040
** 5 0 $\bar{9}$	1.7643	2	8.17	4.087	0.3520	0.100665	-0.3520	0.100665
** 5 0 $\bar{4}$	1.7579	1	4.21	4.205	0.0012	0.000000		
** 4 0 $\bar{2}$	1.4864	2	3.32	1.662	1.7565	0.155781	-1.7565	0.155781
** 4 2 $\bar{2}$	1.5521	2	5.74	2.872	1.0009	0.089314	-0.8008	0.145720
2 1 $\bar{1}$	3.4692	3	21.05	7.018	-0.0197	0.004153		
4 1 $\bar{1}$	2.6041	3	17.80	5.935	0.0843	0.005590		
6 1 $\bar{1}$	1.6771	2	7.61	3.804	0.0471	0.000854		
2 0 0	5.3961	7	56.71	8.102	0.0000	0.053650		
4 0 0	2.0433	4	9.67	2.418	0.0000	0.008834		
0 1 1	2.0013	2	10.33	5.163	-2.8792	0.007648	2.9336	0.023768
** 1 1 8	2.7441	2	10.27	5.135	0.7022	0.026699	-0.6204	0.078028
Cycle Fig. of Merit		0.046856.						
7 HOL Fig. of Merit		0.366513.						

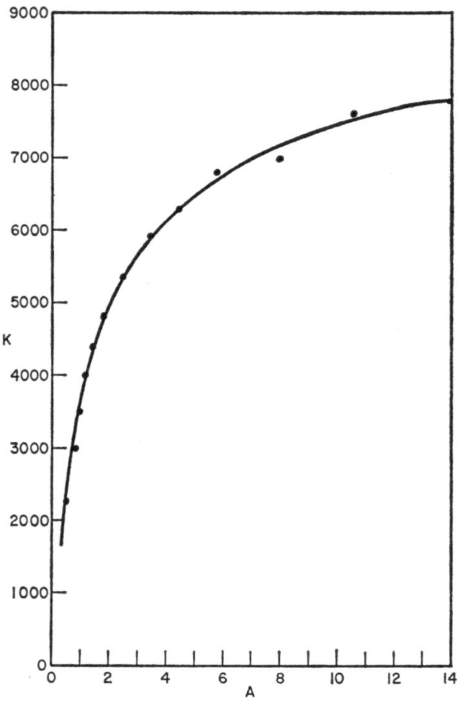

Fig. 1. *The dependence of K on A.*

5.3. *The calculation of the initial phases by means of least-squares*

Once the values of the required 1805 structure invariants were found as described in § 5.2, the values of the individual phases were computed by the method of least-squares (§ 4). The results of the first five cycles of this iteration are shown in Table III. Columns 1, 2, 4, and 5 are self-explanatory. Column 3 gives the number of contributors in $\sum_{\mathbf{k}} A_{\mathbf{k}}$ and the entries in Column 6 are the values of the phases $\varphi_{\mathbf{h}}$ (in radians) which minimize the weighted sum of squares, Φ, in Eq. 4.3. The residuals of Column 7 are the minimum values of Φ. In those cases that Φ has two relative minima, the corresponding second values of the phase, $\varphi_{\mathbf{h}}$, and the residual, minimum value of Φ, are listed in Columns 8 and 9 respectively. If it should happen that the residual listed in Column 7 is not significantly less than the corresponding residual of Column 9, then it may be necessary to permit the two possible values for $\varphi_{\mathbf{h}}$, as explained earlier. The five phases in the basic set were recomputed in the later cycles as contributors to $\sum_{\mathbf{k}} A_{\mathbf{k}}$ for these phases became available. The asterisks preceding Column 1 indicate the new phases computed for the first time in the corresponding cycle.

The computed value of the structure invariant

$$\cos(\varphi_{011}+\varphi_{\bar{2}00}+\varphi_{2\bar{1}\bar{1}})=0{\cdot}9629$$

was the first one which differed from ± 1. This invariant, together with the computed value of unity for $\cos(\varphi_{011}+\varphi_{\bar{4}00}+\varphi_{4\bar{1}\bar{1}})$, led, in Cycle 3 of Table III, to the two possible values $-2{\cdot}8982$, $+2{\cdot}8991$ for φ_{011} with identical residuals (except for round-off errors). The program arbitrarily selected one of these two values (actually the one corresponding to the 'smaller' residual) thus fixing the enantiomorph. All other phases were then uniquely determined.

The cycle figure of merit shown in Table III is simply the root-mean-square value of the residuals in that cycle. The values of the two-dimensional phases φ_{h0l} are known to be either 0 or π. However, this information was not used in the computation of their values. Thus it was in order to compute an *H0L* figure of merit for each cycle. This figure of merit is the average value of the magnitudes of the deviations between the computed values of the φ_{h0l} and the nearer of 0, π. The integer preceding *H0L* is the number of contributors to this average. Evidently the smaller the figures of merit the greater the expectation that the phase determination is correct. Experience shows that cycle figures of merit of the order of 0·18 and *H0L* figures of merit of the order of 0·3 are acceptable.

As indicated in Table III, the least-squares iterative process described earlier was not rigidly adhered to in computing the phases for estriol. Instead, sandwiched between those cycles in which new phases were computed (the odd-numbered cycles), there occur refinement cycles (the even-numbered ones) in which phases determined earlier were recomputed by means of the same least-squares technique (assuming as known all phases determined in the previous cycle). Other modifications of

Table IV. Values (in radians) of 60 representative phases as computed in cycles 17–29.

$\vec{h}$ \ cycle	29	28	27	26	25	24	23	22	21	20	19	18	17
1 0 $\bar{9}$	0.00	0.00	0.00	-0.16	-0.54	0.44	0.00	0.00	0.00	0.13	0.00	0.00	0.00
1 1 $\bar{9}$	2.52	2.27	2.51	2.38	2.45	2.38	2.46						
1 2 $\bar{9}$	-2.75	-2.95	-2.73	-2.86	-2.60	-2.78	-2.22	-2.80	-2.22	-2.77	-2.50	-2.86	-2.59
3 0 $\bar{9}$	0.00	0.00	0.00	0.00	0.00	0.11	0.00	-0.20	0.00	0.00	0.00	0.00	0.00
3 1 $\bar{9}$	2.27	2.19	2.28	2.21	2.18	2.25	2.28						
3 2 $\bar{9}$	-2.66	-3.03	-2.60	-3.00	-2.63	-2.95	-2.33	-2.88	-2.39	3.01	-2.79	3.00	-2.62
3 4 $\bar{9}$	1.64	0.49	1.82	0.41	2.13	0.38	2.46	0.16	2.82	0.01	2.74	0.04	2.72
5 0 $\bar{9}$	0.00	0.00	0.00	0.00	0.00	0.00	0.00	0.00	0.00	0.01	0.00	0.00	0.00
5 2 $\bar{9}$	-2.51	-3.10	-2.44	-3.11	-2.62	3.11	-2.68	3.10	-2.76	2.90	-3.06	3.10	-2.82
7 0 $\bar{9}$	0.12	0.00	-0.08	0.00	0.10	0.00	-0.06	0.00	0.17	0.00	-0.38	0.00	0.00
1 1 $\bar{8}$	-2.88	-3.00	-2.89	-2.93	-2.87	-3.03	-2.59	-2.66	-2.85	-2.61	-2.66	-2.66	-3.04
1 11 $\bar{8}$	-2.73	-2.79	-2.67	-2.75	-2.77	-2.82	2.72						
3 1 $\bar{8}$	-2.42	-2.50	-2.46	-2.49	-2.31	-2.08	2.53	-2.23	2.44	-2.18	2.54	-2.34	2.51
4 5 $\bar{8}$	-1.43	-1.91	-1.37	-2.01	-1.29	-1.84	-1.50	-1.85	-1.29				
5 1 $\bar{8}$	-2.27	-2.82	-2.22	-2.74	-2.25	-2.82	-1.92	-2.81	2.46	-3.07	-1.95	-2.96	2.47
5 3 $\bar{8}$	-1.61	-0.84	-1.79	-0.91	-1.67	-0.97	-1.81	-0.96	-1.76	-0.96	-1.57	-0.71	-1.50
6 4 $\bar{8}$	0.82	0.59	0.79	0.44	0.76	0.69	0.75						
2 6 $\bar{7}$	1.56	1.47	1.46	1.53	1.45	1.52	1.41	1.56	1.44	1.55	1.15		
4 4 $\bar{7}$	-1.83	-1.72	-1.78	-1.81	-1.62	-1.98	-2.53	-1.89	-2.46	-1.96	-2.23	-1.88	-2.24
4 5 $\bar{7}$	-1.54	-1.40	-1.51	-1.43	-1.45	-1.51	-1.42						
4 6 $\bar{7}$	1.27	1.35	1.19	1.37	1.24	1.28	1.16	1.18	1.20	1.30	1.16		
6 4 $\bar{7}$	-2.18	-1.55	-1.95	-1.62	-2.09	-1.70	-2.56	-1.94	-2.27	-1.98	-2.40	-1.90	-2.26
6 5 $\bar{7}$	-1.43	-1.52	-1.40	-1.55	-1.33	-1.51	-1.58	-1.75	-1.39				
6 6 $\bar{7}$	0.88	1.02	0.92	0.98	0.86	0.86	1.09	0.84	1.42	1.07	1.34		
2 3 $\bar{6}$	-2.57	-3.07	-2.64	-2.93	-2.83	-2.91	-2.66	-2.72	-2.66	-2.71	-2.57	-2.93	-2.74
2 4 $\bar{6}$	-1.19	-1.15	-1.24	-1.18	-1.22	-1.31	-1.61	-1.37	-2.16				
2 5 $\bar{6}$	0.76	0.90	0.79	0.93	0.77	1.12	0.79	1.59	0.72	1.58	0.77	1.64	0.58
2 6 $\bar{6}$	1.63	1.83	1.61	1.82	1.68	1.76	0.80	2.50	0.27				
2 7 $\bar{6}$	-1.27	-1.68	-1.27	1.59	-1.42	-1.51	-1.53	-1.75	-1.12				
4 3 $\bar{6}$	-2.70	2.96	-2.79	3.04	-2.76	-1.96	-2.56	-1.86	-2.62	-1.84	-2.24	-1.68	-1.93

the same general technique suggest themselves. For example, instead of picking up new phases at the rate 1, 3, 5, . . . it may, on occasion, be desirable to modify this sequence. Thus in one application, it was necessary to pick up only one new phase during each of the first three cycles before accelerating the pick-up rate. Again, instead of ignoring the fact that the phases φ_{h0l} must be 0 or π, one could, in each cycle, have forced the values of these phases to the appropriate one of 0, π. Thus, in Cycle 5 of Table III, $\varphi_{50\bar{9}}$ was computed to be 0·3520 and this value, rather than 0, was tentatively accepted. Clearly the value of 0 for this phase was strongly indicated and the program could easily be modified to permit this option. Of course, if this is done, then there would be no point in computing an *H*0*L* figure of merit.

A total of 29 cycles, of which Table III shows the first 5, were carried out. Of these, 12 were cycles during which new phases were determined, 11 were refinement cycles sandwiched between the others, and the last 6 cycles were final refinement cycles during which no new phases were determined. In Table IV are shown the values of 60 representative phases (in radians) as computed in cycles 17–29 inclusive. For most vectors **h** the values of the φ were reasonably stable and the convergence to their respective limiting values acceptable. However, in some cases, *e.g.* $34\bar{9}$, $52\bar{9}$, $45\bar{8}$, $51\bar{8}$, $53\bar{8}$, stable values of the phases were not attained. (It has been found that averaging the computed value of a phase with its value as determined in the previous cycle tends to improve the convergence behavior and this practice is therefore recommended. While this averaging technique was not employed in the structure determination for estriol, it has subsequently been incorporated into the computer program and was successfully used in two later structure determinations.) Of the total of 149 phases obtained in this way, the values of 103 were deemed acceptable and were used in the tangent formula, Eq. 4.10, in order to obtain final values for all the 1023 permissible phases.

Not only has the estriol structure been found in the manner described, but two others, 5β-androstan-3α, 17β-diol ($C_{19}H_{32}O_2$) and 5α-androstan-3β-ol-17-one, both crystallizing in the space group $P2_1$ and each having one molecule in the asymmetric unit, have also been solved in the same way. In addition the structure of 17β-(trimethylsiloxy) androst-4-en-3-one

Table IV (Concl.)

h \ cycle	29	28	27	26	25	24	23	22	21	20	19	18	17
$2\ 9\ \bar{1}$	2.57	2.61	2.66	2.72	2.72	2.67	2.10	2.72	2.99	2.59	-3.14		
$2\ 11\ \bar{1}$	-0.46	-0.54	-0.49	-0.52	-0.58	-0.30	-0.70						
$4\ 1\ \bar{1}$	0.08	0.09	0.06	0.16	0.03	0.05	0.12	0.25	0.07	0.20	0.09	0.35	0.10
$4\ 2\ \bar{1}$	1.53	1.51	1.56	1.52	1.63	1.44	1.43	1.46	0.75				
$4\ 11\ \bar{1}$	-0.15	-0.50	-0.17	-0.30	-0.38	-0.04	-0.10	-0.06	0.03				
$6\ 1\ \bar{1}$	0.18	0.14	0.21	0.17	0.14	0.28	0.33	0.40	0.17	0.40	0.22	0.48	0.11
$6\ 2\ \bar{1}$	1.42	1.29	1.37	1.30	1.46	1.26	0.83						
0 10 0	-0.56	-0.51	-0.50	-0.45	-0.66	-0.65	-0.70	-0.72	-1.14	-0.85	-0.95		
0 12 0	2.92	2.85	2.91	2.87	2.71	2.60	-2.96	2.14	3.10				
2 0 0	0.09	0.14	0.12	-0.02	0.02	0.01	-0.07	0.12	-0.11	0.02	-0.01	-0.01	0.00
2 1 0	0.48	0.32	0.51	0.40	0.63	0.55	1.07						
2 12 0	2.85	2.63	2.81	2.78	2.80	3.12	2.75						
4 0 0	-0.02	-0.01	0.02	0.01	0.00	0.00	0.01	-0.17	-0.02	-0.01	0.31	0.17	-0.47
4 2 0	-2.34	2.75	-2.16	2.73	-2.07	2.79	-1.98	2.67	-2.01	2.51	-1.87	2.59	-1.86
4 4 0	1.71	2.70	1.75	2.67	1.86	2.43	1.75	2.21	1.72	2.25	1.58	2.01	1.54
0 1 1	-2.31	-3.02	-2.30	-3.07	-2.36	-2.95	-2.63	-3.08	3.04	-3.05	3.10	-2.94	2.68
0 9 1	-0.38	-0.22	-0.34	-0.43	-0.18	-0.32	0.06	-0.31	0.03	-0.12	0.33		
0 2 2	-1.54	-0.68	-1.27	2.95	-1.27	2.58	-1.40	2.62	2.16	2.41	-1.08	2.83	2.09
1 5 2	-1.99	-2.16	-2.04	-2.11	-1.88	-2.10	-1.69	-2.33	-2.01	-2.34	-2.38		
1 6 2	2.28	2.14	2.31	2.09	2.28	2.05	1.61	2.00	1.37				
1 7 2	0.75	0.90	0.73	1.05	0.84	0.97	2.12	1.80	1.86	1.83	1.59		
3 4 2	-2.28	-2.25	-2.32	-2.28	-2.30	-2.29	-2.10	-1.79	-2.16	-2.57	-2.12	-2.53	-1.90
3 5 2	-2.10	-1.96	-2.02	-1.89	-1.94	-1.91	-1.76	-2.01	-1.81	-2.05	-1.95	-2.18	-2.06
3 7 2	0.82	0.99	0.84	1.10	0.96	1.31	1.78	1.44	1.60	1.25	1.48		
5 4 2	-2.20	-1.94	-2.20	-2.04	-2.12	-2.03	-2.13	-2.54	-2.12	-2.69	-2.02		
5 5 2	-2.28	-1.94	-2.28	-1.85	-2.41	-1.93	-2.21	-1.62	-1.75	-1.66	-1.58		
5 7 2	0.64	0.80	0.70	0.82	0.80	1.17	1.08	1.02	0.98				
1 0 3	-3.13	2.18	-3.13	2.44	3.11	2.93	-2.99	-1.80	-2.95	-1.90	-3.02		
1 2 3	2.14	1.69	2.08	1.62	1.97	1.42	1.66	0.71	1.23				
1 3 3	-2.69	3.05	-2.68	-3.12	-2.83	-3.01	-2.48	-2.70	-2.51	-2.60	-2.64	-2.64	-2.53

($C_{22}H_{36}O_2Si$), space group $P2_12_12_1$, has also been determined recently in this way by Charles Weeks (private communication).

Appendix

The conditional probability distribution of $\cos(\varphi_{\mathbf{h}}+\varphi_{\mathbf{h}}+\varphi_{-\mathbf{h}-\mathbf{k}})$

Assume that the vector **h** is fixed and that **k** ranges uniformly throughout reciprocal space. It turns out that the conditional probability distribution of the random variable $X=\cos(\varphi_{\mathbf{h}}+\varphi_{\mathbf{k}}+\varphi_{-\mathbf{h}-\mathbf{k}})$, given that $|E_{\mathbf{k}}|$ and $|E_{\mathbf{h}+\mathbf{k}}|$ have specified, fixed values, depends only on

$$A=\frac{2}{N^{1/2}}|E_{\mathbf{h}}E_{\mathbf{k}}E_{\mathbf{h}+\mathbf{k}}| \qquad (1)$$

and is therefore denoted by $P_A(x)$. By means of an analysis which is too lengthy to be included here (see Cochran, 1955, for a related distribution) it can be shown that

Table V. For fixed **h** the conditional probability that

$\cos(\varphi_{\mathbf{h}}+\varphi_{\mathbf{k}}+\varphi_{-\mathbf{h}-\mathbf{k}})>x$,

given $A=\frac{2}{N^{1/2}}|E_{\mathbf{h}}E_{\mathbf{k}}E_{\mathbf{h}+\mathbf{k}}|$

A \ x	0.995	0.980	0.955	0.921	0.878	0.825	0.765	0.698	0.622
0.0	0.0318	0.0637	0.0955	0.1273	0.1592	0.1910	0.2228	0.2546	0.2865
0.5	0.0493	0.0984	0.1469	0.1948	0.2418	0.2876	0.3321	0.3753	0.4169
1.0	0.0682	0.1358	0.2020	0.2663	0.3282	0.3871	0.4428	0.4951	0.5437
1.5	0.0864	0.1715	0.2542	0.3332	0.4078	0.4773	0.5410	0.5990	0.6510
2.0	0.1028	0.2036	0.3005	0.3919	0.4764	0.5532	0.6218	0.6821	0.7344
2.5	0.1174	0.2319	0.3408	0.4421	0.5339	0.6154	0.6863	0.7465	0.7970
3.0	0.1303	0.2568	0.3761	0.4852	0.5823	0.6665	0.7375	0.7962	0.8435
3.5	0.1420	0.2792	0.4072	0.5227	0.6235	0.7088	0.7788	0.8347	0.8783
4.0	0.1528	0.2996	0.4352	0.5557	0.6590	0.7443	0.8124	0.8650	0.9046
4.5	0.1627	0.3183	0.4606	0.5853	0.6900	0.7744	0.8400	0.8892	0.9248
5.0	0.1720	0.3357	0.4839	0.6119	0.7173	0.8004	0.8631	0.9086	0.9404
5.5	0.1808	0.3519	0.5055	0.6361	0.7416	0.8228	0.8824	0.9243	0.9526
6.0	0.1891	0.3673	0.5255	0.6582	0.7634	0.8424	0.8988	0.9371	0.9621
6.5	0.1971	0.3818	0.5443	0.6785	0.7829	0.8595	0.9126	0.9477	0.9697
7.0	0.2047	0.3956	0.5619	0.6973	0.8005	0.8745	0.9245	0.9563	0.9757
7.5	0.2120	0.4087	0.5785	0.7146	0.8165	0.8878	0.9346	0.9635	0.9804
8.0	0.2190	0.4213	0.5941	0.7307	0.8310	0.8995	0.9432	0.9694	0.9843
8.5	0.2258	0.4334	0.6089	0.7457	0.8442	0.9099	0.9507	0.9744	0.9873
9.0	0.2324	0.4449	0.6229	0.7597	0.8562	0.9192	0.9571	0.9785	0.9897
9.5	0.2387	0.4560	0.6363	0.7727	0.8672	0.9274	0.9627	0.9819	0.9917
10.0	0.2449	0.4668	0.6490	0.7849	0.8773	0.9347	0.9675	0.9848	0.9933
10.5	0.2509	0.4771	0.6611	0.7963	0.8865	0.9413	0.9717	0.9872	0.9946
11.0	0.2568	0.4871	0.6726	0.8070	0.8950	0.9471	0.9753	0.9892	0.9956
11.5	0.2625	0.4968	0.6836	0.8171	0.9027	0.9524	0.9784	0.9909	0.9964
12.0	0.2680	0.5061	0.6942	0.8265	0.9099	0.9570	0.9811	0.9923	0.9971
12.5	0.2735	0.5152	0.7042	0.8354	0.9165	0.9613	0.9835	0.9935	0.9976
13.0	0.2788	0.5240	0.7139	0.8438	0.9225	0.9650	0.9856	0.9945	0.9981
13.5	0.2840	0.5325	0.7232	0.8517	0.9281	0.9684	0.9874	0.9954	0.9984
14.0	0.2890	0.5408	0.7321	0.8592	0.9333	0.9715	0.9889	0.9961	0.9987
14.5	0.2940	0.5489	0.7406	0.8662	0.9381	0.9742	0.9903	0.9967	0.9990
15.0	0.2989	0.5567	0.7488	0.8728	0.9425	0.9767	0.9915	0.9972	0.9992

Table V. (Concl.)

A \ x	0.540	0.454	0.362	0.267	0.170	0.071	-0.416	-0.801	-0.990
0.0	0.3183	0.3501	0.3820	0.4138	0.4456	0.4775	0.6366	0.7958	0.9549
0.5	0.4570	0.4953	0.5321	0.5671	0.6005	0.6323	0.7696	0.8795	0.9743
1.0	0.5887	0.6301	0.6679	0.7024	0.7337	0.7620	0.8685	0.9364	0.9869
1.5	0.6973	0.7381	0.7737	0.8048	0.8317	0.8548	0.9307	0.9693	0.9939
2.0	0.7792	0.8170	0.8486	0.8749	0.8965	0.9143	0.9654	0.9860	0.9973
2.5	0.8384	0.8720	0.8990	0.9203	0.9371	0.9502	0.9833	0.9939	0.9989
3.0	0.8809	0.9099	0.9322	0.9491	0.9617	0.9711	0.9921	0.9974	0.9995
3.5	0.9114	0.9361	0.9542	0.9672	0.9766	0.9832	0.9963	0.9989	0.9998
4.0	0.9335	0.9542	0.9687	0.9787	0.9855	0.9901	0.9983	0.9996	0.9999
4.5	0.9498	0.9670	0.9785	0.9861	0.9910	0.9942	0.9992	0.9998	1.0000
5.0	0.9619	0.9761	0.9852	0.9909	0.9944	0.9965	0.9996	0.9999	1.0000
5.5	0.9710	0.9826	0.9897	0.9940	0.9965	0.9979	0.9998	1.0000	1.0000
6.0	0.9778	0.9873	0.9928	0.9960	0.9978	0.9988	0.9999	1.0000	1.0000
6.5	0.9830	0.9906	0.9950	0.9973	0.9986	0.9993	1.0000	1.0000	1.0000
7.0	0.9869	0.9931	0.9965	0.9982	0.9991	0.9996	1.0000	1.0000	1.0000
7.5	0.9899	0.9949	0.9975	0.9988	0.9994	0.9997	1.0000	1.0000	1.0000
8.0	0.9922	0.9962	0.9982	0.9992	0.9996	0.9998	1.0000	1.0000	1.0000
8.5	0.9940	0.9972	0.9988	0.9995	0.9998	0.9999	1.0000	1.0000	1.0000
9.0	0.9953	0.9979	0.9991	0.9996	0.9998	0.9999	1.0000	1.0000	1.0000
9.5	0.9964	0.9985	0.9994	0.9998	0.9999	1.0000	1.0000	1.0000	1.0000
10.0	0.9972	0.9989	0.9996	0.9998	0.9999	1.0000	1.0000	1.0000	1.0000
10.5	0.9978	0.9992	0.9997	0.9999	1.0000	1.0000	1.0000	1.0000	1.0000
11.0	0.9983	0.9994	0.9998	0.9999	1.0000	1.0000	1.0000	1.0000	1.0000
11.5	0.9987	0.9995	0.9998	0.9999	1.0000	1.0000	1.0000	1.0000	1.0000
12.0	0.9990	0.9996	0.9999	1.0000	1.0000	1.0000	1.0000	1.0000	1.0000
12.5	0.9992	0.9997	0.9999	1.0000	1.0000	1.0000	1.0000	1.0000	1.0000
13.0	0.9994	0.9998	0.9999	1.0000	1.0000	1.0000	1.0000	1.0000	1.0000
13.5	0.9995	0.9999	1.0000	1.0000	1.0000	1.0000	1.0000	1.0000	1.0000
14.0	0.9996	0.9999	1.0000	1.0000	1.0000	1.0000	1.0000	1.0000	1.0000
14.5	0.9997	0.9999	1.0000	1.0000	1.0000	1.0000	1.0000	1.0000	1.0000
15.0	0.9998	0.9999	1.0000	1.0000	1.0000	1.0000	1.0000	1.0000	1.0000

$$\left.\begin{aligned} P_A(x) &= \frac{\exp(Ax)}{\pi I_0(A)\sqrt{1-x^2}} && \text{if } |x|<1\,, \\ &=0 && \text{if } |x|>1\,, \end{aligned}\right\} \qquad (2)$$

where I_0 is the Bessel function of imaginary argument. Since, as explained in § 3, this distribution is needed in order to calculate the structure invariants, the values of the cumulative distribution,

$$F_A(x)=\int_x^1 P_A(x)\,dx\,, \qquad (3)$$

have been calculated for $|x|<1$ and are shown in Table V. The entries in this table then are the conditional probabilities that $\cos(\varphi_{\mathbf{h}}+\varphi_{\mathbf{k}}+\varphi_{-\mathbf{h}-\mathbf{k}})>x$, given A. As described earlier, the sliding scale factor K of Eq. 3.2 is so determined as a function of A that the distribution of the resulting computed values of the $\cos(\varphi_{\mathbf{h}}+\varphi_{\mathbf{k}}+\varphi_{-\mathbf{h}-\mathbf{k}})$ agrees with the distribution shown in Table V.

Acknowledgment

The author acknowledges his grateful indebtedness to Mr. C. T. Lu and Dr. Dorita Norton, who supplied the experimental data, and to Miss Janet Fisher and Mr. Harrison Hancock, who carried out the extensive computer programming required for the evaluation of the phases for estriol.

References

Cochran, W. (1955). *Acta Cryst.* **8**, 473.
Hauptman, H. (1963). *The Encyclopedia of X-Rays and Gamma Rays*, p. 749. New York: Reinhold Publ. Corp.
Hauptman, H. (1964). *Acta Cryst.* **17**, 1421.
Hauptman, H., Fisher, J., Hancock, H. & Norton, D. (1969). *Acta Cryst.* B**25**, 811.
Hauptman, H. & Karle, J. (1953). *Solution of the Phase Problem.* I. *The Centrosymmetric Crystal.* A.C.A. Monograph No. **3**.
Hauptman, H. & Karle, J. (1956). *Acta Cryst.* **9**, 45.
Karle, J. & Hauptman, H. (1956). *Acta Cryst.* **9**, 635.
Karle, J. & Hauptman, H. (1957). *Acta Cryst.* **10**, 515.
Karle, J. & Hauptman, H. (1958). *Acta Cryst.* **11**, 264.
Vaughan, P. A. (1958). *Acta Cryst.* **11**, 111.

Discussion

BEURSKENS: In view of the computing time involved, is it not best to find a sufficient number of centrosymmetric projection phases and then continue with the tangent formula?

HAUPTMAN: Yes, that is a good suggestion. Dr. Stewart has recently done something similar to this for centrosymmetric structures. However, it is still necessary to find several three-dimensional phases before the tangent formula can be used.

A6. PHILOSOPHIES AND UNIQUE FEATURES OF PROGRAMS FOR PHASE DETERMINATION

Automatic Phase Determination for Centrosymmetric Structures

F. R. AHMED

Biochemistry Laboratory, National Research Council of Canada, Ottawa 7, Ontario, Canada.

The NRC crystallographic programs for the IBM/360 system include a program in FORTRAN IV, originally written by Hall (1966) and modified by Ahmed (1968), for automatic execution of all operations between the stages of data reduction and E-map calculation as outlined by Karle & Karle (1966). The program proceeds in steps as follows:

1. Estimates the overall B_{iso} and scale from the Wilson plot;
2. Calculates $|E_{\mathbf{h}}|$ and their statistical distribution, then sorts those with $|E|$ more than about 1·5 in descending order of $|E|$;
3. Prepares for each reflexion, **h**, a list of the pairs **k** and **h**$-$**k** satisfying the Σ_2 relationships provided that $|E_{\mathbf{k}}|, |E_{\mathbf{h}-\mathbf{k}}| \geq (\sim 2{\cdot}0)$;
4. Selects the origin-defining reflexions, assigns the symbols as needed, applies the Σ_2 relationships to determine the signs of the reflexions in two main stages: first, for $|E| \geq (\sim 2{\cdot}0)$, then for the weaker ones. It then sorts the signed reflexions in the order required for the Fourier program.

The user is allowed, if he chooses, to feed in his own values for B_{iso}, scale, origin-defining reflexions, signs of the symbols or of any reflexion.

The discussion here is concerned with the execution of the first stage of step (4). The input tape for this stage contains:

A. **h**, $|E_{\mathbf{h}}|$ down to the limit $|E| \simeq 1{\cdot}5$;

B. For each **h**, the list of pairs **k**, $|E_{\mathbf{k}}|$, **h**$-$**k**, $|E_{\mathbf{h}-\mathbf{k}}|$ without repetition of any triplet **h**, **k**, **h**$-$**k**;

C. The quantity $\sigma_3\sigma_2^{-3/2}|E_{\mathbf{h}}E_{\mathbf{k}}E_{\mathbf{h}-\mathbf{k}}|$ for each triplet, where $\sigma_n = \sum_i Z_i^n$ and the sum is over all atoms in the unit cell. This quantity is related to the probability of the determined sign of E as in the expression

$$P_+(\mathbf{h}) \simeq \tfrac{1}{2} + \tfrac{1}{2} \tanh \left(\sigma_3\sigma_2^{-3/2}|E_{\mathbf{h}}| \sum_{\mathbf{k}} E_{\mathbf{k}}E_{\mathbf{h}-\mathbf{k}}\right) \qquad (1)$$

Procedure for acceptance of tentative signs

The basis of the procedure is to be very critical in accepting newly determined signs in the early stages, and to relax this condition in steps as the search for new signs progresses.

The program executing the first stage of step (4), which is the sign determination for the strongest $|E|$'s, is set to handle up to 400 such reflexions, to introduce a maximum of four symbols (A, B, C, D), and to employ as many as 16 probability limits of decreasing values. Instead of computing the full probability expression (1), the program evaluates only the sums $\sum_{\mathbf{k}} \sigma_3\sigma_2^{-3/2}|E_{\mathbf{h}}|E_{\mathbf{k}}E_{\mathbf{h}-\mathbf{k}}$. Sixteen separate counters are assigned to each of the 400 reflexions for accumulation of $\sum_{\mathbf{k}}$ for the 32 possible signs ± 1, $\pm A$, $\pm B$, $\pm AB$, $\pm C$, ..., $\pm ABCD$. In each counter, the distinction between the + and $-$ alternatives is achieved by adding the contribution to the counter for the + alternative, and subtracting it for the $-$.

At the start, all counters are set to zero. When the origin-defining reflexions are selected by the program, their E's are assumed to be +ve and the corresponding ± 1 counters are set arbitrarily to +10. The search for new signs is then started, employing the highest specified $\sum_{\mathbf{k}}$ limit (limit 1). When two of the signed reflexions produce an indicated sign for a third reflexion, the contribution $\sigma_3\sigma_2^{-3/2}|E_{\mathbf{h}}E_{\mathbf{k}}E_{\mathbf{h}-\mathbf{k}}|$ is added to the ± 1 counter of that reflexion when the indicated sign is +ve, or subtracted if the sign is −ve. If the |sum| in this counter is higher than limit 1, the sign is accepted tentatively, otherwise the sign is left as undetermined. The search for new signs with limit 1 is repeated every time one or more new signs are accepted until it becomes impossible to determine any more new signs. At this stage, limit 2 is introduced and the search is repeated as described for limit 1, then when the search is exhausted limit 3 is introduced and again the search is repeated.

The first symbol (A) is then assigned to one of the top 5 to 10 unsigned reflexions, chosen to have the longest list of pairs $(\mathbf{k}, \mathbf{h}-\mathbf{k})$, and the $\pm A$ counter of that reflexion is set to +10. The signing process is repeated, accumulating $\sum_{\mathbf{k}}$ in the ± 1 and $\pm A$ counters, and testing them against limits 4, 5, then 6 as previously described. When this search is completed symbol B is introduced, if necessary; the $\sum_{\mathbf{k}}$ contributions are accumulated in the ± 1, $\pm A$, $\pm B$, $\pm AB$ counters; and the tests are carred out with limits 7, 8, then 9. Similarly, the process is repeated with symbols C and D if needed. Finally, all the accumulated sums are tested against the minimum limit (limit 16). Suitable numerical values for the limits, as applied in solving successfully a triclinic structure with $\sigma_3\sigma_2^{-3/2} = 0{\cdot}157$, are listed in Table I.

Table I. Limits for acceptance of tentative signs

Limits	Numerical values	P (minimum)
1, 2, 3	8·0, 7·5, 7·0	1·0000
4, 5, 6	6·5, 6·0, 5·5	1·0000
7, 8, 9	5·0, 4·5, 4·0	0·9996
10, 11, 12	3·5, 3·0, 2·5	0·9933
13, 14, 15	2·4, 2·2, 2·0	0·9820
16	1·8	0·9734

Determination of the symbols

In general, more than one of the 16 counters of a given reflexion will be non-zero, thus indicating that the sign of that reflexion may be identified as ± 1, $\pm$ a symbol, or $\pm$ a symbol product. Assuming the phase determination is correct, then all indications for the same reflexion must be equivalent. Thus the values of the symbols can be determined either directly or in terms of other symbols. For example, symbol A can be determined as + or − if a sufficient number of reflexions have

Table II. Pairs of counters producing the signs of symbols

Symbol	Pairs of counters							
A	1, *A*	*B*, *AB*	*C*, *AC*	*BC*, *ABC*	*D*, *AD*	*BD*, *ABD*	*CD*, *ACD*	*BCD*, *ABCD*
B	1, *B*	*A*, *AB*	*C*, *BC*	*AC*, *ABC*	*D*, *BD*	*AD*, *ABD*	*CD*, *BCD*	*ACD*, *ABCD*
AB	1, *AB*	*A*, *B*	*C*, *ABC*	*AC*, *BC*	*D*, *ABD*	*AD*, *BD*	*CD*, *ABCD*	*ACD*, *BCD*
C	1, *C*	*A*, *AC*	*B*, *BC*	*AB*, *ABC*	*D*, *CD*	*AD*, *ACD*	*BD*, *BCD*	*ABD*, *ABCD*
AC	1, *AC*	*A*, *C*	*B*, *ABC*	*AB*, *BC*	*D*, *ACD*	*AD*, *CD*	*BD*, *ABCD*	*ABD*, *BCD*
BC	1, *BC*	*A*, *ABC*	*B*, *C*	*AB*, *AC*	*D*, *BCD*	*AD*, *ABCD*	*BD*, *CD*	*ABD*, *ACD*
ABC	1, *ABC*	*A*, *BC*	*B*, *AC*	*AB*, *C*	*D*, *ABCD*	*AD*, *BCD*	*BD*, *ACD*	*ABD*, *CD*
D	1, *D*	*A*, *AD*	*B*, *BD*	*AB*, *ABD*	*C*, *CD*	*AC*, *ACD*	*BC*, *BCD*	*ABC*, *ABCD*
AD	1, *AD*	*A*, *D*	*B*, *ABD*	*AB*, *BD*	*C*, *ACD*	*AC*, *CD*	*BC*, *ABCD*	*ABC*, *BCD*
BD	1, *BD*	*A*, *ABD*	*B*, *D*	*AB*, *AD*	*C*, *BCD*	*AC*, *ABCD*	*BC*, *CD*	*ABC*, *ACD*
ABD	1, *ABD*	*A*, *BD*	*B*, *AD*	*AB*, *D*	*C*, *ABCD*	*AC*, *BCD*	*BC*, *ACD*	*ABD*, *CD*
CD	1, *CD*	*A*, *ACD*	*B*, *BCD*	*AB*, *ABCD*	*C*, *D*	*AC*, *AD*	*BC*, *BD*	*ABC*, *ABD*
ACD	1, *ACD*	*A*, *CD*	*B*, *ABCD*	*AB*, *BCD*	*C*, *AD*	*AC*, *D*	*BC*, *ABD*	*ABC*, *BD*
BCD	1, *BCD*	*A*, *ABCD*	*B*, *CD*	*AB*, *ACD*	*C*, *BD*	*AC*, *ABD*	*BC*, *D*	*ABC*, *AD*
ABCD	1, *ABCD*	*A*, *BCD*	*B*, *ACD*	*AB*, *CD*	*C*, *ABD*	*AC*, *BD*	*BC*, *AD*	*ABC*, *D*

sums accumulated in their ± 1 and $\pm A$ counters, provided a large majority of these indications results in a consistent value for A. Also, the same result can be obtained from the pairs of counters $\pm B$ and $\pm AB$ of the reflexions, whether B is known or not. In total, there are 8 such pairs of counters which can lead directly to the sign of each of the symbols and symbol products, as listed in Table II. The program scans all 8 pairs of counters of all reflexions in order to determine the value of each symbol or symbol product. Finally, it checks identities such as

$$\text{sign}(A) \,.\, \text{sign}(B) \equiv \text{sign}(AB)$$

and informs the user whenever contradictions are detected.

References

Ahmed, F. R. (1968). Revisions to the Symbolic Addition Program. Unpublished.

Hall, S. R. (1966). *World List of Crystallographic Computer Programs.* 2nd ed., Appendix p. 52, Program 537.

Karle, J. & Karle, I. L. (1966). *Acta Cryst.* **21**, 849.

Discussion

DEWAR: How much computer time is involved in a typical phase determination?

AHMED: About 15–20 minutes on an IBM 360/50 for approximately 400 phases.

ROLLETT: Do you select the reflections to be used as symbols at the start of the process?

AHMED: No, a new symbol is selected only when the current phase extensions have stopped and some phases have still not yet been determined.

SORTE - A Program to Aid in the Implementation of the Symbolic Addition Method for the Direct Determination of Centrosymmetric Crystal Structures

ALLAN L. BEDNOWITZ

IBM Watson Research Center, Yorktown Heights, New York.

In the early work using the symbolic addition technique (Karle & Karle, 1963) manual methods were used extensively. A Σ_2 listing was produced by computer, but the symbolic representations for the phases were introduced after manual inspection of the listings. It is entirely conceivable that many centrosymmetric phase determinations can be carried out by manual iteration. However, once past the initial stage, the necessary manipulations increase rapidly.

In order to minimize the tedious and repetitive aspects of the manual methods, the phase determination procedure was partially automated by the program SORTE (Bednowitz & Post, 1966; Okaya & Bednowitz, 1967). The choice of origin-defining reflections and the symbol assignments is left to the discretion of the user. In addition, the solution of a set of unknown signs still requires manual intervention between passes through the computer.

The input to the program consists of a deck containing reflection indices and the estimated *E*-factor for each reflection. *E*-factors less than 1·0 are usually not required for the analysis. The initial input deck may (as a rule of thumb) contain only those *E*-factors greater than 1·7. The input deck need not be in any order, the program will order the reflections in the following manner:

$$H_i \leq H_{i+1},\ K_i \leq K_{i+1},\ L_i \leq L_{i+1}\ . \quad (1)$$

The reflection indices are packed into one variable,

$$\mathrm{HKL}(I) = 10000{*}h + 100{*}(k+50) + l + 50\ , \quad (2)$$

and ordered in increasing values of HKL. This simplifies the sorting and speeds the search procedures. The index, *I*, is used as a pointer to all the data associated with a unique set of packed indices. The output of the first run consists of a Σ_2 list, where the defining relation used is,

$$sE_{\mathbf{H}} \sim s \sum_{\mathbf{k}} E_{\mathbf{K}} E_{\mathbf{H}-\mathbf{K}}\ . \quad (3)$$

H is termed the primary reflection, **K** is the secondary, and **H**−**K** is the interacting reflection. The output listing is ordered on decreasing magnitude of the primary reflection. A minimum primary *E*-factor of 1·7 is usually sufficient for the initial listing.

On the basis of the Σ_2 listing, arbitrary signs (+ or −) must be assigned to three reflections in the *P*1 case (primitive orthorhombic or lower symmetry). The three reflections are chosen to satisfy several requirements:

A. They should, if possible, interact with each other.

B. They should, if possible, have a relatively large set of Σ_2 interactions.

C. They should be among the largest *E*-factors.

On occasion, one or more of these criteria may have to be relaxed in order to obtain an initial set of signs. This is one of the reasons manual initiation of the symbolic addition procedure is retained with SORTE.

It may also be necessary to assign a symbolic phase to one or more reflections at this point in order to initiate the procedure. Any symbolic assignments should, if possible, also satisfy the requirements A, B, and C. It may not be necessary to assign more than one or two symbols at this point. The fewer symbols assigned, the fewer decisions to be made later in the analysis. Again, this is added incentive for initiating the symbolic addition procedure by hand.

Starting with the assigned phases and symbols, the starting set is increased judiciously with the aid of the basic sign relationship of Eq. (3). The expression 'judiciously' can best be explained using a plot of probability contours for the basic sign relation. The probability function in Fig. 1 has been simplified with the assumption that the atoms have equal scattering power. The contours are plotted on a semi-log

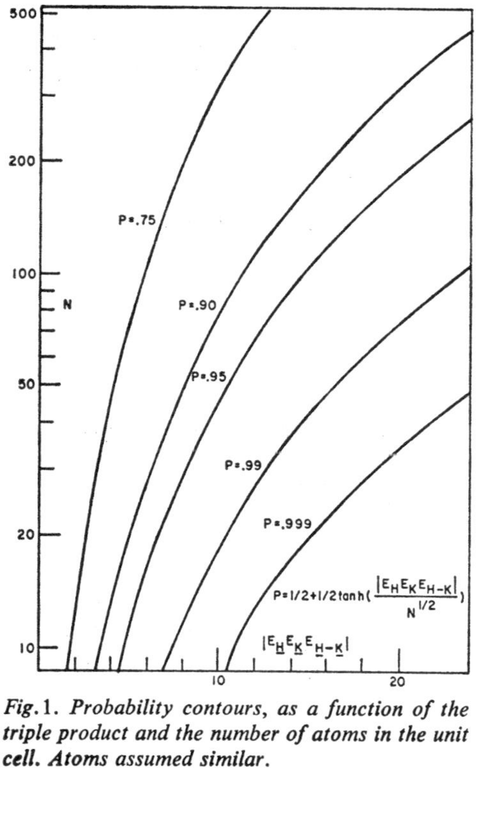

Fig. 1. Probability contours, as a function of the triple product and the number of atoms in the unit cell. Atoms assumed similar.

grid with the number of atoms in the unit cell on the vertical log scale and the triple product as abscissa. For example, consider a molecule of *p*-azoxyanisole ($C_{14}H_{14}N_2O_3$) which has 19 'heavy' atoms with four molecules per unit cell, for a total $N=76$. To maintain a reliability level of 95% for the basic sign relation, a triple product of ~12·0 would be required. For a level of 99% reliability the triple product should be ~20·0. At least a 95% level or higher should be maintained in the initial manual stage.

Table I contains a worksheet showing the first few steps for manually initiating the symbolic addition procedure. The example (*p*-azoxyanisole) crystallizes in the space group $P2_1/n$ which has the phase relation of $E_{hkl} = -E_{h\bar{k}l}$ when $h+k+l = 2n+1$. It is this type of additional phase relation which usually facilitates the assignment of signs to the symbolic phases in monoclinic or higher symmetry crystals. These relations are introduced directly into the SORTE program by modifying a few statements in the subroutines GROUP and INDEX.

As shown in Table I, eleven additional phases were determined from the assignment of three positive signs and a symbol *A*. When no new reliable phases were indicated from the Σ_2 listings, a symbol, *B*, was assigned to E_{310}. This resulted in the generation of 14 additional phases with high reliability as indicated by the triple products. However, there still seemed to be a set of *E*-factors with large magnitudes which did not interact with the set of determined phases. In an attempt to find a pathway into the undetermined set, a symbol *C* was assigned to E_{741}. At the same time it was noted that several interactions pointed to the equivalence of the symbol *A* and the negative phase.

Replacing *A* by $-$ and introducing symbol *C* still left a set of large *E*-factors with unassigned phases. Therefore, the symbol *D* was assigned to E_{339} enabling the derivation of 8 more phases. At this point almost all of the largest E-factors had a symbol assigned and it was evident that several primary phases could be obtained by more than one interaction. The seven defined and 35 derived phases were used as input to SORTE for an iterative procedure based on Eq. (3).

The assigned symbols were punched on the same deck used as input for the first pass through the computer. In the second pass each interaction in the Σ_2 list was re-examined by the program. If the symbolic phases of both the secondary and interacting E-factor had been previously assigned, and if the triple product was larger than the minimum limit, a subroutine PHASE was called. This subroutine determines the symbolic product of the two phases and adds the double product [right-hand side of Eq. (3)] to a sum associated with the derived symbol. Given a primary E-factor, and several interactions with differing symbolic phase indications, PHASE sets up a double product sum for each symbol.

The probability relation applicable to the Σ_2 relationship is,

$$P_+(E_\mathbf{H}) = \tfrac{1}{2} + \tfrac{1}{2}\tanh\left(\frac{\sigma_3}{\sigma_2^{3/2}}\,|E_\mathbf{H}|\sum_\mathbf{K} E_\mathbf{K}E_{\mathbf{H}-\mathbf{K}}\right) \tag{4}$$

where $\sum_\mathbf{K} E_\mathbf{K}E_{\mathbf{H}-\mathbf{K}}$ is the double product sum based on known signs for all E-factors in the set $\mathbf{K}$. This relation in a slightly modified form is used to decide the 'correctness' of a symbolic phase.

For a primary E-factor the 'dominant' symbol is defined as the symbol whose double product sum is greater than the sum of the double products associated with all other symbols. The difference between this dominant sum and the sum of double products from all other symbols (R_s) is a worst case estimate of $|\sum_\mathbf{K} E_\mathbf{K}E_{\mathbf{H}-\mathbf{K}}|$. The probability relation (4) can thus be modified to,

$$P_s(E_\mathbf{H}) \geq \tfrac{1}{2} + \tfrac{1}{2}\tanh\left(\frac{\sigma_3}{\sigma_2^{3/2}}\,|E_\mathbf{H}|R_s\right), \tag{5}$$

where $P_s(E_\mathbf{H})$ is the probability that the symbol s is correct. This expression allows a reasonably quantitative relation to be used for the decisions performed by SORTE.

Table I. SORTE – initial worksheet

p-azoxyanisole $a = 10{\cdot}98$, $b = 8{\cdot}09$, $c = 14{\cdot}99$, $\beta = 107{\cdot}3°$, S.G. $= P2_1/n$, $E_{hkl} = -E_{h\bar{k}l}$, $h+k+l = 2n+1$

( ) – indicates defined symbol

H	$\lvert E_\mathbf{H}\rvert$	s(H)		K	s(K)	H–K	s(H–K)	Triple product
5 6 $\bar{1}$	3·92	(+)	*ugu*					
2 1 0	3·89	(+)	*gug*					
5 1 $\bar{1}\bar{2}$	2·10	(+)	*uug*					
2 2 0	4·20	(*A*)	*ggg*					
3 4 $\bar{1}$	2·02	*A*		$\bar{2}$ $\bar{2}$ 0	(*A*)	5 6 $\bar{1}$	(+)	33·2
4 3 0	2·59	*A*		2 1 0	(+)	2 2 0	(*A*)	42·3
7 4 $\bar{1}$	2·20	*A*		2 $\bar{2}$ 0	(*A*)	5 6 $\bar{1}$	(+)	36·2
7 7 $\bar{1}$	3·20	+		2 1 0	(+)	5 6 $\bar{1}$	(+)	48·8
0 5 11	2·06	+		5 6 $\bar{1}$	(+)	$\bar{5}$ $\bar{1}$ 12	(+)	17·0
7 5 $\bar{1}$	1·85	−		2 $\bar{1}$ 0	(−)	5 6 $\bar{1}$	(+)	28·2
9 3 $\bar{1}$	1·83	−*A*		2 $\bar{1}$ 0	(−)	7 4 $\bar{1}$	*A*	16·5
1 4 $\bar{1}\bar{2}$	2·59	−*A*		5 1 $\bar{1}\bar{2}$	(+)	$\bar{4}$ 3 0	−*A*	14·1
5 8 $\bar{1}$	2·00	−		$\bar{2}$ 1 0	(−)	7 7 $\bar{1}$	+	24·0
3 7 $\bar{1}$	2·57	−		7 4 $\bar{1}$	*A*	$\bar{4}$ 3 0	−*A*	14·6
3 2 $\bar{1}\bar{2}$	1·75	−		3 7 $\bar{1}$	−	0 $\bar{5}$ $\bar{1}\bar{1}$	+	9·2

If the right-hand side of Eq. (5) is found to be greater than a minimum limit, SORTE will assign the symbolic phase s to $E_{\mathbf{H}}$. The procedure is repeated for each primary reflection greater than the primary E-factor limit. The procedure can be recycled several times in order to assign as many new phases as possible.

The output of this second pass consists of a summary listing indicating the double product sum for each symbol found in the primary interaction list (see Table II) and

Table II. Sample from the summary output of SORTE for Pass 2

Cycle	H	Symbol assigned	$\lvert E_{\mathbf{H}}\rvert$	Interaction summary N(DP SUM)s	N(DP SUM)s
3	7 2 $\overline{11}$	D	2·39	1 (6·47)B	2 (6·89)D
3	2 5 8		2·03	1 (5·34)$-B$	1 (8·66)$-D$
3	1 3 $\overline{13}$		1·92	1 (5·67)$-D$	1 (5·38)$-B$
3	2 5 $\overline{10}$	$-D$	1·82	2 (12·47)$-B$	5 (19·50)$-D$
3	6 4 $\overline{11}$	BD	1·79	2 (11·15)BD	1 (2·93)+
3	0 5 10	D	1·78	3 (15·56)D	2 (7·28)B
3	9 4 $\bar{4}$	D	1·74	5 (25·21)D	1 (8·87)B
3	10 0 0		1·74	4 (18·91)BD	2 (11·40)+
3	3 1 11	$-B$	1·72	2 (7·72)$-B$	
3	6 4 $\bar{2}$		1·68	3 (18·98)BD	5 (27·20)+
3	2 1 $\overline{14}$	$-D$	1·64	2 (11·99)$-D$	
3	4 6 4		1·63	1 (4·92)D	

This is only a sample of the indications for $D \equiv B$.

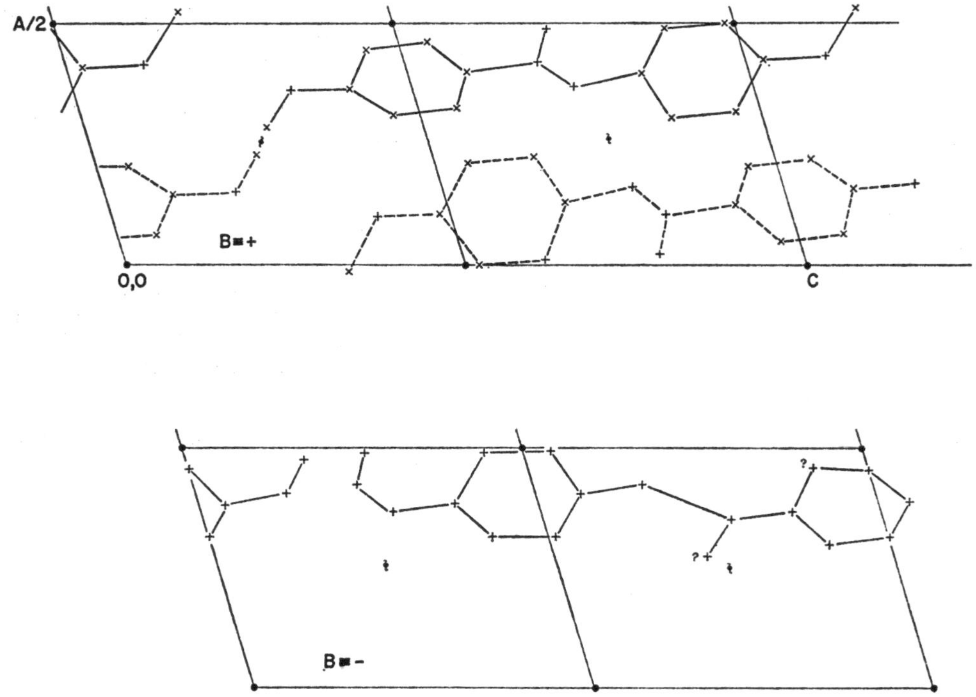

Fig. 2. Projection (down 'b' axis) of the E-map peak positions for p-azoxyanisole where the last symbol was assigned the values: $B \equiv +$ and $B \equiv -$.

a punch card deck containing the E-factors with their current symbolic phase assignments. For our example, *p*-azoxyanisole, the second pass contained a large number of indications that D equivalent to B and that C could be replaced by $-BD$ or its equivalent, $-$.

The symbol equivalent for C and D were input to SORTE for the third pass. The program automatically reassigned symbolic designations and repeated the procedure of pass 2 with lower limits on the E-factors of the primary and secondary reflections. The subsequent output gave several indications that B was negative. On the fourth pass an attempt was made to assign signs for all the E-factors down to a magnitude of 1·0.

An E-map was prepared and a model constructed using the peak positions in the Fourier. Fig. 2*b* shows the projection of this set of peaks down the '*b*' axis. The whole molecule did not appear, some of the peaks (indicated on the projection) were low and seemed almost to blend into the background ripple. An attempt was made to refine this structure by iterative structure factor and difference Fourier calculations. The agreement factor could not be reduced below 40%.

Backtracking and repeating the fourth pass with B set to $+$ resulted in a set of peaks from the E-map as shown in Fig. 2*a*. The nineteen heavy atoms showed up as the nineteen highest non-equivalent peaks in the E-map. The structure was refined and the expected bond distances and angles resulted.

On rechecking the symbolic addition procedure it was clear, from hindsight, that the indications for $B \equiv -$ were rather weak, but, the indications for $B \equiv +$ were no better. Also, in most low symmetry structures there is a tendency for all symbols to indicate $+$ when the defining signs are all positive.

Positive indications are the only ones possible in the triclinic case. In higher symmetry crystals negative indications result from the phase relations between equivalent reflections. This difficulty *cannot* be avoided by using negative signs in the origin definition. If the first E-map does not clearly show the structure, it is often worthwhile to try several other combinations of signs for the symbolic phases.

References

Bednowitz, A. L. & Post, B. (1966). *Acta Cryst.* **21**, 566.

Karle, J. & Hauptman, H. (1953). *Solution of the Phase Problem.* I. *The Centrosymmetric Crystal.* A.C.A. Monograph No. **3**.

Karle, I. L. & Karle, J. (1965). *Acta Cryst.* **16**, 969.

Okaya, Y. & Bednowitz, A. L. (1967). *Acta Cryst.* **22**, 111.

Discussion

BERMAN: I notice the phases you used in the beginning of the determination were for *hk*0, and *hk*12 reflections. Do you think if you had taken the symbol D earlier, the solution would have been more rapid?

BEDNOWITZ: No, I think it would have been about the same.

A System of Direct Methods Programs

ROBERT B. K. DEWAR

Information Science Center, Illinois Institute of Technology, Chicago, Illinois, 60616 *U.S.A.*

A series of programs has been developed for the application of the symbolic addition method of phase determination for both centrosymmetric and noncentrosymmetric space groups as described by Karle & Karle (1966). These programs are described individually.

FAME

FAME calculates a Wilson plot and from this determines overall scale and thermal parameters. Normalized structure factors are then calculated using appropriate symmetry factors.

Two options are available: the generated E's can be rescaled so that $\langle E^2 \rangle = 1$. This is useful in normalizing probabilities. In addition, there is an option to renormalize subgroups of various parities so that the relation $\langle E^2 \rangle = 1$ holds for each parity group. This has proved useful in cases of pseudo-symmetry, particularly pseudo-centering.

For the centrosymmetric case, FAME also has provision for selecting a starting set of symbols for symbolic addition. Three criteria are applied:

A. A variety of parity groups should be represented.

B. Large E values are preferable.

C. Chosen reflections should interact with other large E values.

No origin-determining phases are chosen at this point. This process is deferred till after the symbolic addition. At the present time, the choice of starting reflections in the noncentrosymmetric case is left to the user.

MAGIC
(centrosymmetric case)
MAGIA
(non-centrosymmetric case)

These programs apply the symbolic addition procedure. The process is iterative; and once a phase has been determined, it is never redetermined. Our experience has shown that recomputation of determined phases would not be worthwhile. A variety of criteria including the probability (MAGIC) and variance (MAGIA) are used to determine the acceptance of new phases. The levels of these criteria are automatically manipulated by the program to give satisfactory progress.

During the symbolic addition process, indications of the values of symbols and combinations of symbols are collected, but no attempt is made to interpret these indications at this stage.

One interesting observation in the centrosymmetric case (but *not* in the noncentrosymmetric case) is that very small E's can be successfully determined when the process is automated. For example, in one case, where only 350 reflections were observed, a reflection with $E = 0{\cdot}205$ was successfully phased at a probability level of 0·99 from 205 contributors, none of whose individual probabilities exceeded 0·55. It thus appears that the probability formula works well in such extreme cases.

LINK, LINKA

These programs process MAGIC and MAGIA output respectively. In the centrosymmetric case, the origin is specified at this point. The motive in deferring the origin choice is to avoid using reflections to determine the origin which subsequently interact to a small extent. For the symbols remaining, all possible combinations of values are examined and a consistency index calculated. In the centrosymmetric case, this is a probability weighted count of contradictions. In the noncentrosymmetric case, the index is the product of the probability densities as calculated from the variances. For output convenience, the log of this quantity is displayed. A specified number of the 'best' sets are then output.

SYMPL, SYMPLA

These programs convert the symbolic phases to absolute phases and organize the resulting data. In the centrosymmetric case, input to a Fourier program is prepared. For the noncentrosymmetric case, input to TANG is generated. There is a provision at this stage for rejecting phases which, in retrospect, were not reliably determined.

TANG

This is an elaborate program for application of the tangent formula, either in conjunction with the previously described programs, or separately.

The tangent formula (Karle & Hauptman, 1956) has been given, variously, in two forms:

$$\tan \varphi_{\mathbf{h}} = \frac{\sum_{\mathbf{k}} |E_{\mathbf{k}} E_{\mathbf{h}-\mathbf{k}}| \sin (\varphi_{\mathbf{k}} + \varphi_{\mathbf{h}-\mathbf{k}})}{\sum_{\mathbf{k}} |E_{\mathbf{k}} E_{\mathbf{h}-\mathbf{k}}| \cos (\varphi_{\mathbf{k}} + \varphi_{\mathbf{h}-\mathbf{k}})} \quad (1)$$

and

$$\tan \varphi_{\mathbf{h}} = \frac{\langle E_{\mathbf{k}} E_{\mathbf{h}-\mathbf{k}} \sin (\varphi_{\mathbf{k}} + \varphi_{\mathbf{h}-\mathbf{k}}) \rangle_{\mathbf{k}}}{\langle E_{\mathbf{k}} E_{\mathbf{h}-\mathbf{k}} \cos (\varphi_{\mathbf{k}} + \varphi_{\mathbf{h}-\mathbf{k}}) \rangle_{\mathbf{k}}} \quad (2)$$

The first of these corresponds to a vector sum, the second to a vector average. The predicted phases are the same. However, the implied values of the E_c (E-calculated) are different. This program calculates both values; those derived from Eq. 2 agree better with the observed E.

The following criteria are applied in admitting new phases:

A. Maximum oscillation of value (a history of determined phases is maintained).

B. Minimum value of ECALCA from Eq. 2.

C. Minimum value of ECALCS from Eq. 1.

D. Minimum value of the ratio ECALCA/EOBS.

E. Minimum number of contributors.

F. Maximum estimated error (standard deviation). This is estimated statistically from the vector distribution which may optionally be displayed in histogram form. Note that this measure is quite different from the analytically determined variance (Karle & Karle, 1966).

At the end of each cycle, a series of graphs showing phase shifts as a function of various parameters is displayed.

An important feature of TANG is the ability to hold phases constant; such phases only enter on the right-hand side of the calculations. There is also provision for editing out reflections and maintaining a variety of data files.

Conclusion

The centrosymmetric programs are usually run through automatically with no intervention. In the great majority of cases, a successful unique solution is obtained in a few minutes of computer time.

The noncentrosymmetric programs are somewhat less successful. In some cases, the procedure is as smooth as the centrosymmetric determination. In other cases, several different starting sets must be tried. It also appears to be useful to introduce auxiliary phase information (*e.g.* from the Σ_1 formula). Hopefully, further experience with the program will lead to suitable programmable algorithms for selection of the starting set which is certainly the critical step.

References

Karle, J. & Hauptman, H. (1956). *Acta Cryst.* **9**, 635.

Karle, J. & Karle, I. L. (1966). *Acta Cryst.* **21**, 849.

Discussion

SOLTZBERG: In MAGIC, special classes of reflections (such as $h0l$ in $P2_1/c$, 2nd setting) appear to be determined redundantly from, say, a pair of reflections and then from the pair related by symmetry to the first. Does this give an incorrectly large value for the probability?

DEWAR: No. The symmetry-related reflections must be taken into account to give the correct probability. In high symmetry space groups you do get double and quadruple indications of the same triplet and you must retain these.

NICHOLS: You mentioned that a number of structures had been solved with your programs. Have you noticed certain types of analyses give particular trouble?

DEWAR: There are several structures that we have had difficulty with. These are mainly $P\bar{1}$ and large equal atom cases.

Direct Phasing by Repeated Tangent Iteration

S. R. HALL

Department of Energy, Mines and Resources, Ottawa, Canada.

There have been a number of programs designed for the computer application of statistical structure factor relationships to crystal structure analysis. The majority employ techniques based on the sigma-2 (Hauptman & Karle, 1953) and associated tangent relationships (Karle & Hauptman, 1956). In these routines the basic processes of normalization and generation of sigma-2 relationships are similar and it is the method of applying these relationships, particularly in the initial stages, which differs considerably. This is not surprising as it is at this point in the analysis that there is no one procedure to follow which will absolutely ensure a structural solution. It is a difficult task indeed to set out a basic algorithm which will handle even the simplest type of determination. Programmers have tended therefore to concentrate on the more repetitious parts of the direct phasing process and leave the decision-making, such as required in the initial selection of starting phases, to the user. Some programs designed for the solution of centrosymmetric structures have, however, reduced and sometimes eliminated user intervention to a large extent. Several such programs, employing variations on the symbolic addition procedure of Karle & Karle (1966), are described in this text.

In the case of noncentrosymmetric determinations the initial phasing process is further complicated by the reduced choice of phases which may be conveniently used to specify the origin and enantiomorph of the structure. Generally these analyses require not only an increased number of symbol phases to be assigned in a symbolic addition procedure, but the subsequent allocation of numerical values to the symbols is more difficult due the unrestricted nature of the phases.

The programs to be discussed here attempt to overcome this problem by avoiding the use of symbol phases as much as possible. It has been found that a surprisingly limited set of numeric phases is often sufficient to initiate a phasing process based on repeated tangent iteration. This approach is attractive not only for its simplicity, and therefore ease in coding, but because the phase extension and refinement process take place simultaneously. In the tangent iteration process the E-threshold of the sigma-2 triplets is gradually lowered in steps thus ensuring that the most statistically reliable reflections are evaluated and refined before extending their influence to the more numerous reflections with lower E-values.

The limited set of numeric phases necessary for initiating this process is generally obtained through the specification of the cell origin and enantiomorph. Several other values may be added either to provide ready connection to all index parities or to improve phase extensions in the higher E-values. These are obtained by the allocation of a single symbol phase and the limited use of symbolic addition procedure to obtain numeric values, usually by association.

Description

The programs which apply the repetitive

tangent iteration procedure have been designed for use with the XRAY67 programming system (Stewart, 1967) and expand on concepts of earlier routines (Hall, 1968). The programs are applicable to all centrosymmetric and noncentrosymmetric spacegroups and have been coded in a basic form of FORTRAN IV which is acceptable to a number of computers with the necessary *memory overlay* and *re-read* features.

The programs are in the form of three separate overlay substructures which have the following basic functions:

The calculation of normalized structure factors (referred to as *NORMSF*).

The generation of sigma-2 relationships for all E-values above a specified minimum and the preparation of starting phases (referred to as *SIGMA*).

The extension and refinement of phases by repetitive tangent iteration (referred to as *TANGENT*).

Subsequent calculation of the E-map is performed by existing routines in the XRAY67 system.

The functions of each of these three programs can be best described in terms of the operation of each of the overlay subroutines.

NORMSF

The five subroutines in this program perform the following operations:

1. Enter optional parameter cards and store necessary reflection data from binary data tape.

2. Sum values of $(m|F^2_{obs}|)$, $(m\varepsilon \sum_i^N f_i^2)$ and $\sin^2\theta/\lambda^2$ into 19 overlapped $\sin^2\theta/\lambda^2$ ranges, which each occupy equal volume in reciprocal space. A least-squares line and a non-linear curve are fitted to the plot of $\log_e[\langle m|F^2_{obs}|\rangle/\langle m\varepsilon \sum_i^N f_i^2\rangle]$ versus $\sin^2\theta/\lambda^2$. The former provides the overall isotropic temperature factor and overall F_{obs}-scale.

3. Calculate isotropic and anisotropic corrections to the linear and non-linear values. The anisotropic values are estimated from a Wilson-type plot using h^2; hk; hl; k^2; kl; l^2 rather than $\sin^2\theta/\lambda^2$ (Maslen, 1968) to give the values $\Delta\beta_{11}$; $\Delta\beta_{12}$; $\Delta\beta_{13}$; $\Delta\beta_{22}$; $\Delta\beta_{23}$; $\Delta\beta_{33}$; respectively. This routine is repeated until convergence is obtained.

4. Graph the least-squares line and non-linear curve before and after corrections.

5. Calculate normalized structure factors from the linear and non-linear isotropic values, with and without anisotropic corrections, for output on the binary data tape. Summarize the $|E|$ statistics and distribution.

SIGMA

This program is composed of five subroutines which perform the following sequential operations:

1. Generates the phase shifts for all equivalent reflections, including the transformed indices, with $|E|$'s above a specified minimum. Reflections with restricted phase values (say, 0 or π; $\pi/2$ or $3\pi/2$) are also identified.

2. Sorts all accepted reflections in descending magnitude of $|E|$ to form the basic E-list.

3. Searches the E-list for sigma-2 related triplets and stores them on a scratch file.

4. Surveys the triplet list and accumulates the number of triplets r in which each reflection is involved and the value of $\kappa = \sigma_3\sigma_2^{-3/2}E_{\mathbf{h}} \sum^r E_{\mathbf{h}-\mathbf{k}}E_{\mathbf{k}}$. It also stores triplets, such as the sigma-1 kind (Hauptman & Karle, 1953), which aid in the later allocation of starting phases.

5. Prepares alternate sets of phases for starting the tangent iteration process.

These sets are composed of the essential origin and, if necessary, enantiomorph-defining reflections which satisfy the structural invariant conditions of the particular space group. The starting phases are selected according to the criteria which are listed here in order of precedence:

(I) The reflections should appear jointly in 'special' triplets. Special triplets provide phases of reflections which cannot be specified directly but are very important to the phasing process. For example, in the space group $P2_12_12_1$ reflections with general indices ($hkl \neq 0$) are difficult to specify initially due to their unrestricted phase values, yet these reflections are usually in the majority in the *E*-list and form by far the largest number sigma-2 relationships due to their higher multiplicity. It is important therefore that the starting phases should include, if possible, projection reflections ($hkl = 0$) which lead directly to the phases of general reflections. Triplets involving two projections and one general reflection warrant special consideration in this instance and phase allocations are made whenever possible to use these relationships.

(II) The reflections should appear jointly in as many sigma-2 triplets as possible. Phases estimated from direct interactions between the starting set, or even the second or third-order extensions, are usually very reliable.

(III) The reflections should have high κ-values. This ensures that they are involved in a small number of triplets with all high $|E|$'s or a larger number with small values, or both. These reflections are very influencial in the phasing hierarchy and should be used as soon as possible.

(IV) The reflections should have large $|E|$-values.

These selection conditions are not, of course, independent of each other, so that a reflection that satisfies (II) will probably comply with (III) and (IV). It is quite possible, however, that a number of these conditions cannot be fulfilled within the space group requirements. In any case the routine will proceed to make the best possible choice from the eligible reflections.

To ensure ready extension of the phase set in the early stages of the tangent interation process it may be necessary to start with more than the cell-specifying phases. The need for further phases is at present decided on an empirical basis by testing whether the phasing process expands suitably to all possible reflection parities. If it is considered necessary, extra phases are obtained by a simple symbolic addition process. A symbol phase is assigned according selection rules discussed above, although the phase need not be of the restricted type. A limited extension process is carried out and this usually results in the generation of a numerical value which is added to the starting set. If no phase is obtained, another reflection is assigned the symbol and the process is repeated.

In this routine the various structure invariant conditions specifying linear independence, etc., are handled through the available symmetry information. The special functions required in the selection of phases for each particular space group type are provided by a number of separate overlay subroutines. These enable ready changes to be made to an algorithm of a particular space group type if experience so dictates.

TANGENT

This program has three overlay subroutines which perform the following operations:

1. Stores the *E*-list and sigma-2 triplets from the scratch file and accepts the starting phases in any of three forms: (I) from the scratch file as one of the alternate sets prepared in *SIGMA*; (II) from the binary data output tape of the structure

factor program, when $|F_{calc}| > p \cdot |F_{obs}|$ – this enables phase assignments through partial structural knowledge (Karle, 1968); and (III) directly from cards.

2. Extends and refines the structure factor phases by repeated tangent iteration at successively lower E-thresholds (referred to as 'cycles'). This routine incorporates a number of tests which check the reliability of the estimated phases after each iteration step. These ensure that phases of reflections which appear unreliable will not be used repeatedly in the phasing process. Briefly the tests are as follows: (I) The calculated normalized structure factor E_{calc}, estimated from the sum of the squares of the numerator and denominator in the tangent expression, must not exceed a specified minimum. (II) Restricted phases must not deviate from their allowable discrete values by more than a specified value. (III) In three successive iterations each of the incremental changes in phase may not exceed a specified maximum. This is to reduce the effect of individual phase oscillations. (IV) There is an optional test, particularly useful in centrosymmetric space groups, which restrains phases that have not been estimated by more than one triplet. Along with test (I) this can be used to ensure that there are at least two *similar* phase estimates before this value is used in further extensions.

The routine allows for a maximum of twenty iteration steps in each cycle (*i.e.* for a given $|E|$ threshold), however, the next cycle will be initiated automatically when the extension and refinement process has converged for the current phases. This decision is based on the variations of the average value of $\langle|E_{obs}|\rangle/\langle|E_{calc}|\rangle$ for each iteration. A similar test is used to detect overall phase oscillations.

3. Outputs the estimated phases, in the order of the original binary tape, in preparation for the E-map calculation. A summary of the phase distribution is also prepared.

References

Hall, S. R. (1968). *Crystallographic Programs for the PDP-6 Computer* UWAC-17. University of Western Australia.

Hauptman, H. & Karle, J. (1953). *Solution of the Phase Problem.* I. *The Centrosymmetric Crystal.* A.C.A. Monograph No. **3**.

Karle, J. (1968). *Acta Cryst.* **B24**, 182.

Karle, J. & Hauptman, H. (1956). *Acta Cryst.* **9**, 635.

Karle, J. & Karle, I. L. (1966). *Acta Cryst.* **21**, 849.

Maslen, E. N. (1968). *Acta Cryst.* **22**, 945.

Stewart, J. M. (1967). *X-Ray 67 Program System for X-ray Crystallography.* Computer Science Centre, University of Maryland, Tech. Report 67–58.

Discussion

DEWAR: Have you had difficulty using the tangent formula with heavy atom structures with phases reverting to their heavy atom values?

HALL: No.

ROLLETT: When this does happen, does it mean that the heavy atom would be sufficient to solve the structure anyway?

HALL: Yes, possibly it does. However, I think that direct phasing procedures such as this should still be the first step in a structural solution, and one should revert to some other technique only in the event of failure.

MASLEN: The trouble with the heavy atom method is that in space groups such as $P2_1$ and $P2$ there is an arbitrary y-coordinate, and one hopes to have a method which will avoid this difficulty.

HOPPE: Have you tried a solution process involving a permutation of starting phases and then using some criteria to judge which is best?

HALL: No, we have found in most determinations only one phase, other than those specifying the origin, need be included in the starting set. This value is obtained directly as described in the lecture.

WEBER: I have found in the repeated application

of the tangent formula, especially in the space group $P2_1$, that a consistent set of phases is obtained, but the phases have inherent pseudo-symmetry such that each class of reflections is effectively centrosymmetric. This results in a single large peak in the E-map. Have you had this experience?

HALL: I have seen this effect but it was due to incorrect definition of the enantiomorph. However, a $P2_1$ structure recently solved by Dr. Brisse at NRC, using the repeated tangent iteration method, gave all phases as multiples of $\pi/4$. This approximation was, however, sufficient to solve the structure.

DEWAR: Our experience is similar to Dr. Weber's, and I am sure this phenomenon is related to what sometimes happens in the heavy atom case. In both cases we have found it helpful to fix certain phases in the tangent refinement.

Phase Determination as Applied in the XRAY67 System

J. M. STEWART

Computer Science Center, University of Maryland, U.S.A.

The success of the methods of phase determination based upon the symbolic addition procedure has given rise to the search for suitable algorithms for programming the method. In our attempt to add this capability to our computing system, XRAY67, we have developed a set of codes which seek a solution or 'solutions' directly from the observed data. The algorithm employs a strategy slightly different from the typical symbolic addition procedure. We believe that the method is particularly suitable for carrying out the symbolic addition procedure with a computer. The work is due to the efforts of Dr. R. V. Chastain, Dr. E. G. Boonstra, Dr. H. L. Ammon, Dr. J. R. Holden, Mr. C. Dickinson, and myself.

The overall procedure may be given as the following series of steps:

1. Preliminary data treatment and formation of E values.
2. Sigma-two generation.
3. Formation of equations.
4. Sorting of equations on the basis of probability.
5. Reduction of equations to fewest possible symbols.
6. Definition of unknown symbols.
7. Verification and modification of equations utilized in the solution.
8. Preparation of the E-map.

The preparation of the E-values is carried out by minimizing the value of the summation of $(E^2-1)^2$ over all reflections with the average value of all the E's constrained to be 1·00. The summation is carried out for various physically reasonable values of B and x (*e.g.* for an

$$E^2 = \frac{k^2 F^2_{\text{obs}}}{e^{-2B\sin^x\theta/\lambda^x} \sum_{j=1}^{n} f_j^2}$$

organic compound 2·0 to 4·0; for an inorganic compound 1·5 to 2·5) with the additional condition that the value of k is adjusted such that the average value of E is forced to be 1·000. The final values of B and x are found by interpolation from the table of summations of (E^2-1) which are formed. Note that the procedure is purely digital and is therefore very well adapted to computer use. It gives satisfactory results which are in good agreement with those obtained from a Wilson procedure or the K curve method. The final values of K and B agree well with the final least-squares scale and temperature parameters. The method works best when all unique data within a sphere are collected and used, including the 'unobserveds'. The link of XRAY67 which carries out this calculation is DATFIX.

The sigma-two procedure is carried out as a separate step. The process is considered to be a 'once only' calculation just as in the formation of the E-values. In this process all the possible relationships for which $\mathbf{h}_1+\mathbf{h}_2+\mathbf{h}_3=0$ for E-values above an arbitrary limit are searched out and stored. Great pains need be taken to conserve storage and to enhance speed. To this end, the symmetry operations of the space group are used to conserve storage and only those reflections of an asymmetric set are kept in tables in the immediate access store. Thus, for purposes of the code, the search is over all n reflections where $\mathbf{h}_i$ is

$$\mathbf{h}_i + \mathbf{h}_j R_k + \mathbf{h}_l R_m = 0$$

is over all n reflections where $\mathbf{h}_i$ is the ith reflection, $\mathbf{h}_j$ is the ith, $i+1$, $i+2$, . . ., to nth, l ranges over i, $i+1$, . . ., n and k and m range over all possible rotation matirces of the space group under consideration.

In other words, each reflection in turn is combined with itself and all beyond it in the table (including their symmetrically equivalent reflections) and the resulting reflections and their symmetric equivalents are sought in the table. In such a procedure, the whole emphasis of the code must be on 'failure' since, typically, for a 6000 reflection structure in a monclinic space group with E-values above 1·5, there will be 3000 useful interactions, while there are 3 million possible interactions. The necessity then is to define an algorithm which terminates the fruitless searches in as few steps as possible. This may be accomplished by means of pointer tables and tables of maximum and minimum indices within the asymmetric set. The techniques employed will be discussed in other sessions. The link of XRAY67 which carries out this process is SIGMA2.

At the end of processing, the program has produced a list (in the centrosymmetric case) of product equations of the type $s_{\mathbf{h}_1} s_{\mathbf{h}_2} s_{\mathbf{h}_3} = \pi$ with probability P. In our method of procedure $\mathbf{h}_1$, $\mathbf{h}_2$, and $\mathbf{h}_3$ are all in the supplied asymmetric set of reflections. The factor π is the resultant phase shift which makes the condition $\mathbf{h}_1 + \mathbf{h}_2 + \mathbf{h}_3 = 0$ hold. It must be remembered that for any space group with translational symmetry, the application of a given rotation matrix R may mean a corresponding shift in phase for the related reflection obtained. In the generation process $\mathbf{h}_2$ and $\mathbf{h}_3$ have both been operated upon by all the possible symmetry operations of the space group being treated and the two which have given rise to a useful relationship may also give rise to a phase shift. In the centrosymmetric case, this phase shift will be only zero or π radians.

Each reflection is assigned a symbol as its phase. These are the integers 1, 2, . . ., n, in order of decreasing E-value. Thus, the highest E valued reflection is given symbolic phase '1' and so on down the line. This again is done for programming reasons since it obviates the necessity for equivalence tables as the symbol is the index of the position of the reflection in stored tables.

Next, all the relationships generated directly from SIGMA2 are loaded in the search for a solution for the phases of the m largest reflections. In this loading process, all relationships are taken where $\mathbf{h}_1$ and $\mathbf{h}_2$ are for reflection less than or equal to the m phases sought. $\mathbf{h}_3$ may have any value at all. We then distinguish two types of relationships of essentially equivalent usefulness: relationships of the first kind, where $\mathbf{h}_3$ is less than or equal to m; relationships of the second kind which are obtained by the elimination of a common $\mathbf{h}_3$ from two of the originally loaded relationships. Consider the following typical example in which the top 9 phases are to be solved for:

The loaded data consist of a list, portions of which are shown below.

Symbolic phase	Symbolic phase	Symbolic phase	π	P
1	4	7	−1	0·999
:	:	:	:	:
2	3	4	−1	0·992
:	:	:	:	:
2	3	8	+1	0·990
:	:	:	:	:
5	6	12	−1	0·980
:	:	:	:	:
5	7	8	+1	0·970
:	:	:	:	:
5	8	13	−1	0·950
:	:	:	:	:
6	7	13	+1	0·945
:	:	:	:	:

From the relationships shown, one may obtain the following 'relationship of the second kind' from the elimination of reflection 13 between the last two shown relationships 5*6*7*8 = −1 (0·945). We arbitrarily, after some trials, used the lower probability from the two combined equations.

The resulting list of relationships of the first and second kind are then sorted in order of decreasing probability. This list of equations is the one from which the solution, or solutions, is sought. It should be noted that there are many more equations than unknowns in a typical list of this type if the probability cut-off is set low (*e.g.* 0·75). Because of the sorting procedure, setting it low is not of great concern since the relationships are used from high to low values of probability and only those required for a solution are actually used. All the rest are used to corroborate or refute the ones actually used in the solution.

The actual programming of the solution is of interest because it has four important characteristics:

1. It reduces the set of symbols to a minimum.
2. It leaves a complete history of the relationships used to find each phase.
3. It permits testing the validity of the *relationships* used in the solution for each phase.
4. It is computer-oriented procedure.

A vector and a diagonal matrix are set up along with 'pedigree' tables. The vector is reserved for the relative phase as ± (the product of all the π's used) with zero used to indicate no relationships known. The pedigree information shows all the relationships from the list which are actually required in the solution of a given phase. Consider again the nine phase example cited above.

Initially,

	rel phase	1	2	3	4	5	6	7	8	9
1	0	1								
2	0	0	1							
3	0	0	0	1						
4	0	0	0	0	1					
5	0	0	0	0	0	1				
6	0	0	0	0	0	0	1			
7	0	0	0	0	0	0	0	1		
8	0	0	0	0	0	0	0	0	1	
9	0	0	0	0	0	0	0	0	0	1

When the first relationship (1*4*7 = −1) is treated, the vector and matrix are changed to be as follows:

	rel phase	1	2	3	4	5	6	7	8	9
1	0	1								
2	0	0	1							
3	0	0	0	1						
4	0	0	0	0	1					
5	0	0	0	0	0	1				
6	0	0	0	0	0	0	1			
7	−1	1	0	0	1	0	0	0		
8	0	0	0	0	0	0	0	0	1	
9	0	0	0	0	0	0	0	0	0	1

In the pedigree for phase 7 is stored the fact that it depends on relationship No 1. being correct.

The second relationship (2*3*4 = −1) results in the following change shown in two steps, first:

	rel phase	1	2	3	4	5	6	7	8	9
1	0	1								
2	0	0	1							
3	0	0	0	1						
4	−1	0	1	1	0					
5	0	0	0	0	0	1				
6	0	0	0	0	0	0	1			
7	−1	1	0	0	1	0	0	0		
8	0	0	0	0	0	0	0	0	1	
9	0	0	0	0	0	0	0	0	0	1

Then because 4 has been reduced, all phases dependent upon 4 may be reduced by a column search so that this results:

	rel phase	1	2	3	4	5	6	7	8	9
1	0	1								
2	0	0	1							
3	0	0	0	1						
4	−1	0	1	1	0					
5	0	0	0	0	0	1				
6	0	0	0	0	0	0	1			
7	+1	1	1	1	0	0	0	0		
8	0	0	0	0	0	0	0	0	1	
9	0	0	0	0	0	0	0	0	0	1

Next pedigree of phase 4 is started and that of phase 7 updated to show their dependence upon relationship two. At the end of the scan of all the relationships, the matrix will have been reduced to the fewest possible 'symbols' in terms of all the available relationships. This usually results in a possible solution depending upon the space group.

Once the origin definition has been made and any additional symbols have been fixed for difficult cases, the whole solution may be substituted into *all* the relationships in the sorted list. In every case, the substitution either agrees with the phases found or disagrees. When there is agreement, we consider that by use of the pedigrees of the phases a count of accordant relationships may be formed. When there is disagreement, we consider that each relationship that figured in the solution is 'discrepant' from the standpoint of this particular relationship. At the end of the survey, one has counts of the number of times each of the relationships actually used in the solution appear accordant and discrepant in terms of all the known relationships. Any such relationship may then be 'condemned' or removed from the solution process.

The steps of picking out the appropriate relationships, forming relationships of the second kind, sorting them, reducing to the fewest possible symbols, defining the origin if possible, checking the relationships used in the solution, and producing a set possible phases is carried out for the centrosymmetric case with the XRAY67 link PHASE.

The process is quite successful and completely automatic for space groups like $P2_1/c$ or *Pbca*. For $P\bar{1}$ it virtually always gives several possibilities and the various *E* maps must be searched for the correct solution. In at least one case (Sax *et al.*, 1969) the pedigree has eliminated incorrect relationships and led to a solution when other procedures have failed.

Reference

Sax, M., Pletcher, J., Yoo, C. S. & Stewart, J. M. (1969). *Acta Cryst. Suppl.* A**25**, 202.

Special Considerations in Applying the Procedure for Phase Determination

J. KARLE

U.S. Naval Research Laboratory, Washington, D.C.

Although quite general procedures have been developed for phase determination, any particular crystal may present circumstances which require special understanding and consideration.

A commonly recurring misconception is that the correct physical solution will be obtainable from that set of phases which obeys phase determining formulas such as sigma-2 or the 'sum of angles' formula with the least number of exceptions, 'optimum internal consistency'. It is immediately apparent, of course, that the set of all zero phases satisfies this internal consistency criterion and, with rare exception, it does not correspond to the physical solution. It often occurs that the next most consistent set also does not lead to the physical solution, although the correct structure is always found from a phase set having very high consistency. In a successful phase determination by the symbolic addition procedure, the possible violations of internal consistency are contained among the limited number of ambiguous alternatives implied by the various possible values of the symbols. In developing computer programs, it is necessary to allow for the occasional violation of the phase determining relations and to not rely too heavily on optimum internal consistency. Otherwise it will be necessary to expect a certain number of failures.

One measure of internal consistency is the number of phases in a phase determination which could not be established because of contradictory indications from the contributors to the phase determining formula. In this regard, a good example showing the limitations of the internal consistency criterion is the structure investigation of jamine (Karle & Karle, 1964). Besides the case of all phases equal to zero, another set of phases led to the acceptance of all phase indications. There were also 5 sets, each leading to from 11 to 16 cases of unacceptable phase indications. One of these latter sets corresponded to the correct answer. Similar observations have been made in other laboratories. Among other manifestations of inconsistency encountered in our laboratory, it was found that a probability of over 0·9 for a relationship among letters in the investigation of auramine perchlorate (Fratini, Karle & Karle, 1967) was incorrect. Significant inconsistencies were also observed in many other studies, such as for example with a derivative of 1,4 benzodiazepin-2-one (Karle & Karle, 1967) and β-fumaric acid (Bednowitz & Post, 1966).

It is often possible to determine the structure of a disordered crystal by direct phase determination if the disorder is quite regular and recognizable among the numerous additional peaks generated in a Fourier map. Such was the case in the investigation of a dimerized product of hexafluorobutadiene (Karle, Karle, Owen & Hoard, 1965). The phase determination, however, manifested one unusual feature which is statistically very unlikely and worth mentioning. For approximately the first one hundred largest $|E|$ values, all the phases were zero, although there is no atom at the origin. The phase balance of roughly one-half zero and one-half π was attained by proceeding with the phase determination to smaller $|E|$ values.

When an insufficient number of phases has been determined before the computation of an E-map, spurious peaks are likely to appear, often occuring as a displaced image of part of the correct structure. This can confuse the interpretation of the Fourier map, for example for structures composed of long hydrocarbon chains or multiple rings. Such problems are very easily avoided by extending the set of phases to lower $|E|$ values.

There are many space groups, of which $P\bar{1}$ is an example, in which the assignment of signs to specify the origin cannot lead through the sigma-2 relation directly to a negative sign for a structure factor. This has to be accomplished through the assignment of an unknown symbol. If only one unknown symbol were specified to carry through the phase determination, it would generally have to represent a negative sign. A similar circumstance arises with the sum of angles formula for noncentrosymmetric crystals. For many space groups, phases other than zero or π are obtained through the specification of unknown symbols which may take on a variety of phase values.

Several computer programs for phase determination involve the use of basic sets of numerical phases for immediate application, in the case of noncentrosymmetric crystals, with the tangent formula. The basic sets are composed of phases assigned to specify origin and enantiomorph and reference frame, if appropriate, plus a small number of additional phases which are permitted to assume a variety of possible values. The correct solution is to be found among the sets of phases generated from each of the basic sets. Such programs have often met with success and it is of interest to make some general comparisons with the initial use of symbols. The assignment of numerical values to basic sets affords the opportunity to immediately carry out a simple routine on a computer. There is apparently a limit to the number of phases which can be used in the basic numerical sets, since too many members in such sets would generate too many alternatives to consider. The objective in employing symbols is to afford the opportunity to readily utilize a larger basic set. The symbol is carried through the phase determination as a representative of a range of possible values. In the course of the determination, relations often develop among the symbols or their values are limited in such a way that the number of alternatives which needs to be considered is small and readily handled. A computing program which can combine the ease of utilizing numerical values in a basic set with the opportunity to have a large set, as afforded by the use of symbols, should be very useful for noncentrosymmetric crystals.

There has been a large number of computing programs written for the purpose of carrying through phase-determining procedures. Some programs are undoubtedly quite good, in view of the successes which have been achieved. They are not however of equivalent quality as regards capacity and the decision-making processes which are required. The clear implication is that the ultimate capability of a method for phase determination is not necessarily measured by the capability of a particular computing program. If an initial attempt does not meet with success it is worth trying again with a new set of assignments for a basic set of phases.

It is quite worthwhile to carry out the initial stages of a phase determination by hand. The process is not time-consuming and is quite instructive to new students of the subject. In our laboratory, phase determinations using the symbolic addition procedure have always been initiated by hand and then continued by computing machine after some tens of phases have been evaluated. For some non-centrosymmetric crystals it has been necessary to restart the procedure as many as four times before a solution was obtained, but generally it has not been necessary to start more than twice. With care and attention to detail, it is possible to have a high degree of success with non-centrosymmetric crystals.

References

Bednowitz, A. & Post, B. (1966). *Acta Cryst.* **21**, 566.
Fratini, A. V., Karle, I. L. & Karle, J. (1967). *Appl. Opt.* **6**, 2091.
Karle, I. L. & Karle, J. (1964). *Acta Cryst.* **17**, 1356.
Karle, I. L., Karle, J., Owen, T. B. & Hoard, J. L. (1965). *Acta Cryst.* **18**, 345.
Karle, J. & Karle, I. L. (1967). *J. Amer. Chem. Soc.* **89**, 804.

Discussion

J. Karle: I would like to make a few additional comments in closing. Firstly, one must be aware of statistical relationships, such as Σ_1, Σ_3, and $B_{3\cdot 0}$ types, as these sometimes provide important additional information to the phasing process. Secondly, the advent of fast computers and accurate diffractometer data has enabled a number of the other inequality relationships, derived twenty years or so ago, to become practical. Dr. Tsoucaris and his colleagues have been able to make use of 16th and 20th order determinants in a phasing process. The higher order determinants can be expected to be quite powerful. One would have hesitated to have used these in the past with inaccurate data because of their unacceptable stability. Dr. Tsoucaris has also found a mathematical principle which enables a set of correct phases to be selected. If one uses the various phase possibilities in a high order determinant, the result of interest will be positive or zero. However, Dr. Tsoucaris has found that the most probable set of signs to be used with components of the determinant gives the determinant its maximum value. This approach has already succeeded in the solution of a $P\bar{1}$ structure.

TOPIC B

AUTOMATIC INTERPRETATION OF VECTOR MAPS

CHAIRMAN: C. E. NORDMAN

Patterson Analysis by Superpositions, Symmetry, and Peak Evaluation*

R. A. JACOBSEN

Institute for Atomic Research and Department of Chemistry, Iowa State University, Ames, Iowa U.S.A.

There exists a variety of methods for deconvolution or partial deconvolution of the Patterson function. These methods can be subdivided into two general classes: those making detailed use of chemical knowledge and those using chemical knowledge only indirectly. It is the latter approach that I shall discuss here. No chemical knowledge will be directly used, but we often rely on other information, primarily of the space group type, to aid us in the deconvolution of the Patterson function in a superposition type of procedure.

Patterson superposition method

A brief algebraic analysis of the Patterson superposition approach will be considered first. Assume that N atoms are arranged in a unit cell at positions $\mathbf{A}_1, \mathbf{A}_2, \ldots, \mathbf{A}_N$. The Patterson function can be thought of, in the point atom approximation, as the vector set $(\mathbf{A}_j - \mathbf{A}_i)$, in which i and j can take on any value from 1 to N. Now the Patterson superposition method, as it is usually employed, involves first the selection of a suitable peak from this map. Then the Patterson is shifted by this vector and compared with the original unshifted Patterson, using a suitable comparison function – usually a minimum function. Thus, mathematically, we select a vector $(\mathbf{A}_\beta - \mathbf{A}_\alpha)$, add this to the original set, obtaining $(\mathbf{A}_j - \mathbf{A}_i) + (\mathbf{A}_\beta - \mathbf{A}_\alpha)$, and compare the results with the original set $(\mathbf{A}_j - \mathbf{A}_i)$. Assuming $(\mathbf{A}_\beta - \mathbf{A}_\alpha)$ is unique, *i.e.* the peak selected represents a single interaction, the subsets $(\mathbf{A}_j - \mathbf{A}_\alpha)$, $j=1, \ldots, N$, and $(\mathbf{A}_\beta - \mathbf{A}_i)$, $i=1, \ldots, N$ are common. However, any other vectors which represent cases where other vector triangles, involving two vectors from the original set and the selected shift vector, could form would also be present. The two subsets represent, respectively, an image of the atomic arrangement in the unit cell with atom α at the origin and an inverted image of the unit cell structure with atom β at the origin. It can be readily shown that these two images are related by a center of symmetry one-half way along the shift vector. Therefore, if atoms α and β are related by a center of symmetry in the unit cell, one superposition is in theory sufficient to determine the atomic arrangement.

In the noncentrosymmetric case, a second peak can be selected as long as it belongs to either of the subsets mentioned above, and the superposition process is repeated. If, for example, $(\mathbf{A}_\gamma - \mathbf{A}_\alpha)$ were selected, only the subset $(\mathbf{A}_j - \mathbf{A}_\alpha)$ would remain. For further details the reader should consult some of the original papers on the superposition method (Wrinch, 1939; Buerger, 1950, 1951, 1959; Beevers & Robertson, 1950; Clastre & Gray, 1950; Garrido, 1950, 1951; McLachlan, 1951).

With point atoms very few accidental triangles would be found. In practice, however, we are not dealing with point atoms and the Patterson peaks have appreciable width, as shown in Fig. 1, for

* Work was performed at the Ames Laboratory of the U.S. Atomic Energy Commission. Contribution No. 2573.

the zero section of a Patterson for 1,5-gluconolactone. The width of these peaks can result in the possible formation of several accidental triangles. To reduce the peak width, it is necessary to employ sharpened coefficients. We compute sharpened Pattersons in our laboratory by replacing $|F_{hkl}|^2$ by $[\frac{1}{6}+(\sin\theta/\lambda)^2]$ $(|F_{hkl}|^2/\hat{f}^2)\exp(2B-B')\sin^2\theta/\lambda^2$ where $\hat{f}=\Sigma f_j/\Sigma Z_j$ and the first term represents a derivative sharpening contribution (Jacobson, Wunderlich & Lipscomb, 1961). One must always strike a suitable balance between sharpening the peaks and introducing ripples due to termination of series effects. Fig. 2 shows the same zero section of a Patterson map for 1,5-gluconolactone but now computed with sharpened coefficients. It should be compared with Fig. 1. Note the greatly improved resolution and the appearance of additional peaks that were indiscernible in the unsharpened map.

Patterson superposition program

Our Patterson superposition program was written in FORTRAN IV for the IBM 360/65 with four basic requirements in mind: (1) reasonably small computer core allocations, (2) high speed execution, (3) use with Patterson or symmetry maps,

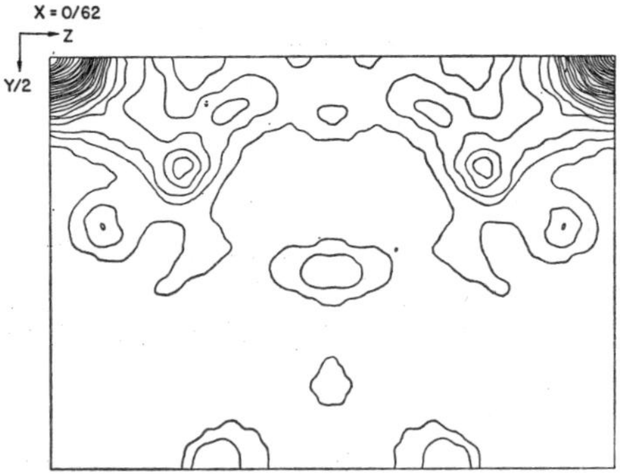

Fig. 1. The zero section of the unsharpened Patterson function for 1,5-*gluconolactone (orthorhombic,* $P2_12_12_1$).

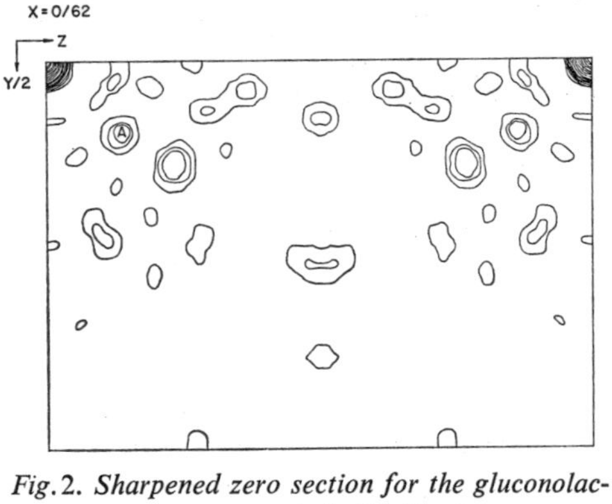

Fig. 2. Sharpened zero section for the gluconolactone, computed with $B'=3$.

and (4) weighting, origin shift, and multiple superposition options.

The program makes extensive use of disk memory in a direct access mode and operates such that two sections, each as a two-dimensional array, are available in the computer core. One of these is a section of a sharpened Patterson map; the other may be a layer of a symmetry map, a Patterson, or a previous superposition map. Arrays for all sections are stored on magnetic disks using two byte words and are read in the direct access mode. By using two input units, one for each array and the macro's FIND and READ, input delays are virtually eliminated. The programs calculate all superpositions (1 to 10) for a given layer of the resultant map before proceeding to the next layer. Another feature of the program is interpolation in either of the two dimensions of the array. Speed can be best illustrated by an example. Four superpositions were carried out on a recent run on a map of grid size $60\times100\times32$. The necessary core allocation was 70K bytes; of this, the program occupied 46K bytes. The central processor time was only 4·6 min.

Success using this approach depends to a large extent on the initial selection of peaks for the superpositions. One obviously wants to select a peak of low multiplicity to obtain a minimum number of images of the structure, a peak that is well resolved, *i.e.* one where the investigator can be confident as to the center of the peak and a peak in an appropriate position on the map. By this last is meant that for the first superposition or so, it is wise to avoid those lines or sections that contain a large number of peaks, since one wishes to have the least number of possible extraneous peaks in the resultant map while still retaining an image of the structure.

A peak labeled *A* in Fig. 2 was selected for a superposition shift vector and the result is shown in Fig. 3. (Of course, in practice you would probably avoid the selection of a peak in this zero section. This peak is selected here for ease of illustration.) A second superposition was then carried out using the peak labeled *B* in Fig. 3; the result is shown in Fig. 4. Although the fragment of the molecule lying in this section can certainly be seen, it is also evident that there are still appreciable numbers of extra peaks present after these two superpositions. It must be recognized that even with sharpened Patterson functions, appreciable numbers of coincidences will occur and, of course, great care must be exercised to avoid the selection of any extraneous peaks for the second or later superpositions. We therefore make extensive use of weights and symmetry in the superposition procedure whenever possible to aid in the selection of the appropriate peaks (Jacobson, 1966).

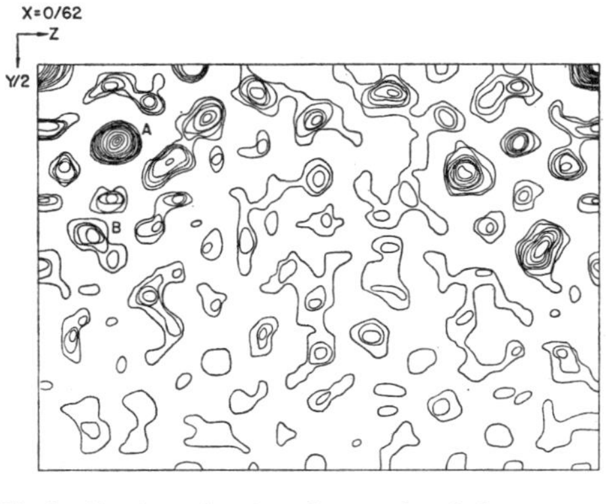

Fig. 3. Section showing the result of first superposition on gluconolactone using peak A in Fig. 2 (using finer contour divisions).

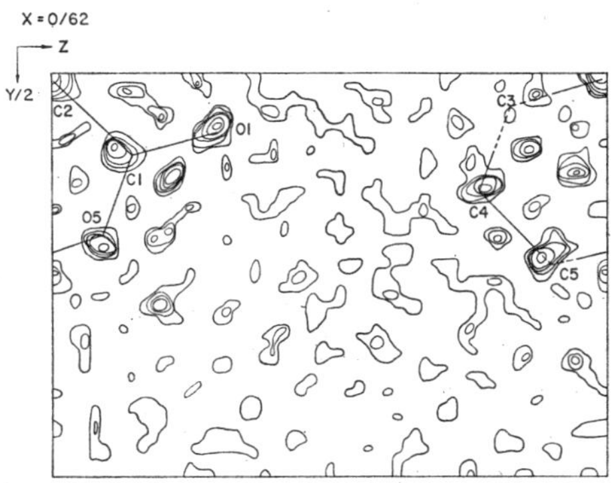

Fig. 4. Section showing the result of a second superposition using peak B in Fig. 3.

Symmetry restrictions

Space group symmetry provides, in general, some very useful restrictions on the peaks appropriate for the superposition. Consider again the 1,5-gluconolactone structure, space group $P2_12_12_1$. Because of the equivalent positions in this space group, there must be vectors at $(\frac{1}{2}-2x, 2\bar{y}, \frac{1}{2})$, $(2\bar{x}, \frac{1}{2}, \frac{1}{2}-2z)$, and $(\frac{1}{2}, \frac{1}{2}-2y, 2\bar{z})$ in Patterson space for every atom at (x,y,z) in electron density space. Therefore we compute a map, commonly called a symmetry map (Mighell & Jacobson, 1963; Simpson, Dobrott & Lipscomb, 1965; Hamilton, 1965), which assigns to the point (x,y,z) a value obtained essentially by taking the minimum of the values at the three positions in Patterson space indicated above (or the square root of this minimum value). If the asymmetric unit of the Patterson is not too large, the computation is quite straightforward, since everything can be kept in core. In practice, we modify the procedure slightly. Since we are using sharpened Pattersons and since every other point in the Patterson would be skipped due to the form of the vector expressions above, we select the maximum of the point and the four other points immediately surrounding this point in this case. A section of the symmetry map which was computed using this approach is shown in Fig. 5. One must remember that thus far no origin has been specified. Therefore, because of the orthorhombic symmetry, not one image but 64 images of the unit cell structure appear on this map. However, to isolate one image, we can select one peak and superimpose the sharpened Patterson on this point and all symmetry-related points in the symmetry map. Fig. 6 shows the results of one set of superpositions of this type using the peak marked A in Fig. 5. A second peak can be selected (B) and a second set of superpositions carried out (Fig. 7). If one compares this with Fig. 4, it is evident that fewer extraneous peaks are present and the atomic positions are reasonably easy to discern in this map. Also, this procedure can be repeated a few times with different starting peaks and the results can then be correlated. Such an approach was successfully used in the structure determination of the gluconolactone. Of course, one could alternatively carry out superpositions guided by, but not directly involving, the symmetry map.

Fig. 5. Section of the symmetry map computed for gluconolactone.

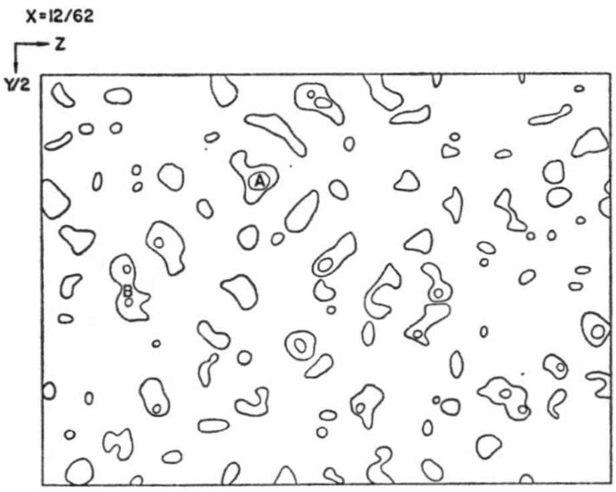

Fig. 6. Section showing results of four superpositions on the symmetry map using peak A (Fig. 5) and its three related positions.

The discriminator function

One of the major difficulties with a superposition method is the incomplete deconvolution of the associated maps, *i.e.* there are still appreciable numbers of extraneous peaks left after the series of superpositions have been carried out. It would be useful to have some test that could be applied to a structure when only a fraction of the atoms have been found by the superposition method or any other method would give a clear indication of whether the atom positions are indeed correct. A limiting case would be, of course, some type of positional parameter discriminator function which could be used to test each new atom as it is added to the partial model. In order to develop such a function, we have proceeded as follows. Let

$$S = \int |P_o - P_c| dV$$

where P_o is the observed Patterson function and P_c is the calculated Patterson for that fragment of the structure which is being tested. A necessary condition for the fragment of the structure to be correct is that S be a minimum, since incorrect atoms would in general give rise to peaks in the calculated Patterson where there are no peaks in the observed Patterson and hence increase the value of S. A discriminator value for the atom or fragment can then be defined as

$$D = \frac{\Delta S_t - \Delta S_a}{\Delta S_t}$$

where ΔS_t is the value of the change in S that should occur when atoms are in their theoretically correct positions and ΔS_a is the change that is actually observed. Letting N equal the number of atoms in the unit cell and n and n' represent the numbers of atoms in the two S calculations ($n > n'$),

$$\Delta S_t = [(\overset{N}{\Sigma} Z_i)^2 - (\overset{n'}{\Sigma} Z_i)^2] - [(\overset{N}{\Sigma} Z_i)^2 - (\overset{n}{\Sigma} Z_i)^2]$$

$$\Delta S_a = S_{n'} - S_n$$

where the Z's are the atomic numbers of the atoms involved; $S_{n'}$ and S_n are computed by a numerical integration procedure.

To improve the sensitivity of this approach, *i.e.* to make it not only a necessary but also a sufficient condition for the correctness of the partial structure, we use sharpened coefficients in the calculation of S and hence in the calculation of the discriminator.

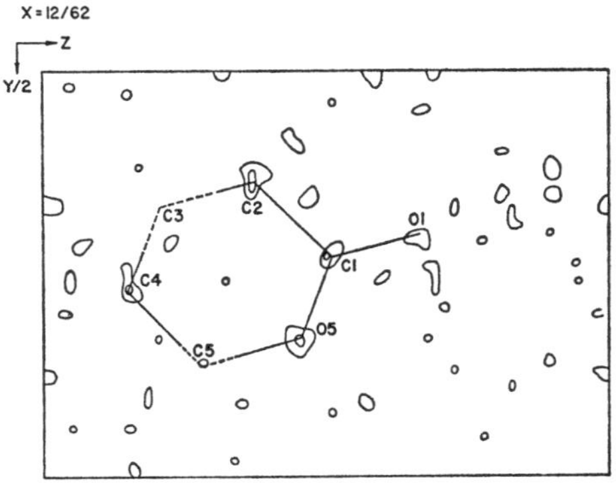

Fig. 7. Section showing results of eight superpositions using peaks A and B (Fig. 6) and their symmetry-related positions.

To exemplify the power and usefulness of this function, consider again the 1,5-gluconolactone example. In Table I are listed some of the results we have obtained in this case. The positions chosen were selected from those peaks remaining after the superposition procedure had been applied. Note that the correct atoms gave significantly lower values than did the incorrect positions, thus readily distinguishing those accidental peaks that remained in the map. It would be very difficult to distinguish these positions using the conventional discrepancy factor, R. Also note that the discriminator value is relatively insensitive to the number of atoms in the correct fragment although ΔS

Table I. Discriminator values for various molecular fragments

n-1 fragment	nth atom	R	ΔS_t	D_n
O 1	O 5	0·68	3072	0·08
O 1	O*	0·72	3072	0·82
O 1, O 5	C 1	0·64	3648	0·13
O 1, O 5	C*	0·64	3648	0·68
O 1, O 5, C 1	C 5	0·61	4800	0·08
O 1, O 5, C 1	C*	0·62	4800	0·47
O 1, O 5, C 1, C 5	C 4	0·58	5952	0·09
O 1, O 5, C 1, C 5	C*	0·58	5952	0·42
O 1, O 5, C 1, C 5, C 4	C 3	0·54	7104	0·03
O 1, O 5, C 1, C 5, C 4	C*	0·56	7104	0·50
O 1, O 5, C 1, C 5, C 4, C 3	C 2	0·51	8256	0·08
O 1, O 5, C 1, C 5, C 4, C 3	C*	0·52	8256	0·49

* Indicates atom position selected from different spurious peaks on resultant superposition map.

is strongly dependent on this variable. Further details and results will soon be reported (Hackert & Jacobson, 1970).

Acknowledgments

The assistance of M. L. Hackert and C. R. Hubbard is gratefully acknowledged. Liberal use has been made of figures and results of their unpublished work.

References

Beevers, C. A. & Robertson, J. H. (1950). *Acta Cryst.* **3**, 164.
Buerger, M. J. (1950). *Acta Cryst.* **3**, 87.
Buerger, M. J. (1951). *Acta Cryst.* **4**, 531.
Buerger, M. J. (1959). *Vector Space.* New York: John Wiley.
Clastre, J. & Gray, R. (1950). *Compt. Rend.* **230**, 1876.
Garrido, J. (1950). *Compt. Rend.* **230**, 1878.
Garrido, J. (1951). *Compt. Rend.* **231**, 297.
Hackert, M. L. & Jacobson, R. A. (1970). *Acta Cryst.* To be published.
Hamilton, W. C. (1965). *Acta Cryst.* **18**, 866.
Jacobson, R. A. (1966). *Trans. Amer. Cryst. Assoc.* **2**, 39.
Jacobson, R. A., Wunderlich, J. A. & Lipscomb, W. N. (1961). *Acta Cryst.* **14**, 598.
McLachlan, D. (1951). *Proc. Nat. Acad. Sci. U.S.* **37**, 115.
Mighell, A. D. & Jacobson, R. A. (1963). *Acta Cryst.* **16**, 443.
Simpson, P. G., Dobrott, R. D. & Lipscomb, W. N. (1965). *Acta Cryst.* **18**, 169.
Wrinch, D. (1939). *Phil. Mag.* (7) **27**, 98.

Discussion

SCOTT: Could you remark on how to choose the variables in your sharpening function?

JACOBSON: We compute the Patterson with 3 or 4 different values of B' and when the peaks start to shift, we know B' is too small. The idea is to use a large enough value of B' to avoid severe rippling.

YANNONI: In the procedure shown in your Table I, is there any indication of trouble if a spurious peak is chosen at the start?

JACOBSON: Once the fragment has reached reasonable size, the D value should certainly tell you if one or more atoms are in spurious positions.

Mind 3: A FORTRAN Patterson Superposition Program for the IBM 1130

NADRIAN C. SEEMAN

Department of Biochemistry and the Crystallography Laboratory, University of Pittsburgh, Pittsburgh, Pa., U.S.A.

The hands-on approach to doing Patterson superpositions on a small computer (Corfield, 1965) has in the past proved remarkably successful in the solution of relatively small light atom crystal structures (*e.g.*, Kim & Rosenstein, 1967; Kim, Jeffrey, Rosenstein & Corfield, 1967; Sax, Desiderato & Dakin, 1969; Seeman, McGandy & Rosenstein, 1968). With this fact in mind, MIND 3 was written for the 16K IBM 1130 in the hope that the same success would be noted in solving the larger light atom problems now under investigation.

Although the program permits the sum and maximum functions, it was designed for use with the minimum function. Upon determining the absolute position of N (usually two or three) atoms in the cell by use of an N-atom Symmetry Minimum Function (Corfield, 1965; Corfield & Rosenstein, 1966), the 0,0,0 vector is discarded entirely and the origin of the Patterson is placed at presumed atomic positions. The persistence of these positions as peaks throughout many minimum functions implies that they are real atomic locations. This method of course implies the presence of a highly positive origin in the Patterson used by the program. Otherwise the atomic positions would be washed out and there would be no real check on the authenticity of the assumed coordinates. Therefore a synthetic origin must be supplied with an E^2-1 Patterson, although this would in general have a smaller volume than the normal Patterson origin. Thus, the procedure is reduced to finding a new peak (atom) in the minimum function, putting it into a new minimum function in all positions implied by the space group symmetry, and seeing if it remains or is wiped out. This practice of putting in all coordinates related by space group symmetry constrains any prospective model to the symmetry of the space group and similarly imposes the symmetry of the space group on the minimum function. If all peak positions chosen are real atomic sites, they should remain as peaks because the overlap of true interatomic vectors will keep them from disappearing. The inverse of this statement is of course equally true: false peaks should sooner or later disappear because the vectors are not in the Patterson. The absolute height of a given true peak, however, will eventually decrease to the height of a single interatomic vector as one performs higher and higher order minimum functions, approaching the number of atoms in the unit cell. The advantage of coupling this procedure with the hands-on approach is that any potential atomic position can be tested, or refined a grid point or so, much more rapidly than normal computer center turnaround times permit. It takes perhaps five minutes to calculate the section of interest. This time period compares very favorably to the two or more hours the average computer center demands.

The program is designed to give the user maximum resolution in his Patterson. In order to maximize the amount of core available for grid arrays, all permitted subroutines have been placed on 'local' (*i.e.*, stored on disk until needed by the program). The grid is at the discretion of the user, as is the number of bits carried per grid point. Only the asymmetric unit of the Patterson is required for symmetries orthorhombic and below. 9600 grid points per section and 1·3 million overall are permitted for the asymmetric unit. The latter restriction is caused by the limited size of the disk, even though grid points are usually packed four to a sixteen-bit word. Besides saving core, packing the grid points minimizes disk I/O time. Symmetry generation of parts of the Patterson not in the original asymmetric unit is also facilitated, since four grid points are handled as one. Naturally the user does not want a Patterson with grid points ranging in value from −500 to 1000 set on a scale from 0 to 15. However, the only region of real interest, in terms of a minimum function, in such a Patterson of a light-atom structure, is that portion between, say, −50 and 100 which probably contains all the single interatomic vectors and many multiple vectors as well. In a heavy-atom structure, the absolute position of the 'interesting region' would probably be higher, but the principle remains the same. The program overflows and underflows those peaks which are above and below the user-chosen cutoffs. This does not decrease the resolution of the Patterson significantly.

In the interest of speed and saving core, the entire program has been written in integer arithmetic. The IBM 1130 has no floating point hardware and stores only 24 mantissa bits in a two-word floating point number. As a result, besides being faster, it is more efficient to store four grid points in a single integer word than six in the words required by a floating point number. Due to the slow speed of card-reading, the Patterson is read onto disk and packed only once, after which time the Patterson is manipulated exclusively from the disk. On a reasonably large Patterson (150,000 grid points per asymmetric unit) this saves about an hour of execution time. A single digit alphameric punchout also speeds up the output on the column-at-a-time 1442 Model 6 card read/punch. Punching 3600 grid points takes about two minutes as single contiguous digits, and the time is proportional to the number of columns punched.

If the user has a low-order minimum function, M_n, of which he is relatively confident and wishes to use it as the basis of a new minimum function, M_{m+n}, he can save time by calculating the M_m and then taking the minimum of the two functions through an auxiliary program. The time required is then of the order of an M_{m+1}, rather than that of an M_{m+n}. This is an extremely useful technique if one is using high-order minimum functions (n greater than five or six times the number of general positions in the space group). Other auxiliary programs do peak searches and glean off all high ground for transforming the minimum function (Corfield & Rosenstein, 1966). Calculating an *E*-map with the phases of structure factors determined from the positions of any and all high ground in the minimum function has been a useful technique in the selection of new peaks for succeeding minimum functions. This *E*-map is considerably sharper, with respect to peak height, than the minimum function. Peaks which are in high ground and are not old atoms are excellent choices for new atoms to be tested.

References

Corfield, P. W. R. (1965). *IBM 1620 Programs for the Solution of the Patterson Function by Superposition Methods.* Technical Report, Crystallography Laboratory, University of Pittsburgh, Pittsburgh, Pa.

Corfield, P. W. R. & Rosenstein, R. D. (1966). *Trans. Amer. Cryst. Assoc.* **2**, 17.

Kim, S. H., Jeffrey, G. A., Rosenstein, R. D. & Corfield, P. W. R. (1967). *Acta Cryst.* **22**, 733.

Kim, S. H. & Rosenstein, R. D. (1967). *Acta Cryst.* **22**, 648.

Sax, M., Desiderato, R. & Dakin, T. W. (1969). *Acta Cryst.* **B25**, 362.

Seeman, N. C., McGandy, E. L. & Rosenstein, R. D. (1968). Paper presented at *Amer. Cryst. Ass.*, Summer Meeting, Buffalo, N.Y.

Patterson Function Interpretation by Transformation into Reciprocal Space

PATRICK TOLLIN

Carnegie Laboratory of Physics, University of Dundee, Scotland.

In any Patterson function interpretation method using a known rigid group the three angles required to describe its orientation and the three coordinates required to specify the position of the group are determined in sequence. In the present method for determining these six parameters, the process of identifying the vector set of the group, within the Patterson function, is transformed into operations in reciprocal space.

The sequence of steps in the method is shown schematically in Fig. 1. Each box in Fig. 1 represents one program in a series written in ALGOL for the Elliot 4130 computer. It is often the case that the rigid group is planar and, if it is so, a significant saving in computing time can be achieved. The determination of the orientation of a planar group can be broken down into two stages: the determination of the two angles which specify the orientation of the plane of the group, and the determination of the azimuthal angle defining the orientation of the group in that plane. In the present method, therefore, different paths are chosen depending on whether the group is planar or non-planar. In order to give some idea of the computing involved, typical times for each stage have been quoted for the 4130 computer which has a 2 μsec cycle time. The structure determination of ninhydrin ($C_9H_6O_4$), although rather simple since all but two of the atoms lie in a plane, has been used as an example. This structure was first solved (Tollin, 1963) using some of the techniques described here in an earlier form, and since then an independent and more accurate determination has been performed (Medrud, 1969). Much of the programming of this method has been performed by Dr D. W. Young and Mr A. R. I. Munns in this laboratory and the author wishes to acknowledge their collaboration in this work.

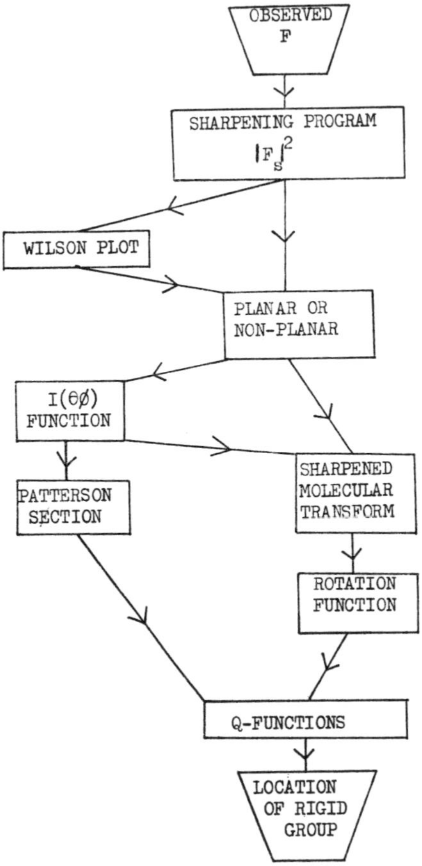

Fig. 1. Schematic diagram indicating the series of programs which perform the procedure described in the paper.

The non-planar group

The rotation function, defined by Rossmann & Blow (1962) as

$$R(\theta_1\theta_2\theta_3)=\int_u P_2(\mathbf{x}_2)P_1(\mathbf{x}_1)d\mathbf{x}_1$$

measures the degree of coincidence within the volume u when the Patterson function P_1 is rotated on the Patterson function P_2. Any point $\mathbf{x}_1$ in P_1 is related to any other point $\mathbf{x}_2$ in P_2 through the rotation matrix $\mathbf{C}$ by the relation $\mathbf{x}_2=\mathbf{C}\mathbf{x}_1$. The above integral has been shown to reduce to the double summation

$$R=\sum_{\mathbf{p}} |F_{\mathbf{p}}|^2\sum_{\mathbf{h}}|F_{\mathbf{h}}|^2G_{\mathbf{hh}'}$$

where $|F_{\mathbf{p}}|$ and $|F_{\mathbf{h}}|$ are the structure amplitudes of the crystals 2 and 1 respectively. $G_{\mathbf{hh}'}$ is an interference function whose magnitude depends on the volume u and the distance of the non-integral reciprocal space vector $\mathbf{h}'=\mathbf{pC}$, from the reciprocal lattice point $\mathbf{h}$.

In the present application the known group is positioned in an arbitrary cell and sharpened Patterson coefficients for this model are calculated. These coefficients are then used with the observed $|F|^2$ values in the calculation of $R(\theta_1\theta_2\theta_3)$. The programs used have been described previously in some detail (Tollin & Rossmann, 1966). A general search is made first using the Large Terms program and the final accurate determination of the orientation is performed using the general program.

The range of Eulerian angles $\theta_1\theta_2\theta_3$ which must be searched depends on the symmetry in reciprocal space of the $|F_{\mathbf{h}}|$ and the $|F_{\mathbf{p}}|$. A full discussion of this has been given by Tollin, Main & Rossmann (1966) and some observations are made in the Appendix to this paper. It is important to choose a sufficiently large cell for the model, otherwise origin peaks from neighbouring cells may again lead to confusing results. The model should be placed in the cell so that any symmetry it possesses becomes a space group operation. In this way the size of asymmetric unit of the rotation function can be significantly reduced. Also if the model does possess symmetry and it is not used in this way, it will lead to more than one peak in the asymmetric unit; and since these peaks will be related by a nonlinear symmetry operation, they may lead to considerable confusion.

Since, in this application of the rotation function, one has usually a fairly small volume of integration, the problem of the cut-off in the G-table, discussed by Tollin & Rossmann (1966), becomes more acute. It is usually necessary to perform the general search with a rather crude interpolation, say the nearest neighbours to the point $\mathbf{h}'$ only, but to determine the precise orientation one may have to take into account contributions from 100–200 lattice points around the point $\mathbf{h}'$. A typical first search at 10° intervals would require about four hours computing time.

The orientation of the planar group

In order to achieve the separation of variables mentioned above, it is convenient in this case to use the spherical polar angles θ and φ and the azimuthal angle ψ to describe the orientation of the planar group.

The angles θ and φ are determined using the function $I(\theta\varphi)$ defined by Tollin & Cochran (1964). The function $I(\theta\varphi)$ recognizes that the vectors between pairs of atoms in the planar group will all lie in a plane, parallel to the plane of the group in the structure and passing through the origin of the Patterson function. If a disc, comparable in size with the planar group, is placed with its centre at the origin, then the integral of the Patterson function over the disc should have a large value when the disc is in the same orientation as the planar group. The function $I(\theta\varphi)$ is thus defined as

$$I(\theta\varphi) = \int_r P(\mathbf{r})t(\mathbf{r})d\mathbf{r}$$

where $t(\mathbf{r}) = 1$ on the disc and 0 elsewhere. Apart from constant factors the function can be expressed in reciprocal space as

$$I(\theta\varphi) = \sum_{\mathbf{h}} |F(\mathbf{h})|^2 2\pi R^2 \frac{J_1(2\pi RS)}{2\pi RS}$$

where R is the radius of the disc and S is the perpendicular distance of the reciprocal lattice point **h** from the normal to the disc. The function $I(\theta\varphi)$ has a value at every point on a sphere and it has been found convenient to use a Samson-Flamsteed sinusoidal equal area projection to depict the results (Steers, 1949). The range of angles which must be examined depends on the symmetry of the crystal. Fig. 2 shows the results for ninhydrin. The peak maximum occurs at $\theta = 59°$, $\varphi = 136{\cdot}5°$ and the time required for the calculation was six minutes.

The azimuthal angle ψ can be determined in two ways: either by calculating the section through the origin of the Patterson function in the orientation obtained from the $I(\theta\varphi)$ map and finding the best fit of the vector set of the planar group or by using the rotation function. The Patterson section program is straightforward, a typical computing time is six minutes.

In using the rotation function to determine the azimuthal angle, it is convenient to place the model in the *ab* plane of the model unit cell and to use the observed $|F|^2$ values as the coefficient of the first crystal, that is, the $|F_h|^2$, and the model coefficients as the $|F_p|^2$. With this choice all that needs to be calculated is the line the rotation function defined by $\theta_1 = \varphi + \pi/2$ and $\theta_2 = \theta$ with θ_3 running from 0–2π or over a more restricted range if the model possesses symmetry. For ninhydrin the required line is $\theta_1 = 226{\cdot}5°$ and $\theta_2 = 59°$, and this line is shown in Fig. 3*a*. The corresponding rotation of the model on itself is shown in Fig. 3*b*. From the comparison of these we conclude that the required rotation is $\theta_3 = 115°$. Fig. 3*c* shows a line section at right angles to the line $\theta_1 = 226{\cdot}5°$, $\theta_2 = 59°$. This section shows the height of the peak relative to the background level of the rotation function. The

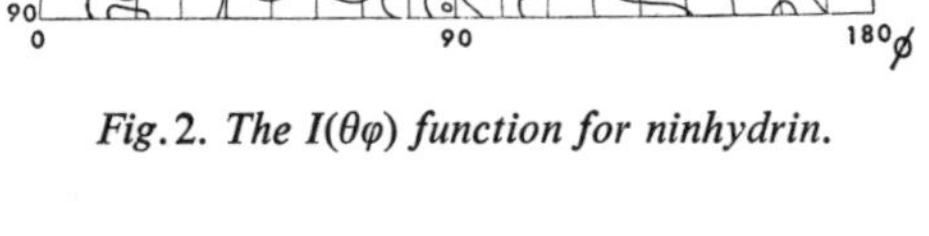

Fig. 2. The $I(\theta\varphi)$ function for ninhydrin.

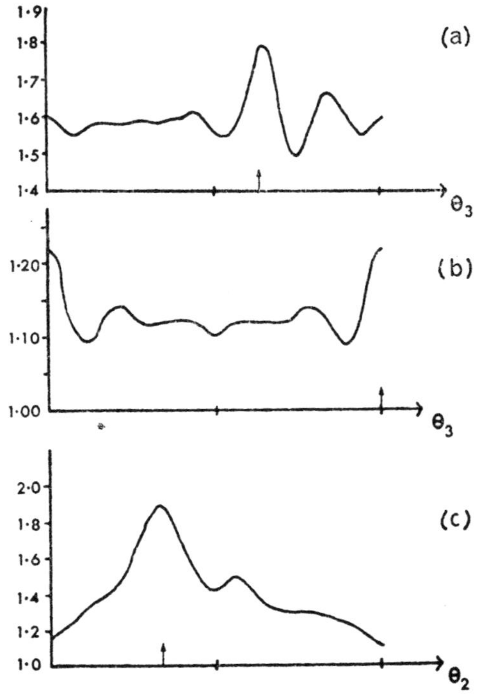

Fig. 3. The rotation function for ninhydrin. (a) The line $\theta_1 = 226{\cdot}5$, $\theta_2 = 59°$: the model against the structure. (b) The line $\theta_1 = 0$, $\theta_2 = 0$: the model against itself. (c) The line $\theta_1 = 226{\cdot}5°$, $\theta_3 = 115°$.

time involved was about one hour using a 3 2 2 grid.

Once the vectors $\mathbf{r}_j$, which define the relative positions of the n atoms of the group, are known, the parameters which remain to be determined are the three coordinates specifying the vector $\mathbf{R}_0$ which defines the position of one of the atoms with respect to the origin of the unit cell. Whichever method is used to solve this translation problem, it is always possible, except for the space group $P\bar{1}$, to effect separation of variables by determining the position of the group relative to the individual symmetry elements of the space group in turn.

The translation problem is solved in the present system by using the Q-functions (Tollin & Cochran, 1964; Tollin, 1966). The Q-functions in principle interpret a vector superposition performed using the sum function (Buerger, 1959).

The sum function, $S_n(\mathbf{r})$, is obtained by setting down the origin of the Patterson function at the ends of the vectors $\mathbf{r}_j$ and summing the result, thus $S_n(\mathbf{r}) = \sum_{j=1}^{n} P(\mathbf{r}-\mathbf{r}_j)$.

$S_n(\mathbf{r})$ contains peaks corresponding to all the atoms of the symmetry related molecules in the structure, in their correct positions relative to the arbitrarily chosen origin of the molecule used for the superposition. If a particular symmetry operation of the space group is described by $T(\mathbf{r})$, then for each atom in the group there will be a peak in the sum function at $T(\mathbf{r}_j + \mathbf{R}_0) - \mathbf{R}_0$. The Q-function, defined as the sum over all such positions for all possible values of $\mathbf{R}_0$, can then be written

$$Q(\mathbf{R}_0) = \sum_{j=1}^{n} S_n[T(\mathbf{r}_j + \mathbf{R}_0) - \mathbf{R}_0] \, .$$

The sum function has been used because it allows the Q-functions to be expressed as

$$Q(\mathbf{R}_0) = \sum_h |F(\mathbf{h})|^2 \sum_{j,j'=1}^{n} \cos \{2\pi\mathbf{h}.[\mathbf{r}_j + \mathbf{R}_0 - T(\mathbf{r}_{j'} + \mathbf{R}_0)]\}$$

and the expressions for particular symmetry operations have been given by Tollin (1966).

The translation function proposed by Crowther & Blow (1967) has recently (Tollin, 1969*a*) been shown to be virtually identical to the Q-functions. As a result the paper by Crowther & Blow provides an alternative description of the Q-functions in terms of a search through the Patterson function for a fit of the cross-vector between the known molecule and one related to it by a symmetry operation.

A number of modifications to the Q-functions can be made. If the group contains atoms of markedly different atomic number, it is useful to replace the sum function by the α-synthesis (Ramachandran & Raman, 1959). Provided the absolute scale of the diffraction data is known, it is possible to remove either the vectors between atoms in the known group or the origin peak from the Patterson function before the superposition is performed. Fig. 4*a* shows the function $Q(X_0Z_0)$ for ninhydrin and this contains examples of the false peaks mentioned previously (Tollin, 1966). Fig. 4*b* shows that removal of the origin peak from the Patterson function removes these false peaks. Although

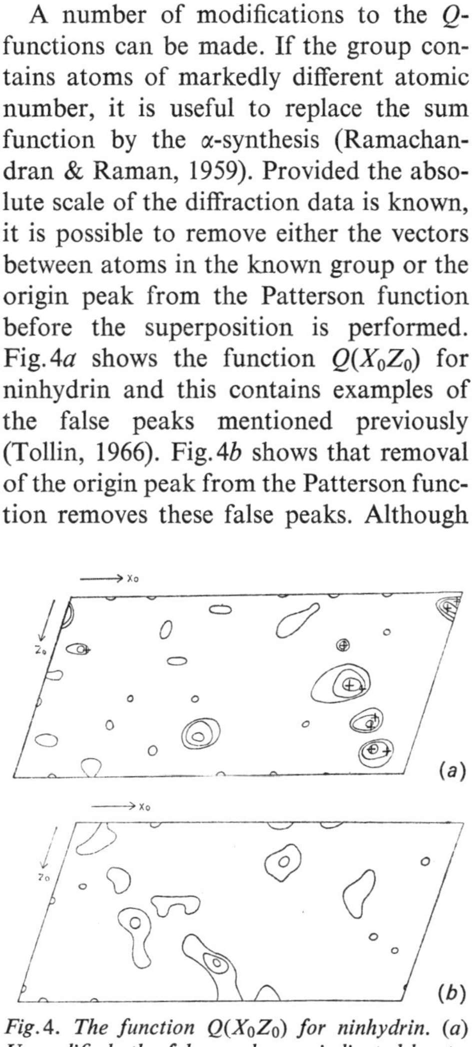

Fig. 4. The function $Q(X_0Z_0)$ for ninhydrin. (a) Unmodified: the false peaks are indicated by +. (b) With origin peak removed.

for a small molecule these false peaks can be predicted *a priori* and lead to no confusion of interpretation, it has recently been shown (Tollin, 1969*b*) that the application of the Q-function to a protein structure requires the removal of the origin peak.

Some examples of the application of these methods have been given by Munns & Tollin (1969), Young, Tollin & Sutherland (1968), and Young, Tollin & Wilson (1969).

Acknowledgements

I wish to thank Mr A. R. I. Munns for helping with the calculations for the ninhydrin example and, along with Dr H. R. Wilson and Dr D. W. Young, for helpful discussions.

Appendix

Symmetry

Whatever method is used to determine the orientation of a known group in the crystal, because of the symmetry of the diffraction data and the symmetry, if any, of the group, it is not necessary to explore all possible angular coordinates describing the orientation of the group. It is also important to know how many different orientations of the model correspond to the same structure lest confusion in the interpretation of the resulting maps ensues. A detailed and theoretical discussion of the effects of the symmetry of the structure and the model has been given by Tollin, Main & Rossmann (1966). It is not intended to reproduce this here but to give some useful practical results.

There are many sets of angular coordinates which can be used to describe the orientation of a rigid body. Tollin, Main & Rossmann (1966) point out that a discussion of symmetry is simplified if one chooses to use the Eulerian angles $\theta_1\theta_2\theta_3$ and the discussion here will be restricted mainly to these. There are a number of different definitions of the Eulerian angles, depending on the sense of the rotations, the choice of axes of rotation, and whether the rotation is an active or a passive one. The definition which we have used is that given by Goldstein (1951).

There are four types of symmetry element produced in $\theta_1\theta_2\theta_3$-space. Because there are always two descriptions of a rotation the operation

$$(\theta_1\theta_2\theta_3)=(\pi+\theta_1,\ -\theta_2,\pi+\theta_3) \qquad (1)$$

is always present. Since the first rotation is performed about the X_3-axis an n-fold rotation axis along **c** produces the operation

$$(\theta_1\theta_2\theta_3)=(2\pi/n+\theta_1,\theta_2,\theta_3) \qquad (2)$$

which causes a reduction in the unit cell size so that the unique portion occurs for $0\leq\theta_1\leq\dfrac{2\pi}{n}$. In addition, if n is even, this operation will combine with (1) to convert it from an n-glide to a c-glide plane. The presence of a two-fold axis along b produces the operation

$$\theta_1\theta_2\theta_3=\pi-\theta_1,\pi+\theta_2,\theta_3 \qquad (3)$$

Finally, the three-fold axis along [111] in the cubic groups, and any symmetry axes not along the X_1, X_2 or X_3 directions, give rise to the nonlinear symmetry operations mentioned by Tollin, Main & Rossmann (1966).

The expressions (2) and (3) assume that the rotation axes occur in the model being rotated. In the notation used to describe the rotation function, the model corresponds to P_1. If the axes occur in the structure, the corresponding operations are

$$(\theta_1\theta_2\theta_3)=\left(\theta_1,\theta_2,\ \frac{2\pi}{n}+\theta_3\right)$$

and

$$\theta_1\theta_2\theta_3=(\theta_1,\pi+\theta_2,\pi-\theta_3)\ .$$

For any particular problem these symmetry operations combine to form a space group in $\theta_1\theta_2\theta_3$ space.

In using the rotation function to determine ψ for a planar group, it is more convenient to choose the model as P_2. Changing the order of the model and structure corresponds to the transformation $(\theta_1\theta_2\theta_3) \rightarrow (-\theta_3-\theta_2-\theta_1)$. A definition of the Eulerian angles different from that given by Goldstein may mean that a similar transformation of the symmetry elements is required. For example, a popular definition of the angles follows that given by Margenau & Murphy (1943) and since their definition corresponds effectively to choosing the first rotation about $X1$, the unique rotation axes should be chosen along **a** rather than along **c**, as is conventional.

References

Buerger, M. J. (1959). *Vector Space.* New York: John Wiley.

Crowther, R. A. & Blow, D. M. (1967). *Acta Cryst.* **23**, 544.

Goldstein, H. (1951). *Classical Mechanics*, p. 107. Cambridge, Mass.: Addison-Wesley Press

Margenau, H. & Murphy, G. M. (1943). *The Mathematics of Physics and Chemistry*, p. 272. Princeton, N. J.: Van Nostrand.

Medrud, R. C. (1969). *Acta Cryst.* B**25**, 213.

Munns, A. R. I. & Tollin, P. (1969). *Nature,* **222,** 1170.

Ramachandran, G. N. & Raman, S. (1959). *Acta Cryst.* **12**, 957.

Rossmann, M. G. & Blow, D. M. (1962). *Acta Cryst.* **15**, 24.

Steers, J. A. (1949). *The Study of Map Projections.* London: University of London Press.

Tollin, P. (1963). Ph. D. Thesis. University of Cambridge.

Tollin, P. (1966). *Acta Cryst.* **21**, 613.

Tollin, P. (1969*a*). *Acta Cryst.* A**25**, 376.

Tollin,P. (1969*b*). *J. Mol. Biol.* **45**, 481.

Tollin, P. & Cochran, W. (1964). *Acta Cryst.* **17**, 1332.

Tollin, P., Main, P. & Rossmann, M. G. (1966). *Acta Cryst.* **20**, 404.

Tollin, P. & Rossmann, M. G. (1966). *Acta Cryst.* **21**, 872.

Young, D. W., Tollin, P. & Sutherland, H. (1968). *Acta Cryst.* B**24**, 161.

Young, D. W., Tollin, P. & Wilson, H. R. (1969). *Acta Cryst.* B**25**, 1423.

Discussion

R. HUBER: Could you state the advantage of using two steps in determining the orientation; (A) varying two Eulerian angles, then (B) a one-dimensional rotation, instead of a 3-D rotation?

TOLLIN: The computation is more economical this way.

BURNETT: If there is a known symmetry element in a structure, then the use of the spherical polar coordinates in the rotation function will isolate only the symmetry operation of interest and save computing time.

TOLLIN: I agree.

Programmed »Faltmolekül« Method

R. HUBER

Abteilung für Röntgenstrukturforschung am Max-Planck-Institut für Eiweiss- und Lederforschung, München und Physikalisch-Chemisches Institut der Technischen Hochschule München. Abteilung für Strukturforschung, München, West Germany.

A number of solutions have been proposed for determining the orientation and translation parameters of a known molecule or molecular fragment in the unit cell. These methods can be divided into two types: one operating in reciprocal space (Rossmann & Blow, 1962; Tollin & Cochran, 1964; Crowther & Blow, 1967), the other in real space (Hoppe, 1957; Nordman & Nakatsu, 1963; Huber, 1965; Braun, Hornstra & Leenhouts, 1969). The formalism of the two types of methods is in principle interconvertible. The differences and relative merits seem to reside mainly in the different criteria of fit, calculation procedures, and approximations which have to be used.

The purpose of this article is to describe the programmed Faltmolekül (convolution molecule) method, a Patterson search method operating in real space (Hoppe, 1957; Huber, 1965; Hoppe & Paulus, 1967).

The convolution molecule

The Faltmolekül or convolution molecule will be referred to as *CM*. The electron density in the unit cell ϱ can be described as the sum of the electron densities of the individual molecules

$$\varrho = \sum_{k}^{K} \varrho_M(\mathbf{E}_k \,.\, \mathbf{r}_k) \qquad (1)$$

ϱ_M is the electron density of an individual molecule described from an arbitrary origin and in an arbitrary orientation. ϱ_M is transformed by the matrix $\mathbf{E}_k$ and translated by the vector $\mathbf{r}_k$ to coincide with the actual position of the kth molecule in the unit cell. The summation is to be made over the K molecules in the general equivalent positions of the space group. $\mathbf{E}_1$ and $\mathbf{r}_1$ are the unknown parameters, the other matrices and vectors being related to these by the symmetry operations of the space group.

A relation equivalent to (1) exists connecting the inverted electron density functions ϱ^* and ϱ_M^*

$$\varrho^* = \sum_{k}^{K} \varrho_M^*(\mathbf{E}_k \,.\, \bar{\mathbf{r}}_k) \qquad (2)$$

The Patterson function P is the convolution product $\varrho\varrho^*$

$$P = \varrho\varrho^* = \sum_{k}^{K} \varrho_M \varrho_M^* [\mathbf{E}_k \,.\, 0] \qquad (3a)$$

$$+ \sum_{\substack{k \\ k \neq k'}}^{K} \sum_{k'}^{K} \varrho_M(\mathbf{E}_k) \varrho_M^*(\mathbf{E}_{k'}) \, [\mathbf{r}_k - \mathbf{r}_{k'}] \qquad (3b)$$

The Patterson function, therefore, consists of two kinds of *CM*. The K *CM* of type *A* in Eq. (3) represent the self-vector sets of the K molecules in the unit cell. They are centred in the origin (translation independent) and have the orientations of the molecules derived from them. (We will refer to these *CM* as self-*CM*.) The *CM* of type *B* represent the cross-vector sets. Their *structure* is orientation dependent, as molecules of different orientations are folded (mixed *CM*). They are centred at the vertices of the difference vectors of the particular molecular positions, as shown in Fig. 1. Replacing the electron density function ϱ_M in Eq. (3) by point atoms, it might easily be seen that the translational part in (3) is the vector set of the general equivalent positions.

As an example I will derive the *CM* structure of a crystal structure in space group $P2_1$ containing 2 molecules/unit cell. The two self-*CM* are the self-vector sets of the two symmetrically related molecules. $\mathbf{E}_1$ is an orientation matrix, depending on the three Eulerian angles ψ, θ, φ (defined later). As space group $P2_1$ involves only rotations, $\mathbf{E}_2$ is also an Eulerian matrix of the angles $-\psi, -\theta, 180° + \varphi$ into which the angles ψ, θ, φ are transformed by the operation of the twofold axis about y. As the general equivalent positions in $P2_1$ are x, y, z ($\mathbf{r}_1$) and $-x, \frac{1}{2}+y, -z$ ($\mathbf{r}_2$), the two mixed *CM* are centred at positions $2x$, $\frac{1}{2}$, $2z$ and $-2x$, $\frac{1}{2}$, $-2z$ respectively. (A derivation of the *CM* positions for all triclinic to orthorhombic space groups is given by Will, 1964.)

The most important principles can be summarized as follows:

1. If there are K symmetrically related molecules in the unit cell, the Patterson structure contains K^2 *CM*. Only K of these, however, are independent, $K.(K-1)$ being related by the symmetry of Patterson space.

2. One of these K-independent *CM* is translation independent and centred in the

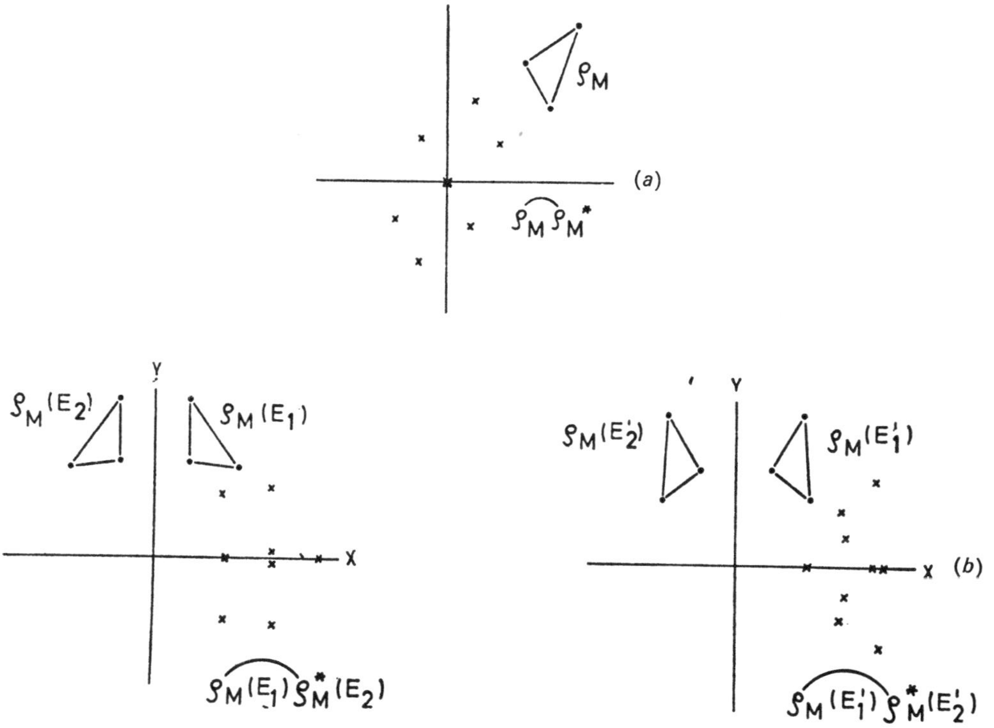

Fig. 1. Two-dimensional point structure ϱ_M and self-CM (Fig. 1a). The self-CM is centred in the origin and has the orientation of the structure derived from it. Its structure is orientation independent.

$\varrho_M(\mathbf{E}_1)$ and $\varrho_M(\mathbf{E}_2)$ are related by a mirror line y (Fig. 1b). The mixed CM $\varrho_M(\mathbf{E}_1)\varrho^*_M(\mathbf{E}_2)$ is centred at the vertex of the difference vector of $\varrho_M(\mathbf{E}_1)$ and $\varrho_M(\mathbf{E}_2)$. It shows the symmetry of the Patterson structure at its position, that is, a mirror line x. Its structure depends on the orientation of ϱ_M as the comparison of the two mixed CM shows.

origin of the Patterson function. $K-1$ are translation dependent.

3. Most *CM* exhibit certain symmetry properties, reflecting the symmetry of the Patterson function at their particular positions (all self-*CM* have a centre of symmetry; the mixed *CM* have a mirror plane perpendicular to their y axis in $P2_1$).

4. If there is more than one molecule in the asymmetric unit, *CM* exist which represent all possible combinations of vector sets between these independent molecules.

5. In centrosymmetric space groups, the *CM* centred in the origin and on Harker sections (Lipson & Cochran, 1957) have multiple weight.

Principle of the calculation

Eq. (3) shows that the *CM* structure depends in principle on six variables ψ, θ, φ, x, y, z (in space group $P2_1$ the variable y is arbitrary).

It is possible, however, to separate the orientation and translation determinations into a two-step operation, the first step being the orientation determination using translation independent self-*CM* centred in the origin. If the orientation is known, the mixed *CM* can be calculated and the translation determined.

Criterion of fit

There are several ways of defining a criterion for the best fit between a *CM* and the Patterson function (Rossmann & Blow, 1962; Nordman & Nakatsu, 1963). As the Patterson function is a superposition of all the *CM* which are positive everywhere, it should be expected that all values of a particular (correctly scaled) *CM* are lower than the corresponding values of the Patterson function if it is in the correct orientation (or translation). Therefore, the following function can be defined

$$R(\psi,\theta,\varphi) = \Sigma(P - \varrho_M \varrho_M^*[\psi,\theta,\varphi])^2 \quad (4)$$

the summation being made over all points of the *CM* which fall on points of the Patterson function lower than the *CM*. This function should be zero at the correct angles ψ, θ, φ, deviations being caused by errors in scaling of the *CM* and Patterson function and in the geometry of the model. An analogous function is defined for the translation determination.

Calculation procedure

1. The *CM* are calculated in two steps.

A. Structure factors are calculated for the two molecules to be folded.

B. The *CM* is then the Fourier synthesis calculated with the product of these structure factors as coefficients. The advantage of this procedure is that the *CM* are directly comparable to the Patterson function if they are calculated using the same scale, overall temperature factor, and resolution.

2. Next the orientation is determined by rotating a self-*CM* about the three Eulerian angles and calculating the value of R for every set of angles. It would be best to use all grid points of the Fourier synthesis of $\varrho_M \varrho_M^*$, but this is too time consuming. Therefore, we only use the peaks of the *CM*, which are generally built up by overlapping vectors. The inclusion of single weight vectors proves to be unnecessary, as their height is in most cases below the level of the Patterson function. As these points generally do not coincide with grid points of the Patterson function, we make a linear interpolation using the four neighbouring points.

3. Having determined the orientation, we transform the model coordinates according to this orientation and calculate the mixed *CM*.

4. These *CM* are translated in the Patterson function and the function R is calculated for every step. No interpolation is necessary as the grids of the Patterson function and of the *CM* have the same orientation.

The calculation of the CM by a structure factor calculation and a Fourier synthesis provides an easy means of subtracting CM which have been determined in their orientation or translation from the Patterson function and so increasing the contrast of the R function for following CM searches.

The coefficients of such a modified Patterson function are $|F_{\mathbf{H}}|^2 - |F_{CM,\mathbf{H}}|^2$, where $F_{CM,\mathbf{H}}$ are the structure factors of a particular correctly oriented and translated CM.

Eulerian angles

In our definition the Eulerian angles ψ, θ, φ produce rotations about Cartesian axes, resulting in a transformation matrix (Eulerian matrix) $\boldsymbol{E}$.

$$\boldsymbol{E} = \begin{vmatrix} \cos\psi & -\sin\psi & 0 \\ \sin\psi & \cos\psi & 0 \\ 0 & 0 & 1 \end{vmatrix} \cdot \begin{vmatrix} 1 & 0 & 0 \\ 0 & \cos\theta & -\sin\theta \\ 0 & \sin\theta & \cos\theta \end{vmatrix} \cdot \begin{vmatrix} \cos\varphi & 0 & \sin\varphi \\ 0 & 1 & 0 \\ -\sin\varphi & 0 & \cos\varphi \end{vmatrix} \quad (5)$$

so that $\mathbf{r}(\psi, \theta, \varphi) = \boldsymbol{E} \cdot \mathbf{r}_0$. $\mathbf{r}_0$ is a vector in the reference system of the model, $\mathbf{r}(\psi, \theta, \varphi)$ is the rotated vector. This corresponds to the following definition of the Eulerian angles:

ψ rotates about z to give $x'y'z'$,
θ rotates about x' to give $x''y''z''$,
and φ rotates about y'' to give $x'''y'''z'''$.

The range of rotations necessary to cover all possible orientations is

ψ	0–360°
θ	0–180°
φ	0–360°

This range is reduced further by symmetries of the CM and the Patterson function. Only the rotational symmetries, however, may be taken into account, as inversion operations cannot be produced by rotations about Eulerian angles. As an example, if the CM has a twofold axis about its y-axis, the relation holds $\mathbf{R}(\psi, \theta, \varphi) = \mathbf{R}(\psi, \theta, \varphi + 180°)$. If the Patterson function has a twofold axis about b (Patterson symmetry in space group $P2_1$ of our example), then $\mathbf{R}(\psi, \theta, \varphi) = \mathbf{R}(-\psi, -\theta, \varphi + 180°)$. These relations may be derived from the matrix $\boldsymbol{E}$ [Eq. (5)] (Tollin, Main & Rossmann, 1966).

In crystal systems with oblique axes, a transformation into a Cartesian system is necessary, so that

$$\mathbf{r}(\psi, \theta, \varphi) = \boldsymbol{A} \cdot \boldsymbol{E} \cdot \boldsymbol{B} \cdot \mathbf{r}_0$$

where $\boldsymbol{B}$ transforms $\mathbf{r}_0$ from the model reference system into a Cartesian system, and $\boldsymbol{A}$ transforms from the Cartesian system into the crystal system.

Programs

With the exception of two short subroutines, all programs are written in Fortran for a 32 K machine. As the Patterson function has to be kept in the core storage, we found it necessary to pack two values in a 36-bit word of storage.

The calculation time (IBM 7090) for an orientation search can be given approximately by the formula $t = nc/350$ (sec) where n is the number of peaks of a particular CM and c is the number of sets of Eulerian angles. A characteristic example with $n = 50$ and $c = 5800$ (ψ, θ, φ, 0–180°, 10° interval) takes $t \sim 14$ min. The translation search is approximately twice as fast. To these numbers, the time for calculating the CM has to be added, *i.e.* about 10 min/CM.

Experience

A number of structures has been solved by this method. Ecdyson (Huber & Hoppe, 1965), γ-Rhodomycinon (Röhrl, 1969), Bicyclo-2,2,1 Heptadien 2,5-Dicarbonsäure 2,3 (Steigemann, 1968), 1,6-Oxido-cyclo-decapentaen (Brodherr, 1969), Bullvalen (Amit, Huber & Hoppe, 1968), Tetramethylcyclohept-azulen (Qasba, Brandl, Hoppe & Huber, 1969), 2,2 Dimethyl-Thioindoxyl-1,1-Dioxyd (Preuss, 1969), 21-Hydroxy-Ecdyson (Dammeier, 1969), Neo-Phorbol (Hoppe, Brandl, Strell, Röhrl, Gassmann, Hecker, Bartsch, Kreibich, Szczepanski, 1967).

Two structure determinations will be discussed, one (γ-Rhodomycinon, Röhrl, 1969) because it contains two molecules, *i.e.* 52 light atoms/asymmetric unit, and the other (Neo-Phorbol) because it shows an interesting combination of *CM* and heavy atom methods.

Fig. 2 shows the known part of the γ-Rhodomycinon molecule which was used for the calculations. It contains 20 atoms out of 27 per molecule. The space group is $P2_1$. The rotational symmetries of the Patterson function and the self-*CM* (two-fold axis along **b** of the Patterson function and 222 symmetry of *CM*) restrict the range of Eulerian angles to

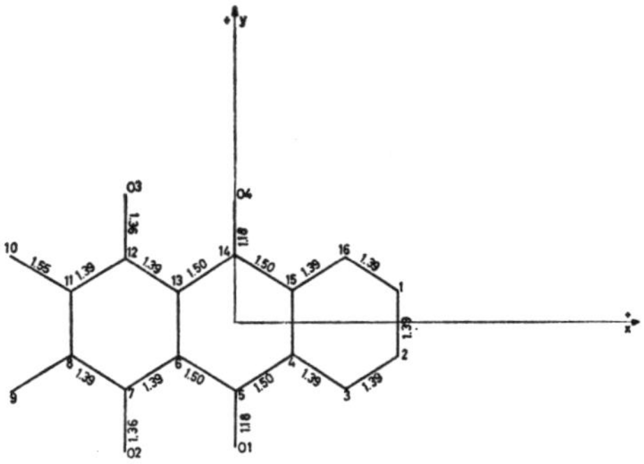

Fig. 2. Model for CM calculations of the γ-Rhodomycinon structure.

ψ	0– 90°
θ	0–180°
φ	0– 90°

Fig. 3 shows the results of the orientational search which showed only one dominant minimum. This result means that both molecules have either an identical orientation or orientations which differ from one another by a rotation of 180° about the z-axis of the model (Fig. 2), as the model symmetry is lower than the *CM* symmetry. The translational search with a mixed *CM* (Fig. 4) showed two minima arising from the solutions of the two independent molecules. The problem of the relative orientation of the two molecules mentioned above and the origin correlation could be

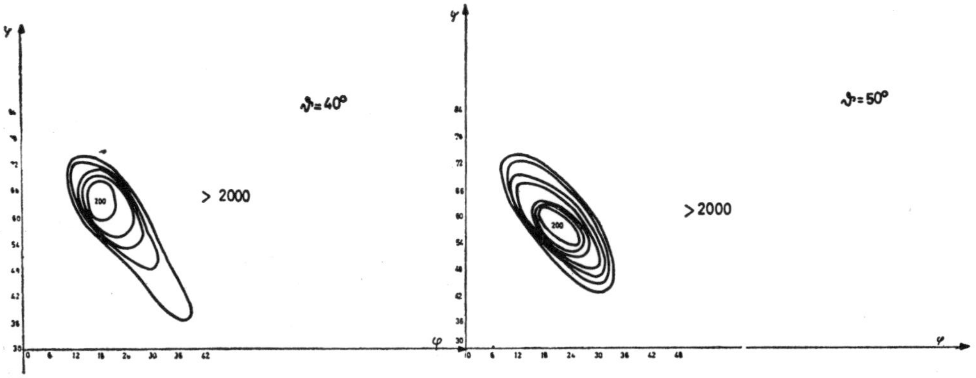

Fig. 3. Orientation determination with the CM of γ-Rhodomycinon. Two sections through the three-dimensional **R** *(ψ, θ, φ) function containing the deepest minimum (height 200) are shown. The mean background is higher than 2000. Scale in degrees.*

solved by a translational search using a *CM* representing the set of vectors between the two independent molecules. Two such calculations have to be made, the molecules arranged in the two different relative orientations. The deeper minimum provides the correct relative orientation and the origin correlation.

Neo-Phorbol (Hoppe *et al.*, 1967) ($C_{13}H_{35}O_9Br$) crystallizes in space group $P2_1$ with one molecule/asymmetric unit. It contains a *p*-bromo-benzoate group. The bromine-bromine vector was clearly visible in the Patterson synthesis. A Fourier synthesis based on the bromine positions alone, which form a centrosymmetric configuration, would show the structure and its centrosymmetric image. We, therefore, made orientational and translational *CM* calculations using the bromobenzoate group. Phasing with this group, which forms a noncentrosymmetric model, leads directly to the structure by successive Fourier syntheses. Fig. 5 represents the translation calculation which showed a subsidiary minimum, as the Br–Br vector dominates in the *CM*.

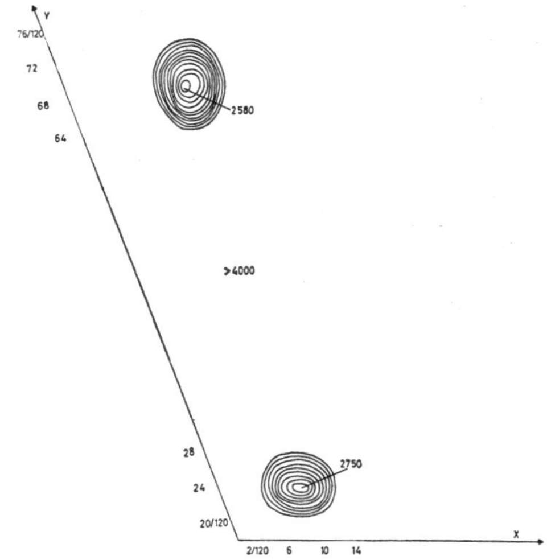

Fig. 5. Translation determination of the p-bromobenzoate group in Neo-Phorbol. The two-dimensional **R** $(x, y, \frac{1}{2})$ *function shows the correct solution* (2580) *and a subsidiary minimum, arising from the dominating* Br–Br *vector (Space group* $P2_1$*, unique axis z).*

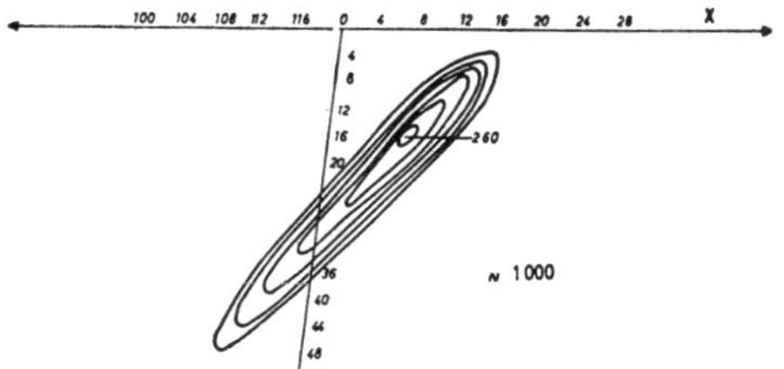

Fig. 4. Translation determination of γ-Rhodomycinon. The two-dimensional function **R** $(x, \frac{1}{4}, z)$ *has two minima (of height* 260 *and* 256*) representing solutions for the two independent molecules (average background about* 1000*). Scale in* $\frac{1}{120}$ *of cell constant (Space group* $P2_1$*, unique axis y).*

References

Amit, A., Huber, R. & Hoppe, W. (1968). *Acta Cryst.* **B24**, 865.

Braun, P. B., Hornstra, J. & Leenhouts, J. I. (1969). *Philips Res. Rep.* **24**, 85.

Brodherr, N. (1969). Diss. TH München.

Crowther, R. A. & Blow, D. M. (1967). *Acta Cryst.* **23**, 544.

Dammeier, B. (1969). Diss. TH München.

Hoppe, W. (1957). *Elektrochem.* **61**, 1076.

Hoppe, W., Brandl, F., Strell, I., Röhrl, M., Gassmann, J., Hecker, E., Bartsch, H. & Szczepanski, Ch. (1967). *Angew. Chem.* **79**, 824.

Hoppe, W. & Paulus, E. F. (1967). *Acta Cryst.* **23**, 339.

Huber, R. (1965). *Acta Cryst.* **19**, 353.

Huber, R. & Hoppe, W. (1965). *Chem. Ber.* **98**, 2403.

Lipson, H. & Cochran, W. (1957). *The Determination of Crystal Structures.* London: Bell.

Nordman, C. E. & Nakatsu, K. (1963). *J. Amer. Chem. Soc.* **85**, 353.

Preuss, L. (1969). Diss. TH München.

Qasba, R., Brandl, F., Hoppe, W. & Huber, R. (1969). *Acta Cryst.* **B25**, 1198.

Röhrl, M. (1969). Diss. TH München.
Rossmann, M. G. & Blow, D. M. (1962). *Acta Cryst.* **15**, 24.
Steigemann, W. (1968). Diplomarbeit TH München.
Tollin, P. & Cochran, W. (1964). *Acta Cryst.* **17**, 1322.
Tollin, P., Main, P. & Rossmann, M. G. (1966). *Acta Cryst.* **20**, 404.
Will, G. (1964). *United States Army Electronics Research and Development Laboratories.* Fort Monmouth, N. J., USA ELRDL Technical Report 2436.

Discussion

BURNETT: Do you refine your model at any stage in the determination or orientation and translation?

R. HUBER: No. We have not found that these methods are very dependent on the accuracy of the model used.

A Program which, Using a Known Part of the Molecule, Produces the Parameters of all Atoms

J. HORNSTRA

Philips Research Laboratories, N. V. Philips' Gloeilampenfabrieken, Eindhoven, The Netherlands.

Crystal structures of organic molecules without heavy atom and without centre of symmetry are difficult to solve. In many cases, however, the conformation of part of the molecule is known and this may be used in a Patterson search method (Nordman & Nakatsu, 1963; Huber, 1965; Braun, Hornstra & Leenhouts, 1969*a*). This paper comments on our programs which have been developed to make this procedure as automatic as possible.

Separate programs are used for different space groups. When a new space group is met, one of the existing programs can easily be modified. Our programs are written in FORTRAN 63 for a CDC 3600 computer.

Preparation of the Patterson function

The sharpening of the Patterson function and its storage in the computer memory depends on the decision as to whether the minimum function, the sum function, or some modification of either will be used as a measure of fit of a vector set.

If no heavy atoms are present, the Patterson function will show a large number of small peaks only. The atoms of C, N, and O are considered to have the same weight and hydrogen atoms are neglected. With medium-sized unit cells (1000 to 2000 Å^3) even a sharpened Patterson function will hardly show any empty space between peaks. As a sharpening function we mostly used multiplication of $|F(hkl)|^2$ by $\sin^2 \theta / f_C(hkl)$. The steroid structures solved by us (Braun, Hornstra & Leenhouts, 1969*b*, 1970) had unit cells of about 1800 Å^3 containing four molecules of about 25 non-hydrogen atoms. Hence, the Patterson function had about 10000 peaks per unit cell. From the shape of the origin peak the radius at half-height of a peak was found to be 0·4 Å. A sphere with this radius has a volume of 0·27 Å^3 and may be considered as the volume of one peak. Since 10000 of these spheres have a volume of 1·5 times the volume of the unit cell, there will be considerable overlap. This explains why attempts to use the unmodified minimum function failed.

The distance between grid points in Patterson space is chosen between 0·15 and 0·20 Å in order to avoid the need for interpolation. Then, the limited capacity of the fast-access memory does not allow the storage of more than two bits per grid point. With a two-bit code, the numbers 0, 1, 2, and 3 represent ranges of values of the Patterson function. These ranges are defined by three levels, called test levels.

Measure of fit

The choice of these test levels and of the measure of fit (*MOF*) was guided by the following reasoning. A vector set generated by the known part of the molecule in a wrong position corresponds to a random sample from the values of the Patterson function. A vector set generated by the correct position corresponds to a biased sample because at the end point of each vector at least one peak will be present. The sum function, or some modification of it, seems to be the most efficient measure of fit for distinguishing correct and incorrect vector sets.

The use of three test levels looks very similar to the use of three percentiles in the estimation of the mean of a sample. According to Dixon & Massey (1951), the latter problem is solved by taking $(P_{17}+P_{50}+P_{83})/3$. Therefore, we determine the cumulative distribution function of the values of the Patterson function and take $(P_{83}-P_{17})/2$ as the spacing between test levels. This quantity is approximately equal to the standard deviation $\sigma(P)$ of the Patterson function if the origin peak is disregarded. The central test level is chosen neither at the mean value of the Patterson function nor at the expected mean of a correct vector set, but half-way between these two. The distance between these two means, the height of a single peak *HSP*, is calculated from the height of the origin peak. If no centre of symmetry is present, this height is divided by the number of atoms in the primitive unit cell. In centrosymmetric structures all vectors between non-equivalent atoms occur twice and single peaks are rare. Then the height of the origin peak divided by one half of the number of atoms in the unit cell is taken as *HSP*. Interference by the presence of the origin peak is avoided by taking the median P_{50} instead of the mean. Hence the central test level is chosen at $P_{50}+HSP/2$ and the others at $\sigma(P)$ above and below it. As a measure of fit we take the sum of the two-bit values of the Patterson function at grid points nearest to the end points of the vectors. This *MOF* is superior to the normal numerical sum function because unwanted effects of very high peaks have been eliminated.

Experience has shown that this choice of the test levels gives a good distinction between correct and incorrect vector sets if the molecules are not planar. With planar molecules several vectors coincide and a better result is obtained if the levels are shifted to higher values, *e.g.* the central test level at $P_{50}+HSP$.

A more detailed statistical analysis of this measure of fit has been given by Braun, Hornstra & Leenhouts (1969*a*). This analysis shows that a good chance of solving the structure requires that the number of atoms in the known part of the molecule should be equal to or greater than the square root of the number of non-hydrogen atoms in the unit cell. This agrees with our experience.

Orientation search

This and the next part of our procedure will not be discussed in detail because they are similar in principle to those described by Huber (1965) and Nordman (1966). These details may be found elsewhere (Braun, Hornstra & Leenhouts, 1969*a*).

With the crystal structures which we solved, the known part of the molecule, called the model, contained 8 to 11 atoms. Hence, the vector set used in the orientation search contained 28 to 55 non-equivalent vectors. Sometimes different vectors of the model coincided, but this was always disregarded. It may be shown that when the sum function is used, the value of the Patterson function should be multiplied by the multiplicity or the weight of the vectors in order to get an optimum result. If only coincidence of vectors with the same weight occurs, this happens automatically when these vectors are treated independently, as in our program.

All significantly different orientations of the vector set are generated by rotation through three angles α, β, and γ. The necessary ranges of these angles are determined by the symmetry of Patterson function and model. The increments by which these angles are changed were usually 9°. It was found that smaller increments did not lead to a more accurate orientation because of the small number of vectors in the vector set. Because the space of angular rotations is degenerate when $\beta = \pm 90°$, the increment of γ is increased in the neighbourhood of this degeneracy.

In this way several thousand orientations are generated. For each of them the measure of fit (*MOF*) of the vector set is calculated by adding the corresponding values of the coded Patterson function, in which 3 corresponds to a low value, 0 to a high value. During this search a limited number (*e.g.* 60) of the best orientations are retained in a sorted list. The *MOF* of each new orientation is compared with the worst of this list and rejected if it is worse. If it is better and the orientation does not lead to unacceptable intermolecular distances, it is added to the list and the worst of the list is rejected. At the end of the orientation search the correct orientation was always found within this list, but rarely as number one. This list is reduced by omitting orientations very close to better ones. As a critical angular distance, we chose 18°. Each of the remaining possible orientations (about 20) is subjected to a translation search, based on the symmetry elements present.

Translation search

The position of the rotated model within the unit cell may be described by a vector with components *a*, *b*, and *c* from the origin to the centre of gravity of the model. In some low-symmetry space groups one or more of these components are indeterminate. The position of the origin with respect to the symmetry elements is taken from the *International Tables for X-ray Crystallography* (1952). Each (twofold) symmetry element gives rise to a new vector set which may be used to determine one or more of these components. A mirror or glide plane yields only one component, a twofold axis or screw axis yields two, and a centre of symmetry yields all three. For efficiency reasons, the symmetry elements are treated separately in this order. Afterwards or during the search with a centre of symmetry, the *MOF*'s are combined and a list of the best combinations is retained. As a combination we use the sum of squares instead of the sum of the *MOF*'s. This is supposed to give a better distinction if the *MOF* is defined in such a way that a low value means a good fit.

By way of example, we shall discuss space group $P2_12_12_1$. The twofold screw axis $(\frac{1}{4}, 0, z)$ transforms x, y, z into $\frac{1}{2}-x, -y, \frac{1}{2}+z$. The positions of the atoms in the rotated model with respect to the centre of gravity being x_i, y_i, z_i, the vector set becomes

$$v_{ij} = (\tfrac{1}{2} + x_i + x_j + 2a,\quad y_i + y_j + 2b,\ \tfrac{1}{2} + z_i - z_j)$$

for all combinations $i \geq j$. The vectors are rounded to the nearest grid point at $a = b = c = 0$. For efficiency reasons, $2a$ and $2b$ are then varied over the entire unit cell by increments equal to an integral number of distances between grid points, usually 0·4 to 0·5 Å. This leads to a *MOF* field *MOF* (a,b). In this space group the combined *MOF* is

$$MOF(a,b,c) = MOF^2(a,b) + MOF^2(b,c) + MOF^2(a,c)$$

The sorted list of the best combined *MOF*'s belonging to one orientation is reduced by intermolecular distance tests and by omitting solutions very near to better ones.

If the model and the structure have a centre of symmetry, the variation of $2a$, $2b$, and $2c$ may be restricted to one-half of the unit cell.

Pseudo-symmetry

In this stage we sometimes found elements of pseudo-symmetry, *e.g.* a pseudo-screw axis parallel to a crystallographic screw axis, as occurs in L-ascorbic acid (Hvoslef,

1968), or a pseudo-centre of symmetry in a centrosymmetric space group, as occurs in canthaxanthin (Bart & MacGillavry, 1968). Both structures were used to test our program. In both cases the translation search yielded the position of the model with respect to the pseudo-element of symmetry.

This may be explained with the two-dimensional example of Fig. 1. The unit cell contains two molecules related by a centre of symmetry, but each molecule has a pseudo-centre of symmetry in itself. Let the model consist of the six-membered ring with three substituents. The orientation is found easily, because the four parts *A*, *B*, *C*, and *D* all contribute to the same vector set. Part *A* is related to *C* and *B* to *D* by crystallographic symmetry and, moreover, part *A* is related to *B* and *C* to *D* by pseudo-symmetry. The vector $\mathbf{r}_i^A - \mathbf{r}_j^B$ coincides exactly with $\mathbf{r}_j^D - \mathbf{r}_i^C$ and roughly with $\mathbf{r}_j^A - \mathbf{r}_i^B$ and $\mathbf{r}_i^D - \mathbf{r}_j^C$, whereas vectors such as $\mathbf{r}_i^A - \mathbf{r}_j^C$ only coincide with $\mathbf{r}_j^A - \mathbf{r}_i^C$. Therefore, the translation search will yield the position with respect to the pseudo-centre with a better *MOF* than the two positions with respect to the crystallographic centres.

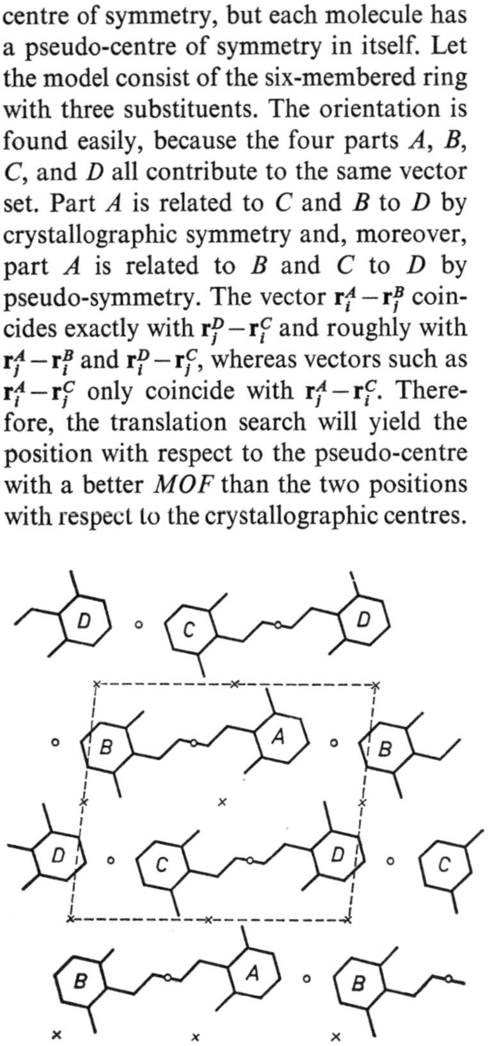

Fig. 1. Two-dimensional case of pseudo-symmetry. Group A is related to B by centres of pseudo-symmetry (open circles), as is C to D. The crystallographic centres (crosses) transfer A into C and B into D.

After the case of pseudo-symmetry has been recognized, one may combine two models related by pseudo-symmetry into a new one and use this new model to find the position of the combination with respect to the true symmetry element. This procedure has been found to work well.

Optimization

Both orientation and translation search are made with large increments in order to save time. Therefore, it is necessary to refine the orientation and translation parameters afterwards in Patterson space. The list of parameter sets to be refined is obtained by transferring a fixed number, *e.g.* 3, of the list of possible translation vectors found for each possible orientation to a new list, which is then sorted with respect to the combined *MOF*. Only a fixed number, *e.g.* 20, of the best of this list are subjected to optimization.

In each cycle each of the six parameters is treated independently. It is varied by amounts equal to one-half of the original increments. From the values of *MOF* $(\alpha - \Delta\alpha)$, $MOF(\alpha)$, and $MOF(\alpha + \Delta\alpha)$ the program decides to change α by $\Delta\alpha/2$, 0, or $-\Delta\alpha/2$. The parameters β, γ, a, b, and c are treated similarly. At the end of the cycle these shifts are applied and a new cycle is started. The procedure ends if four successive cycles do not further improve the *MOF*. Then the entire procedure is repeated once with halved increments. In practically all cases one set of parameters led to a *MOF* significantly lower than all others. This set turned out to be the correct solution.

Atoms outside the model

The search for atoms outside the model will be described in more detail because it contains some new features. The procedure is typical for organic molecules, which do not show large variations of bond angles and bond lengths. This stereochemical knowledge is used to select a small number of possible positions of new atoms. For each of these positions, instead of for all grid points within the asymmetric part of the unit cell, the program calculates the set of vectors to all atoms known at this time and determines the *MOF*. A new atom is added at the position with the best *MOF* if this will not give rise to too short interatomic distances. If it does, the next-best one is considered, etc. In this way, new atoms are added one by one until a given number has been added or until the interatomic distance test does not allow the addition of any more atoms.

The possible positions of new atoms are found by a procedure called model analysis. For each known atom the number of neighbours within 1·7 Å is determined; this is the number of bonds, which may be 1, 2, 3, or 4. If four bonds are found, or three approximately in one plane, this atom does not lead to possible positions of new atoms. If the three bonds are not coplanar, one possible position is determined. The *MOF* is calculated for a position at a given distance in the direction of the fourth corner of a tetrahedron defined by the three other bonds, but also for six other positions at the same distance from the central atom and at distances of 0·15 Å from each other and from the position first mentioned. Of these seven positions, that with the best *MOF* is added to a list of potential atoms.

If two bonds to known atoms are found, three possible positions are determined, one giving a planar configuration and two giving a tetrahedral configuration. For each of these positions the best out of seven is taken as a potential atom, as discussed above.

If the atom has only one bond to a known atom, the possible positions lie on cones corresponding to a list of five bond angles, ranging from 105° to 125°. On each cone 60 possible positions are calculated at the given distance from the central atom. The *MOF*'s of the 300 positions are determined and the three with the best *MOF*'s and at distances of at least 0·7 Å from each other are taken as potential atoms. All procedures are executed for each distance from a list of given bond lengths, usually two.

The 15 best of the potential atoms are stored and sorted in a list and optimized after completion of the model analysis. In this optimization procedure, x, y, and z are varied with steps of 0·075 Å in the positive and negative direction. From the resulting 27 points, the one with the best *MOF* replaces the original one. Finally, the list is sorted again.

Before a new atom from this list is added to the model, it is subjected to interatomic distance tests. It is rejected if the distance to an atom of another molecule is less than a given value, *e.g.* 3·0 Å, or if the distance to an atom of the original model is less than 0·7 Å. If a distance of less than 0·7 Å is found between the new atom and one of the atoms added before, these two atoms are considered to be identical, and the average of their coordinates is taken as the most probable position.

It sometimes happens that near the end of the procedure wrong atoms are added which prevent the addition of correct ones. Because all lists of potential atoms are printed, such errors can easily be corrected afterwards.

Further developments

This procedure is rather slow if the complete model analysis is repeated after the addition of each new atom. Therefore, a partial model analysis has been introduced, by which new positions are calculated only around the atom that has been added and its neighbours. Of all other potential atoms the *MOF* is increased by the contribution of the new atom and those related to it by symmetry. At regular intervals a complete model analysis is made.

Another improvement has been made by the introduction of more chemical knowledge about the molecule into the program. This information is stored in two lists. One list contains all atoms which have at least one bond to a new atom. The first cycles of model analysis are applied to atoms of this list only. The other list contains all atoms that have no bonds to new atoms, but only to known atoms and hydrogen. Atoms of the latter list are always skipped in the model analysis.

These two lists contain information on model atoms only. Similar information on added atoms cannot be given to the computer in this way. Therefore, we are now developing a conversational program with graphic display. In this program each cycle of the model analysis is applied only to atoms indicated by the operator. After this analysis, he determines which atom or atoms have to be added to the model. This version of the program is still in the development stage, but it looks very promising.

Results

Up till now eleven new crystal structures (Braun, Hornstra & Leenhouts, 1969*b*, 1970) have been solved by this method, eight in space group $P2_12_12_1$, one in $P2_1$, one in $P\bar{1}$, and one in $P2_1/c$. If no difficulties occur, the complete search takes about half an hour computer time and the resulting atomic position parameters are sufficiently accurate (errors of about 0·1 Å) to start a converging least-squares refinement. In one case we had trouble because the model of eleven atoms contained two atoms in a wrong position. One of these errors was detected after the search for atoms outside the model and corrected before the refinement. The other wrong atom moved to the correct position in the first cycles of the least-squares refinement. This example shows that, although the presence of wrong atoms in the model decreases the efficiency, it does not always have a disastrous effect. However, if it is known that the position of an atom is doubtful, it should be omitted from the model.

References

Bart, J. C. J. & MacGillavry, C. H. (1968). *Acta Cryst.* B**24**, 1587.

Braun, P. B., Hornstra, J. & Leenhouts, J. I. (1969*a*). *Philips Res. Rep.* **24**, 85.

Braun, P. B., Hornstra, J. & Leenhouts, J. I. (1969*b*). *Philips Res. Rep.* **24**, 427.

Braun, P. B., Hornstra, J. & Leenhouts, J. I. (1970). *Acta Cryst.* To be published.

Dixon, W. J. & Massey, F. J. (1951). *Introduction to Statistical Analysis*, p. 230. New York: McGraw-Hill.

Huber, R. (1965). *Acta Cryst.* **19**, 353.

Hvoslef, J. (1968). *Acta Cryst.* B**24**, 23.

International Tables for X-ray Crystallography (1952). Vol. I. Birmingham: Kynoch Press.

Nordman, C. E. (1966). *Trans. Amer. Crystallogr. Ass.* **2**, 29.

Nordman, C. E. & Nakatsu, K. (1963). *J. Amer. Chem. Soc.* **85**, 353.

Discussion

GASSMANN: Instead of calculating a fine grid and having to use packed words, would it not be simpler to have a coarser grid and interpolate between the points?

HORNSTRA: Interpolation would be a waste of time since it would have to be done very often. Also, in a sharpened Patterson function, a coarse mesh leads to loss of information.

R. HUBER: When you refer back to the Patterson, would it not be more straightforward to use the superposition method instead of searching?

HORNSTRA: We did this at the beginning but found it to be wasteful. The search is limited to the vector set from a small number of reasonable positions, so we calculate the vectors only from the points where a new atom is to be expected. Whether one should use the minimum function or the sum is a matter of taste, but we use the sum function for simplicity. The minimum function would require more test levels and hence more memory space.

NORDMAN: Can you build the full molecule with this method alone?

HORNSTRA: Yes, there may be problems with space groups of low symmetry, but not if there are several symmetry positions to check the new atom. Usually errors cancel out and the determined atoms are within 0·1 Å of their correct positions. After completing the full model we start the L.S. refinement.

HARDGROVE: If you subtract the vectors of the known part of the molecule from the Patterson function, would not this improve the chances of locating the remaining atoms?

HORNSTRA: This is possible and we have tried it, but the success is not great because one can subtract only a very small fraction of all vectors in the Patterson.

Calculation and Use of Vector Overlap Weights in Patterson Search and Refinement

C. E. NORDMAN & J. W. SCHILLING

Department of Chemistry, University of Michigan, Ann Arbor, Michigan, U.S.A.

When part of the structure of a molecule is known, one can approach the solution of the crystal structure by seeking a fit between the Patterson function and the 'Patterson', or vector set, of the fragment. This has been implemented by 'searching' the Patterson function with a partial Patterson computed by a Fourier synthesis (Hoppe, 1957*a,b*; Huber, 1965) and by a seemingly simpler approach, searching with the point vectors generated by the atoms in the fragment. The former approach has an important advantage in that it automatically represents the vectors as finite, generally unresolved peaks. The latter is advantageous in that it permits one to keep easy track of the identities of the vectors which fail to satisfy the Patterson function, and thereby to check on the atoms in the assumed fragment. Errors may be detected and adjustments, additions, and deletions easily made (Nordman & Kumra, 1965; Nordman, 1966).

It is clear that more meaningful criteria of fit can be formulated if the expected heights of Patterson peaks are available in addition to the expected relative positions of these peaks.

A criterion of fit found useful in both vector search and refinement is the 'weighted minimum average' function

$$\min (M)_N = \sum_j P_j / \sum_j w_j$$

where N is the number of search vectors, P_j is the actual value of the Patterson function at the vector point, w_j is the 'weight' or *a priori* expected vector density at the vector point, and the sums include M terms corresponding to the M lowest values of P_j/w_j. If $M=1$, the criterion is a weighted minimum function; if $M=N$, it corresponds to the sum function. This paper is an account of procedures used to calculate peak overlap and total vector densities in the point vector model.

The weight of a vector arises from two sources. The first is the 'intrinsic' weight of the peak, taken proportional to Z_iZ_j, the product of the atomic numbers of the two atoms i and j involved in the vector. The second is the accumulation of overlap from all neighboring vector peaks from which the vector ij is incompletely resolved. Such overlap contributions depend only on the intervector distance r, provided that the vector peaks are spherically symmetrical. The value of the overlap is given by a function $P(r)$ expressing the profile of the overlapping vector peak.

Calculation of peak profile, $P(r)$

Let us assume that the Patterson function, with origin removed, has been sharpened to represent 'point atoms'. That is, we assume that the coefficients have been divided by Σf_i^2 and by the natural overall temperature fall-off, either isotropic or anisotropic. In the anisotropic case the 'point atom' Patterson coefficient is

$$C_o(\mathbf{h}) = Q|F_o(\mathbf{h})|^2 - \sum_i Z_i^2$$

where

$$Q = K^2 \sum_i Z_i^2 / [\sum_i f_i^2 \exp(-\sum_{jk} h_j h_k \beta_{jk})]$$

and K^2 is a constant which places the $|F(\mathbf{h})|^2$ on an absolute scale.

(A Patterson synthesis using these coefficients would produce 'point' peaks if the range in **h** were infinite.)

To allow a choice of sharpening, an exponential damping factor is provided, and the coefficients actually employed are of the form

$$C(\mathbf{h}) = C_o(\mathbf{h}) \exp[-(D/4)s^2]$$

where $s = 2 \sin \theta/\lambda$, and D is a damping parameter, usually chosen in the range 0–5 Å^2.

The function representing a single equal-atom vector peak centered at the origin is then given by

$$P(uvw) = \frac{1}{V} \sum_h \sum_k \sum_l Z^2 \exp[-(D/4)s^2] \times \cos 2\pi(hu + kv + lw)$$

$$\simeq \iiint Z^2 \exp[-(D/4)s^2] \times \cos 2\pi(hu + kv + lw) dV^*$$

where the integration is carried out over the region in reciprocal space for which data are available. Assuming this region to be a sphere of radius $s_m = 2 (\sin \theta/\lambda)_{\max}$, we transform to spherical coordinates and integrate over the two angular variables. The result

$$P(r) = 2Z^2 \left(\frac{1}{r}\right) \int_0^{s_m} \exp[-(D/4)s^2] \times s(\sin 2\pi rs) ds$$

gives the profile of the vector peak as a function of D and s_m. The integral can be evaluated by expanding the integrand into an even power series in s, and integrating term by term. By rearranging the terms, the result can be expressed as a power series in r:

$$P(r) = 4\pi Z^2 s_m^3 \sum_{n=0}^{\infty} A_n r^{2n} \qquad (1)$$

where the coefficients A_n are themselves power series in s_m, namely

$$A_n = [(2\pi s_m)^{2n}/(2n+1)!] \sum_{p=0}^{\infty} t_{np}$$

where

$$t_{np} = \frac{(-1)^{n+p}(Ds_m^2/4)^p}{p!(3 + 2p + 2n)}$$

The peak height $P(0)$ is then given by

$$P(0) = 4\pi Z^2 s_m A_o$$

and the ratio $P(r)/P(0)$ by

$$P(r)/P(0) = 1 + \sum_{n=1}^{\infty} k_n r^{2n} \qquad (2)$$

where

$$k_n = A_n/A_o \, .$$

Data describing several Patterson peak profiles calculated by means of (2) are given in Table I. Since the resolution in reciprocal space is finite, the $P(r)/P(0)$ *vs.* r curves will typically show ripples. We are not interested in the behavior of the curves beyond the first zero, which falls at about 0·6 Å for Cu sphere resolution. For r values less than about 0·6 Å, the series (2) requires about 10 terms for accurate computation.

In any practical application the computation of vector overlap weights is highly repetitive, and the parameters s_m and D are constant. The most efficient way to evaluate the overlap weights is to use Eq. (2) to prepare a table of $P(r)/P(0)$ vs. r, and to recover the weights by table lookup. The routine which calculates the table entries also evaluates a parameter *RMAX* denoting the value of r at which $P(r)/P(0)$ first reaches a specifiable small value, typically 0·02–0·05. Vector peaks separated by more than *RMAX* are considered non-overlapping.

Calculation of vector weights

The program (VEC), which calculates

search vector components and weights, accepts the following input:

- Control parameters, including s_m and D
- Cartesian coordinates and atomic numbers of the atoms in the known group
- Cartesian-to-crystal and crystal-to-Cartesian coordinate transformation matrices
- Euler angles, if applicable
- One or more crystallographic symmetry transformation matrices.

If the vector set for a translation search is being prepared, the input coordinates are rotated through the known Euler angles, transformed to crystal space, and one or more symmetry-related sets of atom coordinates generated. If the vector set for an orientation search is being calculated, the rotation is omitted, and one 'symmetry related' set is generated by the identity operation. In either case the components of all unique vectors from the 'original' to the 'symmetry related' set or sets are calculated.

Alternatively two crystallographically independent sets of atom coordinates can be read in; vectors between these sets can be used to find the relative location of the two groups in the same molecule or asymmetric unit.

To initiate the weight calculation each vector is assigned an 'intrinsic' weight $w^0_{ij} = Z_iZ_j/36$, *i.e.* a non-overlapped weight scaled so that a carbon-carbon vector ($Z_iZ_j = 36$) equals unity. The overlap contributions are now accumulated by a systematic examination of intervector distances in Cartesian space. Each vector ij is in turn assigned the role of 'pivot vector', and its distance from each subsequent vector in the list calculated. For each vector mn found to be at a distance $r < RMAX$ from ij the weight of ij is incremented by $w^0_{mn}[P(r)/P(0)]$ and that of mn by w^0_{ij} $[P(r)/P(0)]$.

In addition, the overlap contributions to ij of Patterson-symmetry replicates of both ij itself and the vectors mn must be included. A distinction is made here between 'fixed' and 'variable' overlap contributions to the weight of a pivot vector ij. The fixed contributions arise from symmetry replicates (of ij or mn) whose distance from ij is constant in a given vector search. For example, in an orientation search any overlap contribution made to ij by the vector related to mn by the center of symmetry at the Patterson origin will be independent of the orientation of the search vector set. Similarly, in an x, z translation search using vectors between sets of atoms related by a 2_1 axis along y, the mirror replicate of mn across the Patterson mirror plane at $v = \frac{1}{2}$ makes a constant overlap

Table I. Calculated Patterson peak profiles

The table entries are the values of r, in Å, for which $P(r)/P(0)$ assumes the value given in the left-hand column. The first two columns of r values correspond approximately to Cu-sphere resolution, the last two to approximately 1·5 Å resolution.

$P(r)/P(0)$	$s_m = 1{\cdot}19$	1·19	0·665	0·665 Å^{-1}
	$D = 0{\cdot}0$	3·0	0·0	8·0 Å^2
1·00	0·000	0·000	0·000	0·000 Å
0·95	0·096	0·103	0·170	0·180
0·90	0·137	0·147	0·243	0·257
0·85	0·169	0·181	0·301	0·318
0·80	0·198	0·212	0·351	0·372
0·75	0·223	0·239	0·397	0·420
0·50	0·336	0·362	0·597	0·635
0·25	0·448	0·487	0·796	0·853
0·00	0·604	0·681	1·074	1·188

contribution to ij. Such fixed overlaps are routinely included in calculating the search vector.

On the other hand, 'variable' overlap weight contributions to ij from Patterson-symmetry replicates of ij or mn depend on the values of the (rotational or translational) search parameters. For example, in the course of an orientation search one or more search vectors may fall near a Patterson mirror plane causing overlap with mirror replicates. The option of including variable weight contributions – at a relatively high cost in execution time – is provided in the search programs themselves.

Practical experiences

The execution time for the VEC program depends strongly on the number of atoms (N) in the known molecular fragment. This is because the program examines all distances between vector points, the number of such intervector distances being approximately proportional to the square of the number of vectors, or to the fourth power of N. For known fragments of moderate size, say $N < 30$, the execution time of VEC is insignificant. For very large molecular fragments this is no longer true. For example, with $N = 60$ the IBM 7090 execution time is approximately 20 minutes. Furthermore, the tables of vector components to be stored become very large.

For even moderately large molecular fragments the total number of independent vectors may be larger than the number of search vectors that could be used without prohibitively increasing the execution time of the Patterson searches. Thus a smaller set of vectors must be selected.

The weights and the intervector distances provide a convenient criterion for selecting the most effective search vectors from a large list of interatomic vectors. The following simple routine (VSEL) is provided for this purpose:

- The vectors are arranged in order of descending weight.
- The first vector in the list is selected as a search vector, and all subsequent vectors which are separated from the search vector by less than a specifiable distance R are removed from the list.
- The next vector among those remaining in the list is selected as a second search vector, all subsequent vectors within a distance R of it are removed, and so on.

The vectors remaining in the set after completion of this selection procedure are the ones of highest weight, subject to the additional condition that no two vector points are separated by less than the distance R. The rationale for this latter condition is that near-coincident vectors in effect sample identical points in Patterson space. Including two or more such vectors in the search conveys no more information than one vector. It merely corresponds to erroneously giving the one vector additional weight, the weight of the vector already having been accounted for by the overlap of its near-coincident neighbors.

Our experience with large search groups has been that even a small number of high-weight vectors tend to be much more effective as search vectors than correspondingly larger numbers of medium or low-weight vectors, assuming for the sake of comparison that the sums of the vector weights in the sets are the same.

Examples of large molecular fragments for which the VEC program has been used, and where very high vector weights are present, are the planar heme group ($N = 33$) and a 4-turn α-helix ($N = 74$) in myoglobin. At 1·5 Å resolution the 10 strongest of 528 intra-heme vectors average 32·0 (C–C vectors) in weight. For the α-helix the 10 strongest (of 2701) vectors average 24·9 in weight. Orientation searches of the Patterson function of myoglobin carried out with a moderate number (30–70) of the highest-weight vectors of each category proved very effective in determining the orientations of the heme group and the directions of the prominent α-helices. By comparison, similar searches with larger numbers of vectors of moderate weight (5·0–6·0) proved much less discriminating.

References

Hoppe, W. (1957*a*). *Acta Cryst.* **10**, 750.
Hoppe, W. (1957*b*). *Z. Elektrochem.* **61**, 1076.
Huber, R. (1965). *Acta Cryst.* **19**, 353.
Nordman, C. E. & Kumra, S. K. (1965). *J. Amer. Chem. Soc.* **87**, 2059.
Nordman, C. E. (1966). *Trans. Amer. Cryst. Ass.* **2**, 29.

Discussion

R. Huber: What was the result of the myoglobin search that you mentioned?

Nordman: The intramolecular search for the orientation of the helices gives unambiguously the five prominent helices. About 70 search vectors are picked in this way using either 1·5 or 2·0 Å data. The translation search is not nearly as clear. The orientation of the heme group plane is very good, and so is the rotation normal to the heme group.

A Fortran Library of Programs for Vector Space Search, Refinement, and Model-Building

JESSE W. SCHILLING

Department of Chemistry, University of Michigan, Ann Arbor, Michigan, U.S.A.

This paper will attempt to describe the operational details of a structure solution carried out by the Vector Search Procedure (Nordman, 1966). For more details and examples, see Schilling (1968).

The Euler angles relate the coordinates of each point in a rigid body in any orientation to those of the same body in a standard orientation, expressed in a fixed Cartesian system (X, Y, Z). The angles correspond to rotations applied in a fixed sequence about Cartesian axes (X', Y', Z') which turn with the body; each rotation is taken about an axis which was moved only by the previous turn. In the ORV program, the rotations A, B, C* are applied clockwise as one looks down the rotation axis towards the origin of a right-handed frame as follows:

A is a turn about the original X axis;
B is a turn about the *rotated* Y axis (Y');
C is a turn about the *rotated* X axis (X').

For a given value of C, the rotations B and A can be visualized as coordinates on the surface of a sphere ('latitude and longitude'), since the axes about which they are made are orthogonal; likewise C and B for given A. Furthermore, the correlation between A and C is large when B is small; when $B=0$, A and C are degenerate.

For a given C, the spherical relationship between A and B is usually given as a stereographic projection. The topography of the sphere then represents values of the image-seeking function and is not meant to be interpreted in terms of any individual vector. One might equally wish to plot the value of C for which the image-seeking function (ISF) has the highest value for a given A and B, except that this function may be multiple-valued. A point on the sphere represents the 'orientation' of the model relative to the chosen 'standard' orientation. This does not have a simple physical relationship to the quantity being plotted, *e.g.*, the value of the lowest vector, since the role of lowest passes unpredictably from one vector to another. Thus, the artificial properties of the stereographic projection must not be literally applied to the search vectors themselves. For example, the latitudes diminish in circumference near the poles, suggesting that larger angular steps may be taken for A when B is small. However, this is true only of the stereographic projection of a single vector; that is, any individual vector which projects primarily onto X' will be relatively insensitive to the A (and C) rotations. Equally well, a vector nearly parallel to Y is insensitive to B. Thus, the area of a minimum function peak may be broadened whenever the lowest vector vertex is near the pole of the sphere, but this may occur at any point on the orientation plot for the *model*. The degeneracy of A and C at low values of B, however, makes it expedient to maintain equal steps in C and take larger steps in A when B is small. This has an effect similar to the maintenance of an equal-area grid near the poles, but the rationalization is quite different.

* These names are used instead of the usual ϕ, Θ, ψ, so that the reader will be forced to refer to the following definitions and avoid confusion with some other, more familiar, set of conventions.

The symmetry of the Patterson will appear in the topography of this sphere if the search is carried out over more than the unique range. For example, since the Patterson is centric, *A* and *B* for a single-vector search are limited to a unique range of 180°. If the symmetry is $2/m$ (with $2 \| y$), then *A* is restricted to 90°; if *mmm*, both *A* and *B* are restricted to 90°. But when several vectors are sampling the Patterson, the symmetry of the set is important. For example, if the set does not possess a center of symmetry, it will not be superimposable upon its center-related image and a larger range in *B* is required. In general, the *C* rotation is about an axis which does not interact with any of the symmetry elements of the Patterson and its range is limited only by the symmetry of the group about the *X* axis. (See Tollin, Main & Rossmann, 1966.)

The superposition process in a digital computer consists of a repeated evaluation of the vectors between the *n* known atoms and a 'roving point' which scans (on a grid) every unique position in the cell. This is equivalent to translating the image of the known atoms and is almost trivially easy to program. At each setting (x,y,z) the vectors have values corresponding to the evaluation of *n* Pattersons (each translated to bring a different atom to the scan origin) at the point which overlies (x,y,z).

A location scan involving *one* symmetry element can be carried out efficiently by deleting any reference to 'atoms'. That is, if the correctly oriented group of *n* atoms is placed anywhere in the unit cell, the vertices of the inter-fragment vectors generated by one symmetry element represent *n* images of the group, each viewed from an atom of the related group. Since the viewing atoms are rigidly connected, the entire set of image-vectors is rigid whatever the location of the original group. Thus, these vectors can be used to find the true location of the group with respect to the given symmetry element by an ordinary superposition in which they are translated through the Patterson. The reader can construct a figure which shows this correspondence for a 2_1 axis along *z*, in which the chemical group is placed at the origin initially and then translated in the positive *x* and *y* directions in steps of δ to $a/2$, $b/2$; it is easy to see that the vectors so generated scan the Patterson in exactly the same way as if the vectors generated at the initial position were translated in steps of -2δ to $-a$, $-b$. This sort of scan does not require repeated recalculation of atom coordinates and generation of vector components. The generalization to several symmetry elements will be mentioned later. Note that the position of the chemical group in the unit cell when the vectors are initially calculated is immaterial, except that the correspondence between steps in the Patterson space vector scan and positions of the group in direct space depend upon this choice. The simple relationship mentioned above obtains when the symmetry elements are placed at the positions they actually have in the unit cell and the group is at the origin.

The advantages of using estimated overlap density to attach 'weights' to the search vectors has been described in a previous paper. Suppose a vector, assumed to have a weight derived only from the atomic numbers of its constituent atoms, falls on high ground (thus supporting the hypothesis of a good fit to the current model); if it can be demonstrated *a priori* that another vector of the model approaches it closely, then a full-peak-height value is insufficient to constitute a good fit. This extends the competition for the role of lowest to vectors which would otherwise be excluded. Briefly, then, the weighting facility is an attempt to take the vector-set of an assumed chemical fragment into a 'partial Patterson'. The weights attached to search vectors enter into the *ISF* as follows:

$$\min(M) = \Sigma P_j / \Sigma w_j \quad (1 \leq j \leq M) \qquad (1)$$

where min (M) is the 'minimum average' function, a figure of merit for a hypothetical structural model. One might at first consider the funtion $\min'(M)=(1/M)\Sigma(P_j/w_j)$ since P_j/w_j, the 'reduced value' of the 'jth lowest' vector, is a meaningful quantity, namely the highest possible value of the figure of merit for that vector (it might be lower if we could augment the weight to account for overlap of vectors involving unknown atoms). However, consider the two cases:

$$\text{A}\quad \left\{\begin{matrix} P=1,2,3,4,5 \\ w=1,2,3,4,5 \end{matrix}\right\} \begin{matrix} \min'(5)=1{\cdot}0; \\ \min(5)=1{\cdot}0 \end{matrix}$$

$$\text{B}\quad \left\{\begin{matrix} P=1,2,3,4,5 \\ w=5,4,3,2,1 \end{matrix}\right\} \begin{matrix} \min'(5)=1{\cdot}74; \\ \min(5)=1{\cdot}0\,. \end{matrix}$$

Case B is clearly a poorer fit than is A. In practice, as long as M is significantly smaller than the total number of vectors, the two functions give the same results.

The 'minimum accumulation functions' of Raman & Lipscomb (1961) are image-seeking functions of a somewhat different variety. They begin with the fact that additional image-vectors, $\mathbf{r}$, may be derived from a known vector $\mathbf{u}$, if the minimum function $\min[P(\mathbf{r}), P(\mathbf{r}+\mathbf{u})]$ is high; they then invert this statement to say that if there are a number of points $\mathbf{r}$ for which the minimum function is high, $\mathbf{u}$ must be a legitimate interatomic vector. The function $MA(\mathbf{u})\equiv\Sigma\min[P(\mathbf{r}), P(\mathbf{r}+\mathbf{u})]$ (where the summation over $\mathbf{r}$ scans the entire unit cell) gives a figure of merit for the likelihood that $\mathbf{u}$ is an interatomic vector, in fact without necessarily referring to the actual value of $\mathbf{u}$ itself. That is, $MA(\mathbf{u})$ is high if the roving point, $\mathbf{r}$, scans a significant number of points which represent other vectors in the same image set.

Applied to the search procedure, this function can be regarded as a way of obtaining confirmation of a proposed partial structure from as yet unknown atoms which are not part of the assumed fragment. Stated in another way, if the n-atom partial structure is true, its image will exist repeatedly in the Patterson, as viewed by each of the additional atoms, known and unknown. Therefore, if the image of the model (n vector vertices) is translated through the Patterson, it will more often fall on high ground (all vectors simultaneously) if the model is true than if it is not. Functions such as min (M) implicitly involve only the n images of the n-atom partial structure as viewed from each of its constituent atoms, *i.e.*, $n(n-1)/2$ non-null vectors; clearly the accumulation function makes use of independent information. In general, however, it is more laborious to compute. For each step in a search it requires $m\,.\,n$ vectors to be evaluated (where m is the number of steps in the grid over which the image of the model is translated), while the ordinary image-seeking functions require $n(n-1)/2$. This gives a computation ratio varying approximately as m/n (and in general $m\gg n$).

(Note that the accumulation function for, say, an orientation search is operationally equivalent to carrying out a superposition on the search group for each orientation.)

The primary objection to this procedure is that the search is necessarily performed with *single* images rather than with *convoluted* fragments. The number of searching points is fewer and they are necessarily all of single weight. In structures large enough for the background to be the equivalent of a single vector peak, such a function would be completely useless; experiments to determine the orientation of an eight-atom fragment in a known structure ($C_{18}H_{14}$, $P2_12_12_1$, $Z=4$) indicate that the MA function, while successful, was less convincing than, say, a min (5) convolution search.

An important distinction between the vector search method and those methods which employ a partial Patterson is that one can make use of the *identities* of the badly fitting vectors. For example,* suppose one atom in a ten-atom search group has been wrongly assumed. Then, in an orientation search, say, 9 of the 45 vectors may have low values at the correct setting and the minimum function will be disappointing. If a range of values of M is used, say min (1), min (5), min (10), etc., a peak can emerge in the higher-order minimum averages. If the lowest vectors at that setting involve the same atom unusually often, one may conclude that further chemical work is necessary, or that the search should be repeated with an alternative fragment.

Implementation

The vector-peak model derived in the previous paper allows a crude way of estimating the minimum requirements in storing the Patterson. For example, we can postulate an expression for the probable number, f, of times one vector will sample an ideal peak within the radius, r, at which it falls to an arbitrary percentage, q, of its full height. We will call f the 'sampling frequency'. Let $q(r) \equiv 100 \cdot P(r)/P(0)$; f may be taken in three dimensions as the volume ratio $(\frac{4}{3}\pi r^3)/V$, where V is the grid volume (Å^3). We may then speak of a limiting value of q, q_{max}, as the highest q for which f is 1·0. The value of r ('limiting radius') which corresponds to the arbitrary parameter q is a function of the degree of sharpening and the adequacy of the data. If the tabulation frequency is given (*e.g.*, one's Fourier program offers only a limited number of choices), we can derive q_{max} and hope to say whether it is adequate; otherwise we can derive the optimum tabulation by choosing a reasonable q_{max}, based perhaps on past experience.

The search method is a serial technique, with each stage dependent upon the accuracy of the preceding one. It is thus important to consider the minimum requirements of range and accuracy in the stored Patterson. The range of values which must be stored is larger if the weights attached to the vectors span a range significantly larger than 1:1. Both range and accuracy depend upon the number of bits used for the values and upon the choice of integer or floating-point representation.

It is probably never necessary to work with vectors whose weight range is greater than 1:8; *i.e.*, lower-weighted vectors will so seldom be included in the *ISF* that they can be ignored. Let us divide the Patterson by $P(0)$, the theoretical full height of a vector between the two lightest non-hydrogen atoms present; values of the *ISF* will then give some indication of how well this model works in a given case. The minimum range of values is $`l' \leq P \leq 8$ where 'l' is a fraction of a single peak height which must be low enough so that the background of the *ISF* is distinguishable from any meaningful peaks. Anything less than 'l' will underflow to zero.

A four-bit *integer* representation gives an absolute range 0, 1–15; if the Patterson is scaled by $(\frac{15}{8})/P(0)$ so that we can just represent the upper limit of the range, the smallest value (represented by a single bit) is $P(0)/1{\cdot}9 = 0{\cdot}53\ P(0)$. If the values were rounded before storing, there is a constant absolute error of $\frac{1}{2}$, or a relative error of 50% in the lowest value; 0·53 $P(0)$ is still too high for such an error to be acceptable – more bits are needed. A four-bit *floating-point* representation requires two bits for the 'characteristic', giving an absolute range of 0, $\frac{1}{2}$–8. The absolute error of the two-bit mantissa is 2^{-3}, giving a maximum relative error (for the smallest value which can be represented by the mantissa, $\frac{1}{2}$) of 25% and a minimum error of 12·5% for values over the entire range of representation. The actual mode used must depend upon the structure and size of the computer in question; if a modest machine is being

* See also 'vector-space refinement' below.

used, it may be necessary to change the mode for different problems. On the IBM 360 the 'natural' choices are four, eight, or sixteen bits (the latter two units are individually addressable in Fortran); on the IBM 704 family, six, nine, and twelve bits.

Subroutine RDOUT is a nuclear routine for storing and retrieving three-dimensional functions (it may be used, for example, to aid in evaluating arbitrary planes through an electron-density function). Retrieval of stored values may be accomplished with eight- or four-point interpolation or with no interpolation (the function value at the nearest tabulated point is returned). RDOUT references the tabulated function as an array of half-words or quarter-words on the IBM 360.

The search programs themselves are modularized for adaptability to different computing environments. The tables of vector components (and other quantities) and the Patterson table are defined in one module so that dynamic storage allocation can be employed or at least so that re-compilation for larger or smaller problems is simple and inexpensive. All output of *ISF* maps is through one subprogram, which contains a provision for up to 12 different *ISF*'s (generally different values of M in Eq. 1) which may be deposited in separate data sets (*e.g.*, disk files) or together on, say, 'the output tape'. Up to 12 'rank-identities' (which vector is lowest, next-lowest, etc.) may also be produced, also on different or the same data sets.

The weighting feature gives rise to an efficiency measure called the 'partial *ISF*'. That is, since the highest-weighted vectors are more likely to appear in the *ISF*, one can specify the list of search vectors in order of decreasing weight and stipulate that a proposed model (*e.g.*, orientation) be rejected if a specified number of vectors are found to have reduced values less than some specified cutoff. In general, it is necessary to read out Patterson values for *all* the search vectors only in the vicinity of a few high peaks in the *ISF*.

The 'proximity feature' is another efficiency device. A 'shell' of high vector density exists in the Patterson near the origin (in organic structures at 1·3–1·8 Å); vectors sampling this region of the Patterson need not be evaluated, since they are more likely to be high whether or not the model is correct.

Steps in an attempted structure solution

1. Generate the Patterson, using tabulation intervals which:

A. make the sampling as nearly isotropic as possible;

B. make the limiting radius for a unit-sampling-frequency as small as possible with the memory capacity available (preferably corresponding to $q_{max}=90\%$ or higher).

The order of the three summations may be chosen in whatever way is most efficient; the DEK program converts it to whatever is the standard form chosen for a particular computer.

2. Generate a symmetry routine for the space group *of the Patterson* to transform any vector components into the particular asymmetric unit which is stored.

3. Generate a model of the search group. If it is three-dimensional, a published crystal structure which contains the same fragment will furnish coordinates as input to a program (MDL) which generates the Cartesian components. The internal X axis must be along the highest-order symmetry axis.

4. Use the VEC program to generate $n(n-1)/2$ intrafragment vector components (Cartesian) and weights (including overlap due to center-related replicates).

Optionally remove those $<2{\cdot}0$ Å in length.

5. Carry out an orientation search over the minimum Euler angle range (ORV program); vectors related to the originals by a center are *not* included. Choose a reasonable number of min (M)'s [including min (1) and up to about 80% of the vectors]; print out the identities of at least the two lowest vectors. Use fine-step scans to establish the exact coordinates of the best peaks.

6. If several symmetry elements are present, the LOC program will carry out simultaneous searches of the type mentioned above and print composite *ISF* maps, accepting as input the Cartesian atom coordinates of the model, a hypothetical set of Euler angles, a specification of the symmetry elements, and the minimum unique search range for each. Or, use the VEC program to generate $n(n+1)/2$ vectors for each symmetry element; the SUP (or any superposition) program can carry out one-, two-, or three-dimensional searches (for planes, axes, or a center, respectively). The *ISF* maps for the separate searches must be 'correlated' by hand; that is, in $P2_12_12_1$, for example, 3 two-dimensional searches (xz, xy, yz) must concur on the best xyz for the molecular location.

7. Choosing the most likely location, generate the entire partial structure (all symmetry-replicates) and carry out a three-dimensional superposition over one asymmetric unit of the cell. Search one or more *ISF* maps for additional atoms. Low-resolution difference Fouriers should be used as a corroborative technique.

Additional procedures

Vector-space refinement

The vector-search technique is quite dependent upon the accuracy of the assumed model and may be impeded by distortions of the assumed geometry due to packing strain or to imprecise knowledge of the stereochemistry, as well as the inevitable accumulation of errors in the determination of the orientation and location of the group. Therefore it is natural to seek a way of improving a partial structure by optimizing its fit to the Patterson.

Unfortunately, the minimum function is discontinuous, since the role of lowest can pass from one vector to another in an unpredictable fashion. However, it is possible to devise a trial-and-error algorithm which finds the gradient of the Patterson function at the site of the lowest vector and shifts the atoms comprising that vector so as to move it to a higher value. Such an algorithm was used with some success by Nordman & Kumra (1965) in solving the structure of Villalstonine, where the group actually found in the Patterson search was chemically different from the assumed search group, being geometrically similar to it except for a distortion involving one atom. In addition, the program played an important role in giving a check on the relative reliability of peaks in the orientation and location searches. Such a method would ideally compensate for the existence of unsuspected distortion or error in the search group and make slight adjustment of the orientational and/or locational parameters.

Although it is true that the Patterson is not uniquely related to the direct structure, it is not likely that a starting model which already fits the Patterson will refine to some homometric mate without passing through unacceptable 'transition states'. A more serious objection is the fact that the presence of artifacts in the Patterson can seriously interfere with such a refinement, especially since it hinges upon the low vector. This could lead to no refinement at all or to more serious consequences if the center of the lowest vector peak has been shifted by the absence of some strong Fourier wave. Consider a vector lying on the slope of a large series-termination ripple near the base of a Patterson peak. The gradient at that point has no necessary relationship to the direction in which the vector should actually be moved. If a shift is made based upon such a false gradient, no progress can be made unless that shift causes the vector in question to fall upon sufficiently high ground so that the role of lowest passes to another vector. In general, the only way in which the first vector can be moved from its stagnant position is to refine upon a vector which has one atom in common with it and which is itself 'refinable'.

In the original program, the refinement typically left many atoms completely unshifted. It was discovered that further improvement was possible by shifting the next lowest vector after the lowest had been optimized; it was sometimes possible to improve the minimum average in this way in spite of a drop in the value of the lowest vector. The obvious next step was to generalize the algorithm to permit a number of the lowest vectors to participate in the refinement. Briefly this is as follows. The role of 'test vector', *i.e.*, the vector whose value is to be maximized, is always held by the vector which occupies a particular *rank* such as lowest (rank 1), next-lowest (rank 2), and so on, regardless of the *identity* of the vector. Of course, as shifts are made, vectors having one atom in common with the test vector will move up or down and the most likely candidate for improving the hypothetical configuration is always the vector occupying the lowest rank which has shown any change (up or down) since the previous cycle. Therefore, whenever a shift applied to the vector of rank 'n' produces an increase in the minimum average, min (M), we shift attention to the first rank lower than n which has changed, if any. Otherwise, we proceed with rank n. Whenever a shift lowers min (M), we reinstate the previous coordinates of the atoms constituting the test vector, find the lowest of all the M ranks which has decreased, establish which atom is in common with the test vector (if not both) and halve the shift on that atom (or both), and try again to shift the vector of rank n. When one or both shifts have become smaller than an arbitrary value, *MIN*, the role of test vector passes to the next higher rank (a new 'phase' of refinement) until the vector of rank M has been tried and abandoned, ending the 'pass'. If the pass just ended was successful (the final min (M) is higher), a 'check pass' is made, starting with rank 1. In most cases, the role of test vector passes upward to M with no improvements in the model; when this happens, the refinement stops. However, it may have happened that a vector became 'refinable' when some other vector of higher rank, with one atom in common, was shifted. If an improvement was registered on the first check pass, another check pass is made. The terminating condition is completion of a pass with no improvement in min (M).

Obviously, this generalized 'round robin' procedure permits more vectors to participate in the refinement and combats the possibility that noise in the Patterson will cause stagnation of the refinement. The more worrisome possiblity of false refinement is ameliorated by an internal consistency condition which considers the relative changes in all vectors having one atom in common with the test vector; also, for the same reasons, shifts which are too large will be more likely to be corrected.

Weighting of vectors due to overlap with others of the known set can be used to detect low vectors which would otherwise be masked. This leads to the observation that the calculated weights should be frequently updated as the refinement proceeds to discourage the tendency of neighboring vectors to 'share' a peak by moving closer together; if the weights are continually revised, they will become higher when this happens and there will be no advantage in such clustering of vectors. Of course, overlap due to vectors of the unknown part of the structure will still handicap the algorithm. It would be very questionable to permit a vector to refine to a value higher than the 'ideal' single vector peak height, since it could be creeping up an overlapping peak. (The details of this algorithm are given by Schilling, 1968).

Model-testing

In a difficult case, the correct structure may not be apparent by inspection of a Patterson superposition; that is, there are typically very many 'false peaks' and it may be that some correct image-vector peaks are lowered or completely masked by artifacts or errors in the assumed model. In cases like this, one may resort to other methods to help choose among the various candidates, but it is natural to ask whether one can go any further with the Patterson.

One way to proceed by Patterson means is to choose strong, chemically reasonable peaks from the *NS*-fold superposition, one at a time, submit these to vector-space refinement to optimize the fit to the Patterson, then do an $NS+1$-fold superposition upon the resulting partial structure, and repeat the process. This was done by Nordman & Kumra (1965) in solving the structure of Villalstonine. It is evident that this is likely to be a time-consuming process, since there will be a number, *NT*, of possibilities for the first additional atom and each successively higher-order superposition will have a like number of candidates for the choice, and so on. One might then for expediency add a number of candidates from the first superposition, each of which is chemically reasonable taken alone but not necessarily compatible with all the other candidates, and perform vector refinement on the entire $(NA+NT)$-atom structure. The atoms least likely to be correct are those which are involved in the lowest vectors and so can be deleted before performing a higher-order superposition or difference Fourier.

The 'model-testing program' (MDL) originated as an effort to automate the above-outlined procedure. One submits to the program a list of *NS* assumed atoms, which form the nucleus of a model which is to be built up by adding, one at a time or in groups, atoms from a list of *NT* hypothetical sites. The assumed group is refined initially to obtain optimized coordinates and a value of min (M) for use as a guide in choosing additional atoms. Normally the 'test atoms' are then added one at a time and refined with the assumed group (taken together, the 'test group'). This is called 'mode 1'. The optimized coordinates and final value of min (M) for each are retained and finally the atom with the highest such value is added to form an 'augmented group'. If there are several equally qualified candidates, *i.e.*, if several of the 'best' test atoms reached values of min (M) which are so similar that a choice is difficult, the program enters 'mode 2'. All the leading candidates are added and refined together. If, as usually happens, min (M) is lower than for the assumed group, then the program begins deleting first one then the other of the pair of atoms involved in the lowest vector (unless one of these is an assumed atom). Whichever most improves min (M) by its absence is deleted from the model and this process is continued until min (M) reaches a value which is equal to or greater than an arbitrary minimum cutoff. Then mode 1 is entered again to test whatever atoms are left in the test list, and so on.

The entire procedure is similar to a series of superpositions of successively higher order $(NA+1,\ NA+2,\ \ldots)$ evaluated at a limited number of points rather than over the entire asymmetric part of the cell, *i.e.*, at points which were initially high. (It is, in fact, similar to a limited application of Raman & Lipscomb's (1961) GMA procedure.) However, aside from the trivial increase in convenience, the procedure described here is different in two respects: first, in the opportunity provided to make adjustments in the internal geometry of the growing model and, second, in the increased discrimination in choosing correct hypothetical atom sites due to the examination of vectors between test atoms themselves.

This procedure cannot stand alone as a model-building technique since it does not possess the power to suggest hitherto unsuspected atom sites. It should be most powerful when combined with difference-Fourier techniques following the original superposition. In addition to the fact that superpositions tend to be more time-consuming than low-resolution Fourier syntheses for higher symmetries (and more so as the number of assumed atoms increases), they are also not capable of demonstrating false atoms in the assumed model: the best that can be said is that min (M) is capable of indicating additional atoms in spite of an error so long as M exceeds the number of vectors systematically in error due to the false atoms.

The simplest criterion of chemical reasonability, *i.e*, distance-of-approach, is invoked here whenever a vector is encountered which is shorter than a specified distance, D. Both, one, or neither of the atoms involved may be *assumed* atoms and the situation may not occur until after many cycles of refinement. In mode 1, the test atom is disqualified if at the end of refinement there is a short vector in the model (whether or not the test atom itself is involved in any short vector); in mode 2 where all added atoms are open to challenge, any atoms from the test list which are involved in such short vectors are removed from the augmented group and restored to the test list for subsequent testing in the next mode 1 pass.

References

Hoppe, W. (1957). *Acta Cryst.* **10**, 750.

Nordman, C. E. (1966). *Transact. Amer. Crystallogr. Ass.* **2**, 29.

Nordman, C. E. & Kumra, S. K. (1965). *J. Amer. Chem. Soc.* **87**, 2059.

Raman, S. & Lipscomb, W. N. (1961). *Z. Krist.* **116**, 314.

Schilling, J. W. (1968). Ph. D. Thesis, University of Michigan.

Tollin, P., Main, P. & Rossmann, M. G. (1966). *Acta Cryst.* **20**, 404.

Discussion

GASSMANN: Why do you not use shell Patterson functions, since you know the length of the vectors and their orientation and you can save 50% of the space?

SCHILLING: This would be very useful for an orientation search, but reading out a Patterson in this form would not be as straightforward.

TOPIC C

UTILIZATION OF ANOMALOUS DISPERSION AND ISOMORPHOUS REPLACEMENT FOR ELUCIDATION OF LARGE MOLECULES

CHAIRMAN: G. KARTHA

Anomalous Dispersion in Diffraction Studies: An Introducion

BY Y. OKAYA

Department of Chemistry, State University of New York, Stony Brook, New York, U.S.A.

Anomalous dispersion, once considered a source of nuisance, has attained a deserving importance in diffraction studies. In this brief note, I wish to make some introductory remarks on the theory and applications of the anomalous dispersion methods. Since the majority of the diffraction studies involving the use of anomalous dispersion are concerned with X-ray diffraction, this report will devote most of its space to X-ray crystallographic studies. However, the effects are also important in neutron and electron diffraction phenomena, and many interesting problems in these areas have been encountered. I will touch upon these problems.

In what follows, the notations summarized in Table I, will be used in mathematical descriptions.

Freidel's rule and the effect of anomalous scatterers

When there is no anomalous scattering in a crystal, the atomic scattering factors, f, are real for all atoms. The so-called structure factor

$$F_{\mathbf{H}}=\sum_{j} f_j \exp 2\pi i(\mathbf{H} \,.\, \mathbf{r}_j)$$

derives the intensity of reflection $I_{\mathbf{H}}$ or the weight at a reciprocal lattice point,

$$I_{\mathbf{H}}=KF_{\mathbf{H}} \,.\, F^*_{\mathbf{H}}.$$

It is quite obvious from the real character of f that $F_{\mathbf{H}}=F^*_{\bar{\mathbf{H}}}$ and

$$I_{\mathbf{H}}=I_{\bar{\mathbf{H}}}\,.$$

This relationship, which is valid for centrosymmetric as well as noncentrosymmetric crystals, is called Friedel's rule. It states that, when there is no effect of anomalous dispersion, X-rays are scattered with equal intensity from the forward side and the backward side; or, the diffraction adds a center of symmetry to the distribution of weights in reciprocal lattice points, the center being at the origin of reciprocal space. It must be pointed out that this statement is not equivalent to the often-cited misstatement that, in the absence of an anomalous scatterer, the diffraction effects add centers of symmetry to the structure. It was clearly shown that the statistical distribution of the weights at reciprocal lattice points (or the intensity distribution) is different for centro- and noncentrosymmetric crystals. Also for

Table I

$\mathbf{H}$	(h,k,l)	a reciprocal lattice point
$\mathbf{r}$	(x,y,z)	a point in real space
$\mathbf{u}$	(u,v,w)	a point in vector space
$\mathbf{b.H}$	$ha* + kb* + lc*$	a reciprocal lattice vector from the origin to $\mathbf{H}$
F_H	F_{hkl}	structure factor
$f=f'+if''$	f',f''	real and imaginary parts of atomic scattering factor
$\varrho(\mathbf{r})$	$\varrho(x,y,z)$	the electron density
$\bar{\mathbf{H}}$	$(\bar{h},\bar{k},\bar{l})$	a bar on any symbol designates the negative value

certain space groups, the observed systematic absences can be used as unambiguous proof for the presence or absence of centers of symmetry.

Now consider the case when atoms in a crystal are anomalous scatterers. The atomic scattering factors are now, in general, complex.

$$f = f' + if'', \; f \neq f^*$$

Under this condition, and if the crystal has no center of symmetry,

$$F_{\mathbf{H}} \cdot F^*_{\mathbf{H}} \neq F_{\bar{\mathbf{H}}} \cdot F^*_{\bar{\mathbf{H}}}, \; I_{\mathbf{H}} \neq I_{\bar{\mathbf{H}}},$$

and so Friedel's rule is no longer valid.

It should be pointed out that if a crystal is composed of one kind of atom, whether anomalous or not, Friedel's rule is again valid for noncentrosymmetric crystals. In this sense, it is important to note that the effect of anomalous dispersion is indicative of the differences between **H** and $\bar{\mathbf{H}}$ only by the interference between an anomalous scatterer and other scatterers, anomalous or not, of a different kind.

Table II. Equality relations among reflections

Point groups	(h,k,l)	$\neq$	$(\bar{h},\bar{k},\bar{l})$	Laue symmetry $(h,k,l)=(\bar{h},\bar{k},\bar{l})$
1	hkl		$\bar{h}\bar{k}\bar{l}$	i
2	$hkl, \bar{h}k\bar{l}$		$\bar{h}\bar{k}\bar{l}, h\bar{k}l$	$2/m$
m	$hkl, h\bar{k}l$		$\bar{h}\bar{k}\bar{l}, \bar{h}k\bar{l}$	$2/m$
$mm2$	$hkl, \bar{h}kl, h\bar{k}l, \bar{h}\bar{k}l$		$\bar{h}\bar{k}\bar{l}, h\bar{k}\bar{l}, \bar{h}k\bar{l}, hk\bar{l}$	$2/m\ 2/m\ 2/m$
222	$hkl, \bar{h}\bar{k}l, \bar{h}k\bar{l}, h\bar{k}\bar{l}$		$\bar{h}\bar{k}\bar{l}, hk\bar{l}, h\bar{k}l, \bar{h}kl$	$2/m\ 2/m\ 2/m$
$\bar{4}$	$hkl, \bar{k}h\bar{l}, \bar{h}\bar{k}l, k\bar{h}\bar{l}$		$\bar{h}\bar{k}\bar{l}, k\bar{h}l, hk\bar{l}, \bar{k}hl$	$4/m$
4	$hkl, \bar{k}hl, \bar{h}\bar{k}l, k\bar{h}l$		$\bar{h}\bar{k}\bar{l}, k\bar{h}\bar{l}, hk\bar{l}, \bar{k}h\bar{l}$	$4/m$
$\bar{4}2m$	$hkl, khl+\bar{4}$		$\bar{h}\bar{k}\bar{l}, \bar{k}\bar{h}\bar{l}+\bar{4}$	$4/m\ 2/m\ 2/m$
$\bar{4}m2$	$hkl, kh\bar{l}+\bar{4}$		$\bar{h}\bar{k}\bar{l}, \bar{k}\bar{h}l+\bar{4}$	$4/m\ 2/m\ 2/m$
$4mm$	$hkl, \bar{h}kl+4$		$\bar{h}\bar{k}\bar{l}, h\bar{k}\bar{l}+4$	$4/m\ 2/m\ 2/m$
422	$hkl, \bar{k}\bar{h}\bar{l}+4$		$\bar{h}\bar{k}\bar{l}, khl+4$	$4/m\ 2/m\ 2/m$
3				
primitive	hkl, ihl, kil		$\bar{h}\bar{k}\bar{l}, \bar{i}\bar{h}\bar{l}, \bar{i}\bar{k}\bar{l}$	$\bar{3}$
rhombohedral	hkl, klh, lhk		$\bar{h}\bar{k}\bar{l}, \bar{k}\bar{l}\bar{h}, \bar{h}\bar{k}\bar{l}$ (cyclic permutation)	
$3m1$	$hkl, \bar{k}\bar{h}l+3$		$\bar{h}\bar{k}\bar{l}, kh\bar{l}+3$	$\bar{3}\ 2/m$
$31m$	$hkl, khl+3$		$\bar{h}\bar{k}\bar{l}, \bar{k}\bar{h}\bar{l}+3$	
$3m$ (rhombohedral)	$hkl, khl+$cyclic		$\bar{h}\bar{k}\bar{l}, \bar{k}\bar{h}\bar{l}+$cyclic	
312	$hkl, \bar{k}\bar{h}\bar{l}+3$		$\bar{h}\bar{k}\bar{l}, khl+3$	$\bar{3}\ 2/m$
321	$hkl, kh\bar{l}+3$		$\bar{h}\bar{k}\bar{l}, \bar{k}\bar{h}\bar{l}+3$	
32 (rhombohedral)	$hkl, \bar{k}\bar{h}\bar{l}+$cyclic		$\bar{h}\bar{k}\bar{l}, khl+$cyclic	
6	$hkl, ihl, kil,$ $\bar{h}\bar{k}l, \bar{i}\bar{h}l, \bar{k}\bar{i}l$		$\bar{h}\bar{k}\bar{l}, \bar{i}\bar{h}\bar{l}, \bar{k}\bar{i}\bar{l},$ $hk\bar{l}, ih\bar{l}, ki\bar{l}$	$6/m$
$\bar{6}$	hkl, ihl, kil $hk\bar{l}, ih\bar{l}, ki\bar{l}$		$\bar{h}\bar{k}\bar{l}, \bar{i}\bar{h}\bar{l}, \bar{k}\bar{i}\bar{l},$ $\bar{h}\bar{k}l, \bar{i}\bar{h}l, \bar{k}\bar{i}l$	$6/m$
622	$hkl, kh\bar{l}+6$		$\bar{h}\bar{k}\bar{l}, \bar{k}\bar{h}l+6$	$6/m\ 2/m\ 2/m$
$6mm$	$hkl, khl+6$		$\bar{h}\bar{k}\bar{l}, \bar{k}\bar{h}\bar{l}+6$	$6/m\ 2/m\ 2/m$
$\bar{6}m2$	$hkl, \bar{k}\bar{h}l+\bar{6}$		$\bar{h}\bar{k}\bar{l}, kh\bar{l}+\bar{6}$	$6/m\ 2/m\ 2/m$
$\bar{6}2m$	$hkl, khl+\bar{6}$		$hkl, \bar{k}\bar{h}\bar{l}+\bar{6}$	
23	same as 222+cyclic permutation			$2/m\ \bar{3}$
432	same as 422+cyclic permutation			$4/m\ \bar{3}\ 2/m$
$4\bar{3}m$	same as $42m$+cyclic permutation			$4/m\ \bar{3}\ 2/m$

General remarks. The second setting is used for the monoclinic system. For the tetragonal system, observe the effect of diagonal transformation on $\bar{4}2m$ and $\bar{4}m2$. For the trigonal system, if the rhombohedral space groups are described by the hexagonal settings, use the relationships for $3m1$ or 321.

The true point group symmetry and the Laue symmetry

When there exist noticeable effects of anomalous dispersion, it is possible to make an unequivocal determination of the point group of a crystal. The equivalent conditions found in the weights at reciprocal lattice points, or the intensities of diffracted beams, depend on point-group symmetry and, in certain symmetry groups, on axial settings. They do not reflect space group or translational symmetry. The relationship for all noncentrosymmetric point groups are given in Table II. For crystals without anomalous scatterers, the inequality sign between the **H** and $\bar{\mathbf{H}}$ sides is changed into an equality sign. This results in some degeneracies as shown in the last column of the Table which lists the corresponding Laue symmetry groups.

It seems worthwhile to mention here that for a noncentrosymmetric crystal with anomalous scatterers, a Laue symmetry group having lower symmetry than the true group might be assigned erroneously. This might affect the estimate of standard deviations of intensity measurement.

Anomalous dispersion as a physical phenomenon

The treatment by Hönl (1933) of the theory of anomalous dispersion is now a classical and famous one. He has treated the required correction to the real part and the imaginary component of the atomic scattering factor as functions of the ratio of the incident wavelength to the absorption edge of the atom. The result can be found in the treatise by R. W. James (1958). Dauben & Templeton (1955) and Cromer (1965) have prepared extensive tables for these corrections.

When the accuracy of intensity measurement was relatively low, such as obtained by visual recording of photographic films, it was customary to use as large f'' values as possible. This requirement imposed many restrictions on the choice of anomalous scatterers and X-ray target materials. At the same time, errors due to the absorption effect became a critical issue. However, with increasing improvement in the accuracy of intensity measurements, especially by the use of counter techniques, we can make use of relatively small f'' values, thus eliminating the absorption problem. Now, one can find many examples of light atoms being used as anomalous scatterers. The absolute configurations of optically active silicon compounds (Okaya & Ashida, 1966) were determined by using Si atoms as anomalous scatters. It was recently reported that the effect of oxygen dispersion can be sufficient ($f'' \simeq 0{\cdot}02$) for deciding the absolute configurations of organic molecules (Hope & de la Camp, 1969).

In contrast to the X-ray anomalous dispersion which results in relatively small imaginary components, the neutron dispersion effect can exhibit far larger imaginary values when proper choices of wavelength and isotopes are made. This phenomenon might be useful especially when a crystal contains only light atoms which have anomalously scattering isotopes (Li, B, D, etc.) or when a crystal contains too many light atoms compared to dispersing heavy atoms. An excellent use of the neutron dispersion method can be found in the absolute configuration determination of a product of enzymatic deuteration (Johnson, Gabe, Taylor & Rose, 1965).

A similar situation concerning large imaginary components is also found in electron diffraction, especially for low energy electron diffraction. Future developments on the theory and applications of dispersion techniques on surface crystallography are expected.

Complex vector space

The electron density function for a crystal

can be expressed as the Fourier series

$$\varrho(\mathbf{r}) = \sum_{\mathbf{H}} F_{\mathbf{H}} \exp(-2\pi i \mathbf{H}\cdot\mathbf{r})$$

Now, $\varrho(\mathbf{r})$ is a complex function. The convolution of $\varrho(\mathbf{r}+\mathbf{u})$ and $\varrho^*(\mathbf{r})$, called the Patterson function, is

$$P(\mathbf{u}) = \sum_{\mathbf{H}} |F_{\mathbf{H}}|^2 \exp(-2\pi i \mathbf{H}\cdot\mathbf{u})$$

It is well known that for crystals with only normal scatterers, $P(\mathbf{u})$ is real and also centrosymmetric.

When anomalous dispersion occurs,

$$|F_{\mathbf{H}}|^2 + |F_{\overline{\mathbf{H}}}|^2 = 2\sum_{j,k} (f'_j f'_k + f''_j f''_k) \times \cos 2\pi[\mathbf{H}\,.\,(\mathbf{r}_j - \mathbf{r}_k)]$$

$$|F_{\mathbf{H}}|^2 - |F_{\overline{\mathbf{H}}}|^2 = 2\sum_{j,k} (f'_j f''_k - f''_j f'_k) \times \sin 2\pi[\mathbf{H}\,.\,(\mathbf{r}_j - \mathbf{r}_k)]$$

Inserting these expressions into the general formula of the Patterson function

$$P(\mathbf{u}) = P_c(\mathbf{u}) - iP_s(\mathbf{u})\,,$$

where

$$P_c(\mathbf{u}) = \sum_{\mathbf{H}} |F_{\mathbf{H}}|^2 \cos 2\pi\mathbf{H}\,.\,\mathbf{u}\,,$$

$$P_s(\mathbf{u}) = \sum_{\mathbf{H}} |F_{\mathbf{H}}|^2 \sin 2\pi\mathbf{H}\,.\,\mathbf{u}\,.$$

The imaginary component of the Patterson function, $P_s(\mathbf{u})$, is an antisymmetric function and the usefulness of this function to crystal structure and absolute configuration determinations was first pointed out by Okaya, Saito & Pepinsky (1955). Extensive analysis of the $P_s(\mathbf{u})$ was given in a later article (Okaya & Pepinsky, 1960).

Aid to the phase problem

In addition to the use of $P_s(\mathbf{u})$, anomalous dispersion is used extensively in solving the phase problem. Since these subjects will be discussed in the subsequent papers of this session, the author wishes to make no further remarks on the phase problem.

It might be appropriate to issue a general warning to those who believe the results of absolute configuration determinations. For establishment of absolute configuration, it is essential to hold strictly to one coordinate system (*e.g.* right-handed) throughout the experiments and subsequent computations. Any departure from this discipline might result in erroneous results. (The fact that even numbers of errors give the correct answer is not an excuse for lack of proper attention. For the absolute configuration there is no intermediate answer; either the result is correct or incorrect.) Unfortunately, the author has spotted a number of examples with questionable determinations.

Anomalous dispersion, absolute configuration, and the physical and chemical significance

The first example of absolute configuration determination was made about forty years ago on zincblende, ZnS, by Nishikawa & Matsukawa (1928) and by Coster, Knol & Prins (1930). After twenty years of neglect, the method was revived in 1951. Bijvoet and coworkers (Bijvoet, 1949; Peerdman, van Bommel & Bijvoet, 1951) determined the absolute configuration of the (+)-tartrate ion. The L-configuration of natural amino acids was then established using isoleucine by Trommel & Bijvoet (1954); the configuration was further substantiated by the determination of the L-aspartate ion by Doyne, Pepinsky & Watanabe (1957) using the $P_s(\mathbf{u})$ function. At the end of the 1960's, there are many dozens of absolute configuration determinations of organic molecules and it is not possible to make a list in this short contribution.

The first determination of the absolute configuration of optically active complex-ion compounds is that of D-$[Co(en)_3]^{+++}$ by Saito, Nakatsu, Shiro & Kuroya (1955). In 1969, one can again find as many absolute configuration determinations in complex-ion compounds as in organic molecules. For metal organic compounds, the determinations of optically active $SiH(CH_3)(C_6H_5)$ alpha-$C_{10}H_7$ and corresponding Ge compounds (Okaya & Ashida 1966; Okaya, 1967) were made.

Among optically active crystals and minerals, the following examples might be worth mentioning as earlier examples; ullmanite (NiSbS) by Takeuchi (1957), alpha-quartz (deVries, 1958), and CdS by neutron diffraction.

Extensive studies on the absolute configurations of various molecules before and after stereospecific reactions answer without ambiguity the question of retention or inversion of the configurations. The results are helpful in elucidating many key reactions, especially in many new systems of compounds whose chemistry is still at an initial stage of understanding. The determinations for many organic molecules and diastereoisomers are of fundamental importance in successful detections of extraterresterial life.

In order to obtain unequivocal relations between crystal structures and polar phenomena such as piezoelectricity, pyroelectricity, ferroelectricity, and others, it is essential to establish the absolute configurations of crystal structures showing such phenomena under various conditions. The author should add a few examples here; the piezoelectric deformations of KDP structures and the absolute configurations (Unterleitner, 1962); relationship between the morphology and the structures of a semiconductor, InSb (Warekois & Metzger, 1960); the polarization and absolute direction of atomic shifts in a ferroelectric, $(glycine)_3$ sulfate (Unterleitner, 1962); and perfection studies of GaAs surfaces (Cole & Stemple, 1962).

This study was supported in part by the U.S. Air Force Cambridge Research Laboratory, contract no. AFCRL-F19(628)–69–C–0082.

References

Bijvoet, J. M. (1949). *Koninkl. Ned. Akad. Wetenschap.* **52**, 313.

Cole, H. & Stemple, N. R. (1962). *J. Appl. Phys.* **33**, 2227.

Coster, D., Knol, K. S. & Prins, J. (1930). *Z. Physik.* **63**, 345.

Cromer, D. T. (1965). *Acta Cryst.* **18**, 17.

Dauben, C. H. & Templeton, D. H. (1955). *Acta Cryst.* **8**, 841.

Doyne, T., Pepinsky, R. & Watanabe, T. (1957). *Acta Cryst.* **10**, 438.

Hönl, H. (1933). *Z. Physik.* **84**, 1.

Hope, H. & de la Camp, U. (1969). *ACA Meeting Abstracts,* H-5, *Seattle, Wash.*

James, R. W. (1958). *Optical Principles of the Diffraction of X-rays.* London: Bell.

Johnson, C. K., Gabe, E. J., Taylor, M. R. & Rose, I. A. (1965). *J. Amer. Chem. Soc.* **87**, 1802.

Nishikawa, S. & Matsukawa, K. (1928). *Proc. Imperial Acad. Japan*, **4**, 96.

Okaya, Y. (1967). *ACA Meeting Abstracts.*

Okaya, Y. & Ashida, T. (1966). *Acta Cryst.* **20**, 461.

Okaya, Y. & Pepinsky, R. (1960). *Computing Methods and the Phase Problem in X-ray Crystal Analysis.* Oxford: Pergamon Press.

Okaya, Y., Saito, Y. & Pepinsky, R. (1955). *Phys. Rev.* **98**, 1857.

Peerdman, A. F., van Bommel, A. J. & Bijvoet, J. M. (1951). *Proc. Roy. Soc. Amsterdam*, **B54**, 16.

Saito, Y., Nakatsu, K., Shiro, M. & Kuroya, H. (1955). *Acta Cryst.* **8**, 729.

Takeuchi, Y. (1957). *Mineral. J.* **2**, 90.

Trommel, J. & Bijvoet, J. M. (1954). *Acta Cryst.* **7**, 703.

Unterleitner, F. (1962). Thesis, Department of Physics, Pennsylvania State University.

Vries, A. de (1958). *Nature*, **181**, 1193.

Warekois, E. P. & Metzger, P. H. (1960). *J. Appl. Phys.* **30**, 960.

Use of Isomorphous Series and Anomalous Scattering Data in Structure Determination of Large Molecules

GOPINATH KARTHA

Roswell Park Memorial Institute, Buffalo, New York, U.S.A.

To date, most of the structure investigations of complicated molecules forming noncentrosymmetric crystals have been successfully carried out by the use of the so called *indirect* methods of solving the phase problem. By *indirect*, as opposed to direct, methods I mean those which involve the use of certain known structural features and their scattering behaviour in deriving either partial or complete information about the phase angles. The simplest and most obvious of such indirect methods is the heavy atom method. Here, one or a few atoms of high atomic numbers are incorporated in the unit cell of the crystal. The positions of these heavy atoms are easily determined by Patterson methods and hence the phase angles of their contributions to the reflections calculated. The basic assumption of the heavy atom method is that the phases of their contribution are close enough to the true phases of the reflections so that they, in conjunction with the measured amplitudes, provide a good enough approximation to the actual structure to permit successful refinement procedures. How good this approximation is, of course, depends on the nature and positions of the heavy atoms as well as the total number and types of atoms in the rest of the structure. The striking success of this assumption in solving molecules of moderate complexity (even around a hundred atoms per molecule) is now well established.

However, some of the molecules being studied now, like proteins, contain a thousand or more atoms per molecule. In such cases, the scattering from a few heavy atoms may be only a small fraction of the scattering from the total structure and the basic assumption of the heavy atom method may no longer give a valid starting-point for further refinement. In the case of proteins, progress in the solution of the phase problem by these indirect methods came through the use of the isomorphous replacement (IR) (Green, Ingram & Perutz, 1954) and anomalous scattering (AS) methods. In the application of these methods, to be considered in the rest of this section, we shall assume that we are referring to the diffraction of X-rays by crystals, even though many of the results derived are also applicable to the case of diffraction of neutrons by ordered nuclei in the crystals.

Isomorphous replacement (IR) method

This method has in many ways close similarity to the heavy atom method. The known structural feature used in solving the phase problem is the information about the change in scattering amplitude of the total structure caused by changes in the scattering properties of a few atoms in known positions. The principle is illustrated in Fig. 1.

The crystal gives a scattering amplitude $|F_P|$ with phase angle α_P. Additional scattering material in known positions in the isomorphous crystal results in a total scattering amplitude $|F_{PH}|$ and phase angle α_{PH}. These amplitudes are experimentally measured, while the phases are not. Further, the vector $\mathbf{F}_H$ due to the additional scattering material in the second crystal has the magnitude $|F_H|$ and phase angle α_H. In vector notation we write

$$\mathbf{F}_{PH} = \mathbf{F}_P + \mathbf{F}_H \tag{1}$$

and

$$\alpha_P = \alpha_H + \varphi \tag{2}$$

In triangle OLM we know the length of the three sides. From this we write

$$|F_{PH}|^2 = |F_P|^2 + |F_H|^2 - 2|F_H|\,|F_P| \cos\theta$$

and

$$-\cos\theta = \cos(\pi - \varphi) = [|F_{PH}|^2 - |F_P|^2 - |F_H|^2] \div 2|F_H|\,|F_{PH}| \tag{3}$$

Note that Eq. (3) gives only the cosine of the angle φ, and φ may be positive or negative creating a twofold ambiguity for the angle. In other words, it is possible to construct another vector solution for the problem by reflecting the diagram in Fig. 1 about the $\mathbf{F}_H$ vector, giving an alternate orientation for the $\mathbf{F}_P$ vector. These two possible solutions for α_P are symmetrical about α_H, the phase of the replaceable group. Hence we write

$$\alpha_P = \alpha_H \pm \varphi \tag{4}$$

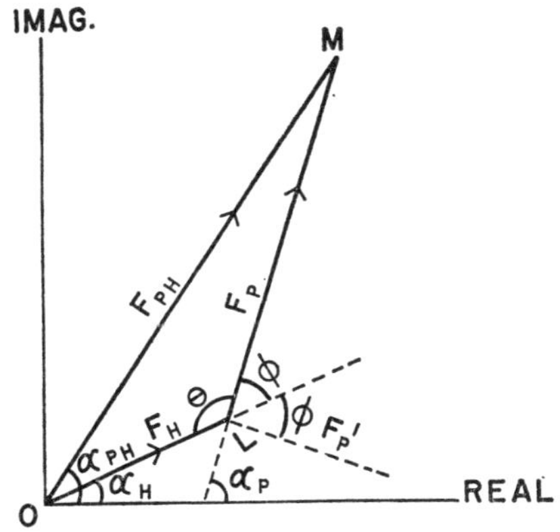

Fig. 1. Vector representation of $\mathbf{F}_H$ *(heavy atom),* $\mathbf{F}_P$ *(protein), and* $\mathbf{F}_{PH}$ *(protein + heavy atom). If* α_H *is known, the directions of* $\mathbf{F}_P$ *and* $\mathbf{F}_{PH}$ *can be deduced, giving two ambiguous solutions arranged symmetrically about* $\mathbf{F}_H$.

where φ corresponds to the smallest positive angle satisfying Eq. (3).

In the centrosymmetric case no phase ambiguity exists since $\varphi = 0$ or π and in either case $-\varphi = +\varphi$ and both solutions in Eq. (4) for α_P coincide. For noncentrosymmetric crystals this ambiguity can be resolved by having another isomorphous crystal with the replaceable atoms in different positions. Using data from this crystal, we get another possible pair of values for α_P which are symmetrical about another value α_H. Then, one of the values of α_P, the true one, should be the same in the two substitutions and this solves the phase problem. This is the basis of the multiple isomorphous replacement (MIR) method and Harker (1956) has given a geometrical construction for evaluating the phase angles from the measured amplitudes of the isomorphs and a knowledge of the replaceable atom configurations. The MIR method has been the basis of all successful protein structure investigations so far. But, the requirement that the crystals must have substituents at different sites and still be isomorphous has limited its application in the study of ordinary organic structures. However, in spite of the phase ambiguity with the single isomorphous series data, valuable information regarding the structure could still be obtained (Kartha, 1961) from an electron density map computed by including both ambiguous phase solutions of the single isomorphous series method.

In using Eqs. (3) and (4) to evaluate α_P we need both magnitude and direction of $\mathbf{F}_H$. These are given by the usual structure factor expression

$$\mathbf{F}_H(\mathbf{H}) = \sum_j f(j) \exp 2\pi i(\mathbf{H} \cdot \mathbf{r}_j) \tag{5}$$

where $f(j)$ is the change in scattering power due to the replacement at $\mathbf{r}_j$ and the summation extends over all regions where the scattering power changes. If we know the types and sites of the replaceable atoms, we can calculate their contribution $\mathbf{F}_H$ to every reflection from Eq. (5).

Anomalous scattering (AS) method

Even though the effects of anomalous scatterers on the intensities of crystal reflections have been known for a long time (Coster, Knol & Prins, 1920), its use was mostly in the determination of the absolute configurations of structures that have been determined by standard methods. This involved only comparison of related pairs of $F(hkl)$ and $F(\bar{h}\bar{k}\bar{l})$ reflections for which anomalous scattering effects caused significant intensity differences to allow a choice between the two possible configurations. Okaya, Saito & Pepinsky (1955) pointed out the possibility of more extensive use of the anomalous scattering effect as another powerful '*indirect*' method in obtaining phase information. Different ways of utilizing this effect have been proposed (Okaya & Pepinsky, 1961; Ramachandran & Raman, 1959); and recently this method, in conjunction with the isomorphous series method, has been of great value in protein structure investigations. It is possible that the AS method, using both X-rays and neutrons, will find much wider application in structure analysis when the ease and accuracy of measurements of reflection intensities are increased, resulting in a corresponding increase in the significance of the anomalous scattering measurements.

The anomalous scattering effects arise from the fact that, in general, the atomic scattering factor $f(j)$ in Eq. (5) is a complex number, although the imaginary component is negligible in most cases. However, for certain wavelengths and atoms, these components become appreciable and give a scattering component $\mathbf{F}''_H$ for $\mathbf{F}_H$ which has a phase advance of $\pi/2$ over the component $\mathbf{F}'_H$ that is in phase with the incident beam. This out-of-phase component has the effect of making the structure amplitudes $F(hkl)$ and $F(\bar{h}\bar{k}\bar{l})$ [hereinafter referred to as $F(+)$ and $F(-)$] from the front and back surfaces of atomic planes different for noncentrosymmetric crystals, in apparent violation of Friedel's law.

The vector diagram illustrating this is shown in Fig. 2. Here, the crystal, in addition to the normal scatterers, also contains a few atoms with non-negligible anomalous scattering components for the wavelength employed. For these atoms, the atomic scattering factor $f(j)$ is expressed as a complex number.

$$f(j)=f'(j)+if''(j) \tag{6}$$

where now f' denotes the atomic scattering component that is in phase with, and f'' the component that is $\pi/2$ in advance of, the phase of the forward scattering from the atom. These two components have been tabulated for many atoms (Dauben & Templeton, 1956) for the common wavelengths used in X-ray diffraction studies.

These anomalous scattering differences are, in general, small. However, these measurements are the differences of two measurements on the same crystal and are capable of good accuracy. As will be seen later, the measurement of these differences

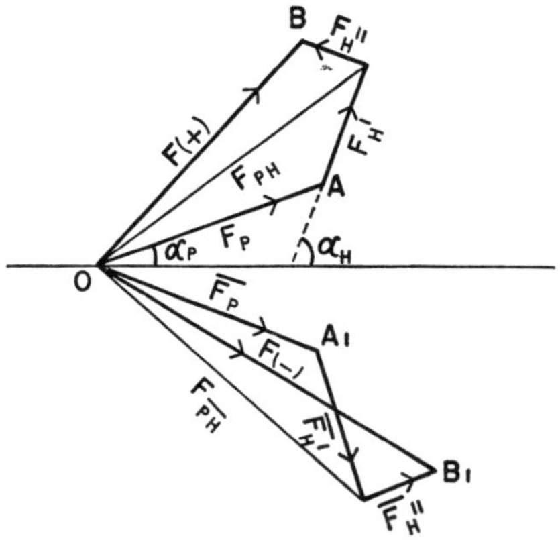

Fig. 2. Relation between $|F(+)|$ and $|F(-)|$ for the hkl and $\bar{h}\bar{k}\bar{l}$ reflections due to the anomalous component $\mathbf{F}''_H$.

gives us the phase difference between the 'normal' and the 'anomalous' scattering components. The result is similar to what one obtains in the case of the IR method and also has an ambiguity. But, luckily, the nature of the ambiguities in the IR and AS cases are different and together they lead to a unique solution of the phase problem. Hence, it is now recognized that it is a combination of the IR and AS methods that gives us the most powerful tool for the 'indirect' solution of the phase problem in the case of large noncentrosymmetric structures like proteins. In what follows, I discuss what appears to me the most direct way of using this information, first for establishing the parameters that lead to a detailed description of the vector $\mathbf{F}_H$ and then the phase angle α_{PH} of the total crystal.

Heavy atom vs. isomorphous replacement method

In any 'indirect' method of phase determination, we assume that we know the positions and nature of the scattering centers and hence the vector $\mathbf{F}_H$ in Eq. (5). In most medium-sized structures, this in itself does not present any serious problem, as these are fairly heavy atoms and their locations can be derived more or less easily from three-dimensional vector maps. However, one may come across special situations in the study of smaller molecules where the extra labour of data collection and the techniques detailed here may pay ample dividends.

For the successful application of the heavy atom method, as a very rough guide, we should have

$$\sum_{1}^{H} f_H^2(j) \simeq \sum_{1}^{L} f_L^2(j)\,. \qquad (7)$$

Here $f(j)$ represents the scattering factor of atom j and the summation on the left corresponds to the heavy marker atoms while the sum on the right is over all remaining light atoms in the unit cell. In large molecules, L may be of the order of thousands and, even with the heaviest atoms available, one has to have a very large number of them in specific and known positions to satisfy Eq. (7) for the usual heavy atom method to be effective. This situation creates practical difficulties in the application of the heavy atom method, so that, to date, the successful protein structure investigations have relied on the multiple isomorphous replacement method. In principle, the heavy atom method and the isomorphous series method are very similar and the basis of deducing the structure from measured intensities are almost identical in the two cases. In theory, at least, the heavy atom method has the same limitations and ambiguities that one faces with the IR method. The main advantage of the IR method over the heavy atom method is in lifting the limitations imposed by Eq. (7) that the heavy atom scattering should be of the same order as the scattering from all the remaining atoms. For the success of the IR method, we only need that the intensity difference produced by the replaceable atom should be significantly higher than the experimental errors in the intensity measurements. Hence, the advantage of the IR method over the heavy atom method is in the accuracy with which the light atoms can be determined for a given accuracy in intensity measurements. Viewed in this way, the advantage of the IR method over the heavy atom method, even in the case of molecules of medium complexity, is easily appreciated.

Multiple isomorphous replacement (MIR) method

In the MIR method for determining noncentrosymmetric structures, the phases of reflections are evaluated by preparing a series of isomorphous derivative crystals, each one containing a small group of heavy atoms attached at different but specific sites. Then the phase evaluation for the reflections can be divided into a number of steps as follows:

1. Determination of the relative arrangements of the heavy atoms in a given derivative crystal.

2. Correlating the positions of the heavy atoms to the same origin in the unit cell of the crystal.

3. Refinement of the heavy atom positional, occupancy, and thermal parameters for accurate evaluation of the vector $\mathbf{F}_H$ and

4. Evaluation of the phase angles of the parent crystal reflections using all the above information.

An actual case of protein structure investigation may necessitate going through these various steps in a series of refinement cycles before satisfactory convergence is obtained. In our experience, during all these stages of the application of the MIR method to protein structure studies, additional information from the anomalous scattering behaviour of the heavy atoms is of great value. Proper combination of the multiple isomorphous replacement and anomalous scattering (MIRAAS) methods not only helps in confirming and complementing the results in a very satisfactory manner, but actually overcomes many undesirable features that crop up when either method is used alone.

We shall, hence, assume that in the MIR method, for all the heavy atom derivatives, the experimental data for any reflection $S(hkl)$ consists, not only of the structure amplitude $|F_P|$ and $|F_{PH}|$, but also the inverse reflection $|F(-S)|$ of the heavy atom derivative, *i.e.* $|F_{\overline{PH}}|$ of Fig. 2.

Many methods have been suggested for the use of the experimental data to compute maps that bring out the heavy atom-heavy atom $(H-H)$ vectors with the least undesirable background. Some of these are considered below.

H–H vector maps from isomorphous series data

From the locations and nature of the heavy atom substitutions, we can compute the difference vector $\mathbf{F}_H$ (Fig. 1) between the scattering vectors $\mathbf{F}_P$ and $\mathbf{F}_{PH}$. Conversely, knowing the vectors $\mathbf{F}_P$ and $\mathbf{F}_{PH}$, we can calculate the difference vector $\mathbf{F}_H$ which leads directly to the heavy atom positions, leaving aside errors due to lack of isomorphism. However, the experimental data consist only of the amplitudes $|F_P|$ and $|F_{PH}|$. We have to determine $\mathbf{F}_H$ and the location of the heavy atoms from this information.

From Fig. 1, considering the vector triangle OLM, we get

$$\begin{aligned}|F_{PH}| &= |F_H| \cos \mathrm{LOM} + |F_P| \cos \mathrm{OML} \\ &= |F_H| \cos(\alpha_{PH}-\alpha_H) \\ &\quad + |F_P| \cos(\alpha_{PH}-\alpha_P)\end{aligned}$$

Rewriting

$$\cos(\alpha_{PH}-\alpha_P) = 1-2\sin^2[(\alpha_{PH}-\alpha_P)/2]$$

we get the following expression for the amplitude difference

$$\begin{aligned}|F_{PH}|-|F_P| &= |F_H| \cos(\alpha_{PH}-\alpha_H) \\ &\quad -2|F_P| \sin^2\left(\frac{\alpha_{PH}-\alpha_P}{2}\right)\end{aligned} \tag{8}$$

We see that this amplitude difference is related to the amplitude $|F_H|$ of the heavy atom scattering vector in a complicated way. We shall denote this measured amplitude difference between the heavy atom isomorph and parent crystals by $\Delta|F|_{\mathrm{iso}}$ so that,

$$\begin{aligned}\Delta|F|_{\mathrm{iso}} &= |F_H| \cos(\alpha_{PH}-\alpha_H) \\ &\quad -2|F_P| \sin^2\left(\frac{\alpha_{PH}-\alpha_P}{2}\right)\end{aligned} \tag{9}$$

The second term on the right hand side of Eq. (9) is usually small compared to the first since $(\alpha_{PH}-\alpha_P)$ is, in general, small and we can write to a very good first approximation

$$\Delta|F|_{\text{iso}} = |F_{PH}| - |F_P| \simeq |F_H| \cos(\alpha_{PH} - \alpha_H) \quad (10)$$

The term $\Delta|F|_{\text{iso}}$ may be positive or negative. Vector maps computed with the square of these amplitude differences have been used by many investigators (Perutz, 1956; Kartha, Bello, Harker & De Jarnette, 1963), both in two and three dimensions, to obtain information regarding the *H–H* vectors. The usefulness of such maps is seen by studying the coefficients used in computing them. From Eq. (10),

$$(\Delta|F|_{\text{iso}})^2 \simeq |F_H|^2 \cos^2(\alpha_{PH} - \alpha_H) = \tfrac{1}{2}|F_H|^2 + \tfrac{1}{2}|F_H|^2 \times \cos 2(\alpha_{PH} - \alpha_P) \quad (11)$$

The first term on the right of Eq. (11) is the appropriate coefficient for computing the Patterson vector map of the heavy atom structure and gives positive peaks of height $(\frac{1}{2}) f_{Hi} f_{Hj}$ at positions $\pm(\mathbf{r}_{H_i} - \mathbf{r}_{H_j})$. The second term, however, leads to a confused background which depends, not only on the heavy atom configuration, but also on the protein density distribution. These background features, which may be either positive or negative peaks, tend to mask the features of the *H–H* vectors we are trying to establish from these maps. Undoubtedly, for locating anything more than very simple heavy atom arrangements, it is a definite advantage to eliminate the undesirable background features represented by the second term in Eq. (11).

H–H vectors from anomalous scattering data

If we assume that the heavy atoms are also anomalous scatterers whose scattering factor $\mathbf{f}(j)$ has an imaginary component, as in Eq. (6), then the structure factors of the direct and inverse reflections (Fig. 2) can be written as

$$\mathbf{F}(+) = \mathbf{F}_{PH}(+) + i\mathbf{F}''_H(+) \quad (12)$$

$$\mathbf{F}(-) = \mathbf{F}_{PH}(-) + i\mathbf{F}''_H(-) \quad (13)$$

where $\mathbf{F}_{PH}(+)$ and $\mathbf{F}_{PH}(-)$ are the scattering vectors assuming that all atoms have their out-of-phase components $f''(j)$ equal to zero. The term $\mathbf{F}''_H$ represents the out-of-phase component introduced by the anomalous scattering from the heavy atoms. We thus have,

$$\mathbf{F}'_H(+) = \sum_{j=1}^{H} f'(j) \exp(2\pi i \mathbf{S} \cdot \mathbf{r}_j) \quad (14)$$

$$\mathbf{F}''_H(+) = \sum_{j=1}^{H} f''(j) \exp(2\pi i \mathbf{S} \cdot \mathbf{r}_j) \quad (15)$$

Even though not essential for the usefulness of the AS method, additional simplicity can be obtained for these equations by considering the special case where all atoms are of similar type, having the same value of (f''/f'). In such a case, the vector $i\mathbf{F}''_H$ will be proportional in length to the $\mathbf{F}'_H$ vector and in advance of α_H by phase angle $\pi/2$. This is expressed by relationship

$$\mathbf{F}''_H = i(f''/f')\mathbf{F}'_H = ik_1 \mathbf{F}'_H \quad (16)$$

Remembering that $\mathbf{F}_{PH}(+)$ and $\mathbf{F}_{PH}(-)$ are related by mirror symmetry about the real axis, we can write down from Fig. 2

$$|F(+)|^2 = |F_{PH}|^2 + |F''_H|^2 - 2|F_{PH}|\,|F''_H| \cos(\pi/2 + \alpha_H - \alpha_{PH})$$

and

$$|F(-)|^2 = |F_{PH}|^2 + |F''_H|^2 + 2|F_{PH}|\,|F''_H| \cos(\pi/2 + \alpha_H - \alpha_{PH})$$

To a good approximation, we may replace $|F_{PH}|$ by $\frac{1}{2}[|F(+)| + |F(-)|]$ and, thus, we get the expression for the amplitude difference between $\mathbf{F}(+)$ and $\mathbf{F}(-)$ as

$$|F(+)|-|F(-)|=2|F''_H| \sin(\alpha_{PH}-\alpha_H) \quad (17)$$

For the case of similar anomalous scatterers, using Eq. (16), we can write down

$$k[|F(+)|-|F(-)|]= |F_H| \sin(\alpha_{PH}-\alpha_H) \quad (18)$$

where $k=(f'/2f'')$ and $|\mathbf{F}'_H|=|\mathbf{F}_H|$ due to the relative smallness of k compared to unity. By denoting the left side of Eq. (18), which is a scaled value of the amplitude difference between $\mathbf{F}(+)$ and $\mathbf{F}(-)$, by $\Delta|F|_{\text{ano}}$, we get

$$\Delta|F|_{\text{ano}}=|F_H| \sin(\alpha_{PH}-\alpha_H) \quad (19)$$

Rossman (1961) suggested the use of maps computed with the squares of the amplitude differences between $\mathbf{F}(+)$ and $\mathbf{F}(-)$ for locating the positions of anomalous scatterers. By squaring both sides of Eq. (19), we get

$$(\Delta|F|_{\text{ano}})^2=|F_H|^2 \sin^2(\alpha_{PH}-\alpha_H) = \tfrac{1}{2}|F_H|^2-\tfrac{1}{2}|F_H|^2 \times \cos 2(\alpha_{PH}-\alpha_H) \quad (20)$$

The first term of the right gives rise to the Patterson vector map of the anomalous scatterers, whereas the second term gives rise to unwanted positive and negative peaks masking the *H–H* vectors we are investigating.

Combination of IR and AS data to obtain H–H vectors

A glance at Eqs. (10) and (19) immediately reveals the complementary features of the information we get about $\mathbf{F}_H$ from the IR difference $\Delta|F|_{\text{iso}}$ and the AS difference $\Delta|F|_{\text{ano}}$. Whereas $\Delta|F|_{\text{iso}}$ gives the component of $\mathbf{F}_H$ along $\mathbf{F}_{PH}$, $\Delta|F|_{\text{ano}}$ gives the component of $\mathbf{F}_H$ perpendicular to $\mathbf{F}_{PH}$. Between the two we get the length of the heavy atom scattering vector $\mathbf{F}_H$ which can be used in computing a vector map giving only *H–H* vectors. This is also seen by inspection of the two vector maps computed using coefficients in Eqs. (11) and (20). In both, the first terms giving the *H–H* peaks are the same. However, considering the second terms, we see that every unwanted feature in one coincides with a corresponding feature of opposite sign in the other. Consequently, a synthesis with coefficients $(\Delta|F|_{\text{iso}})^2+(\Delta|F|_{\text{ano}})^2)$ gives a map where the *H–H* vectors of the two contributing maps add up while the positive and negative background features exactly cancel each other.

Application in practical cases

In our studies on the structure of the protein ribonuclease, we have used similar maps to establish the positions of heavy atoms in the derivative crystals with satisfactory results. Ribonuclease is a protein of molecular weight around 14,000 and it contains more than 1,000 C, N, and O atoms per molecule. The crystal under study belonged to space group $P2_1$ with two molecules in a unit cell of dimensions $a=30{\cdot}1$ Å, $b=38{\cdot}1$ Å, $c=53{\cdot}3$ Å and monoclinic angle $\beta=105{\cdot}7°$. The first successful heavy atom derivative obtained was a platinum derivative crystal. Fig. 3(*a*), (*b*), (*c*) show sections of three-dimensional vector maps at $y=\frac{1}{2}$ showing the Harker peaks between the screw-related atoms. The coefficients used in Figs. 3(*a*) and 3(*b*) were the squares of

$$\Delta|F|_{\text{iso}}=|F_{PH}|-|F_P| \quad (22)$$

and

$$\Delta|F|_{\text{ano}}=k[|F(+)|-|F(-)|]\,. \quad (23)$$

Both these maps give major peaks in the same positions. These maps, along with the corresponding section at $y=0$, can be explained by assuming two major heavy atom substitutions per asymmetric unit, where both kinds of atoms have approximately the same y parameter. This agreement is very satisfying in view of the fact that, while map 3(*a*) used data from the free protein as well as the derivative, map 3(*b*) is based only on the reflection amplitudes from the heavy atom derivative crystals. Further, the atomic parameters obtained from the two maps are in very good agreement. This suggests, that, as far as simple configurations of one or two heavy atoms per asymmetric unit are concerned, either the $(\Delta|F|_{iso})^2$ or the $(\Delta|F|_{ano})^2$ maps by themselves could yield satisfactory heavy atom positions in practical cases of protein investigations. In the case of exceptionally good isomorphism, even small changes in scattering caused by replacement of one atom by another of slightly higher atomic number – say, phosphorus by arsenic – can be easily established

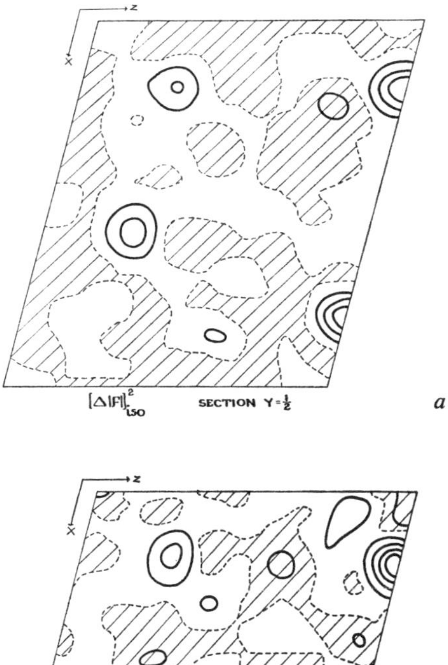

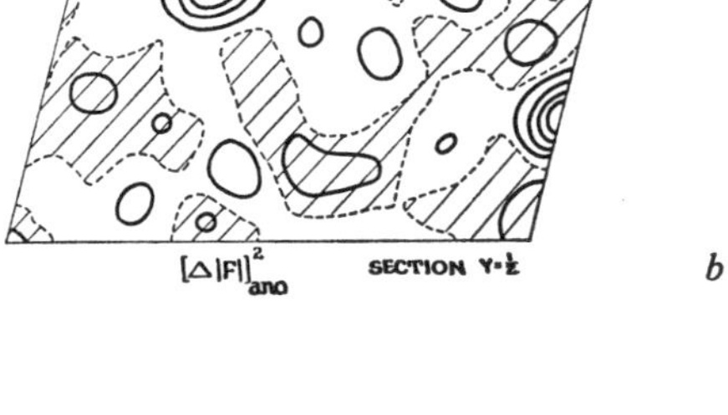

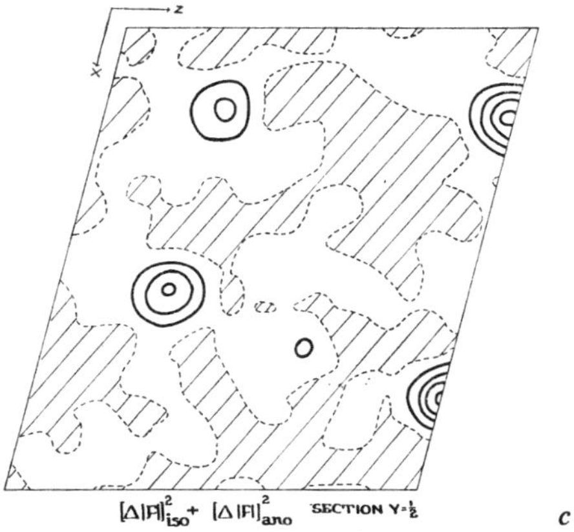

Fig. 3. *Sections of the three-dimensional Patterson function at* $y=\frac{1}{2}$ *for the cis-diglycine platinum derivative of ribonuclease, with* 3 Å *data* (2400 *independent reflections*). (*a*) *The coefficients are* $(\Delta|F|_{iso})^2$ (*caused by isomorphous substitution*); *two of the main peaks are Harker peaks and the third is a non-Harker peak.* (*b*) *The coefficients are* $(\Delta|F|_{ano})^2$ (*caused by anomalous scattering*). (*c*) *The coefficients are* $(\Delta|F|_{iso})^2+(\Delta|F|_{ano})^2$.

(Kartha, Bello & Harker, 1968) from the $(\Delta|F|_{iso})^2$ maps.

More detailed study of these maps reveals other positive and negative excursions in both maps and also the fact that positive features in one correspond to negative features in the other, as predicted by the theory. Hence, a combined map, shown in Fig. 3(*c*), was made using coefficients:

$$(\Delta|F|_{iso})^2+(\Delta|F|_{ano})^2 \simeq |F_H|^2 \qquad (24)$$

Here, $\Delta|F|_{ano}$ coefficients in Eq. (24) are obtained by suitably scaling the anomalous scattering amplitude differences $(|F(+)|-|F(-)|$ by a scaling factor k. These combined maps, in addition to reinforcing features of the individual maps revealing heavy atom vectors, are also relatively free from other background peaks. Similar maps combining the data from IR and AS have been used with other derivatives of ribonuclease, as well as in other proteins (Matthews, 1966). The usefulness of some of these techniques in locating the heavy atom positions has also been established in cases of molecules of smaller size.

Correlation of heavy atom positions in different isomorphous derivatives

The relative arrangements of the heavy atoms in a single derivative can be established from vector maps using the methods described above. However, in many space groups, the choice of origin is not unique in structures derived from vector maps. The application of the MIR method requires that all the heavy atom coordinates are given with the same origin of coordinates fixed with respect to the protein molecule. Hence, it is necessary to refer all these heavy atom coordinates to a common origin in the chosen unit cell. The problem may be further complicated in some space groups, like $P2_1$ and $P4$, where the choice of origin may also involve a variable parameter. Various methods are available for doing this. Here again, it is seen that by combining data from IR and AS methods, we get vector maps that are least subject to unwanted background.

However, the most direct method for solving the origin correlation problem is to avoid it by establishing the heavy atom positions, if possible, in second and later heavy atom derivatives, with respect to the origin chosen for the first derivative. This is easily achieved, with the single isomorphous replacement (SIR) method, in conjunction with the anomalous scattering information. In this, the protein phase angles α_P are established without ambiguity from the single derivative. This is possible, in principle, since $\Delta|F|_{\text{iso}}$ and $\Delta|F|_{\text{ano}}$ give the components of the $\mathbf{F}_{PH}$ vector along two directions, giving a complete description of the $\mathbf{F}_{PH}$ vector, including its phase α_{PH}, from a knowledge of the $\mathbf{F}_H$ vector. Conversely, once we have a set of starting phases, α_P, for the protein with respect to a given origin – which refers to the first $\mathbf{F}_H$ vector – a complete description of another $\mathbf{F}_H$ vector in another derivative is easily obtained. From Eqs. (10) and (19) we have,

$$\Delta|F|_{\text{iso}} = \text{Real part of } |F_H| \exp i(\alpha_{PH_2} - \alpha_{H_2})$$

$$i\Delta|F|_{\text{ano}} = \text{Imaginary part of } |F_H| \exp i(\alpha_{PH_2} - \alpha_{H_2})$$

and hence

$$\mathbf{F}_{H_2} = (\Delta|F|_{\text{iso}} - i\Delta|F|_{\text{ano}}) \exp(+i\alpha_{PH_2})$$

If we assume that, for the majority of reflections, the phase angles of $\mathbf{F}_{PH}$ and $\mathbf{F}_P$ are not widely different, we get

$$\alpha_P \simeq \alpha_{PH1} \simeq \alpha_{PH_2}$$

and

$$\mathbf{F}_{H_2} \simeq [\Delta|F|_{\text{iso}} - i\Delta|F|_{\text{ano}}] \times \exp(+i\alpha_P) \quad (25)$$

Eq. (25) shows that for any heavy atom derivative, a map can be computed with coefficients on the right of Eq. (25), which can be split into $\Delta|F|_{\text{iso}}$ and $\Delta|F|_{\text{ano}}$ parts. These are components of $\mathbf{F}_H$ parallel and perpendicular to the phase of α_P of the protein and may be termed π and σ components respectively. The map so obtained with knowledge of α_P, as well as the π and σ components of $\mathbf{F}_H$, gives us the locations of the heavy atoms in the derivative in question. These positions are given with respect to an origin fixed in the protein and used for evaluating the protein phases with the first derivative data. From Eq. (25), it can also be shown that maps computed using either $\Delta|F|_{\text{iso}}$ coefficient (π component) or the $\Delta|F|_{\text{ano}}$ coefficient (σ component) alone still yields heavy atom positions with the correct origin but on a somewhat fluctuating background. However, these maps, being electron density maps, give us the simple fundamental set of heavy atoms, in contrast to the vector sets obtained earlier. Therefore, these maps usually cause no difficulties in interpretation. In practice, the initial protein phases obtained by the SIR method, even in the absence of the anomalous scattering method, or vice versa, are usually satisfactory as starting phases for correlating origins of heavy atom configurations in different derivatives. As data from more than one derivative become available, estimates of α_P become better. The difference electron density maps computed using these refined phases are usually satisfactory, not only in correlating the origins, but also in refining the positional and occupancy parameters, picking up heavy atom substitutions of lower occupancy, and showing cases of departures from perfect isomorphism in the parent protein. Because of the simplicity and directness of the difference electron density method of avoiding the correlation problem, other vector methods are not considered here any further except to mention that independent checks can be obtained by computing a correlation vector map if one so desires.

However, if two sets of phase angles, α_{P1} and α_{P2}, are obtained for the protein, with heavy atoms referred to different origins, then an origin correlation function (Kartha, 1967) can be computed with coefficients

$$|F_P|\omega \times \exp(i\alpha_{P1}) \times \exp(-i\alpha_{P_2})$$

which gives a single large positive peak, correlating the second origin with the first. Such a map may be useful in comparing the results of two independent phase angle determinations of the same protein at a resolution that is insufficient to establish the relative origins by comparison of features of the low resolution electron density model based on those phases.

Refinement of heavy atom parameters

For accurate evaluation of the protein phase angles by the MIRAAS method, it is essential to have an accurate estimate of the substituted heavy atom parameters. Assuming that the scattering amplitudes from the various derivatives have been placed on the same scale, the main parameters to be refined for each derivative are:

1. Number and locations of the heavy atom substitutions,
2. their thermal parameters, and
3. occupancy factors.

As in classical crystallography, two main methods are used for parameter refinement. The choice of the method depends on the status of the existing list of parameters and computational convenience. However, in the early stages of refinement, the first method appears preferable, even if it is less convenient from the computational point of view. The two methods are discussed below.

A. Fourier refinement

If a set of initial protein phases is available, even though they may still be subject to appreciable errors, an electron density representation for the heavy atoms can be obtained by the difference-Fourier method indicated in Eq. (25). The refinement of the heavy atom parameters can be achieved through alternate cycles of difference-Fourier maps and protein phase re-evaluation with refined parameters. As this is dealt with in greater detail in a later section on phase angle evaluation in proteins by Dr. B. W. Matthews, this method of parameter refinement is not discussed further here.

B. Least squares refinement

Even after obtaining approximate parameters for the heavy atoms, it is not possible to perform the usual least-squares refinement of these parameters against any observed data in the case of the IR method. This is due to the fact that we need information about the protein phase angles α_P in order to find a measured quantity $|F_H|_{obs}$, against which the calculated $|F_H|_{calc}$ can be compared. Hence, in some of the earlier attempts at heavy atom parameter refinement (Hart, 1961) only the centrosymmetric reflections were used. In such a case

$$|F_H|_{obs} = ||F_{PH}| \pm |F_P|| \qquad (26)$$

and the sign giving the smaller difference between $|F_H|_{obs}$ and $|F_H|_{calc}$ is chosen. This method, in addition to using only a very small part of the data, also has the disadvantage of being unable to refine some of the parameters, such as the y parameters in space group $P2_1$.

Here again, by combining IR and AS data, it is possible to provide the unknown phase angle information. With these, we can evaluate the magnitude of the heavy atom vector $\mathbf{F}_H$ that plays the part of the quantity $|F_H|_{obs}$ to be compared with the calculated length of heavy atom vector $|F_H|_{calc}$. From Fig. 1 we can write

$$\begin{aligned} |F_H|^2 &= |F_{PH}|^2 + |F_P|^2 - 2|F_{PH}|\,|F_P| \times \cos(\alpha_{PH} - \alpha_P) \\ &= (|F_{PH}| - |F_P|)^2 + 2|F_{PH}|\,|F_P| \times \{1 - \cos(\alpha_{PH} - \alpha_P)\} \\ &= (\Delta|F|_{iso})^2 + 2|F_{PH}|\,|F_P| \times [1 - \{1 - \sin^2(\alpha_{PH} - \alpha_P)\}^{1/2}] \qquad (27) \end{aligned}$$

The first term $\Delta|F|_{iso}$ is a measured quantity. It is the second part that involves the phase angles α_P and α_{PH}. However, from Eq. (19) we get

$$\sin(\alpha_H - \alpha_{PH}) = -(\Delta|F|_{ano})/|F_H| \qquad (28)$$

and further from Fig. 1

$$\sin(\alpha_{PH} - \alpha_P)/|F_H| = \sin(\alpha_H - \alpha_{PH})/|F_P| \qquad (29)$$

Combining Eqs. (28) and (29), we get

$$\sin(\alpha_{PH} - \alpha_P) = -(\Delta|F|_{ano})/|F_P| \qquad (30)$$

Eq. (30) gives $\sin(\alpha_{PH} - \alpha_P)$ in terms of measured quantities $\Delta|F|_{ano}$ and $|F_P|$. From Eq. (18), it is seen that $\Delta|F|_{ano}$ is obtained by rescaling the difference between $|F(+)|$ and its Friedel conjugate $|F(-)|$ by a constant factor $k = (f'_H/2f''_H)$. In practice, it is customary to consider this scaling factor k also as a variable parameter, to be adjusted for the best overall fit between the observed and calculated data.

From Eqs. (27) and (30) we get the length of the heavy atom scattering vector $|F_H|$ in terms of experimentally measured quantities, except for an ambiguity in the sign of $\cos(\alpha_{PH} - \alpha_P)$. If we make the reasonable assumption that in the protein derivatives, $(\alpha_{PH} - \alpha_P)$ rarely exceeds $\pi/2$, then the positive square root is always implied in Eq. (27). This gives us an equation

$$\{|F_H|_{obs}\}^2 = (\Delta|F|_{iso})^2 + 2|F_{PH}|\,|F_P| \times [1 - \{1 - (\Delta|F|_{ano}/|F_P|)^2\}^{1/2}] \qquad (31)$$

giving the length of the heavy atom vector in terms of measured quantities. In most cases of proteins, we can further make the assumption that $|F_H|$ is small compared to $|F_P|$ and $|F_{PH}|$ and for all practical purposes replace $|F_{PH}|$ by $|F_P|$ in Eq. (31). This gives the simpler expression

$$(|F_H|_{obs})^2 \simeq (\Delta|F|_{iso})^2 + (\Delta|F|_{ano})^2 \qquad (32)$$

derived earlier in Eq. (24). In an actual case of a protein derivative studied, the use of the simpler form (Eq. 32) for $|F_H|_{obs}$ for refinement purposes gave the same results as the more accurate expression given in Eq. (31).

Some results of least squares refinement in ribonuclease

In the study of the protein ribonuclease (Kartha, 1965) the six anisotropic thermal parameters were refined in addition to the three positional parameters for the heavy atoms. The occupancies of the heavy atoms were refined by the study of the heavy-atom electron-density difference-Fourier maps, computed with the last set of protein phases calculated by the MIRAAS method. A number of alternating cycles of phase evaluation and Fourier refinement of heavy atom parameters were performed. In computing the difference electron-density maps for parameter refinement, maps with the π and σ components of $\mathbf{F}_H$, as well as combined maps, were calculated, as indicated in an earlier section.

Two types of least-squares refinement were carried out. In the first, $\Delta|F|_{ano}$ was assumed zero in evaluating $|F_H|_{obs}$, which is then composed of only the π component. In the second refinement, $\Delta|F|_{ano}$ was also included, giving us the full length for the heavy atom vector.

Table I gives the results of four cycles of least-squares runs for the heavy atom parameters in five derivative crystals. Some comments on the results in this case may be worthwhile.

The discrepancies between the observed and calculated amplitudes are indeed very large compared to the usual standards applied in small-structure analysis. This is mainly due to the lack of perfect isomorphism between the parent protein and the derivative crystal, and also to the fact that $|F_H|_{obs}$ is computed from small differences in large measured quantities, and are thus liable to larger percentage errors. However, it is seen that the R values are consistently 10 to 15% lower when the data from the anomalous scattering were also included in the refinement. Further, when the number of parameters refined is small compared to the number of reflections, as in the present case, both refinements gave practically the same positional parameters. This was repeatedly tested by deliberately shifting the input co-ordinates from their correct positions and then subjecting them to the least squares refinement with both sets of data. It was noticed that with shifts less than one-third the data resolution both refinements converged at about the same rate to the correct value. However, the thermal parameters obtained with the set which included the anomalous scattering data converged faster and appeared more reasonable.

Concluding remarks

In order to reconstruct the electron-density distribution from measured diffraction data, we need the relative phases of the reflections. This information is not present in the data. In other words, we need, not the length of the $\mathbf{F}$ vector, but its components A and B along the real and imaginary axes for evaluating the density distribution. However, there is no sanctity about the real and imaginary axes normally used in forming these components, π and σ (or A' and B') of $\mathbf{F}$ along any two non-collinear axes should be equally satisfactory, provided their directions are known. In fact, it is not even necessary to have these axes fixed for all $\mathbf{F}$ vectors

Table I. Discrepancy factors in heavy atom parameters refinement in ribonuclease

Ribonuclease data to 4 Å was used. Set 1 did not include the anomalous scattering information in evaluating $|F_{obs}|$ although the set 2 did. R values are given as percentages. Each derivative contained two major heavy atom sites per molecule. Over 1000 reflections within the 4 Å sphere were included in this refinement.

Derivative no.	R with set 1, π component only. No. of cycle				R with set 2, Both π and σ components. No. of cycle			
	1	2	3	4	1	2	3	4
1	63·2	59·6	53·1	52·9	56·9	41·7	36·4	36·3
2	61·2	60·5	53·3	53·3	56·4	49·8	43·0	42·8
3	57·6	57·4	52·7	52·8	49·6	47·7	39·4	39·4
4	67·6	63·6	60·1	59·6	63·2	49·7	48·0	47·3
5	63·5	59·9	53·2	53·0	59·8	45·6	42·0	41·8

provided they vary from reflection to reflection in a known manner. Knowledge of π and σ components parallel to these axes gives us a full description of **F** and enables us to obtain A and B components required in computing the electron-density maps.

Looking at the phase problem this way, it is immediately seen from Eqs. (10) and (19) that the isomorphous series amplitude difference $\Delta|F|_{\text{iso}}$ gives the (π) component of $\mathbf{F}_{PH}$ along $\mathbf{F}_H$ while the anomalous scattering difference $\Delta|F|_{\text{ano}}$ gives the (σ) component of the same $\mathbf{F}_{PH}$ perpendicular to $\mathbf{F}_H$. Further, $\Delta|F|_{\text{iso}}$ and $\Delta|F|_{\text{ano}}$ together also provide a complete description of the vector $\mathbf{F}_H$ for all the reflections. Hence we can write

$$\alpha_{PH} = \alpha_H + \omega$$
$$A_{PH} = |F_{PH}| \cos(\alpha_{PH} + \omega)$$
$$B_{PH} = |F_{PH}| \sin(\alpha_{PH} + \omega)$$

where ω ($= \alpha_{PH} - \alpha_H$) is the phase advance of $\mathbf{F}_{PH}$ over $\mathbf{F}_H$ and is obtained from Eqs. (10) and (19). The components A_{PH} and B_{PH} are the coefficients used in Fourier synthesis.

Thus, in addition to establishing the $\mathbf{F}_H$ vector completely, the isomorphous series and anomalous scattering differences also give the two components of $\mathbf{F}_{PH}$; π and σ. The π component is along the in-phase component $\mathbf{F}'_H$ of $\mathbf{F}_H$ and the σ component is along the out-of-phase component $\mathbf{F}''_H$ of $\mathbf{F}_H$. These two directions are not always mutually perpendicular if there are different types of anomalous scatterers, but this in itself presents no basic problems. The isomorphous pair differences give $\cos(\omega)$ and hence two ambiguous values $\pm\omega$; while the anomalous scattering differences provide $\sin(\omega)$ and hence two ambiguous values ω and $\pi - \omega$. Together they provide an unambiguous solution for ω, the phase advance of $\mathbf{F}_{PH}$ over $\mathbf{F}_H$. If the $\mathbf{F}_H$ vector is known for each reflection by a complete knowledge of the heavy atom arrangement, *including its absolute configuration*, then here we have a complete and straightforward experimental solution of the phase problem.

Acknowledgements

The discussion of the isomorphous series and anomalous scattering method in structure analysis of large molecules reported here is based on the research and publications of many workers in this field, even though viewed from a somewhat personal point of view. To all these sources my thanks are due. No attempt has been made here to make an exhaustive survey of the literature on the subject, and the bibliography quoted is far from complete. To my colleagues, Professor David Harker and Dr. R. Parthasarathy my thanks are due for many stimulating discussions on the topics detailed here. It is also a pleasure to record many thanks for the support of this work by the National Science Foundation, National Institute of Health, and the New York State Department of Health.

References

Blow, D. M. (1958). *Proc. Roy. Soc.* **A247**, 302.
Bragg, W. L. (1958). *Acta Cryst.* **11**, 70.
Coster, D., Knol, K. S. & Prins, J. (1920). *Z. Physik.* **63**, 345.
Dauben, C. H. & Templeton, D. H. (1956). *Acta Cryst.* **8**, 685.
Dickerson, R. E. (1964). *The Proteins,* Vol. II. Ed. Neurath. New York: Academic Press.
Green, D. W., Ingram, V. M. & Perutz, M. F. (1956). *Proc. Roy. Soc.* A**225**, 287.
Harker, D. (1956). *Acta Cryst.* **9**, 1.
Hart, R. G. (1961). *Acta Cryst.* **16**, 1196.
Kartha, G. (1961). *Acta Cryst.* **14**, 680.
Kartha, G. (1965). *Acta Cryst.* **19**, 883.
Kartha, G. (1967). Unpublished.
Kartha, G. (1969). Under preparation.
Kartha, G., Bello, J., Harker, D. & Dejarnette, F. E. (1963). *Aspects of Protein Structure,* ed. Ramachandran. New York: Academic Press.
Kartha, G., Bello, J. & Harker, D. (1968). *Structural Chemistry and Molecular Biology.* San Francisco: W. H. Freeman.

Kartha, G. & Parthasarathy, R. (1965). *Acta Cryst.* **18**, 745.
Kartha, G. & Parthasarathy, R. (1965). *Acta Cryst.* **18**, 749.
Mathews, B. W. (1966). *Acta Cryst.* **20**, 230.
Okaya, Y. & Pepinsky, R. (1961). *Computing Methods and the Phase Problem.* London: Pergamon Press.
Okaya, Y., Saito, Y. & Pepinsky, R. (1955). *Phys. Rev.* **98**, 1857.
Perutz, M. F. (1956). *Acta Cryst.* **9**, 867.
Phillips, D. C. (1966). *Advanc. Struct. Res.* **2**, 75.
Ramachandran, G. N. & Raman, S. (1959). *Acta Cryst.* **12**, 957.
Rossmann, M. G. (1960). *Acta Cryst.* **13**, 221.
Rossmann, M. G. (1961). *Acta Cryst.* **14**, 383.
Steinrauf, L. (1963). *Acta Cryst.* **16**, 317.

Discussion

HARDGROVE: How are the roles of occupancy factor and the temperature factor kept separate?

KARTHA: This is not an easy task as we find strong correlations between the two. One possibility may be study of variation of residual as a function of scattering angle. We have not been too critical about the individual effects as long as their combined effect on scattering change can be predicted reasonably well.

Determination and Refinement of Phases for Proteins

B. W. MATTHEWS

Institute of Molecular Biology, University of Oregon, Eugene, Oregon, U.S.A.

The number of proteins whose structures have been 'solved' by X-ray methods is now rapidly approaching a dozen. In every case the structure determination has been achieved through application of the method of 'isomorphous replacement' with heavy atoms, first applied to protein crystals by Green, Ingram & Perutz (1954). The idea that the phases of X-ray reflections from a protein crystal could be obtained by the introduction into the crystal of heavy atoms is not new, being suggested by J. D. Bernal in 1939. (Sir Lawrence Bragg, 1965, has given a fascinating account of the first stages in the X-ray analysis of proteins.) Nevertheless it has taken a number of years for the method of isomorphous replacement to evolve into the working technique which is of such prominence in crystallographic studies of proteins. The isomorphous replacement method was used as early as 1927 by Cork in studying the alums. Bokhoven, Schoone & Bijvoet (1951) subsequently extended the method to the study of a noncentrosymmetric projection of strychnine sulphate, using what would now be termed the method of single isomorphous replacement. They also suggested that by using a double isomorphous replacement a unique phase determination would be obtained, even for noncentrosymmetric reflections. The details of the double (or multiple) isomorphous replacement method were worked out by Harker (1956) who introduced the very useful concept of phase circles. Another contribution which has been of great practical value, and which will provide the basis for much of the subsequent discussion, is the method introduced by Blow & Crick (1959) for the treatment of errors in the isomorphous replacement method. In addition to the determination of protein phases by the method of substitution with heavy atoms; it has been shown possible to supplement this information by utilizing the anomalous scattering of the substituent atoms (Bijvoet, 1954; Okaya & Pepinsky, 1960; Ramachandran & Raman, 1956). The first application of the anomalous scattering method to protein crystallography was by Blow (1958) who used the anomalous scattering of the iron atoms to determine phase information for a noncentrosymmetric projection of horse oxyhaemglobin. Subsequent development and application of the method to protein crystallography will be discussed later.

The isomorphous replacement method

Consider a protein crystal with corresponding structure factors $\mathbf{F}_P(\mathbf{h})$ and an isomorphous heavy atom derivative, *i.e.* a modified crystal in which heavy atoms occupy specific sites throughout the crystal, but which is in all other respects identical with the unsubstituted 'parent' crystal. (*Note*: Structure *amplitudes* are indicated by italic type, *e.g.*, F_P, and *vectors* by bold-face type, *e.g.*, $\mathbf{F}_P$.) Let the structure factors of the isomorph be $\mathbf{F}_{PH}(\mathbf{h})$, and of the heavy atoms $\mathbf{F}_H(\mathbf{h})$. In practice one can measure the structure *amplitudes* F_P and F_{PH}, and it is desired to obtain from these observable quantities the value of the phase angles $\varphi(\mathbf{h})$ of $\mathbf{F}_P(\mathbf{h})$ so that a Fourier synthesis of the protein structure may be calculated. It will be assumed that the positions and occupancy of the sites of heavy atom binding have been determined as accurately as possible (G. Kartha, preceding paper).

From the heavy atom parameters, the corresponding structure factor $\mathbf{F}_H(\mathbf{h})$ is

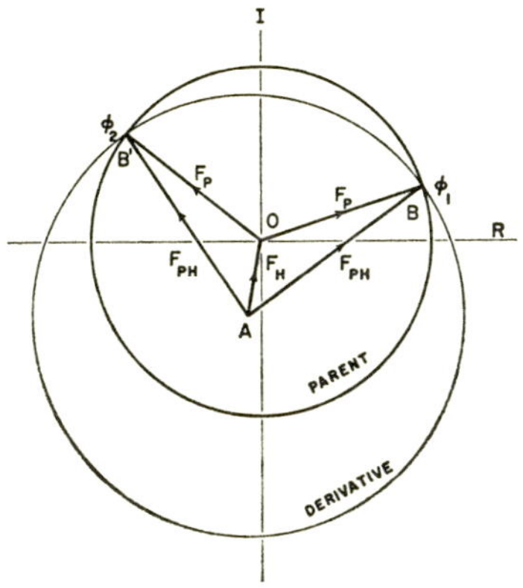

Fig. 1(*a*). *Harker construction for a single isomorphous replacement.* φ_1 *and* φ_2 *are the 'most probable' phases for* F_P.

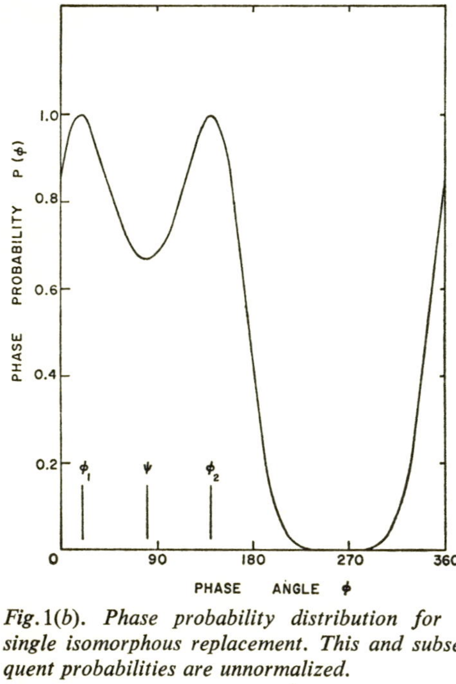

Fig. 1(*b*). *Phase probability distribution for a single isomorphous replacement. This and subsequent probabilities are unnormalized.*

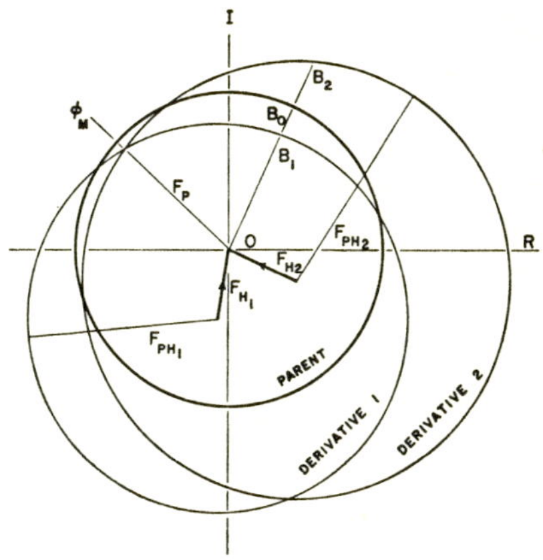

Fig. 2(*a*). *Harker construction for a double isomorphous replacement.* $F_P = 540$, $F_{PH1} = 680$, $F_{PH2} = 730$, $F_{H1} = 230$, $\psi_1 = 80°$, $F_{H2} = 250$, $\psi_2 = 156°$. φ_M *is the 'most probable' phase for* F_P.

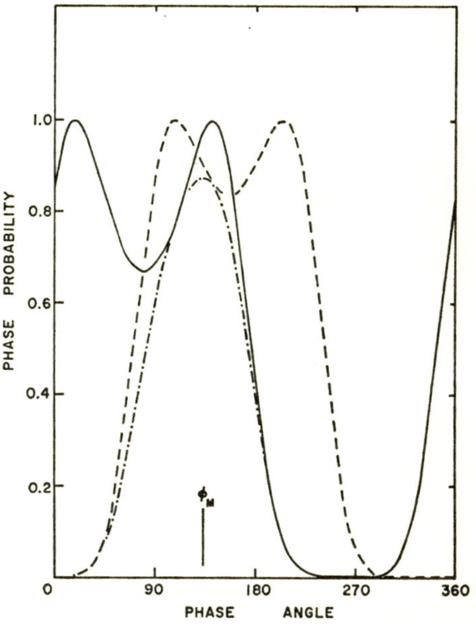

Fig. 2(*b*). *Phase probability distribution corresponding to the double isomorphous replacement shown in Fig.* 2(*a*). *The curve for derivative* 1 *is solid, for derivative* 2 *is broken, and for the combined distribution is drawn as a chain line.* $E_1 = E_2 = 100$. *For this phase determination the most probable phase* $\varphi_M = 134°$, *the best phase* $\varphi_B = 127°$, *and the figure of merit* $m = 0{\cdot}82$.

calculated. To determine φ, the phase of $\mathbf{F}_P(\mathbf{h})$, we construct a set of phase circles, as proposed by Harker (1956). From a chosen origin O [Fig. 1(*a*)] the vector OA is drawn equal to $-\mathbf{F}_H$. Circles of radius F_P and F_{PH} are then drawn about O and A respectively. The intersections of the phase circles at B and B' define two possible phase angles for F_P. Note that the angles are symmetrical about $\mathbf{F}_H$. This ambiguity may in principle be resolved in two ways (A) by using a second heavy atom isomorphous derivative or (B) by utilizing the anomalous scattering effects for the first isomorph. In practice it may be desirable to use a combination of both techniques.

The method of multiple isomorphous replacement

The phase information provided by a second isomorph is illustrated in Fig. 2(*a*). In theory the three phase circles will intersect at a point, and the phase ambiguity will be resolved. In practice there will be errors in the observed amplitudes F_P and F_{PH}, and in the heavy atom parameters (and thus in $\mathbf{F}_H$). Also the isomorphism may be imperfect. As a result the intersections of the phase circles will not coincide exactly. Another complication arises from the fact that for reflections where $\mathbf{F}_H$ is small, the phase determination will become indeterminate. A method of taking all these factors into account has been proposed by Blow & Crick (1959).

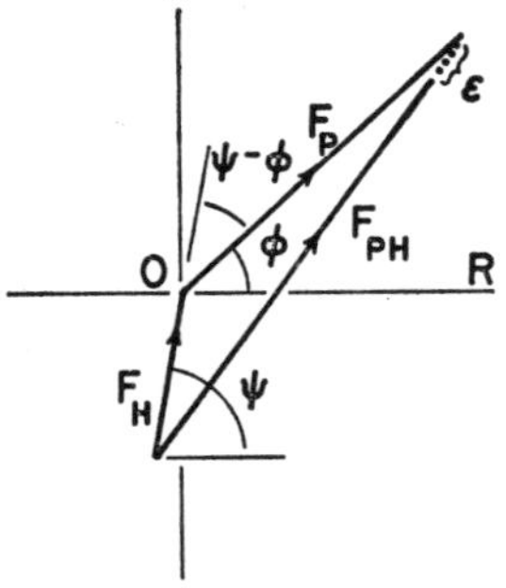

Fig. 3. Vector diagram illustrating the lack of closure, ε, of an isomorphous replacement phase triangle.

The method, or variants of it, has become virtually the standard method for phase determination of proteins, and will therefore be discussed in some detail.

The method of Blow and Crick

Blow and Crick point out that in practice the phase angle φ can never be determined with complete certainty. Rather, there is a finite probability that any arbitrary phase angle may be the correct one. Consider the vector diagram shown in Fig. 3, in which $\mathbf{F}_H$ is known and we wish to determine the probability $P(\varphi)$ that the arbitrary phase angle φ is the correct phase of $\mathbf{F}_P$. Strictly, one should allow for the possibility of errors in $\mathbf{F}_H$, F_P, and F_{PH}, and should consider the probability that the vector $\mathbf{F}_P$ occupies all possible positions in the Argand diagram. However, Blow and Crick suggest that the analysis may be considerably simplified by assuming that F_P and $\mathbf{F}_H$ are known accurately and that all the error lies in the observation of F_{HP}. In other words it is assumed that the vector $\mathbf{F}_P$ must lie on the circle of radius F_P, and the probability distribution of F_P is evaluated as a function of φ only.

For an arbitrary phase angle φ, the phase triangle (Fig. 3) will not close exactly. If we define F_C to be the vector sum of $\mathbf{F}_H$ and $F_P \exp(i\varphi)$, then the lack of closure of the phase triangle is given by

$$\varepsilon = F_C - F_{HP} \tag{1}$$

Clearly

$$F_C = \{F_P^2 + F_H^2 + 2F_P F_H \cos(\psi - \varphi)\}^{1/2}$$
$$= \{F_P^2 + F_H^2 + 2F_P \times (A_H \cos\varphi + B_H \sin\varphi)\}^{1/2} \tag{2}$$

where

$$\mathbf{F}_H = F_H \exp(i\psi) = A_H + iB_H \tag{3}$$

Following Blow and Crick, if E is the r.m.s. error associated with the measurements, and the distribution of error is assumed to be Gaussian, then the probability $P(\varphi)$ of the phase φ being the true phase is

$$P(\varphi)=N \exp(-\varepsilon^2/2E^2) \quad (4)$$

where N is a normalizing factor such that the sum of all probabilities is unity, *i.e.*

$$\int_0^{2\pi} P(\varphi)d\varphi=1 \quad (5)$$

The unnormalized probability distribution corresponding to Fig. 3 [and Fig. 1(*a*)] is shown in Fig. 1(*b*). The two most probable phase angles ($\varphi=\varphi_1$, and $\varphi=\varphi_2$) are the alternative phases of F_P for which the phase triangle is closed.

The individual probability distributions $P_2(\varphi)$, $P_3(\varphi)$, ... for the other isomorphs are derived in an analogous manner and may be combined to give the overall probability distribution

$$P(\varphi)=\Pi P_i(\varphi)=N \exp[-(\Sigma\varepsilon_i^2/2E_i^2)] \quad (6)$$

The joint probability distribution corresponding to Fig. 2(*a*) is shown in Fig. 2(*b*), and in this case the most probable phase is that which simultaneously best fits the observed data for the two isomorphous derivatives.

The main objection which may be made to the Blow and Crick treatment is that it assumes that there is no error in F_P. In an attempt to remove this objection Cullis, Muirhead, Perutz, Rossmann & North (1961*a*) proposed a method in which the lack of closure ε for each derivative (including the parent) is evaluated not from the 'error-free' circle of radius F_P, but rather from the 'mid-points' of all the observations. The method is illustrated by Fig. 2(*a*), in which the lack of closure for the ith derivative is defined to be $[OB_i(\varphi)-OB(\varphi)]$, where OB is the mean of all the OB_i, including the parent. From a computational point of view the method suffers from the disadvantage that reflections for which the origin of the Argand diagram lies outside one or more of the phase circles have to be treated as a special case. Reference can be made to Cullis *et al.* (1961*a*) and to Holmes (1965) for details. With the possible exception of structure determinations where data are available for a large number of isomorphs (*e.g.* more than four), the method of Cullis *et al.* will give results very similar to that of Blow and Crick, and most workers prefer to use the original Blow-Crick formulation.

The best Fourier

A protein crystallographer desires to obtain a Fourier synthesis which he can interpret in terms of an atomic model of the structure. It is by no means obvious what form the Fourier coefficients should take to facilitate this goal. One synthesis which could be calculated is the 'most probable Fourier', obtained by choosing the value of $F_P(\mathbf{h})$ for each reflection which corresponds to the highest value of $P(\varphi)$. Blow and Crick point out that although this Fourier is the most likely to be correct, it has certain disadvantages. In the first place it might tend to give too much weight to certain phases and, in the second place, for cases where $P(\varphi)$ is bimodal there is a strong chance of making a large error in the phase angle. Blow and Crick suggest that in cases such as this a compromise is needed, and that the centroid of the phase probability distribution provides just the required compromise. They show that the corresponding synthesis is the 'best Fourier', which is defined to be *that Fourier transform which is expected to have the minimum mean square difference from the Fourier transform of the true F's when averaged over the whole unit cell.*

Comparisons of the 'best' and 'most probable' Fouriers have been made by Dickerson, Kendrew & Strandberg (1961) and by Cullis *et al.* (1961*b*). The respective Fourier syntheses were found to be very similar, although in both cases the 'best' Fourier was the one preferred.

The centroid of the phase probability distribution may be defined as a point on the phase diagram with polar coordinates (mF_P, φ_B) where φ_B is the 'best' phase angle. Taking the weighted mean of all possible values $F_P \exp(i\varphi)$ for the protein structure factor, we have

$$mF_P \exp(i\varphi_B) = \frac{\int F_P \exp(i\varphi)P(\varphi)d\varphi}{\int P(\varphi)d\varphi} \tag{7}$$

Thus

$$m \cos \varphi_B = \frac{\int P(\varphi) \cos \varphi \, d\varphi}{\int P(\varphi) \, d\varphi} \tag{8}$$

and

$$m \sin \varphi_B = \frac{\int P(\varphi) \sin \varphi \, d\varphi}{\int P(\varphi) \, d\varphi} \tag{9}$$

The quantity m, which acts as a weighting factor for F_P, is called the 'figure of merit' of the phase determination. Its magnitude, between 0 and 1, is a measure of the reliability of the phase determination. Dickerson, Kendrew & Strandberg (1961) show (following Blow and Crick) that the mean square error in electron density of the 'best, Fourier synthesis is simply related to the figures of merit by the equation

$$\overline{\Delta\varrho^2} = \frac{2}{V^2} \sum_{h=0}^{\infty} \sum_{k=-\infty}^{\infty} \sum_{l=-\infty}^{\infty} \{F_P^2(\mathbf{h})[1-m^2(\mathbf{h})]\} \tag{10}$$

This expression has been tested recently in a practical case (Cohen, Matthews & Davies, 1969). For a 5·5 Å resolution electron density map of γ-chymotrypsin, the root mean square error in electron density calculated from Eq. (10) was 0·077 e/Å^3. This agreed reasonably well with a value of 0·094 e/Å^3 derived independently from comparisons with α-chymotrypsin, suggesting that Eq. (10) can be used with confidence, provided the figures of merit are reasonably realistic.

Anomalous scattering

All heavy atoms, and in particular those used in preparing heavy atom isomorphs, give rise to significant anomalous scattering for Cu *Kα* radiation. The atomic scattering factor can be expressed as

$$\mathbf{f} = f_0 + \Delta f' + i\Delta f'' = f' + if'' \tag{11}$$

where f_0 is the normal scattering factor far from an absorption edge and $\Delta f'$ and $\Delta f''$ are the correction terms which arise due to dispersion effects. The quantity $\Delta f'$, in phase with f_0, is usually negative, and $\Delta f''$, the imaginary part, is always $\pi/2$ ahead of the phase of the real part ($f_0 + \Delta f'$). It may be noted that by using different wavelengths, the change in ($f_0 + \Delta f'$) is equivalent to a change in scattering power of the heavy atom and produces intensity differences similar to a normal isomorphous replacement, except that in this case the isomorphism is *exact*. However, the smallness of $\Delta f'$ and the inaccuracies introduced through the necessity of making measurements at different wavelengths have to date made it impossible to determine protein phases by this method of 'anomalous scattering *without phase change*' (Ramaseshan, 1964).

On the other hand, it has been shown feasible to determine the phases of protein crystals by the method of anomalous scattering *with phase change*. From here on we shall be discussing only measurements of this type, *i.e.* made with a single wavelength, and shall refer to this technique as the anomalous scattering technique. The anomalous scattering of a heavy atom is always considerably less than the normal scattering (for Cu *Kα*, $2f''/f'$ ranges from about 0·24 to 0·36), but there are several factors which tend to offset this disadvantage (*e.g.* see Blow, 1958; North, 1965).

Theory of anomalous scattering

Suppose that two isomorphous crystals are differentiated by N heavy atoms, of position $\mathbf{r}_n$, and scattering factor

$$\mathbf{f}_n = f'_n + if''_n \quad (12)$$

Then for the reflection (hkl) the calculated structure factor of the N atoms is

$$F_H(\mathbf{h}) + iF''_H(\mathbf{h}) = \sum_{n=1}^{N} f'_n(\mathbf{h}) \exp(2\pi i\mathbf{h}\cdot\mathbf{r}_n) + i\sum_{n=1}^{N} f''_n(\mathbf{h}) \exp(2\pi i\mathbf{h}\cdot\mathbf{r}_n) \quad (13)$$

On resolving $\mathbf{F}_H$ into real and imaginary components we have

$$\mathbf{F}_H(\mathbf{h}) = A_H(\mathbf{h}) + iB_H(\mathbf{h}) \quad (14)$$

where

$$A_H(\mathbf{h}) = \sum_{n=1}^{N} f'_n(\mathbf{h}) \cos(2\pi\mathbf{h}\cdot\mathbf{r}_n) \quad (15a)$$

and

$$B_H(\mathbf{h}) = \sum_{n=1}^{N} f'_n(\mathbf{h}) \sin(2\pi\mathbf{h}\cdot\mathbf{r}_n) \quad (15b)$$

Similarly

$$\mathbf{F}''_H(\mathbf{h}) = A''_H(\mathbf{h}) + iB''_H(\mathbf{h}) \quad (16)$$

where

$$A''_H(\mathbf{h}) = -\sum_{n=1}^{N} f''_n(\mathbf{h}) \sin(2\pi\mathbf{h}\cdot\mathbf{r}_n) \quad (17a)$$

and

$$B''_H(\mathbf{h}) = \sum_{n=1}^{N} f''_n(\mathbf{h}) \cos(2\pi\mathbf{h}\cdot\mathbf{r}_n) \quad (17b)$$

If the parent crystal contains anomalously scattering atoms, then the contribution of these atoms should be included in the summations for A''_H and B''_H, but not, of course, for A_H and B_H. If the heavy atoms are all of the same type, *i.e.* they all have the same ratio of f'_n/f''_n $(=k)$, then F_H and F''_H are orthogonal and $F''_H = F_H/k$.

The relation between the structure factors of the reflection (hkl) and its Friedel mate ($\bar{h}\bar{k}\bar{l}$) is illustrated in Fig. 4(a). The situation can be conveniently represented [Fig. 4(b)] by reflecting the ($\bar{h}\bar{k}\bar{l}$) diagram through the real axis onto the (hkl) diagram. In cases such as this where Friedel's law breaks down, we shall refer to the difference $\Delta_{PH} = (F_{PH+} - F_{PH-})$ as the Bijvoet difference, or simply the anomalous scattering difference. The Harker phase circles corresponding to Fig. 4(b) are shown in Fig. 5. It will be seen that as in the case of single isomorphous replacement, so with the anomalous scattering data alone, there is an ambiguous phase determination. In the absence of error, the

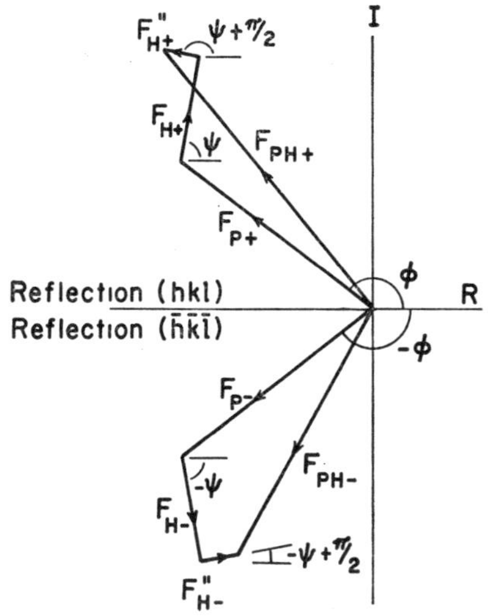

Fig. 4(a). Vector diagrams illustrating anomalous scattering for the reflections (hkl) and ($\bar{h}\bar{k}\bar{l}$).

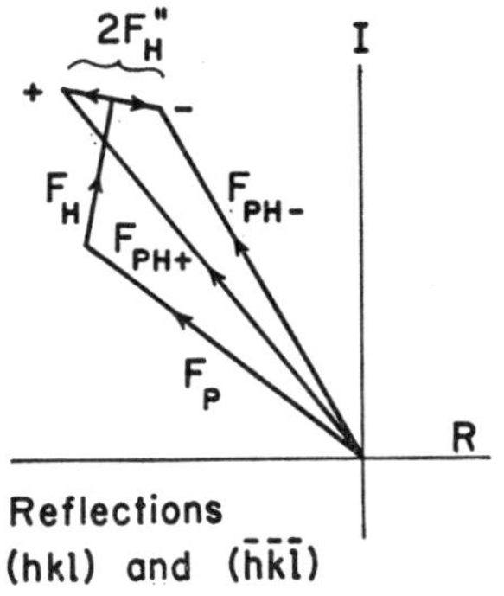

Fig. 4(b). Combined vector diagram for reflections (hkl) and ($\bar{h}\bar{k}\bar{l}$).

three phase circles (Fig. 5) will meet at a point, resolving the phase ambiguity and giving a unique solution for the phase of $\mathbf{F}_P$. The isomorphous replacement method gives phase information symmetrical about the vector $\mathbf{F}_H$, whereas the anomalous scattering phase information for F_{PH} is symmetrical about $\mathbf{F}''_H$, which, for heavy atoms of the same type, is at right angles to $\mathbf{F}_H$. In other words the two methods exactly complement each other, one method providing exactly that information which is not given by the other. The same type of complementarity may be used in locating the position of heavy atoms in protein crystals (Kartha & Parthasarathy, 1965*a*, 1965*b*; Matthews, 1966*a*).

In practice it is difficult to measure the small anomalous difference $(F_{PH+}-F_{PH-})$ accurately, especially for a protein crystal, and it is therefore unwise to attempt an exact solution for the protein phase. Rather, it is preferable to determine the phase angles by a probability method similar to the Blow-Crick method for isomorphous replacement. This approach was first used by Blow & Rossmann (1961).

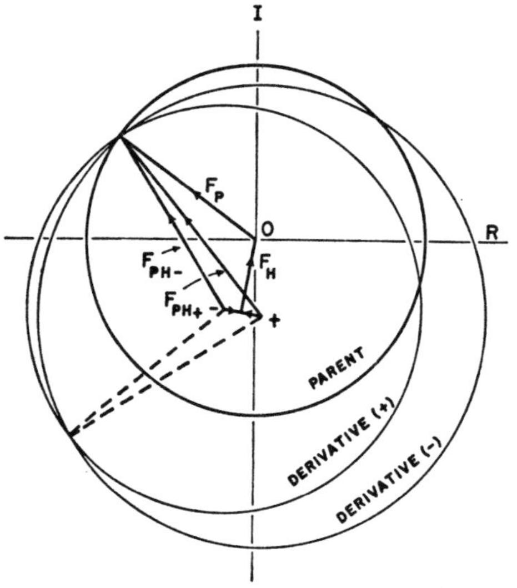

Fig. 5. Harker construction for a single isomorphous replacement with anomalous scattering, in the absence of errors.

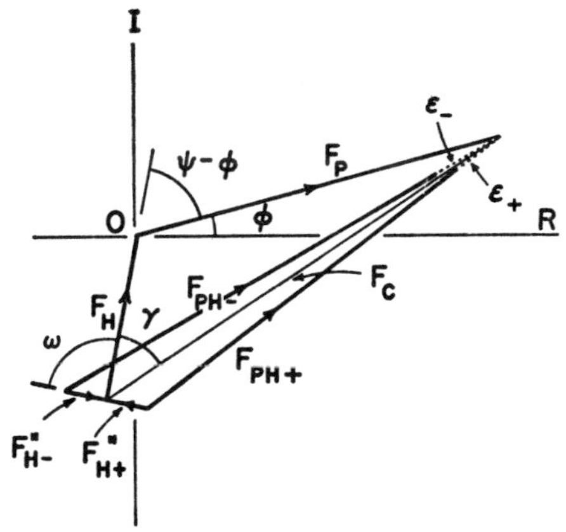

Fig. 6. Vector diagrams illustrating lack of closure in the anomalous scattering method.

Subsequently North (1965) showed how the method could be modified to take proper account of the lower errors in the anomalous scattering measurements. North's approach has been extended by Matthews (1966*b*).

The phase probability distribution for anomalous scattering

From Fig. 6 it can be seen that the most probable phase angle will be the one for which $\varepsilon_+=\varepsilon_-$. At any other phase angle there will be an 'anomalous scattering lack of closure' which we define to be $(\varepsilon_+-\varepsilon_-)$.

In practice the anomalous scattering effects are almost always small compared with the total protein scattering, and it can be assumed that $F''_H << F_{PH}$. It then follows from Fig. 6 that the anomalous scattering lack of closure is

$$\varepsilon_+-\varepsilon_-= F_{PH-}-F_{PH+}+2F''_H\cos(\gamma+\omega) \quad (18)$$

Thus, if the r.m.s. error in $(\varepsilon_+-\varepsilon_-)$ is E', and the distribution of error is assumed to be Gaussian, then from measurements of anomalous scattering the probability $P_{\text{ano}}(\varphi)$ of phase φ being the true phase of $\mathbf{F}_P$ is

$$P_{\text{ano}}(\varphi) = N \exp\left\{ -\frac{1}{2E'^2}[-\Delta_{PH} + 2F''_H \cos(\gamma + \omega)]^2 \right\} \quad (19)$$

N is the usual normalizing factor, $\Delta_{PH} = (F_{PH+} - F_{PH-})$, and the expression on the right can be conveniently evaluated by using the following relationships (Matthews, 1966*b*)

$$\begin{aligned} \sin\gamma &= [F_P(B_H \cos\varphi - A_H \sin\varphi)]/(F_C F_H) \\ \cos\gamma &= [F_H^2 + F_P(A_H \cos\varphi + B_H \sin\varphi)]/(F_C F_H) \quad (20) \\ \sin\omega &= (A_H B''_H - A''_H B_H)/F''_H F_H \\ \cos\omega &= (A_H A''_H + B_H B''_H)/F''_H F_H \end{aligned}$$

An example of an anomalous scattering phase probability distribution $P_{\text{ano}}(\varphi)$ is shown by the dotted curve in Fig. 7. The asymmetry of the distribution arises from the fact that $P_{\text{ano}}(\varphi)$ is the phase probability distribution for F_P rather than that of F_{PH} which would be symmetrical about the phase of F''_H. The overall probability distribution obtained by combining the anomalous scattering data with the previous isomorphous replacement data [Fig. 1(*b*)] is given by

$$P(\varphi) = N P_{\text{iso}}(\varphi) P_{\text{ano}}(\varphi) \quad (21)$$

and is illustrated in Fig. 7.

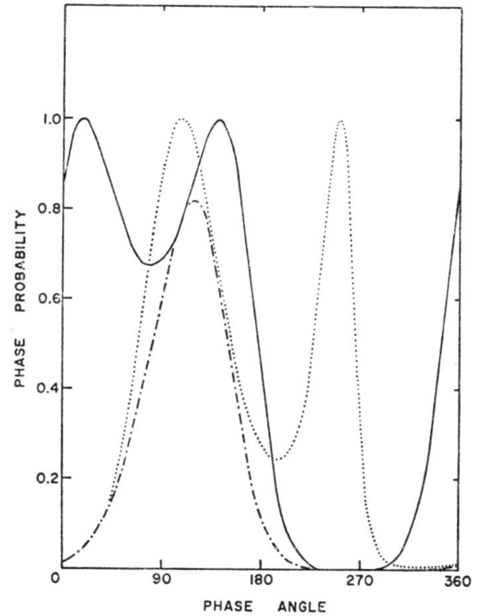

Fig. 7. Combination of isomorphous replacement and anomalous scattering phase probabilities for a single isomorphous replacement. $P_{\text{iso}}(\varphi)$ *drawn solid,* $P_{\text{ano}}(\varphi)$ *dotted, and the combined probability distribution is drawn as a chain line.* $F_P = 540$, $F_{PH+} = 700$, $F_{PH-} = 660$, $F_H = 230$, $F''_H = 62$, $\psi = 80°$, $E = 100$, $E' = 50$. *For this phase determination the most probable phase* $\varphi_M = 120$, *the best phase* $\varphi_B = 113°$, *and the figure of merit* $m = 0.82$.

Anomalous scattering without isomorphous replacement

The treatment outlined above of phase determination by anomalous scattering assumed that data were available for a parent crystal devoid of anomalous scatters and an anomalously scattering isomorphous heavy atom derivative. It is not uncommon that the 'native' protein contains atoms which scatter anomalously (*e.g.* the iron atoms in haemoglobin, or the zinc atom in carboxypeptidase) and in such cases use can be made of this effect.

The expression for the phase probability distribution can either be derived from first principles or alternatively can be obtained from Eqs. (19) and (2), by taking the limits as $F_H \to 0$, and $F_C \to F_P$. This leads to the result that

$$P_{\text{ano}}(\varphi) = N \exp\left\{ -\frac{1}{2E'^2}[-\Delta_P + 2(A''_H \cos\varphi + B''_H \sin\varphi)]^2 \right\} \quad (22)$$

where $\Delta_P = (F_{P+} - F_{P-})$, and E' is the r.m.s. error in the determination of the parent anomalous scattering difference.

The determination of absolute configuration

It should be noted that in calculating phases from anomalous scattering data it is first necessary to determine the coordinates of the heavy atoms *in their absolute configuration.* For the isomorphous replacement method, the choice of a wrong hand for the heavy atom arrangement will result in an electron density map which is the mirror image of the correct map. Such an error is easily rectified by restacking the electron density sections. In contrast, if the wrong hand is used in the anomalous scattering method, the resultant electron density map will in general bear no relation to the correct electron density. This is illustrated by Figs. 8(*a*) and 8(*b*) (Matthews, 1966*a*). The maps not only define the correct enantiomorph, but also determine the positions of the different heavy atoms relative to a common origin.

Refinement of the phase angles

To obtain the most accurate phase angles it is necessary to obtain the best possible values for the heavy atom parameters. Kartha (preceding paper) has discussed methods of refining the heavy atom parameters for a single derivative. If two or more isomorphs are available, it is possible to use their combined phase information to assist in the respective refinements of the heavy atom parameters of each derivative in turn. The method was first introduced by Dickerson, Kendrew & Strandberg (1961), and has since become increasingly popular (*e.g.*, see Kraut, Sicker, High & Freer,

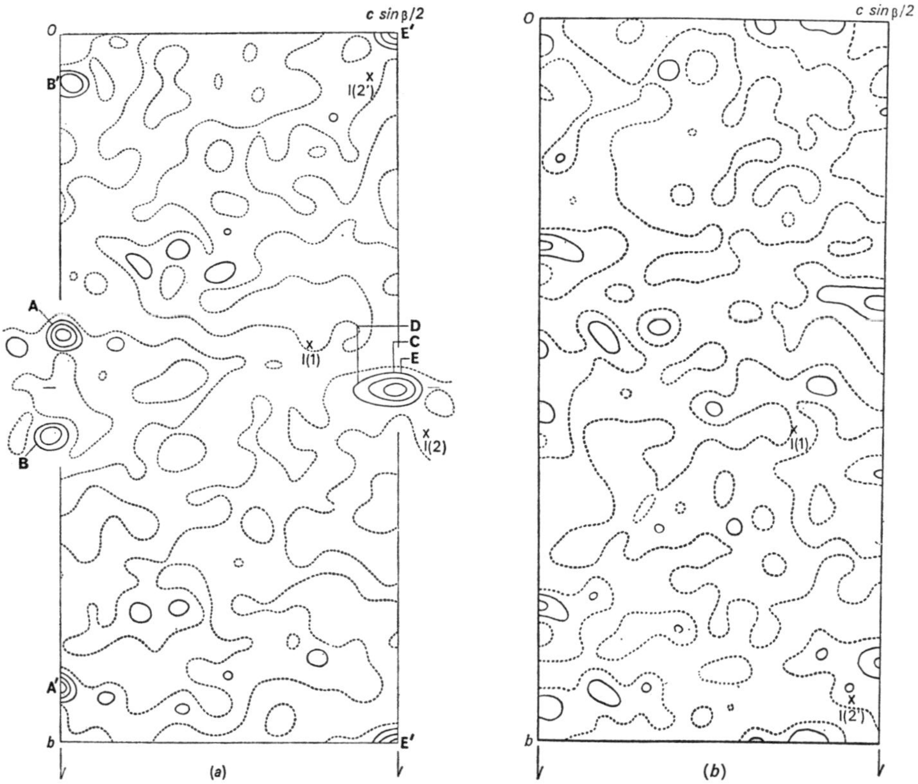

Fig. 8. *Alternative* (0*kl*) *Fourier syntheses illustrating the determination of the absolute configuration of the heavy atom arrangement in α-chymotrypsin. In both cases the amplitudes are* $(F_{PtCl_4}-F_{Native})$ *and the phase angles were determined from the combination of isomorphous replacement and anomalous scattering data from a single isomorph with two iodine atoms* I(1) *and* I(2) *in the asymmetric unit. The binding sites for the* $PtCl_4$ *group are labelled A, B, C, D, E. (From Matthews,* 1966*a.)* (*a*) *Phase angles determined with the iodine atoms in their correct absolute configuration.* (*b*) *Phase angles obtained assuming the alternative configuration.*

1962; Lipscomb, Coppola, Hartsuck, Ludwig, Muirhead, Searl & Steitz, 1966; Dickerson, Weinzierl & Palmer, 1968).

Briefly, the method is as follows. From the approximate heavy atom parameters, phases φ of F_P are calculated. On the basis of these phase angles the 'expected' magnitude F_{Cj} of the scattering amplitude of the jth heavy atom derivative is calculated for each reflection (Eq. 2). The corresponding residual for this reflection (c.f. Fig. 3) is

$$\varepsilon_j = (k_j F_{PHj} - F_{Cj}) \qquad (23)$$

where k_j is a scaling factor. Similar residuals are obtained for the other reflections, and the desired shifts in the heavy atom parameters for derivative j are those which minimize the weighted sum over all reflections of the squares of the residuals ε_j. The normal equations are

$$[\mathbf{A}_{mn}]\,[\Delta \mathbf{x}_m] = [\mathbf{b}_n] \qquad (24)$$

where m and n denote the parameters being refined, $\Delta \mathbf{x}_m$ is the shift in the mth parameter, and the matrix coefficients a_{mn} and b_n are given by

$$a_{mn} = \sum_{\mathbf{h}} \omega \left(\frac{\partial F_{Cj}}{\partial x_m}\right)\left(\frac{\partial F_{Cj}}{\partial x_n}\right)$$

$$b_n = \sum_{\mathbf{h}} \omega \,(F_{PHj} - F_{Cj}) \left(\frac{\partial F_{Cj}}{\partial x_n}\right) \qquad (25)$$

At the end of this cycle of least-squares refinement the phases can be recalculated with the improved heavy atom parameters, and the least-squares process repeated for the same or a different heavy atom derivative. In practice a very considerable economy in time can be achieved by repeating several cycles of least-squares refinement, before recalculating the parent phases. Blow (1969, private communication), and Cohen (1969, private communication) have found it satisfactory to repeat four cycles of least-squares refinement for all derivatives before repeating the phase calculation.

In any refinement of the heavy atom parameters it is desirable to check progress with some type of Fourier method. The synthesis most commonly calculated is the difference Fourier synthesis, with coefficients

$$m(F_{PH} - F_P) \exp(-i\varphi) \qquad (26)$$

This synthesis has the disadvantage that it is not very useful in checking the accuracy of the heavy atom parameters. Also, in syntheses of this type it is difficult to locate light atoms in the immediate vicinity of heavy atoms. A better synthesis, which does not have these disadvantages, can be obtained by using the coefficients

$$m(F_{PH} - F_C) \exp(-i\varphi_C) \qquad (27)$$

where m is the figure of merit of the phase determination, F_C is the amplitude of resultant of $\mathbf{F}_P$ and $\mathbf{F}_H$ (Eq. 2), and φ_C is the phase of $\mathbf{F}_C$. These coefficients correspond to the lack of closure vector, ε, in Fig. 3, weighted according to the figure of merit of the current phase determination. This synthesis, which might be called a 'residual' Fourier (c.f. Eqs. 1 and 23) is expected to reveal the difference between the true heavy atom electron density differentiating a parent and its isomorph and the approximation of that electron density in terms of the current heavy atom parameters. In addition, any heavy atom sites not accounted for should appear as positive peaks, while spurious sites will be indicated by negative regions. In the absence of all errors, the 'residual' electron density would be identically zero. 'Residual' syntheses helped considerably in the refinement of the heavy atom parameters for α-chymotrypsin (Blow & Matthews, to be published).

Some practical considerations in phase determination

Centrosymmetric reflections

In computing phases of noncentrosymmetric reflections, it is usual to calculate the phase probability distribution at intervals of 10° or 15° in φ. A similar treatment of the centrosymmetric reflections, although formally incorrect, may be used to find the 'best' phase, but the figure of merit will be an underestimate of the true value.

A convenient method of programming phase determination for centrosymmetric reflections is to use Eqs. (4), (8), and (9), but calculate the probabilities at intervals of 180° in φ (Generally $\varphi=0$ and $\varphi=\pi$, although other values are necessary for some space groups). It can easily be verified that this is equivalent to using Eq. (19) of Blow and Crick for centrosymmetric reflections.

Calculation of heavy atom structure factors

The estimation of the 'atomic scattering factor' of a heavy atom is in general not trivial, partly because the 'atom' may in fact be a complex ion, and also because one should take into account any electron density displaced by the heavy atom. In practice it has been found satisfactory to simply regard the heavy atom groups as point atoms, and to include the effective atomic scattering factor as part of the exponential temperature factor. (To a resolution of 2 Å the atomic scattering factor of mercury is equal, within about 2%, to $80{\cdot}0 \exp[-4{\cdot}0 \sin^2\theta/\lambda^2]$.) For lysozyme, Blake, Koenig, Mair, North, Phillips & Sarma (1965) were able to use the unusually high proportion of centrosymmetric reflections to estimate the appropriate scattering factor curve for each heavy atom.

Determination of E and E′

Where centrosymmetric data are available, E may be estimated from the relation

$$E=\langle |F_{PH}-F_P|-F_H\rangle \qquad (28)$$

where the brackets $\langle\,\rangle$ denote the root mean square value. It is well known that, in using this equation, errors will be introduced by 'crossover' terms in which $\mathbf{F}_{PH}$ and $\mathbf{F}_P$ have the opposite sign. However, such cases are rare, and a more important source of error arises from reflections for which the experimental inaccuracies are such that $(\mathbf{F}_{PH}-\mathbf{F}_P)$ and $\mathbf{F}_H$ have the *opposite* sign, *i.e.* reflections for which the observed data give a wrong sign determination. This effect will in general become serious at higher angles of diffraction where E, $|F_{PH}-F_P|$, and F_H are often of the same order of magnitude.

Where data are available for two or more heavy atom isomorphs, it is best to calculate the phase angles of the centrosymmetric reflection using preliminary values of E, and then to determine more accurate E values for each derivative from the relation

$$E=\langle \varepsilon(\varphi_M)\rangle \qquad (29)$$

where $\varepsilon(\varphi_M)$ is the lack of closure at the most probable phase φ_M, and the summation includes only the centrosymmetric reflections. In so far as the most probable phases are the correct phases, the determination of E from Eq. (29) will be free of the errors incurred in using Eq. (28).

For space groups which include no centrosymmetric reflections, the determination of E is more difficult, especially when only one or two heavy atom derivatives are available. If more isomorphs can be used, a reasonable value can be obtained for the most probable phase φ_M, and in so far as this is the true phase, the value of E is given by the r.m.s. value of the lack of closure of the isomorphous replacement triangle (Fig. 3) at the overall most probable phase; *i.e.* Eq. (29) is used and *all* reflections are included.

Similarly, for the anomalous scattering data, E' can be estimated from the r.m.s. value of the lack of closure of the anomalous scattering data at the most probable phase, *i.e.*

$$E'=\langle \varepsilon_+(\varphi_M)-\varepsilon_-(\varphi_M)\rangle \qquad (30)$$

Alternatively, if the parent protein does not include anomalous scatterers, E' may be estimated from the observed differences between the amplitudes of Friedel related reflections, *i.e.*

$$E' = \langle F_P(\mathbf{h}) - F_P(\bar{\mathbf{h}}) \rangle \quad (31)$$

Such an estimate does not take into account possible errors in the heavy atom parameters and assumes that the quality of the parent and derivative data is the same.

It may be noted that as additional heavy atom derivations are included in the phase determination, and the values of φ_M approach their true values, so the values of E and E' determined from Eqs. (29) and (30) will increase toward their correct values.

R factors

To assess the usefulness of a given heavy atom isomorph, several 'R factors' have been proposed. The one used most commonly was introduced by Cullis *et al.* (1961*a*):

$$R = \frac{\Sigma||F_{PH} - F_P| - F_H|}{\Sigma|F_{PH} - F_P|} \quad (32)$$

where the summations are over the centrosymmetric reflections.

A second R value, useful for noncentrosymmetric data, was proposed by Kraut *et al.* (1962), and has been widely adopted:

$$R_K = \frac{\Sigma|\varepsilon(\varphi_M)|}{\Sigma F_{PH}} \quad (33)$$

For good derivatives R has been observed to be around 40% or less, and R_K around 10% or less (Dickerson *et al.*, 1968).

Another ratio which has been found to be useful in practice as an indicator of the quality of phase determination is the quantity

$$R_{EF} = \frac{\{\Sigma\varepsilon^2(\varphi_M)\}^{1/2}}{\Sigma F_H} = E/\bar{F}_H \quad (34)$$

where the bar denotes the mean value. The analogous quantity for anomalous scattering data is

$$R'_{EF} = E'/2\bar{F}'' \quad (35)$$

Experience with α-chymotrypsin (Blow & Matthews, to be published) suggested that when R_{EF} exceeded unity, *i.e.* E was greater than the mean heavy atom scattering, the resultant phase information was of marginal benefit. Some theoretical justification of this empirical result can be given by the following argument. It can be shown (Matthews, unpublished) that for a centrosymmetric reflection with heavy atom scattering, F_H, the probability $P(F_H)$ of the phase determination being correct is

$$P(F_H) = \tfrac{1}{2} + \tfrac{1}{2}\,\mathrm{erf}\,(\sqrt{2}F_H/E) \quad (36)$$

where erf is the error function. Thus if $E/\bar{F}_H = 1{\cdot}0$, then for an 'average' centrosymmetric reflection with $F_H = \bar{F}_H$, the phase determination will be correct 84% of the time; if $E/\bar{F}_H = 0{\cdot}5$, then an average reflection will give the correct phase 98% of the time; but if $E/\bar{F}_H = 1{\cdot}5$, then an average reflection will result in a correct phase determination only 75% of the time. (50% of the phases will, of course, be correct just by chance.)

How many derivatives are necessary?

In theory a unique phase determination can be obtained either from two heavy atom derivatives, or from a single isomorph if anomalous scattering data are included. Recently Herriott, Sieker & Jensen (private communication), using the latter method, were able to determine the structure of the nonhaeme iron protein rubredoxin. Although rubredoxin is a small protein (M.W. ≃ 6000) the structure determination by Jensen and collaborators represents a very considerable technical achievement. To date, all published structure determinations of larger proteins have required data from at least three heavy atom derivatives.

Blow & Rossmann (1961) and Kartha (1961) showed that if the heavy atom arrangement is not centrosymmetric, useful phase information for noncentrosymmetric reflections can be obtained from a single isomorph, even in the absence of measurements of anomalous scattering. In the single isomorphous replacement method, as proposed by Blow and Rossmann, the phase probability $P_{iso}(\varphi)$ (Eq. 4) is calculated, and the centroid of this distribution gives the best phase for F_P. The single isomorphous replacement is

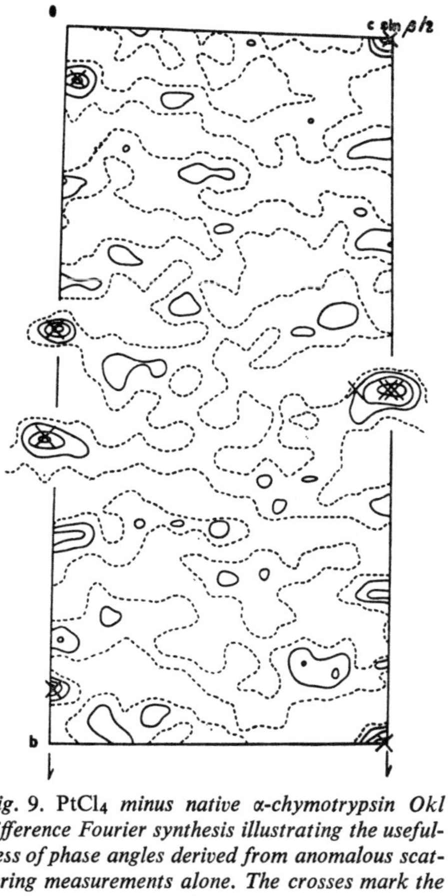

Fig. 9. $PtCl_4$ *minus native α-chymotrypsin Okl difference Fourier synthesis illustrating the usefulness of phase angles derived from anomalous scattering measurements alone. The crosses mark the expected sites for the heavy atom peaks* [*c f. Fig.* 8 (*a*)].

most useful when the first satisfactory isomorph has been found. At this stage preliminary phases can be found using the S.I.R. method, and these may then be used to analyse subsequent derivatives by the powerful difference-Fourier method. The most serious difficulty of the S.I.R. method is due to ghost images which are related to the true electron density by inversion through any partial centres of symmetry in the heavy atom arrangement (Blow & Rossmann, 1961; Kartha, 1961). When using S.I.R. phases to calculate difference Fourier maps, care must be taken to distinguish such ghost images from potential heavy atom binding sites.

By analogy with the method of S.I.R., useful preliminary phases may be obtained from a single set of anomalous scattering measurements. In this case $P_{ano}(\varphi)$ is calculated from Eq. (19) or (22), and the best phase obtained. Application of this method to α-chymotrypsin is shown in Fig. 9. In this case the phases were derived using only the anomalous scattering differences of two iodine atoms in a total molecular weight of 50,000, and the phases were tested by calculating a $PtCl_4$ minus native difference-Fourier map. In the single anomalous scattering method there will be ghost images at exactly the same positions as in the S.I.R. method, but in this case they are negative and therefore would not be confused with heavy atom binding sites.

Acknowledgements

I am grateful to a number of colleagues for their comments and discussion relating to the subject matter of this paper. In particular I would like to acknowledge the contributions of Drs. D. M. Blow, P. B. Sigler, D. R. Davies, and G. H. Cohen. The preparation of this paper was supported in part by the Public Health Service Grant GM 15423.

References

Bijvoet, J. M. (1954). *Nature*, **173**, 888.
Blake, C. C. F., Koenig, D. F., Mair, G. A., North, A. C. T., Phillips, D. C. & Sarma, V. R. (1965). *Nature*, **206**, 757.
Blow, D. M. (1958). *Proc. Roy. Soc.* A**247**, 302.
Blow, D. M. & Crick, F. H. C. (1959). *Acta Cryst.* **12**, 794.
Blow, D. M. & Rossmann, M. G. (1961). *Acta Cryst.* **14**, 1195.
Bokhoven, C., Schoone, J. C. & Bijvoet, J. M. (1951). *Acta Cryst.* **4**, 275.
Bragg, Sir Lawrence. (1965). *Reports on Progress in Physics XXVIII*, 1.
Cohen, G. H., Matthews, B. W. & Davies, D. R. (1969). Submitted to *Acta Cryst.*
Cullis, A. F., Muirhead, H., Perutz, M. F., Rossmann, M. G. & North, A. C. T. (1961*a*). *Proc. Roy. Soc.* A**265**, 15.
Cullis, A. F., Muirhead, H., Perutz, M. F., Rossmann, M. G. and North, A. C. T. (1961*b*). *Proc. Roy. Soc.* A**265**, 161.
Dickerson, R. E., Kendrew, J. C. & Strandberg, B. E. (1961). *Acta Cryst.* **14**, 1188.
Dickerson, R. E., Weinzierl, J. E. & Palmer, R. A. (1968). *Acta Cryst.* B**24**, 997.
Green, D. W., Ingram, V. M. & Perutz, M. F. (1954). *Proc. Roy. Soc.* A**225**, 287.
Harker, D. (1956). *Acta Cryst.* **9**, 1.
Holmes, K. C. (1965). *Computing Methods in Crystallography*, ed. J. S. Rollett. Oxford: Pergamon Press.
Kartha, G. (1961). *Acta Cryst.* **14**, 680.
Kartha, G. & Parthasarathy, R. (1965*a*). *Acta Cryst.* **18**, 745.
Kartha, G. & Parthasarthy, R. (1965*b*). *Acta Cryst.* **18**, 749.
Kraut, J., Sieker, L. C., High, D. F. & Freer, S. T. (1962). *Proc. Nat. Acad. Sci. U.S.* **48**, 1417.
Lipscomb, W. N., Coppola, J. C., Hartsuck, J. A. Ludwig, M. L., Muirhead, H., Searl, J. & Steitz, T. A. (1966). *J. Mol. Biol.* **19**, 423.
Matthews, B. W. (1966*a*). *Acta Cryst.* **20**, 230.
Matthews, B. W. (1966*b*). *Acta Cryst.* **20**, 82.
North, A. C. T. (1965). *Acta Cryst.* **18**, 212.
Okaya, Y. & Pepinsky, R. (1960). *Computing Methods and the Phase Problem in X-ray Crystal Analysis.* London: Pergamon Press.
Ramachandran, G. N. & Raman, S. (1956). *Curr. Sci.* **25**, 348.
Ramaseshan, S. (1964). *Advanced Methods of Crystallography*, ed. G. N. Ramachandran. London: Academic Press.

Discussion

MAILER: Is it possible to use two sources of radiation and determine the heavy atom position from the difference of the anomalous scattering?

MATTHEWS: This is equivalent to isomorphous replacement and is discussed in my paper. It has several disadvantages.

JANSONIUS: In the structure determination of papain (*J. Mol. Biol.* **24**, 449) we have used a different way of finding the correct 'hand' of the structure. We calculated a special kind of difference Fourier, in which the calculated F_H'' values were combined with the corresponding heavy atom phases. This would give a positive peak if the 'hand' is correct and a negative peak otherwise.

MATTHEWS: Yes, I know the method. It has the disadvantage that for coefficients where F_H'' and F_{PH} are small, the experimental inaccuracies may be magnified and lead to large errors in the coefficients. In your calculation did you take this into account, *e.g.*, by setting a lower limit to the amplitude of F_H?

JANSONIUS: Yes, we did.

KARTHA: If data from more than one derivative crystal are available, the hand of heavy atom configuration can be established in many ways. We find that even a figure of merit index shows significant enhancement when the correct hand is used with anomalous scattering measurements. However, with data from only one derivative, choosing the hand is a much more difficult problem.

R. HUBER: Would you agree that the applicability of single isomorphous replacement for a complete structure determination seems not to be proved?

MATTHEWS: With increased experimental accuracy it should be possible to determine the structure of larger proteins from a single isomorphous replacement.

The Influence of the Anomalous Scattering of the Light Atoms on Phase Determinations in X-Ray Studies of Proteins

ADRIAAN DE VRIES

Liquid Crystal Institute, Kent State University, Kent, Ohio, U.S.A.

When the anomalous scattering effect is used in the study of the structures of large molecules, one usually considers only the heavy atoms present in the material. The reason for this is obvious from Table I: the $\Delta f''$ values of the atoms most commonly present (H, C, N, O) are very small compared with those of heavy atoms like I, Pt, and Hg. Blow (1958) has shown, however, that the situation changes somewhat when atoms of medium weight are present, like S or Fe (and S, of course, is quite common in proteins since there are two naturally occurring α-amino acids which contain sulfur). This is already evident from Table I itself, but it becomes even more so when the composition of an actual protein is taken into account (Tables II and III, first three columns of each). With Cr $K\alpha$ radiation the contribution of the S atoms in particular is quite appreciable and with Cu $K\alpha$ radiation the contribution of the Fe atoms is almost equal to that of the Hg atoms. In both cases (columns 3 of Tables II and III) the contribution of the light atoms (C, N, O) is relatively small, but when we use more recent values for $\Delta f''$, the contribution from these atoms increases considerably (columns 5 and 7 of Tables II and III). It seems quite probable, therefore, that the effect of these light atoms is considerably greater than anticipated by Blow. More reliable values of the anomalous scattering factors of these elements are urgently needed to make a more quantitative assessment of their influence.

Table I. Some dispersion corrections ($\Delta f''$) for atomic scattering factors*

Atom	Cr $K\alpha$	Cu $K\alpha$
H	0·0	0·0
C	0·1	0·0
N	0·1	0·0
O	0·2	0·1
S	1·2	0·6
Fe	0·9	3·4
I	13·6	7·2
Pt	15	8
Hg	16	9

* Selected values from the *International Tables for X-ray Crystallography* (1962).

Table II. Anomalous scattering data for Hg-hemoglobin; Cr $K\alpha$ radiation

Components of ½ molecule	Blow–De Vries* $\Delta f''$	Blow–De Vries* $(\Sigma\Delta f''^2)^{1/2}$	Cromer–De Vries† $\Delta f''$	Cromer–De Vries† $(\Sigma\Delta f''^2)^{1/2}$	*International Tables* $\Delta f''$	*International Tables* $(\Sigma\Delta f''^2)^{1/2}$
1580 C atoms	0·01		0·04		0·1	
397 N atoms	0·02	0·8	0·06	2·7	0·1	6·1
439 O atoms	0·03		0·09		0·2	
7 S atoms	1·0	2·6	1·15	3·0	1·2	3·2
2 Fe atoms	0·9	1·3	0·94	1·3	0·9	1·3
0·65 Hg atom	15	9·8‡	12·76	8·3‡	16	10·4‡
all atoms except Hg		3·0		4·3		7·0

* From Blow (1958) and De Vries (1960); the $\Delta f''$ values for C, N, O, and S are estimated from Hönl's results (1933), those for Fe and Hg are from Dauben & Templeton (1955).

† The $\Delta f''$ values for S, Fe, and Hg are from Cromer (1965); the values for C, N, and O are my extrapolations, based on Cromer's data.

‡ Since Hg occupies only one single position, the numbers represented $\Sigma\Delta f''$ rather than $(\Sigma\Delta f''^2)^{\frac{1}{2}}$ values.

The way in which the effect of the lighter atoms (*i.e.* all atoms except the heavy atoms) is taken into account is as follows (De Vries, 1960). Considering a structure containing both heavy atoms and lighter atoms, with a centrosymmetric array of the heavy atoms and a noncentrosymmetric array of the lighter atoms, the following formulas are obtained for the structure factor of the heavy atoms only ($\mathbf{F}_H$), that of the lighter atoms only ($\mathbf{F}_L$), and that of all atoms ($\mathbf{F}$):

$$\begin{aligned}\mathbf{F}_H &= \Sigma_H(f'_j + i\Delta f''_j) \cos \varphi_j = A_H + iA''_H \\ \mathbf{F}_L &= \Sigma_L(f'_j + i\Delta f''_j) \exp(i\varphi_j) \\ &= A_L + iB_L + iA''_L - B''_L \\ \mathbf{F} &= (A_H + A_L - B''_L) + i(A''_H + A''_L + B_L)\end{aligned}$$

In the formulas f'_j includes the real part of the anomalous scattering correction, $\Delta f''_j$ is the imaginary part of this correction, φ_j is the phase angle (all for atom j), each A stands for a $\Sigma f \cos \varphi$, $B = \Sigma f \sin \varphi$, $A'' = \Sigma \Delta f'' \cos \varphi$, and $B'' = \Sigma \Delta f'' \sin \varphi$.

From the formula for $\mathbf{F}$ we find for the Bijvoet difference (the difference between $F^2(hkl)$ and $F^2(\bar{h}\bar{k}\bar{l})$):

$$\Delta F^2 = 4B_LA''_H + 4B_LA''_L - 4B''_LA_H - 4B''_LA_L$$

When the anomalous effect of the lighter atoms is negligible, this formula is reduced to:

$$\Delta F^2 = 4B_LA''_H \tag{1}$$

ΔF^2 can be measured and, the positions of the heavy atoms being known, A''_H can be calculated. Thus from this formula B_L can be calculated and from this in turn the phase angle φ_L of F_L (apart from the ambiguity between φ_L and $\pi - \varphi_L$).

We shall restrict ourselves here to the determination of the sign of B_L, which determines whether φ_L lies between 0 and π or between π and 2π. When we can neglect the contributions from the lighter atoms, this sign can be determined from the relation:

$$\text{sign of } B_L = \text{sign of } \Delta F^2 \times \text{sign of } A''_H. \tag{2}$$

When the anomalous effect of the lighter atoms can no longer be ignored, we have to add to the right-hand side of Eq. (1) the correction term:

$$4C \equiv 4B_LA''_L - 4B''_LA_L - 4B''_LA_H.$$

Because of the presence of this correction term, Eq. (2) will give us the wrong sign of B_L when $|C| > |B_LA''_H|$ when C and $B_LA''_H$ have different signs. Since the positions of the lighter atoms are not yet known at the stage in which use is made of the anomalous scattering effect, we cannot calculate C. We can, however, estimate the probability that Eq. (2) will give the wrong indication for the sign of B_L.

This probability is (De Vries, 1960)

$$P = \frac{2}{\pi\sqrt{2\pi}} \int_0^1 \frac{dx}{1-x^2} \int_{Hx}^{\infty} \exp\left(-\frac{y^2}{2}\right) dy,$$

where $H \equiv |A''_H|(\overline{B''^2_L})^{-1/2}$ = the 'heaviness' of the heavy atoms (for the reflection under consideration) compared with the

Table III. Anomalous scattering data for Hg-hemoglobin; Cu $K\alpha$ radiation

Components of ½ molecule	Blow–De Vries* $\Delta f''$	Blow–De Vries* $(\Sigma\Delta f''^2)^{1/2}$	Cromer–De Vries* $\Delta f''$	Cromer–De Vries* $(\Sigma\Delta f''^2)^{1/2}$	*International Tables* $\Delta f''$	*International Tables* $(\Sigma\Delta f''^2)^{1/2}$
1580 C atoms	0·01		0·02		0·0	
397 N atoms	0·01	0·6	0·03	1·4	0·0	2·1
439 O atoms	0·02		0·05		0·1	
7 S atoms	0·57	1·5	0·58	1·5	0·6	1·6
2 Fe atoms	3·4	4·8	3·45	4·9	3·4	4·8
0·65 Hg atom	8	5·2*	7·04	4·6*	9	5·8*

* See Table II.

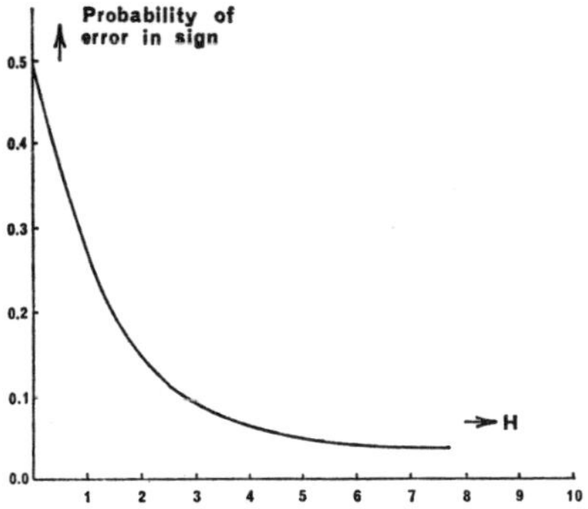

Fig. 1. *Probability of error in the sign of B_L determined from the anomalous scattering effect, when the contribution of the lighter atoms is ignored, plotted against H (the anomalous scattering from the heavy atoms compared to that from the lighter atoms).*

lighter atoms, as far as the anomalous scattering is concerned. The integral over y is tabulated, *e.g.* by Hald (1952) and the integral over x can be evaluated numerically to give P as a function of H (Fig. 1).

From Fig. 1 it can be seen that, in order to obtain results which have a reliability of 90% or more, H has to be at least 2·75. Since $(\overline{B_L''^2})^{1/2} = (\frac{1}{2}\Sigma_L \Delta f''^2)^{1/2} = 2{\cdot}16$, 3·05 or 4·95 respectively, for the three different sets of $\Delta f''$ values for Cr $K\alpha$ radiation (compare Table II), this means that $|A_H''|$ has to be at least 5·9, 8·4, or 13·6, respectively. Comparing these values with the $\Sigma \Delta f''$ values for Hg in Table II (which are maximum values for $|A_H''|$), we find that if either one of the newer sets of data in Table II is correct, a reliability of 90% is not obtainable for any reflection of the Hg-hemoglobin compounds discussed here. This result points once more to the necessity of obtaining more reliable $\Delta f''$ values, especially for the light atoms.

One final comment seems to be in order. If the anomalous scattering of the lighter atoms can be so large as to seriously hinder the phase determination based on considering only the anomalous scattering of the heavy atoms, it should be possible to determine *experimentally* the magnitude of the anomalous scattering from these lighter atoms.

References

Blow, D. M. (1958). *Proc. Roy. Soc.* A **247**, 323.
Cromer, D. T. (1965). *Acta Cryst.* **18**, 17.
Dauben, C. H. & Templeton, D. H. (1955). *Acta Cryst.* **8**, 841.
Hald, A. (1952). *Statistical Tables and Formulas*, p. 34. New York: John Wiley and Sons.
Hönl, H. (1933). *Z. Phys.* **84**, 1.
International Tables for X-ray Crystallography (1962). Vol. III, p. 213. Birmingham: Kynoch Press.
Vries, A. de (1960). *Koninkl. Ned. Akad. Wetenschap., Proc. Ser.* B **63**, 353.

Panel Discussion on Protein Methods

AHMED: When one starts a protein structure, crystals are prepared for the protein without a heavy atom and for at least two heavy atom derivatives. Does one then collect 3-D data from these crystals to 6 Å resolution to start the analysis, or is it enough to start with projection data?

KARTHA: I think we get more positive information from low-resolution 3-D data (to 4 Å) than from high-resolution projection data even if it is centrosymmetric, since these projections are usually along 30 Å or more. Hence, unless experimental set-up dictates it, I prefer to work with 3-D data at lower resolution than with an equal number of 2-D data at higher resolution. With diffractometer data one has more flexibility in choosing the minimum number to be measured. With photographic methods, the 2-D data comes more naturally.

MATTHEWS: Yes, personal preference is a factor, and for multiple-site derivatives, 3-D data may be essential. Usually new derivatives are screened by taking precession photographs, and if intensity differences are apparent, it would be a pity to make no further use of the photograph than a visual inspection. If a derivative is to be used for structure determination at atomic resolution, it will be necessary to carry out some tests with high resolution data.

TOPIC D

LEAST-SQUARES PROCEDURES IN CRYSTAL STRUCTURE ANALYSIS

CHAIRMAN: J. S. ROLLETT

Least-Squares Procedures in Crystal Structure Analysis

J. S. ROLLETT

Oxford University Computing Laboratory, Oxford OX 1 3 PL, England.

In any overdetermined situation there are more observations than parameters available to fit them. The simplest such case is that in which each observation depends linearly on the parameters. Let the parameters be $x_1, \ldots, x_n$ and the observations $b_1, \ldots, b_m$. The 'observational' equations then are

$$\begin{aligned} a_{11}x_1 + \ldots + a_{1n}x_n &= b_1 \\ a_{m1}x_1 + \ldots + a_{mn}x_n &= b_m , \end{aligned} \tag{1}$$

where we are given the b_i and the a_{ij}. If $m<n$ there is no unique solution and we need further observations, or assumptions, to increase the number of equations. If $m=n$, the equations can be solved exactly, but the solution does not give any evidence on the accuracy of the b_i nor of the x_j derived from them. If $m>n$, the situation is overdetermined and we cannot in general satisfy all the equations exactly. We must therefore content ourselves with defining a solution which 'best' fits all of the equations. If we write

$$d_i = -b_i + a_{i1}x_1 + \ldots + a_{in}x_n , \tag{2}$$

then d_i is the error of equation i for given $x_1, \ldots, x_n$. We can set up some function of the d_i and seek to minimize it, defining the best solution as that for the minimum. Many different functions have been proposed, and no single one will be appropriate for all cases. For the method of least-squares the function can be

$$M_2 = \sum_{i=1}^{m} d_i^2 , \tag{3}$$

when all of equations (2) are independent and equally reliable. It may well be, however, that one such equation may be expected to be twice as much in error as the others. The corresponding value of d^2 is likely to be 4 times as large as for the other equations and we should divide this value of d^2 by 4 before adding it into M_2 to obtain the most appropriate figure of merit. The minimization function is then

$$M = \sum_{i=1}^{m} w_i d_i^2 , \tag{4}$$

where w_i is proportional to the reciprocal of the square of the error to be feared in (in statistical terms, the reciprocal of the variance of) the ith equation. It should be stressed that the equations must be independent. If this is not so, we must use $M = \mathbf{d}^T W \mathbf{d}$ where W is the inverse of the covariance matrix of the observational equations. This method has not been used in crystallographic calculations up to now.

A philosophical justification of the method of least-squares is given in Whittaker & Robinson (1937). Here we content ourselves with observing that it is relatively insensitive to gross errors, as compared, say, with a method based on

$$M_1 = \max_i |d_i| . \tag{5}$$

It is also computationally economical and readily yields statistical error estimates. The validity of these is established when the d_i have normal distributions of error; and if m is much larger than n, the reliability of the estimated standard deviations of the x_j is good.

If the function M is at a minimum with respect to x_j, then we have

$$\frac{1}{2}\frac{\partial M}{\partial x_j}=\sum_{i=1}^{m} w_i a_{ij}(a_{i1}x_1+\ldots+a_{in}x_n-b_i)=0 \quad (6)$$

This gives the 'normal equation for x_j',

$$\sum_{i=1}^{m} w_i a_{i1} a_{ij} x_1+\ldots+\sum_{i=1}^{m} w_i a_{ij}^2 x_j+\ldots+\sum_{i=1}^{m} w_i a_{in} a_{ij} x_n=\sum_{i=1}^{m} w_i a_{ij} b_i\,. \quad (7)$$

There is one such equation for each j from 1 to n and so we have n equations in n unknowns, which can be solved exactly for the x_j which minimize M, without the aid of any approximate estimates.

Unfortunately it is exceptional for the experimental observations to depend linearly on the parameters. In the case of a crystal structure analysis in particular, we have typically

$$F_{\text{calc}}(\mathbf{h})=\sum_{i=1}^{n} f_i \exp(2\pi i\mathbf{h}\,.\,\mathbf{x}_i)\,, \quad (8)$$

where f_i is the scattering factor for the ith atom in the unit cell, corrected suitably for vibration effects, $\mathbf{h}$ is the order of diffraction, $\mathbf{x}_i$ is the position of the ith atom and $F_{\text{calc}}(\mathbf{h})$ is the structure factor. We observe $|F_{\text{obs}}(\mathbf{h})|$ and compare it with $|F_{\text{calc}}(\mathbf{h})|$, which varies highly non-linearly with the $\mathbf{x}_i$. The details of this variation are discussed elsewhere, and here we simply show that it is necessary to have a starting approximation to the parameters and that the derivatives of the F_{calc} with respect to the parameters are needed. For generality we will assume that the parameters are $p_1,\ldots,p_n$, which may include positions, vibrations and scalefactors. We choose to minimize

$$M=\sum_{i=1}^{m} w_i(|F_{oi}|-|F_{ci}|)^2. \quad (9)$$

For the trial values $p_{10},\ldots,p_{n0}$ we have $|F_{ci}|=|F_{ci}|_0$ and so for any values of $p_1,\ldots,p_n$ we can write a Taylor series

$$|F_{ci}|=|F_{ci}|_0+\sum_{j=1}^{n}(p_j-p_{j0})\frac{\partial F_{ci}}{\partial p_j}+\text{term including }(p_j-p_{j0})^2+\ldots \quad (10)$$

Provided that (p_j-p_{j0}) is small and the higher derivatives of the $|F_{ci}|$ are not unduly large we can neglect all terms in (10) after those in the first derivatives. We can then obtain equations of the same form as Eq. (1)

$$(p_1-p_{10})\left(\frac{\partial|F_{ci}|}{\partial p_1}\right)_0+\ldots+(p_n-p_{n0})\left(\frac{\partial|F_{ci}|}{\partial p_n}\right)_0=|F_{oi}|-|F_{ci}|_0\,, \quad (11)$$

for the adjustments to $p_1,\ldots,p_n$ which would reduce to zero the discrepancies between $|F_o|$ and $|F_c|$ for the various structure factors. The analogue of Eq. (4) is

$$M=\sum_{i=1}^{m} w_i(|F_{oi}|-|F_{ci}|)^2, \quad (12)$$

and it is readily shown that minimization of M is achieved by solving normal equations of the form

$$\sum_{i=1}^{m} w_i\left(\frac{\partial F_{ci}}{\partial p_1}\right)\left(\frac{\partial F_{ci}}{\partial p_j}\right)(p_1-p_{10})+\ldots+\sum_{i=1}^{m} w_i\left(\frac{\partial F_{ci}}{\partial p_j}\right)^2(p_j-p_{j0})+\ldots+\sum_{i=1}^{m} w_i\left(\frac{\partial F_{ci}}{\partial p_n}\right)\left(\frac{\partial F_{ci}}{\partial p_j}\right)\times(p_n-p_{n0})=\sum_{i=1}^{m} w_i(|F_{oi}|-|F_{ci}|)\left(\frac{\partial F_{ci}}{\partial p_j}\right) \quad (13)$$

Because these equations are derived by ignoring the terms in higher powers of (p_j-p_{j0}) and higher derivatives of $|F_{ci}|$ than the first, they do not generally give an exactly correct result and may actually give a worse set of parameters than those input. Fortunately the linearity of the $|F_c|$ with the parameters is good over a small range of errors of position and both Patterson methods and direct methods of phase determination usually yield positions which are accurate enough to prevent divergence of least-squares refinements. Since the method normally converges, but does not give exact adjustments, it is necessary to repeat the calculations until it is clear that a further cycle of refinement would give insignificant shifts. This can be done conveniently by monitoring the sum of squares of ratios between the shifts and the corresponding standard deviations

$$Q = \sum_{j=1}^{n} [(p_j - p_{j0})/\sigma(p_j)]^2. \qquad (14)$$

We shall see later how the standard deviation $\sigma(p_j)$ can be found.

The primary computing difficulties in forming the normal equations arise from the heavy labour and large storage capacity required for large problems. The formation of a full set of n equations from m structure factors needs $\frac{1}{2}mn(n+3)$ multiplications. If each multiplication requires 10 microseconds (a reasonable figure in 1969), there are 3000 structure factors and 300 parameters, we need about 8 minutes of machine time. Bigger problems are already being solved and many computers are not so fast. Where time must be saved we may resort to approximations which omit the calculation of most of the matrix elements. This may also be forced on us by lack of random access store. Unless the structure factor derivatives are obtained more than once, either by re-calculation or by retrieval from a backing store, we have to scan repeatedly through the $\frac{1}{2}n(n+3)$ normal equation elements. If the main store of the computer is of size 32,768 words, for example, this limits n to 254 or less and the limit is lower if appreciable space must be reserved for the program and other data.

Constraints

It has long been the practice in crystal structure analysis to refine as independent least-squares parameters the position coordinates, the vibrational amplitudes, and sometimes the scale factors of the crystallographically unique atoms. It has been found that both the range of convergence and its speed may be increased in the early stages of an analysis by constraining atomic positions so that they are related by appropriate distances and angles. Where the crystallographic data are of poor quality or few in relation to the number of atoms it may be advisable to retain the constraints in the final refinement. This can also be useful as a way of testing structure models. There are three different ways of applying constraints to a least squares refinement.

The first method is to replace the unconstrained atomic parameters by molecular parameters and to calculate the derivatives of the $|F_c|$ with respect to the molecular parameters from the dependence of the atomic parameters on them. The necessary calculations of derivatives will be discussed fully elsewhere and need not be pursued here. The method has the great advantage that the size of the normal matrix is generally reduced and sometimes greatly so. It is particularly appropriate where the computing equipment is limited or the problem very large (which is practically the same thing). It has therefore been developed particularly for analyses of protein structures. For more general use it has suffered from the drawback that it has not been found easy to produce programs applicable to a wide variety of chemical situations.

The second formal method is that of Lagrangian multipliers. It has the drawbacks of the molecular parameter scheme plus the disadvantage that the normal matrix is bigger, not smaller, than that without constraints. It has therefore not been used for refinement programs. It is a useful general technique, however, and we give here a brief statement of the equations to be solved. If the unconstrained normal equations are

$$\begin{aligned} c_{11}p_1 + \ldots + c_{1n}p_n &= d_1 \\ c_{n1}p_1 + \ldots + c_{nn}p_n &= d_n , \end{aligned} \tag{15}$$

and the parameters $p_1, \ldots, p_n$ must also satisfy the m conditions

$$\begin{aligned} g_{11}p_1 + \ldots + g_{1n}p_n &= h_1 \\ g_{m1}p_1 + \ldots + g_{mn}p_n &= h_m \end{aligned} \tag{16}$$

then we solve the augmented normal equations

$$\begin{aligned} c_{11}p_1 + \ldots + c_{1n}p_n + g_{11}\lambda_1 & \\ + \ldots + g_{m1}\lambda_m &= d_1 \\ c_{n1}p_1 + \ldots + c_{nn}p_n + g_{1n}\lambda_1 & \\ + \ldots + g_{mn}\lambda_m &= d_n \\ g_{11}p_1 + \ldots + g_{1n}p_n &= h_1 \\ g_{m1}p_1 + \ldots + g_{mn}p_n &= h_m . \end{aligned} \tag{17}$$

The $\lambda_1 \ldots \lambda_m$ are called Lagrangian or undetermined multipliers and their values do not in general have any obvious significance. The derivation of Eqs. (17) is beyond our scope here but a rigorous treatment is given by Apostol (1957, pp. 153–157).

The third method has been developed by Waser (1963) for crystal structure refinement. It is essentially different from the other two in that the constraints applied are not exact but approximate. The minimization function of Eq. (12) is changed to

$$M = \sum_{i=1}^{m} w_i(|F_{oi}| - |F_{ci}|)^2 + \sum_{k=1}^{l} w_k[f_k(p_1, \ldots, p_n) - g_k]^2, \tag{18}$$

where f_k and g_k are respectively a function calculated from the parameters and a desired value of this function. The second term on the right-hand side of Eq. (18) can be shown to lead to l extra observational equations

$$\begin{aligned} &\sqrt{w_k}\left(\frac{\partial f_k}{\partial p_1}\right)(p_1 - p_{10}) + \ldots \\ &+ \sqrt{w_k}\left(\frac{\partial f_k}{\partial p_n}\right)(p_n - p_{n0}) \\ &= \sqrt{w_k}\,[g_k - f_k(p_{10}, \ldots, p_{n0})] . \end{aligned} \tag{19}$$

These contribute to the normal equations in just the same way as do Eqs. (11) based on the structure factor agreement. It is straightforward to write a program which generates the normal equation contributions for the commonly required constraints and the size of the normal matrix is not altered by the constraint process. This method therefore gives us a convenient general purpose tool for the application of constraints in a wide variety of contexts. The difficulties are the correct choice of weights, the requirements for the design of the least-squares program, and the need to design the constraints routine to be sufficiently flexible to deal with problems which are likely to arise.

It may be desirable simply to fix certain distances at exact values. In this case all that matters is that each constraint Eq. (19) shall have much higher weight than the total weights of all the structure factor Eqs. (11). This is readily arranged, but difficulties appear if the precaution is carried too far. If the constraint equation has 10,000 times the weight of the structure factor equations combined, its normal equation diagonal term will be approximately four decimal orders of magnitude larger than that from the other equations. If the floating point arithmetic of the machine provides the equivalent of 8 significant decimal figures, the rounding errors will affect the fourth significant figure of the structure factor contribution. Any further increase of weight on the constraint is likely to lead to undesirable rounding errors since the constraints must leave some functions of the parameters free to vary. A normal matrix from constraint equations alone is therefore singular (*i.e.* it yields no unique solution).

It may happen that an interatomic distance is to be constrained to a set value because the X-ray data at hand do not determine the distance as precisely as it can be estimated from the results of previous analyses. These themselves have finite errors and the constraint they provide is, say, a bond-length with an estimated standard deviation. We can then proceed as follows. The weights of the X-ray observations are commonly on an arbitrary scale and a normalization factor is calculated as

$$N^2 = \left[\sum_{i=1}^{n} w_i(|F_{oi}| - |F_{ci}|)^2\right] / (n-s) , \quad (20)$$

where the n values of $|F_o|$ have been fitted with the aid of s parameters. A constraint equation based on a distance d_j with estimated standard deviation $\sigma(d_j)$ is then given weight

$$w_j = N^2/\sigma^2(d_j) . \quad (21)$$

Each of the various types of observational equation, 'prepared' by multiplication by the square root of its weight, then has an expectation r.m.s. error of $Nn/(n-s)$. If the constrained refinement produces a bond-length differing from d_j by several times $\sigma(d_j)$, we can deduce that the data at hand are in conflict with the constraint assumption we have made. This is a valuable safeguard against imposing constraints which are not really valid.

The contributions to the normal equations from constraints of this kind can be included, and the details of the process altered, in a flexible way provided that the subroutine which accumulates the X-ray contributions is separated from that which solves the equations. If the system is not designed in this way, then it may not be possible to amend the constraint system without detailed knowledge of the other routines, and an unwelcome rigidity is likely to be imposed on the facilities provided. Similar remarks apply to the provision of facilities for weighting the X-ray observations themselves. It is best if the weights can be tabulated by an independent subroutine before the main calculation begins. That subroutine can then be altered without deep knowledge of the system when a new weighting scheme is desired.

The constraint of bond lengths has been mentioned. Bond angles can be constrained by fixing distances between next nearest neighbours. Groups of atoms can be forced to remain planar (*e.g.* benzene rings), but it is difficult to do this by constraining distances within the plane. A convenient device is to use a null-atom (of zero scattering power) some 10 Å out of the plane and to constrain its distances from the atoms of the planar group. The position coordinates of the null-atom should be refined so that the position of the plane is not undesirably fixed.

It is advisable to allow for the prescription of distances between atoms related by symmetry operations (including unit cell translations) to those in the parameter list. The obvious drawback of failing to do this is that all atoms in the parameter list may have to be transferred to a single molecule. A less obvious difficulty is that of dealing with a ring surrounding a centre of symmetry or a bond straddling a mirror plane.

A scheme of this type has been developed by Ford & Rollett (unpublished) and has proved extremely convenient. It can, for example, easily deal with an aliphatic side chain of a large molecule in which bond lengths and angles are fixed but dihedral angles may vary considerably. No special programming is required, however complicated the geometry of the chemical situation. A further advantage of this method is that it has been extended to deal with isotropic and anisotropic vibrations. The parameters for these not uncommonly vary by impossibly large amounts from one atom to the next in a large structure and often correspond to imaginary vibration amplitudes when no constraints are applied. The isotropic vibration parameters of neighbouring atoms can be constrained to agree, or to differ by a quantity derived from a rigid body motion analysis. The same treatment can be given to corresponding anisotropic vibration parameters of neighbouring atoms. A more sophisticated approach has proved slightly more successful and easier to apply in practice. If two atoms are joined by a covalent bond then the difference between their vibration amplitudes in the bond direction cannot exceed the amplitude of the bond-stretching vibration. For a carbon-carbon single bond this can be estimated as 0·06 Å at room temperature, using a typical force constant value. We can therefore assign a standard deviation of 0·0036 Å^2 to the difference between the temperature factor parameters of the atoms in the bond direction and this constraint need not be relaxed because of any rigid body librations of the molecule as a whole. It is unfortunate that bond-angle bending motions cannot be treated so simply, but they would, in any case, give much weaker constraints.

If the two bonded atoms have coordinates differing by (x_1, x_2, x_3) and if the distance between them is d, where

$$d^2 = \sum_{i=1}^{3} \sum_{j=1}^{3} g_{ij} x_i x_j \,, \tag{22}$$

because of the (possibly non-orthogonal) unit cell geometry, then we have

$$h_i = \sum_{j=1}^{3} g_{ij} x_j / d \,, \tag{23}$$

as the ith component of a reciprocal vector of unit length in the bond direction. The mean square amplitude of vibration in this direction of an atom with vibration parameters U_{kl} is then U_b, where

$$U_b = \sum_{k=1}^{3} \sum_{l=k}^{3} U_{kl} h_k h_l \,, \tag{24}$$

and where the temperature factor for a reflection (h_1, h_2, h_3) is written in the form

$$T = \exp -\left(\sum_{i=1}^{3} \sum_{j=i}^{3} 2\pi^2 U_{ij} h_i h_j a_i^* a_j^* \right) . \tag{25}$$

This enables us to compute the observational equation for the bond stretching constraint between atoms r and s as

$$\sqrt{w_m} \sum_{k=1}^{3} \sum_{l=k}^{3} h_k h_l (\delta U_{klr} - \delta U_{kls}) = \sqrt{w_m} \sum_{k=1}^{3} \sum_{l=k}^{3} h_k h_l (U_{klr} - U_{kls}) \tag{26}$$

where $\sqrt{w_m} = N/0{\cdot}06$ for a carbon-carbon single bond, with corresponding values for bonds with different force constants and reduced masses.

Solution of normal equations

Most programming systems are equipped with built-in or library subroutines for solving simultaneous equations and inverting matrices. The use of such a routine may save a crystallographic programmer a great deal of hard work, but it is likely to be unsatisfactory unless it provides for the special needs of the crystal structure refinement process. The most vital consideration is that a least-squares normal matrix is symmetric and only the $\frac{1}{2}n(n+1)$ elements on and above the leading diagonal need be stored. If the routine which solves the equations requires storage of the n^2 elements of the full matrix then the size of problem which can be handled will be seriously reduced. Fortunately library programs which require the upper triangle alone are commonly available.

A difficulty which is less often handled by standard routines is that there is a substantial risk that the equations may turn out to be numerically singular. This can be avoided in all cases if the facilities for relating atomic parameters to least-squares parameters are well designed, but the situation is so complicated that some mistakes are inevitable. An unsuitable solution routine will then either break down catastrophically or give enormous, meaningless shifts. This wastes the bulk of the computing time for the refinement cycle, which has been spent in forming the normal equations, and may make it difficult to locate the cause of the trouble.

Some common causes of singularity are

1. Attempts to refine parameters fixed by symmetry, *e.g.* the position coordinates of an atom on a symmetry centre.

2. Attempts to refine as independent parameters those which are equal by symmetry, *e.g.* x,y, and z of an atom on a threefold axis in a cubic space group.

3. Attempts to refine position coordinates of all atoms in the structure for the direction of a polar axis.

4. Attempts to refine an overall temperature factor as well as temperature factors for all atoms of the structure.

There are other, less gross, errors which can also lead to singularities. If two atoms differ in position by half the length of the x axis and have equal y and z coordinates, then exclusion of all reflections with h odd from the refinement will cause trouble. So also will refinement of the position of a site whose occupancy (individual scale factor) has fallen to zero.

If the parameter concerned is not determined by the data, a row and column of the normal matrix will be zero apart from rounding error. If two or more parameters should be equal, the corresponding rows and columns will be identical. These difficulties can be handled by means of a suitably modified Gauss or Cholesky routine (see, for example, Rollett, 1965, pp. 16–21, for a basic description of these processes). During the matrix decomposition step of the Cholesky process a singularity will appear as a diagonal element of the reduced matrix which is below a certain absolute value or has fallen to several orders of magnitude below its value in the original matrix. Such criteria have to be used rather than that of an exactly zero element because of rounding errors. The appropriate action is to set the reciprocal of the diagonal element of the matrix L (where $LL^T=A$, the normal matrix) to zero rather than the large value it would otherwise attain. This ensures that the shift and estimated standard deviation of the parameter will be given as zero, which is appropriate if it is to be kept at a fixed value. Where the parameter should have had a shift equal to that in some other, an unbalanced result will appear. It is therefore advisable to write the program so that the parameter yielding the singularity is indicated by a warning message on the output, and it may be good practice to ensure that further refinement is not allowed until the error is corrected, but this is sometimes unduly restrictive.

A more powerful method of dealing with singularities, and also with ill-conditioning of other kinds, is to compute the latent roots and vectors of the normal matrix using the Householder method (see, for example, Rollett, 1965, pp. 57–65, for an introduction to this). If the normal matrix is numerically singular, one or more of the latent roots will be many orders of magnitude smaller than the largest one. The contributions of their latent vectors to the solution are then ignored. Apart from this, if the normal equations, in matrix form, are

$$\mathbf{Ap}=\mathbf{b}\,, \tag{27}$$

and the latent roots λ_i and vectors $\mathbf{v}_i$ of A satisfy

$$\mathbf{Av}_i=\lambda_i\mathbf{v}_i\,, \tag{28}$$

$$\mathbf{v}^T\mathbf{v}_i=1\,, \tag{29}$$

we have

$$c_i=\mathbf{b}^T\,.\,\mathbf{v}_i\,, \tag{30}$$

$$\mathbf{b}=\sum_i c_i \mathbf{v}_i , \qquad (31)$$

and the solution **p** is given by

$$\mathbf{p}=\sum_i (c_i/\lambda_i)\mathbf{v}_i . \qquad (32)$$

Clearly if one of the λ_i is very small, any error in its c_i will be magnified enormously on division by λ_i. Since such very small λ_i have vectors displaying the symmetries of situations arising for atoms at special positions and so on, neglect of the vector contributions concerned preserves the proper symmetry of the solution. In this sense this method is superior to the modified Cholesky process. It is also more difficult to program, but the library Householder routine for most computer systems can be used unmodified, so that the programming labour involved may actually be less than for modified Cholesky. The Householder scheme is particularly at an advantage for the case of a polar axis. The fixing of a particular coordinate by (or for) the Cholesky process removes the singularity but leaves the normal equations ill-conditioned, especially if there is no dominating heavy atom to be fixed. The Householder process fixes the sum of the shifts of all the coordinates and removes the ill-conditioning completely. (A similar advantage could be secured with a suitably designed set of constraints, but this would be much more troublesome to prepare.)

The refinement of atomic positions and vibrations does not usually lead to badly ill-conditioned normal equations. There are some situations in which this can happen, however. An example is the attempt to refine a structure in space group $C2$ by starting from a slight distortion of a model of symmetry $C2/m$. A number of crystallographers have been distressed to find highly erratic shifts and pathologically large standard deviations in such attempts. Examination of the latent roots and vectors of the normal matrix would reveal unusually small roots (with values sensitive to the size of the distortion) for vectors related to the asymmetry introduced. To exclude the contributions of these vectors would largely prevent the distortion from growing, whereas to include the contributions at full value gives erratic results. The remedy is to filter the solution by scaling down the shift contributions of latent vectors progressively as the latent roots decrease in size. The simple expedient of dropping the contributions abruptly at a certain latent root value has, however, been used with considerable success by Diamond (1958, 1966) in his least-squares adjustments of molecular models to guide points from protein Fourier maps, and in studies of graphite.

Semi-diagonal approximations

We have already seen that the full normal matrix for n parameters has $\frac{1}{2}n(n+1)$ unique elements which have in general to be kept in the main store of the computer simultaneously and require $\frac{1}{2}mn(n+1)$ operations for their evaluation (the right-hand sides add n and mn to these figures respectively). However large and fast the computer used, these figures will imply too much storage and too much computing time for the largest problems likely to be met. In these circumstances we can take advantage of the fact that the off-diagonal elements in the normal matrix are mostly small compared with those on the main diagonal, and omit the calculation of most of these elements. Some approximations which have been used are:

1. A block for all positional parameters and a block for all vibrational parameters with the overall scale factor.

2. As for 1, but with the (anisotropic) vibration parameters in two blocks. This requires special treatment of overall scale.

3. One block for the positional parameters for each axial direction and one

block for the isotropic vibration parameters and overall scale factor.

4. One block for all parameters of atoms in, say, each quarter of the structure.

5. One block for all parameters of each atom.

Any such approximation neglects some of the correlations between the effects of different parameter changes on the $|F_c|$. Most of such correlations are quite small but some are not. In particular, the interaction between the overall scale factor and the isotropic part of any temperature factor is so strong that it should never be neglected. The special device which can be used when various temperature factors occur in different diagonal blocks will be discussed elsewhere and need not be described here.

In the earliest stages of a refinement, when the positional shifts are greater than 0·1 Å and the vibrational shifts greater than 0·005 Å^2, the behaviour of the full matrix process is commonly no better than that of an economical block diagonal scheme, because of non-linearities. There is then much to be said for commencing a large analysis with one or two cycles of the block diagonal type to reduce the shifts to the size at which linearity is a good approximation. Once this point has been reached there appears to be little doubt that it is best to compute blocks of normal matrix which are as large as it is possible to store. The time required for this can be economized greatly at a later stage by recalculating the right-hand sides of the normal equations only and assuming that the elements of the normal matrix are the same as those for a previous cycle. The evidence so far accumulated is meagre, but suggests that in the final stages the rate of reduction of shift size with computing time increases steadily as the proportion of the normal matrix computed goes up. Sparks (1961) analysed a number of closely related single atom block approximations applied to anthracene with 64 parameters. He found typically that the latent roots of the product of the inverse of the block diagonal matrix with the full matrix ranged from 0·3 to 1·9, so that the block diagonal approximation would multiply the corresponding vector components of errors by factors varying from 0·7 to −0·9 (unity less the latent root). For a pure diagonal approximation the situation was much worse and errors could oscillate with divergence. Hodgson & Rollett (1963) have analysed the situation further and shown that if the range of such latent roots can be estimated in advance, an acceleration device provides improved convergence. This device involves multiplication of the shifts by a relaxation factor which varies from cycle to cycle and is sometimes less than and sometimes greater than unity. If the sequence of factors is intelligently chosen, the progress of the refinement yields a check on the actual range of latent roots for the approximation used. It will be seen that this is of value in assessing the standard deviations obtained. Hodgson & Rollett assumed a range of latent roots for the atom-block diagonal iteration system of $0{\cdot}2 \leq \lambda_i \leq 2{\cdot}0$. Further experience showed that it was safer to modify this to $0{\cdot}3 \leq \lambda_i \leq 3{\cdot}0$. As a result the recommended sequence of shift factors became 0·61, 0·47, 0·84, 0·40, 1·28, 0·35, 2·08, 0·33, 3·12. (Many other sequences can be generated for other error reduction factors than that of 160 for this process, or for other latent root ranges.) It is prudent to use the shift factors in this sequence but to monitor the behaviour of the shifts at each cycle. If the effect of the process is to yield successive sets of shifts with the same pattern of signs and an approximately constant ratio of sizes, it is then possible that the process can be speeded up by choosing to use the most appropriate shift factor out of turn.

Stonebridge & Rollett (unpublished) have compared block diagonal procedures on isoclovene (with 16 atoms and 48 parameters). None was so fast as the full matrix process with recalculation of right-hand sides only, but the best results were obtained by alternating between two different block designs, one with one block per atom and the other with one block per axial direction. Presumably each design dealt efficiently with the errors left uncorrected by the other.

The implications of all this for the design of programs are clear. It is necessary to provide for as much flexibility as possible in selecting the number and types of parameters to be included in the blocks of normal matrix to be computed. If this is not done, efficiency will be lost and it may be difficult to deal with problems for which the block designs provided will not fit into the storage available.

Estimated standard deviations

A general survey of the theory of statistical error estimates in crystal structure analysis is given in *International Tables*, Vol. II, Section 2·6 (1959). Further accounts, with more recent references, are given by Cruickshank (1965).

We have, as before, the normal equations

$$\begin{aligned} c_{11}p_1 + \ldots + c_{1n}p_n &= d_1 \\ c_{n1}p_1 + \ldots + c_{nn}p_n &= d_n \end{aligned} \qquad (33)$$

which can be inverted to give

$$\begin{aligned} c_{11}^{-1}d_1 + \ldots + c_{1n}^{-1}d_n &= p_1 \\ c_{n1}^{-1}d_1 + \ldots + c_{nn}^{-1}d_n &= p_n . \end{aligned} \qquad (34)$$

We have already given references to methods for obtaining the c_{ij}^{-1}. The derivation of the expressions usually used for estimating the standard deviations of the p_i then depends on the following assumptions:

1. That convergence has been obtained so that all shifts in the p_i are small compared with their standard deviations.

2. That the weights have been correctly chosen to be proportional to the reciprocals of the variances (*i.e.* of the squares of the standard deviations) of the $(|F_{oi}| - |F_{ci}|)$. It must be emphasized that the variances concerned are those from all sources of error, not just the random errors of quantum counting, for example. If this assumption is correct, then no single value of $w_i(|F_{oi}| - |F_{ci}|)^2$ and no class of such values will be significantly greater than the root mean square value.

3. That the errors of the $(|F_{oi}| - |F_{ci}|)$ are not correlated. Unfortunately it is far from being clear that this is so, particularly for absorption errors and other errors which are likely to vary slowly with position in reciprocal space.

Under these assumptions, and the further assumption that the weights are actually equal to the reciprocals of the variances of the $(|F_{oi}| - |F_{ci}|)$, we have for the variance of the ith parameter.

$$\sigma^2(p_i) = c_{ii}^{-1} . \qquad (35)$$

If the weights are not equal, but only proportional, to the reciprocals of variances, then this must be replaced by

$$\sigma^2(p_i) = c_{ii}^{-1} \sum_{i=1}^{n} w_i(|F_{oi}| - |F_{ci}|)^2/(n-s) , \qquad (36)$$

where, again, n values of $|F_o|$ have been fitted with the aid of s parameters. On the same basis we have

$$\mathrm{cov}\,(p_i, p_j) = c_{ij}^{-1} \sum_{i=1}^{n} w_i(|F_{oi}| - |F_{ci}|)^2/(n-s) , \qquad (37)$$

for the covariance of two parameters.

Eq. (36) provides us with a means of estimating the standard deviations of the parameters used in the least-squares refinement. These will normally include the atomic position coordinates and we can estimate their errors from the c_{ii}^{-1}, the diagonal terms of the inverted least-squares normal matrix. Suppose, however, that we wish to estimate a standard deviation for a function of these parameters such as a bond length or a bond angle. We can write this as a (probably non-linear) function of the least-squares parameters

$$l = \text{fn}\,(p_1, \ldots, p_n)\,. \tag{38}$$

This can be differentiated to give

$$dl = \sum_{i=1}^{n} \left(\frac{\partial l}{\partial p_i}\right) dp_i\,. \tag{39}$$

Hence we deduce

$$\sigma^2(l) = \sum_{i=1}^{n} \left(\frac{\partial l}{\partial p_i}\right)^2 \sigma^2(p_i) + \sum_{i=1}^{n} \sum_{j=1}^{n} \left(\frac{\partial l}{\partial p_i}\right) \left(\frac{\partial l}{\partial p_j}\right) \times \text{cov}\,(p_i, p_j) \tag{40}$$

In order to use Eq. (40) we need to have available the full inverse of the matrix of the normal equations. It is desirable that this should be left in a backing store by the subroutine which solves the normal equations so that it can be used by a subsequent accuracy analysis subroutine. Output of the full inverse for more than a small number of parameters is expensive and hand calculation from it is highly laborious.

So far this discussion of standard deviation estimates has been based on a full-matrix scheme. Where a block diagonal approximation has been used, two errors arise. Firstly we cannot calculate the c_{ij}^{-1} correctly from the partial normal matrix. Secondly we do not have all the c_{ij}^{-1} available for use in Eq. (40). Hodgson & Rollett (1963) have given a means of estimating the overall effects of these inaccuracies. They describe an acceleration device for which shifts in successive least-squares cycles are multiplied by a series of different relaxation factors until a designed reduction factor for the indicated shifts has been achieved. The range of relaxation factors actually required gives an estimate of the range of latent roots of the product of the full normal matrix with the inverse of the approximate block diagonal matrix. This range is the range of possible ratios of variances of functions of parameters derived from the two matrices. An example may make this clear. Reference to Table 1 in Hodgson & Rollett (1963) shows that if the process has been followed until shift factors of 1·92 and 0·53 have been applied and the subsequent shifts indicate convergence, then the latent roots are in the range 0·46 to 1·96. If the variance of some function of the parameters is σ^2, and the approximate variance from the block diagonal matrix inverse is s^2, then we have

$$0{\cdot}46 \leq s^2/\sigma^2 \leq 1{\cdot}96\,. \tag{41}$$

These are the extreme bounds on ratios of variances by the two methods for functions of parameters. Ratios for individual parameters are likely to be closer to unity, and s^2/σ^2 for a single parameter cannot be greater than unity.

There remain two serious uncertainties affecting the reliability of error estimates from the least squares normal matrix. The first is that there is doubt whether the errors of the $(|F_o| - |F_c|)$ are really independent. If not, then the present type of weighting system is certainly not appropriate and the error estimates are not to be trusted. The second difficulty is that if the standard deviation of a positional parameter is more than about 0·03 Å it is quite possible for an actual error to be so large that the behaviour of $|F_c|$ will not be effectively linear. In this case the normal distribution does not give true estimates of the probabilities of errors of large size. It would appear to be better to test any given hypothetical structure model not by comparing deviations from it with the estimated standard deviations but rather by applying the R-factor ratio test of Hamilton (1965) to a free refinement and one constrained to fit the model. Operations of this kind on cases with small standard deviations have confirmed the agreement of these two types of test under good conditions, provided that the Hamilton test is applied to the actual function minimized. Application to the unweighted R-factors is likely to give a false indication that a model can be rejected.

Applications of the least-squares method

So far, the method of least-squares has been discussed in relation to the refinement of the parameters in a structure factor calculation. While this is so far the most important application, it is by no means the only one. We mention here some of the other situations in which the method has been used.

Electron density interpolation

This is a convenient example of the use of the method to fit a curve to a set of given points. It is common practice to compute three-dimensional Fourier maps with grid points spaced about 0·25 Å apart. Since spacings between atoms are commonly no more than 1·2 Å, if a particular grid point is locally the highest on an atomic peak, its nearest and next nearest neighbours are nearer to that peak than any other but this need not be true of the points separated from it by the body diagonal of a grid cell. It is therefore prudent to interpolate for the position and height of the maximum by using the 19 points given by the highest point, its 6 nearest and 12 next nearest neighbours. The peak can be expected to be Gaussian in shape and so we can fit the function

$$\ln \varrho(xyz) = a + bx + cy + dz + ex^2 + fy^2 + gz^2 + hyz + kzx + lxy \quad (42)$$

to the 19 values of ϱ given. There are 10 parameters $a, \ldots, l$ so we retain 9 degrees of freedom. Since in general each ϱ value is likely to be equally accurate we have $\sigma(\ln \varrho) = \text{const}/\varrho$. The most reasonable choice of weights is therefore $w_i = \text{const.} \times \varrho^2$ and the observational Eq. (42) is prepared by multiplying it through by ϱ_i. If any of the ϱ_i chances to be less than zero (which can happen if the process is applied in automatic scanning of Fourier maps), then the weight can be zero without inconsistency and the problem of logarithms of negative numbers is avoided. When the values of $a, \ldots, l$ have been determined from the normal equations, the position of the maximum is readily obtained by deriving three simultaneous equations for x, y and z from the conditions

$$\frac{\partial(\ln \varrho)}{\partial x} = \frac{\partial(\ln \varrho)}{\partial y} = \frac{\partial(\ln \varrho)}{\partial z} = 0\,. \quad (43)$$

Some simplification of the normal equations can be obtained by giving equal weight to the $\ln \varrho$ values and working with coordinates of -1, 0, and 1 for the grid points, but this seems unjustified except for hand calculation. If the peak being analysed is sufficiently distorted, it is possible that the function fitted to $\ln \varrho$ may not go to minus infinity at large distances from the origin in all directions. In this case a negative element on the diagonal will be encountered in the pivotal condensation of the three simultaneous equations given by the conditions (43).

Mean plane through a group of atoms

When the position coordinates of a set of atoms are known it is sometimes necessary to calculate the equation of a best plane through them, for example to determine whether or not the individual atoms depart significantly from this plane. We will suppose that there are n atoms with position coordinates referred to axes of any nature

$$X_i, Y_i, Z_i, \; i = 1, \ldots, n \quad (44)$$

and that their positional uncertainties in the axial system chosen are described by known variances and covariances:

$$\sigma^2(X_i),\ \sigma^2(Y_i),\ \sigma^2(Z_i)$$
$$\operatorname{cov}(Y_i, Z_i),\ \operatorname{cov}(Z_i, X_i),\ \operatorname{cov}(X_i, Y_i)\,. \tag{45}$$

We define the best plane as the plane

$$ax+by+cz=d\,, \tag{46}$$

such that the function

$$M=\sum_{i=1}^{n}[w_{11i}(X_i-x_i)^2+\dots+w_{33i}(Z_i-z_i)^2+w_{23i}(Y_i-y_i)\times(Z_i-z_i)+\dots+w_{12i}(X_i-x_i)\times(Y_i-y_i)] \tag{47}$$

is minimized subject to the subsidiary conditions

$$ax_i+by_i+cz_i+d=0\,,\ i=1,\dots,n\,, \tag{48}$$

and

$$a^2+b^2+c^2=1\,. \tag{49}$$

The w_{jki} are the elements of the matrix obtained by inverting the variance covariance matrices of the X_i, Y_i, Z_i. In consequence the solution gives the maximum likelihood that these points deviate from the plane by random error alone. For brevity, we will outline the iterative algorithm for solving the problem without proving it. The proof is a straightforward application of the method of Lagrangian multipliers.

We start with values of a, b, c all equal to unity. These do not satisfy Eq. (49) and the test for convergence of the iteration always fails on the first cycle. We compute

$$w_i=[a^2\sigma^2(X_i)+b^2\sigma^2(Y_i)+c^2\sigma^2(Z_i)+2bc\operatorname{cov}(Y_i,Z_i)+2ca\operatorname{cov}(Z_i,X_i)+2ab\operatorname{cov}(X_i,Y_i)]^{-1} \tag{50}$$

$$\bar{X}=\sum_{i=1}^{n}(w_iX_i)/\sum_{i=1}^{n}(w_i),\ \bar{Y},\ \bar{Z}\ \text{similarly} \tag{51}$$

$$X'_i=X_i-\bar{X},\ Y'_i=Y_i-\bar{Y},\ Z'_i=Z_i-\bar{Z} \tag{52}$$

On the first iteration we set $\Delta_i=0$ for all i, but on later iterations we get

$$\Delta_i=aX'_i+bY'_i+cZ'_i \tag{53}$$

We then form the equations

$$\sum_{i=1}^{n}[w_iX'^2_i-\sigma^2(X_i)w_i^2\Delta_i^2]a+\sum_{i=1}^{n}[w_iX'_iY'_i-\operatorname{cov}(X_i,Y_i)w_i^2\Delta_i^2]b+\sum_{i=1}^{n}[w_iX'_iZ'_i-\operatorname{cov}(X_i,Z_i)w_i^2\Delta_i^2)c=0 \tag{54}$$

$$\sum_{i=1}^{n}[w_iX'_iY'_i-\operatorname{cov}(X_i,Y_i)w_i^2\Delta_i^2)a+\sum_{i=1}^{n}[w_iY'^2_i-\sigma^2(Y_i)w_i^2\Delta_i^2]b+\sum_{i=1}^{n}[w_iY'_iZ'_i-\operatorname{cov}(Y_i,Z_i)w_i^2\Delta_i^2)c=0 \tag{55}$$

$$\sum_{i=1}^{n}[w_iX'_iZ'_i-\operatorname{cov}(X_i,Z_i)w_i^2\Delta_i^2)a+\sum_{i=1}^{n}[w_iY'_iZ'_i-\operatorname{cov}(Y_i,Z_i)w_i^2\Delta_i^2)b+\sum_{i=1}^{n}[w_iZ'^2_i-\sigma^2(Z_i)w_i^2\Delta_i^2)c=0 \tag{56}$$

There is no exact solution of these homogeneous equations unless their determinant is zero. We therefore obtain the best approximate solution by finding the latent vector corresponding to the smallest latent root, normalized so that $a^2+b^2+c^2=1$. If the sum of squares of differences between the old and new values if a, b, and c is small enough we stop (this cannot happen on the first iteration, a limit of 10^{-12} gives six figures stability in the results). Otherwise we substitute the new a, b, and c for the old and repeat the calculation from Eq. (50). Having obtained convergence we can get

$$d=-(a\bar{X}+b\bar{Y}+c\bar{Z}) \tag{57}$$

$$x_i = X_i - w_i \Delta_i [a\sigma^2(X_i) + b \operatorname{cov}(X_i, Y_i) + c \operatorname{cov}(X_i, Z_i)] \quad (58)$$

with similar expressions for y_i and z_i. The points (x_i, y_i, z_i) are those in the plane least significantly different from the (X_i, Y_i, Z_i).

This is the extension to three dimensions of an algorithm for fitting a line to a set of points which has been developed by Cumming & Rollett and tested on data from problems in geological age determination and reaction kinetics.

Automatic solution of crystal structures

Methods have been developed by Ford, Hodgson, Rollett & Stonebridge (unpublished) for automatic solution of crystal structures. A Fourier map can frequently be produced, by Patterson or direct methods, which reveals the sites of some of the atoms of a structure. If a heavy atom is located with certainty, it can be put into the structure factor calculation with full, fixed weight. Other atoms can be put in with occupation numbers (individual scale factors) of zero, and the correctness of the sites tested by refinement of the occupation numbers by least squares. After three cycles the shifts have usually become small and the improved model can be used to phase a new Fourier map which is calculated and interpreted within the computer to give a list of distinct peaks. Peaks which are redundant, as symmetry repeats, or are lower than a height limit set to give a list of the required length, and so on, are eliminated and a new least-squares parameter list is generated. A further major cycle of the process then begins, starting from the occupation numbers given by the previous cycle, plus zero occupations for new sites. The scheme has been highly successful with noncentrosymmetric structures up to about 35 atoms in the asymmetric unit, in which traditional methods were difficult to apply because of false symmetry effects. A typical case is that of a structure in space group $P2_1$ with one unique heavy atom only.

Many problems arise in this application of the least-squares method. The parameter shifts are large, the equations are highly non-linear, and the normal matrix is ill-conditioned in the case that there is partial resolution of false symmetry. The errors of $(|F_o| - |F_c|)$ are dominated at most stages by the defects of $|F_c|$ rather than $|F_o|$ and weighting schemes which take this into account have aided convergence. Schemes of the types $w = |F_c|/(|F_c| + |F_o|)$ and $w = 1/[fT(|F_o|^2 + |F_c|^2)]$ have both proved effective. Weighting of the Fourier maps, in the way proposed by Sim (1961), is also desirable to reduce the ill effects of poor phasing of reflections with $|F_c| < |F_o|$.

The chemical type of an atom which might be carbon, nitrogen, or oxygen can be checked by assigning it a carbon form factor with occupation unity and refining the occupation of a form factor which is equal to the difference between those for carbon and oxygen. In this, as in the original process to establish the sites, it is unusual for all occupation numbers to agree with expected values such as zero or unity within reasonable statistical error bounds. The discrepancies observed are believed to be due to the strong interaction between the occupation numbers and the corresponding individual atomic temperature factors. Attempts at least-squares refinements of both types of parameter together give very poor results, but the situation can be retrieved by the use of 'bond-stretching' constraints on the temperature factors. The variations of occupation from one site to another are then within statistically reasonable ranges, but there is a tendency for the range of carbon atom occupations to be too high. It is believed that this may be connected with difficulties in compensating for the effects of hydrogen atoms. Unfortunately these cannot be placed until the other atomic types are decided (in general) and the common case is that of distinguishing between CH_3, NH_2 and OH.

References

Apostol, T. M. (1957). *Mathematical Analysis.* Reading, Mass.: Addison-Wesley.

Cruickshank, D. W. J. (1965). *Computing Methods in Crystallography*, pp. 99–116. Oxford: Pergamon Press.

Diamond, R. (1958). *Acta Cryst.* **11**, 129.

Diamond, R. (1966). *Acta Cryst.* **19**, 744.

Hamilton, W. C. (1965). *Acta Cryst.* **18**, 502.

Hodgson, L. I. & Rollett, J. S. (1963). *Acta Cryst.* **16**, 329.

International Tables for X-ray Crystallography (1959). Vol. II. Birmingham: Kynoch Press.

Rollett, J. S. (1965). *Computing Methods in Crystallography.* Oxford: Pergamon Press.

Sim, G. A. (1961). *Computing Methods and the Phase Problem in X-ray Analysis.* Oxford: Pergamon Press.

Sparks, R. A. (1961). *Computing Methods and the Phase Problem in X-ray Analysis.* Oxford: Pergamon Press.

Waser, J. (1963). *Acta Cryst.* **16**, 1091.

Whittaker, E. T. W. & Robinson, G. (1937). *The Calculus of Observations.* London: Blackie.

Discussion

JOHNSON: I question the use of the term 'Householder method' for obtaining the principal eigenvalues and eigen-vectors, since this method only converts the matrix to the codiagonal form, and one still has to use some other procedure to get the principal components of the matrix. Also, instead of the term 'filtering' which has crept into crystallography, it would be more advantageous to use the statistical term 'regression on principal components' so that one can get into more rigorous literature on the subject.

ROLLETT: I am using the term 'Householder method' to cover the whole process, but I agree with what you said. There are many names for the 'filtering' process, so one need not object very strongly to this term.

COPPENS: There is an alternative to the linear constraint of the Lagrangian modifier method, described in Hamilton's book, which does not enlarge the matrix but enables the unconstrained parameters to be corrected *a posteriori* for the effect of constraints.

ROLLETT: This was the way in which Waser originally published the approximate constraint process in 1963. It is rather complicated and we haven't actually done it this way.

SPARKS: How does one decide on how much to relax the constraints for the final set of parameters that are published?

ROLLETT: One way is to compare the minimization function of the constrained and the unconstrained refinements using Hamilton's test and to see whether the difference is significant. In that case one leaves the constraints in with full weights up to the end.

SOLTZBERG: Please comment on the relation between the overall temperature factor as determined from the Wilson plot and by least squares. For one of our structures, the B values were 2 and 5 Å^2, respectively; and we wonder whether disorder or some other physical situation could cause it.

ROLLETT: Rogers has discussed the effect of systematic errors on the Wilson plot in *Computing Methods in Crystallography* (1965).

PAWLEY: What are your objections to using the unweighted R for these tests?

ROLLETT: The quantity $\sqrt{w}||F_o|-|F_c||$ is a normal variate with unit standard deviation; so that if you take $\Sigma\, w(|F_o|-|F_c|)^2$ with proper weights, you have what is called a γ variate. We can apply the χ^2 or Hamilton's test to this variate because each term in the sum is of expectation unity. If the weights are left out, the expected values of the terms are no longer unity and may vary from term to term. In cases where a structure has a small number of very strong reflections, then those few reflections will dominate the refinement and the effective number of degrees of freedom is reduced.

Scaling of Intensity Data Collected with Two or More Crystal Settings

R.A. SPARKS

Syntex Research, Palo Alto, California, U.S.A.

The technique for scaling intensity data that will be described here is that of Hamilton, Rollett & Sparks (1965). The method can be applied to any problem where many measurements of the same data are made in sets and an unknown scale factor is required for each set to place all the data onto a common scale. Thus the technique which will be described has been used by Abrahams, Alexander, Furnas, Hamilton, Ladell, Okaya, Young & Zalkin (1967) to compare data collected by various investigators, each using his own diffractometer. The method has also been used to obtain film factors for a set of films for a given layer of intensity data.

The technique described here differs from most others in the literature in that there is one observational equation for each independent measurement. Most of the other methods have one observational equation for ratios of two measurements. These other methods become cumbersome when there are more than two crystal settings.

The Hamilton, Rollett & Sparks method is as follows:

The observational equations are:

$$\varphi_{hl} = \sqrt{W_{hl}}\,d_{hl} = \sqrt{W_{hl}}(F^2_{hl} - G_l F^2_h) \qquad (1)$$

where F^2_{hl} is the observation of the independent reflection F^2_h measured on the lth film, G_l is a scale factor for the lth film, and $\sqrt{W_{hl}}$ is the weight for the observation and is set proportional to the reciprocal value of the estimated standard deviation.

The minimization function is then

$$M = \sum_{h,l} \varphi^2_{hl} = \sum_{h,l} W_{hl} d^2_{hl} \qquad (2)$$

The best least-squares values of the F^2_h's are found from

$$\frac{\partial M}{\partial F^2_h} = 0$$

Thus,

$$F^2_h = \sum_l (W_{hl} G_l F^2_{hl}) / \sum_l (W_{hl} G^2_l) \qquad (3)$$

The values of F^2_h determined by Eq. (3) are substituted into Eqs. (1). Eqs. (1) are now non-linear and linearized as described by Rollett in the preceding paper. Thus, the observations become:

$$\varphi_{hl} \simeq \sqrt{W_{hl}}(F^2_{hl} - G_l F^2_h) + \sum_k (\partial \varphi_{hl} / \partial G_k) \Delta G_k \qquad (4)$$

Approximate values of the G_l are chosen and the ΔG_l are calculated by the usual least-squares method. In matrix notation the normal equations have the form:

$$\mathbf{Ag} = \mathbf{b} \qquad (5)$$

where the elements of the vector **g** are the ΔG_l.

Several methods have been used for solving the above equations. The one chosen by Hamilton, Rollett & Sparks is to choose one scale factor equal to unity and solve for incremental changes in the other scale factors. In order to force the solution to behave properly, it was necessary to introduce the following constraint: set $\Delta G_l = -0{\cdot}7\, G_l$ if the calculated $\Delta G_l \leq -0{\cdot}7\, G_l$.

The new scale factors are used in the next iteration and the iterations continue until all the ΔG_l are less than a certain preset minimum. Convergence properties are important in this problem and three different techniques have been used. One method, due to Monahan, Schiffer & Schiffer (1967), decreases the amount of computation needed for each iteration. However, it tends to increase the total number of iterations required. This algorithm is as follows. Observe that minimizing M with respect to the G_l leads to the following result:

$$\frac{\partial M}{\partial G_l}=0$$

$$G_l=\sum_h(W_{hl}F_{hl}^2F_h^2)/\sum_h(W_{hl}F_h^4) \qquad (6)$$

With trial values of the G_l Eqs. (3) are solved. Then with the resulting values of the F_h^2 Eqs. (6) are solved. These two steps are alternated until on successive cycles the G_l and F_h^2 do not change appreciably.

Another technique for speeding convergence was described by Fox & Holmes (1966). Fox & Holmes observed that the direction of the shift vector is not an optimum one in the Hamilton, Rollett & Sparks method. They therefore used the following method. Let $\boldsymbol{\beta}$ be the matrix of eigen-vectors of $\mathbf{A}$. Then Eq. (5) can be rewritten:

$$(\boldsymbol{\beta}^T\mathbf{A}\boldsymbol{\beta})\boldsymbol{\beta}^T\mathbf{g}=\boldsymbol{\beta}^T\mathbf{b} \qquad (7)$$

$\boldsymbol{\beta}^T\mathbf{A}\boldsymbol{\beta}$ will be a diagonal matrix with one diagonal element equal to zero. Eq. (7) is solved for the vector $\boldsymbol{\beta}^T\mathbf{g}$. The element of $\boldsymbol{\beta}^T\mathbf{g}$ corresponding to the zero diagonal element of $\boldsymbol{\beta}^T\mathbf{A}\boldsymbol{\beta}$ is set equal to zero. The shifts are then calculated and the iteration proceeds as in the Hamilton, Rollett & Sparks method. The Fox & Holmes method takes fewer iterations but requires more arithmetic operations per iteration.

Ford & Rollett (1967) have essentially combined the methods of Fox & Holmes and Monahan, Schiffer & Schiffer. Ford & Rollett replace Eq. (6) with the following:

$$H_l=G_l \qquad (8)$$

$$A_l=\sum_h[W_{hl}F_h^4+W_{hl}^2F_{hl}^2(F_{hl}^2-2F_h^2H_l)]/(\sum_k W_{hk}H_k^2) \qquad (9)$$

$$G_l=H_l+[\sum_h(W_{hl}F_h^2F_{hl}^2)-H_l\sum_h(W_{hl}F_h^4)]/A_l \qquad (10)$$

This substitution is a good approximation to the method of Fox & Holmes. The procedure then is to iteratively calculate Eqs. (3), (8), (9), and (10) until on successive iterations the changes ΔG_l are less than a specified minimum. In addition Ford & Rollett use an acceleration technique with very little extra cost in terms of arithmetic operations. This acceleration method results in convergence properties almost as good as Fox & Holmes. Thus, the method of Ford & Rollett is undoubtedly the least costly in terms of computer time.

Basically, the least-squares scaling techniques described above are methods designed to reduce the total amount of data by combining independent measures of the same variable. It is important to remember, however, that this data reduction gives data which are not now independent. To my knowledge, no one has dealt with the effect that this might have on a crystal structure. Except in peculiar situations (*i.e.* data very weak in certain data sets), these effects would probably be unimportant. If this interdependence is a problem, the scaling factors can be handled as additional parameters in the atomic parameter least-squares programs.

References

Abrahams, S. C., Alexander, L. E., Furnas, T. C., Hamilton, W. C., Ladell, J., Okaya, Y., Young, R. A. & Zalkin, A. (1967). *Acta Cryst* **22**, 1.

Ford, G. C. & Rollett, J. S. (1967). *Acta Cryst.* B**24**, 293.
Fox, G. C. & Holmes, K. C. (1966). *Acta Cryst.* **20**, 886.
Hamilton, W. C., Rollett, J. S. & Sparks, R. A. (1965). *Acta Cryst.* **18**, 129.
Monahan, J. E., Schiffer, M. & Schiffer, J. P. (1967). *Acta Cryst.* **22**, 322.

Discussion

STEWART: Do not these methods also have applications in cases where several crystals are used to collect the full data?

SPARKS: Yes, they do.

ALCOCK: If the scale factors are going to be refined with the other parameters by least-squares, why should we not start with linear scales initially?

SPARKS: It is important to scale the data earlier on, since, as mentioned earlier, the direct methods work more satisfactorily with accurate data.

TOPIC E

LEAST-SQUARES REFINEMENT OF ATOMIC PARAMETERS

CHAIRMAN: D. W. J. CRUICKSHANK

Least-Squares Refinement of Atomic Parameters

D.W.J. CRUICKSHANK

Chemistry Department, University of Manchester Institute of Science and Technology, Manchester M60 1QD, England.

1. Definitions: structure factors and Fourier series

The object of this contribution is to set down the basic equations which must be programmed to refine atomic parameters by the method of least squares.

If anomalous scattering is negligible the structure factor

$$F(hkl) = V \int_0^1 \int_0^1 \int_0^1 \varrho(xyz) \times \exp[2\pi i(hx+ky+lz)]dxdydz \quad (1\cdot 1)$$

$$= A(hkl) + iB(hkl) \quad (1\cdot 2)$$

$$= |F(hkl)| \exp[i\alpha(hkl)] \quad (1\cdot 3)$$

where V is the cell volume, xyz are fractional coordinates, and $\varrho(xyz)$ is the electron density. Note

$$|F(\bar{\mathbf{h}})| = |F(\mathbf{h})| \text{ and } \alpha(\bar{\mathbf{h}}) = -\alpha(\mathbf{h}). \quad (1\cdot 4)$$

By the properties of Fourier series $F(\mathbf{h})$ is exactly equal to the value of the appropriate Fourier coefficient of $\varrho(\mathbf{x})$, so that

$$\varrho(xyz) = \frac{1}{V} \sum_h \sum_k \sum_l F(hkl) \times \exp[-2\pi i(hx+ky+lz)] \quad (1\cdot 5)$$

where the summations in h,k,l each run from $-\infty$ to $+\infty$.

If ϱ is a superposition of r 'atomic' densities (1·1) may be written

$$F(hkl) = \sum_r f_r(hkl) \times \exp[2\pi i(hx_r+ky_r+lz_r)] \quad (1\cdot 6)$$

where the summation is over all atoms in the unit cell and

$$f_r(hkl) = V \int\int\int_{-\infty}^{+\infty} \varrho_r(uvw) \times \exp[2\pi i(hu+kv+lw)]dudvdw \quad (1\cdot 7)$$

is the scattering factor of the atom r whose electron density is ϱ_r. *If the atomic densities are centrosymmetrical* with respect to their origins (x_r, y_r, z_r), the f_r are real, and

$$A(hkl) = \sum_r f_r(hkl) \times \cos[2\pi(hx_r+ky_r+lz_r)] \quad (1\cdot 8a)$$

$$B(hkl) = \sum_r f_r(hkl) \times \sin[2\pi(hx_r+ky_r+lz_r)] \quad (1\cdot 8b)$$

For space groups with a centre of symmetry at the origin $B(hkl)=0$.

2. Effect of atomic vibrations

A. Since the vibration-averaged density is the convolution* of the rest density with the smearing function,

$$f_T(hkl) = f(hkl)q(hkl) \quad (2\cdot 1)$$

where the scattering factors are f_T for the atom in vibration and f for the atom at rest, and q is the transform of the smearing function t.

B. *If an atom vibrates* at any temperature T *in an isotropic harmonic potential,*

* For an atom not subject to libration.

the smearing function $t(\mathbf{x})$ is a Gaussian (Bloch, 1932),

$$t(\mathbf{x})=(2\pi^2U)^{-3/2}\exp(-x^2/2U) \quad (2\cdot2)$$

in which the only parameter is $U=\overline{u^2}$, the mean square amplitude of vibration in any direction.

(The wave function for the ground state of a harmonic oscillator leads to a Gaussian probability density. Excited states do not correspond to Gaussian densities. Bloch showed that, with a Boltzman $[\exp(-E/kT)]$ probability distribution among the energy levels $[E=(n+\frac{1}{2})h\nu]$, the total average density $t(\mathbf{x})$ was Gaussian.)

C. The transform of $t(\mathbf{x})$ is

$$q(s)=\exp(-2\pi^2Us^2) \quad (2\cdot3)$$

where $s=|\mathbf{s}|=2\sin\theta/\lambda$ is the reciprocal radius, or for isotropic vibration

$$q(s)=\exp[-B(\sin\theta/\lambda)^2] \quad (2\cdot4)$$

where $B=8\pi^2U=8\pi^2\overline{u^2}$ is the Debye factor.

The following table illustrates the relation between B, U, and the r.m.s. amplitude

B	U	r.m.s. amplitude
0·2 Å^2	0·0025 Å^2	0·05 Å
1·0	0·013	0·11
3·0	0·038	0·20
10·0	0·127	0·36

A r.m.s. amplitude of 0·05 Å is about the smallest encountered, *e.g.* roughly that of a carbon atom in diamond. Values of B between 3 and 10 are often encountered in organic structure analyses at room temperature. The corresponding r.m.s. amplitudes show that some caution must be exercised (and appropriate corrections made) before claiming to have determined a bond length to an accuracy of, say, 0·01 Å.

3. Anisotropic harmonic vibrations

In an anisotropic harmonic potential field the vibrations are described by a symmetric tensor $\mathbf{U}$, such that the m.s. amplitude in the direction of a unit vector $\mathbf{l}$ is

$$\overline{u^2}=\Sigma\Sigma\, U^{ij}l_il_j=\Sigma\Sigma\, U_{ij}l^il^j \quad (3\cdot1)$$
$$=U^{11}(l_1)^2+2U^{23}l_2l_3+\ldots$$
$$=U_{11}(l^1)^2+2U_{23}l^2l^3+\ldots$$

Here $\mathbf{U}$ has been expressed in two forms, in terms either of its contravariant components U^{ij} or its covariant components U_{ij} with respect to an arbitary set of base axes $\mathbf{a}_i$ $(i=1,2,3)$, whose reciprocal axes $\mathbf{a}^i$ are defined by $\mathbf{a}_i\,.\,\mathbf{a}^j=\delta_i^j$. Similarly l^i and l_i are the contravariant and covariant components of the unit vector

$$\mathbf{l}=l_1\mathbf{a}^1+l_2\mathbf{a}^2+l_3\mathbf{a}^3=l^1\mathbf{a}_1+l^2\mathbf{a}_2+l^3\mathbf{a}_3\,. \quad (3\cdot2)$$

[For details of tensor notation see Patterson (1959) and Cerrini (1970)].

If the axes are particularized by choosing a base set $\mathbf{e}_i$ such that the members of its reciprocal base $\mathbf{e}^i$ are all of unit length and parallel to the crystal reciprocal axes ($\mathbf{a}^*$, $\mathbf{b}^*$, $\mathbf{c}^*$), then for $\mathbf{l}$ parallel to $\mathbf{a}^1$, it has the single non-zero component $l_1=1$, and

$$\overline{u^2}=U^{11}$$

is the m.s. amplitude in the direction parallel to $\mathbf{a}^*$.

At a reciprocal lattice point $\mathbf{s}$ with components (ha^*, kb^*, lc^*) with respect to the unit reciprocal base $\mathbf{e}^i$, the transform of the anisotropic smearing function is

$$q(hkl)=\exp[-2\pi^2(U^{11}h^2a^{*2}+U^{22}k^2b^{*2}$$
$$+U^{33}l^2c^{*2}+2U^{23}klb^*c^*$$
$$+2U^{31}lhc^*a^*+2U^{12}hka^*b^*)]\,. \quad (3\cdot3)$$

Note that this expression has often been written, *e.g.* Cruickshank (1956), with U^{ij} replaced by 'U_{ij}' (but not with any implication that the 'U_{ij}' were the covariant components of $\mathbf{U}$). Although (3·3) is a mixture of notations, the use of U^{ij} rather than 'U_{ij}' serves as a reminder that (3·3) uses the components of $\mathbf{U}$ with respect to the unit $\mathbf{e}^i$ parallel to the crystal reciprocal axes. For non-orthogonal axes the dual axes $\mathbf{e}_i$ will not be of unit length. In tensor notation, the equality $U^{ij}=U_{ij}$ holds only when the base axes are mutually orthogonal and of unit length, $\mathbf{a}_i \cdot \mathbf{a}_j=\mathbf{a}^i \cdot \mathbf{a}^j=\delta_i^j$, so that $\mathbf{a}_i=\mathbf{a}^i$.

On the proper use of the phrase 'thermal vibration ellipsoid' see the paper on representational surfaces for thermal motion by Nelmes (1969).

4. The normal equations of least-squares

The function most commonly minimized is

$$R=\sum_{hkl} w(|F_o|-|F_c|)^2=\Sigma w\Delta^2 \qquad (4{\cdot}1)$$

where the sum is over the set of crystallographically independent planes and the $w(hkl)$ are weights. The lowest standard deviations in the derived parameters are obtained for

$$w(hkl)=1/\sigma^2(hkl) \qquad (4{\cdot}2)$$

where $\sigma(hkl)$ is the standard deviation of $|F(hkl)|$.

Let $p_1, p_2 \ldots, p_n$ be the n parameters occurring in the $|F_c|$ whose values are to be determined. (Note any scale factors of the $|F_o|$ must be determined from the appropriate inverse parameters of the $|F_c|$.) Then, as discussed by Rollett (1970), the $n \times n$ normal equations for small changes ε_i in the parameters p_i are

$$\sum_i c_{ij}\varepsilon_i=d_j \qquad (4{\cdot}3)$$

where

$$c_{ij}=\sum_{hkl} w \frac{\partial|F_c|}{\partial p_i}\frac{\partial|F_c|}{\partial p_j} \qquad (4{\cdot}4)$$

$$d_j=\sum_{hkl} w\Delta \frac{\partial|F_c|}{\partial p_j}. \qquad (4{\cdot}5)$$

It is the normal equations which must be set up and solved to refine a structure.

In place of the function R the function

$$R_2=\sum_{hkl} w_2(|F_o|^2-|F_c|^2)^2 \qquad (4{\cdot}6)$$

is sometimes minimized. The revised normal equations may be obtained from (4·3) by replacing w by $w_2=w/(4|F_o|^2)$, Δ by $(|F_o|^2-|F_c|^2)$, and $\partial|F_c|/\partial p_j$ by $2|F_c|\partial|F_c|/\partial p_j$.

5. Values of $\partial|F_c|/\partial p_j$

We may write

$$|F_c|=A\cos\alpha+B\sin\alpha \qquad (5{\cdot}1)$$

and

$$A=\sum_r\sum_s A_{rs}=\sum_r A_r\,, \qquad (5{\cdot}2a)$$

$$B=\sum_r\sum_s B_{rs}=\sum_r B_r\,, \qquad (5{\cdot}2b)$$

where A_{rs} is the contribution to A from one atom s related to r by symmetry and A_r is the contribution from atom r and its symmetry equivalents. B_{rs} and B_r are similarly defined. If p_j is a parameter of atom r (and its equivalents), it can be shown that

$$\frac{\partial|F_c|}{\partial p_j}=\frac{\partial A_r}{\partial p_j}\cos\alpha+\frac{\partial B_r}{\partial p_j}\sin\alpha\,. \qquad (5{\cdot}3)$$

To allow for an overall scale factor G on the whole set of F_c, a site occupation factor n_r, and anisotropic vibrations, we may write

$$A_{rs}=Gn_rf_{rs}\exp[-2\pi^2(U^{11}{}_{rs}h^2a^{*2}+\ldots)]$$
$$\times\cos[2\pi(hx_{rs}+\ldots)]\,, \qquad (5{\cdot}4)$$

and similarly for B_{rs}, where f_{rs} is the (real) scattering factor for the atom at rest, and $U^{11}{}_{rs}$ and x_{rs} denote parameters of the atom s symmetry-related to r. For isotropic vibrations the exponential becomes

$$\exp[-2\pi^2U_r(2\sin\theta/\lambda)^2]\,. \qquad (5{\cdot}5)$$

For the moment we will assume that the crystal has only translational symmetry, *i.e.* space group $P1$. To emphasize that for the present A_r and B_r each have only one term, we change the suffix to s, *i.e.* A_s, B_s. We now consider some particular values of $\partial A_s/\partial p_j$, etc.

Atomic coordinates

$$\partial A_s/\partial x_s=(-2\pi h)Gnf \exp[\ldots] \times \sin 2\pi[\ldots]=-2\pi h B_s; \quad (5\cdot6a)$$

$$\partial B_s/\partial x_s=2\pi h A_s\,. \quad (5\cdot6b)$$

Atomic isotropic vibrations

$$\partial A_s/\partial U_s=-2\pi^2(2\sin\theta/\lambda)^2 A_s; \quad (5\cdot7a)$$

$$\partial B_s/\partial U_s=-2\pi^2(2\sin\theta/\lambda)^2 B_s\,. \quad (5\cdot7b)$$

Atomic anisotropic vibrations

$$\partial A_s/\partial U^{11}{}_s=-2\pi^2h^2a^{*2}A_s; \quad \partial B_s/\partial U^{11}{}_s=-2\pi^2h^2a^{*2}B_s; \quad (5\cdot8)$$

$$\partial A_s/\partial(2U^{23}{}_s)=-2\pi^2klb^*c^*A_s, \text{ etc.} \quad (5\cdot9)$$

Site occupation factors

$$\partial A_s/\partial n_s=Gf \exp \cos=A_s/n_s; \quad \partial B_s/\partial n_s=B_s/n_s\,. \quad (5\cdot10)$$

Note that these derivatives (5·6) to (5·10) with respect to atomic parameters are all multiples of A_s and B_s. This greatly eases the programming of SFLS calculations.

Overall isotropic vibrations

If the vibration m.s. amplitude U is common to all atoms

$$\partial|F_c|/\partial U=-2\pi^2(2\sin\theta/\lambda)^2|F_c|\,. \quad (5\cdot11)$$

Overall scale factor

If G is the overall scale factor for the set of F_c

$$\partial|F_c|/\partial G=|F_c|/G\,. \quad (5\cdot12)$$

6. Allowance for symmetry in the SFLS equations

Suppose the symmetry is higher than $P1$. Reconsider A_{rs} (5·4) where s is related to r by symmetry. Any symmetry operation is a combination of a 3×3 rotation matrix $\mathbf{R}$ and a translation vector $\mathbf{t}$. Thus the coordinates of s are

$$\mathbf{x}_{rs}=\mathbf{R}\mathbf{x}_r+\mathbf{t}\,. \quad (6\cdot1)$$

For example the equivalent position $(-y, x-y, z+\frac{1}{3})$ generated by a 3_1 operation in the space group $P3_1$ may be written

$$\begin{pmatrix} x_{rs} \\ y_{rs} \\ z_{rs} \end{pmatrix} = \begin{pmatrix} 0 & -1 & 0 \\ 1 & -1 & 0 \\ 0 & 0 & 1 \end{pmatrix} \begin{pmatrix} x_r \\ y_r \\ z_r \end{pmatrix} + \begin{pmatrix} 0 \\ 0 \\ \frac{1}{3} \end{pmatrix}. \quad (6.2)$$

As a consequence of (6·1) the argument of the cos in A_{rs}, apart from the factors 2π, may be written

$$\mathbf{h}\,.\,\mathbf{x}_{rs}=\mathbf{h}\,.\,(\mathbf{R}\mathbf{x}_r+\mathbf{t})=\mathbf{h}_s\,.\,\mathbf{x}_r+\mathbf{t}_s \quad (6\cdot3)$$

where $\mathbf{hR}=\mathbf{h}_s=(h_s, k_s, l_s)$ is a set of 'equivalent indices' and $t_s=\mathbf{h}\,.\,\mathbf{t}$ is a scalar quantity, independent of the particular atom r of the asymmetric unit. For the 3_1 operation above we find

$$h_s=k\,,\ k_s=-h-k\,,\ l_s=l;\ t_s=l/3\,. \quad (6\cdot4)$$

Equivalent indices may be used in exactly the same way in the argument of the anisotropic exponential. Hence

$$A_{rs}=Gn_rf_{rs} \exp[-2\pi^2(U^{11}{}_{rs}h_s^2a^{*2}+\ldots)] \times \cos[2\pi(h_sx_r+\ldots+t_s)] \quad (6\cdot5)$$

and similarly for B_{rs}. Recalling that A_r is defined as the contribution from atom r and its symmetry equivalents, we have for example

$$\partial A_r/\partial x_r=\sum_s -2\pi h_s B_{rs} \quad (6\cdot6)$$

$$\partial A_r/\partial U^{11}{}_r=\sum_s -2\pi^2h_s{}^2a^{*2}A_{rs} \quad (6\cdot7)$$

where the summations are over the set of symmetry operations including the identity operation. Thus the derivatives of A_r and B_r involve summations of multiples of A_{rs} and B_{rs}.

This treatment is suitable for equivalent positions generated by rotations, reflections, screw axes, and glide planes, but to deal with pure translations in non-primitive cells or with a centre of symmetry at the origin it is convenient to introduce an appropriate multiplicity M so that A_{rs} is redefined as

$$A_{rs} = MGn_r f_{rs} \exp[\ldots] \cos[\ldots] \quad (6{\cdot}8)$$

and similarly for B_{rs}. The summations in (6·6), etc. then include only a minimal set of rotations, reflections, screw axes, and glide planes. The B_{rs} are still calculated if there is a centre of symmetry, since they are needed for the derivatives, but we set $B=0$. [(6·8) is inappropriate for, say, planes with $h+k$ odd in a C lattice, since such planes will be systematically absent. We assume no attempt is made to calculate structure factors for such planes.]

In passing we note the general relation

$$F(\mathbf{h}_s) = F(\mathbf{h}) \exp(-2\pi i t_s) \quad (6{\cdot}9)$$

if $\varrho(\mathbf{x}) = \varrho(\mathbf{Rx}+\mathbf{t})$ for all $\mathbf{x}$.

In polar space groups the position of the origin in one or more dimensions is not fixed by reference to symmetry elements. Thus in $P1$ the origin may be chosen arbitarily, in Pm it may be anywhere in the mirror plane, in $P2$ anywhere along the twofold axis. In such cases it is convenient to fix the arbitrariness by reference to one atom, often the heaviest atom of the structure. Thus in $P2$ we can assign one atom to $(x, 0, z)$ and its y coordinate must not be refined (see Rollett, 1970).

7. Atoms in special positions

For atoms in special positions the summations or multiplicity factors involve a restricted set of symmetry operations. Further, some parameters may have fixed values or may be interrelated. Thus for a position such as $(X, 2X, z)$ in a hexagonal space group the independent variable X is a function of the coordinates x and y, so that

$$\partial|F_c|/\partial X = \partial|F_c|/\partial x + 2\partial|F_c|/\partial y \, . \quad (7{\cdot}1)$$

Similar problems occur with the U^{ij} and a comprehensive SFLS program must be able to handle them. The restrictions imposed on the U^{ij} by site symmetry are listed for all sites in each space group by Petersee & Palm (1966). See also Johnson (1970).

As an example, consider the problems which would arise for N_2O_5 (Grison, Eriks & Vries, 1950). The space group is $P6_3/mmc$, no. 194, $Z=2$, and each atom listed below also occurs centrosymmetrically with respect to the origin.

NO_3^- ions: O_1 at $x,2x,\frac{1}{4};\ 2\bar{x},\bar{x},\frac{1}{4};\ x,\bar{x},\frac{1}{4}$.
N_1 at $0,0,\frac{1}{4}$.

NO_2^+ ions: O_2 at $\frac{1}{3},\frac{2}{3},z;\ \frac{2}{3},\frac{1}{3},\frac{1}{2}+z$.
N_2 at $\frac{1}{3},\frac{2}{3},\frac{3}{4}$.

The general position in this space group is 24-fold so that all the above atoms lie at special positions. If there were an atom at a general position, it would be necessary to supply 12 sets of $\mathbf{R}$ and $\mathbf{t}$ (the centre of symmetry generates the other 12 positions). In this problem 4 sets of $\mathbf{R}$ and $\mathbf{t}$ suffice; these can be determined as follows.

The site symmetry of O_1 is mm, and four positions of the general set:

(1) x,y,z; (A) $y-x,y,z$;
(B) $x,y,\frac{1}{2}-z$; (C) $y-x,y,\frac{1}{2}-z$

have coincided for $y=2x$ and $z=\frac{1}{4}$ to give the special position $x,2x,\frac{1}{4}$. $\mathbf{R}$ and $\mathbf{t}$ are therefore not needed for the operations corresponding to (A), (B), (C). The equivalent positions $2\bar{x},\bar{x},\frac{1}{4}$ and $x,\bar{x},\frac{1}{4}$ of O_1 may be regarded as generated by the general members

(2) $\bar{y},x-y,z$, and (3) $y-x,\bar{x},z$

and $\mathbf{R}$ and $\mathbf{t}$ must be supplied for these.

The site symmetry of O_2 is $3m$ and six positions of the general set have coincided. The equivalent position $\frac{2}{3},\frac{1}{3},\frac{1}{2}+z$ of O_2 may be regarded as generated by the general member

$$(4)\quad \bar{x},\bar{y},\tfrac{1}{2}+z\,.$$

The site symmetries of N_1 and N_2 are both $\bar{6}m2$ and twelve positions of the general set have coincided. These atoms require only the **R** and **t** for the identity operation (1).

Thus we need supply **R** and **t** corresponding to only (1), (2), (3), and (4). O_1 needs (1), (2), and (3); O_2 needs (1), and (4); N_1 and N_2 only (1).

In the least-squares calculations we take $\partial|F_c|/\partial X$ for O_1 as expressed by (7·1). O_2 has a straightforward $\partial|F_c|/\partial z$ and all other coordinates in the structure are fixed.

For O_1 the site symmetry causes the principal axes of its **U** tensor to lie along **a**, **b***, and **c***. Thus $2U^{31}=2U^{23}=0$ and $2U^{12}=U^{22}$. Introducing $V^{22}=U^{22}$ as the independent variable we get

$$\partial|F_c|/\partial V^{22}=\partial|F_c|/\partial U^{22} +\partial|F_c|/\partial(2U^{12})\,. \quad (7{\cdot}2)$$

U^{11} and U^{22} are additional independent variables and give normal derivatives.

For O_2, N_1, and N_2 the site symmetries are all at least $3m$, and the only two independent U^{ij} are U^{11} and U^{33}. The restrictions are $U^{11}=U^{22}=2U^{12}$, and $2U^{13}=2U^{23}=0$. The derivative with respect to the independent variable $V^{11}=U^{11}$ is therefore

$$\partial|F_c|/\partial V^{11}=\partial|F_c|/\partial U^{11}+\partial|F_c|/\partial U^{22} +\partial|F_c|/\partial(2U^{12})\,. \quad (7{\cdot}3)$$

8. Anomalous scattering

Ueki, Zalkin & Templeton (1966) have pointed out the serious coordinate errors which can result from the neglect of the imaginary component $\Delta f''$ of the anomalous scattering in the polar point groups. For a thorium compound, space group *Fdd*2, Mo $K\alpha$ radiation, the neglect of $\Delta f''=9$ for thorium caused an error of about 0·05 Å in the z coordinate of thorium relative to the lighter atoms. Cruickshank & McDonald (1967) have shown for polar space groups that if the maximum acceptable error is 0·005 Å, the neglect of $\Delta f''$ is permissible only if the structure contains no elements heavier than oxygen with Cu radiation or sulphur with Mo radiation. Fortunately the inclusion of $\Delta f''$ in SFLS calculations is straightforward.

For a noncentrosymmetric space group with anomalous scattering $|F(\mathbf{h})|\neq|F(-\mathbf{h})|$, and it is sensible to treat **h** and −**h** as a pair in an SFLS program whenever observations of both have been made. For a centrosymmetric space group $|F(\mathbf{h})|=|F(-\mathbf{h})|$, but the phase $\alpha(\mathbf{h})=\alpha(-\mathbf{h})$ is no longer restricted to 0 or π.

For atoms with anomalous scattering the real dispersion correction $\Delta f'$ will be assumed to be included in the ordinary f. For the imaginary component $i\Delta f''$ we define structure factor contributions,

$$C_{rs}=MGn_r\Delta f''_{rs}\exp[\ldots]\cos[\ldots], \quad (8{\cdot}1$$

$$D_{rs}=MGn_r\Delta f''_{rs}\exp[\ldots]\sin[\ldots] \quad (8{\cdot}2)$$

which are analogous to (6·8). We thus obtain

$$F(\mathbf{h})=\sum_r\sum_s[(A_{rs}-D_{rs})+i(B_{rs}+C_{rs})] =(A-D)+i(B+C) \quad (8{\cdot}3)$$

where the summation for C_{rs} and D_{rs} will extend only over the anomalous atoms. In terms of $A_{rs}=A_{rs}(\mathbf{h})$, $A=A(\mathbf{h})$, etc., the expression for $F(-\mathbf{h})$ is

$$F(-\mathbf{h})=\Sigma\,\Sigma\,[(A_{rs}+D_{rs})-i(B_{rs}-C_{rs})] =(A+D)-i(B-C)\,. \quad (8{\cdot}4)$$

The extra calculations for $F(-\mathbf{h})$ are therefore trivial.

The phase angles are given by

$$\tan\alpha=(B+C)/(A-D)$$
and
$$\tan\alpha(-\mathbf{h})=-(B-C)/(A+D)\,. \quad (8\cdot5)$$

For a centrosymmetric structure $\tan\alpha=\tan\alpha(-\mathbf{h})=C/A$.

The LS calculations are governed by the following analysis,

$$|F|=(A-D)\cos\alpha+(B+C)\sin\alpha\,. \quad (8\cdot6)$$

Hence the derivative with respect to any parameter p is

$$\frac{\partial|F|}{\partial p}=\left(\frac{\partial A}{\partial p}-\frac{\partial D}{\partial p}\right)\cos\alpha+\left(\frac{\partial B}{\partial p}+\frac{\partial C}{\partial p}\right)\sin\alpha \quad (8\cdot7)$$

since the terms arising from the differentiation of $\cos\alpha$ and $\sin\alpha$ cancel.

If x_r is an atomic parameter and h_s is the equivalent index for symmetry position s, we note that

$$\left.\begin{aligned}\partial A_{rs}/\partial x_r&=-2\pi h_s B_{rs}\,;\\ \partial B_{rs}/\partial x_r&=2\pi h_s A_{rs}\,;\\ \partial C_{rs}/\partial x_r&=-2\pi h_s D_{rs}\,;\\ \partial D_{rs}/\partial x_r&=2\pi h_s C_{rs}\,.\end{aligned}\right\} \quad (8\cdot8)$$

If we define

$$G_{rs}=(A_{rs}-D_{rs})\cos\alpha+(B_{rs}+C_{rs})\sin\alpha\,, \quad (8\cdot9)$$

$$J_{rs}=(B_{rs}+C_{rs})\cos\alpha-(A_{rs}-D_{rs})\sin\alpha\,, \quad (8\cdot10)$$

we then have

$$\partial|F_c|/\partial x_r=\sum_s -2\pi h_s J_{rs}\,, \quad (8\cdot11)$$

$$\partial|F_c|/\partial U^{11}{}_r=\sum_s -2\pi^2 h_s{}^2 a^{*2} G_{rs}\,, \quad (8\cdot12)$$

$$\partial|F_c|/\partial n_r=\sum_s G_{rs}/n_r\,. \quad (8\cdot13)$$

These show that in the anomalous case the calculations of the LS contributions for a single plane $\mathbf{h}$ are similar to the non-anomalous case for all stages subsequent to the calculation of G_{rs} and J_{rs}.

For the plane $-\mathbf{h}$ a new set of G_{rs} and J_{rs} is calculated from

$$G_{rs}=(A_{rs}+D_{rs})\cos\alpha(-\mathbf{h})-(B_{rs}-C_{rs})\sin\alpha(-\mathbf{h})\,, \quad (8\cdot14)$$

$$J_{rs}=(B_{rs}-C_{rs})\cos\alpha(-\mathbf{h})+(A_{rs}+D_{rs})\sin\alpha(-\mathbf{h})\,. \quad (8\cdot15)$$

The remainder of the calculation is exactly as for the plane $\mathbf{h}$ even to the use of indices $\mathbf{h}_s$ and not $-\mathbf{h}_s$. For noncentrosymmetric anomalous cases with both $|F_o(\mathbf{h})|$ and $|F_o(-\mathbf{h})|$ provided, the calculation time for the SF part is practically the same as when one alone is provided, but the time for the LS part is doubled. For centrosymmetric space groups $B=D=0$, and the terms in B_{rs} and D_{rs} must be omitted from G_{rs} (8·9) and those in A_{rs} and C_{rs} from J_{rs} (8·10).

The use of the quantities G_{rs} and J_{rs} is also convenient in the non-anomalous case as it enables the structure-factor derivatives to be calculated without the explicit use of (5·3). This aspect is discussed in Cruickshank *et al.* (1961), where Table 2 also gives a flow diagram for an SFLS calculation.

9. Scale factor refinement

Since by definition $F(000)$ is equal to the number of electrons in the unit cell, it is natural to consider that the scale of the $|F_o|$ should be adjusted so as to correspond to $|F_o(000)|$ equal to this value. From this point of view no scale factor should arise in the F_c. This view is proper but so far as any one cycle of the least-squares process is concerned no tampering with the data ($|F_o|$) is allowed and the appropriate inverse scale factor must be determined for the F_c. When the cycle of calculation has been completed the F_c may be freed of the scale factor and the $|F_o|$ put on the desired scale.

A number of cycles of refinement are usually necessary to achieve convergence of the parameters. In the first cycle $G=1$, so that no multiplication by G is necessary in the calculation of A_{rs}, B_{rs}. We may keep this simplification in subsequent cycles if the scale factor is transferred to the $|F_o|$ after each cycle in the manner suggested above and if we reset $G=1$ at the start of each cycle.

If $|F_c|$ denotes a calculated F without a scale factor, then the definition of Δ given in § 4 corresponds to $\Delta=|F_o|-G|F_c|$. If we let $a=1/G$ and now redefine $\Delta=a|F_o|-|F_c|$, since this is the 'natural' value, we can express the function minimized as

$$\begin{aligned} R &= \Sigma w(|F_o|-G|F_c|)^2 \\ &= (1/a)^2\Sigma w(a|F_o|-|F_c|)^2 \\ &= (1/a)^2\Sigma w\Delta^2 \,. \end{aligned}$$

Note therefore that though R should converge to a minimum in successive cycles, $\Sigma w\Delta^2$ may not because a may change from cycle to cycle. Since $\Sigma w|F_o|^2$ is a constant for the given set of data, the function

$$R' = \frac{\Sigma w\Delta^2}{a^2(\Sigma w|F_o|^2)} = \frac{\Sigma w\Delta^2}{\Sigma w(a|F_o|)^2}$$

is also minimized in the refinement. The square of this function has a value similar to that of the usual discrepancy factor $\Sigma\Delta/\Sigma(a|F_o|)$. Thus $(R')^{1/2}$ is a useful index for comparing the degree of refinement of different structures.

In the correct procedure when G is treated as a parameter in the refinement equations, the least-squares condition $\partial R/\partial G=0$ corresponds to $\Sigma w\Delta|F_c|=0$. The incorrect procedure treating a directly as a parameter corresponds to $\Sigma w\Delta|F_o|=0$. These two conditions are different, so that two different results are achieved. With Δ defined as in the last paragraph, the correct procedure makes $\Sigma w\Delta^2$ converge to a minimum in successive cycles. The incorrect procedure yields a low value of a, the $|F_o|$ scale, and this is accompanied by high values of the vibration parameters.

In general the normal equations are valid only for small changes in the parameters, since $|F_c|$ is not a linear function of the parameters. However, since $|F_c|$ is a linear function of G, the derivatives of $|F_c|$ with respect to G beyond the first are zero, and the normal equations are valid for large changes of scale, provided the changes in the other parameters are small.

Rollett (1970) has discussed the semi-diagonal approximations which may be used when the calculation of the full normal matrix is impracticable. Here we note only that the estimation of the scale change from a 1×1 equation is completely unsatisfactory. Because of the very high correlation of the scale with the vibration parameters, the scale must be determined from at least a 2×2 equation for the scale and the (possibly dummy) overall isotropic vibration parameter. Such a 2×2 equation gives a good estimate for the change of scale. The effect of the scale change upon individual atomic vibration parameters derived from, say, 9×9 or 4×4 atom-parameter matrices can be allowed for by augmenting the individual ΔU^{ij} or ΔU by $(\Delta\bar{U}_2-\Delta\bar{U}_1)$ where $\Delta\bar{U}_2$ is the change of overall vibration from the 2×2 equation and $\Delta\bar{U}_1$ is from the 1×1 equation for the dummy overall vibration parameter.

10. Least-squares treatment of batch or layer-line scaling by minimization of R'

In § 9 we supposed that a single scale factor might be required for the set of $|F_o|$. We now consider the problem where a number of scale factors are required for the data, and each $|F_o(\mathbf{h})|$ is observed on one or more 'layer lines'. In case a scale is truly common to several layers or relates to only part of a layer, we shall speak of batches i. In accordance with the fundamental principle of least-squares the observations are regarded as given, and all the parameters are formally applied to the $|F_c|$. We call the scale factors for the different batches G_i, so that ideally we wish to achieve $|F_o(\mathbf{h})|_i = G_i|F_c(\mathbf{h})|_i$. The sum in the minimization is over all $\mathbf{h}$ and i (and is thus more extensive than usual).

The LS equations are simplified if we adopt the legitimate device, mentioned in the second paragraph of § 9, of applying the inverse scale factors to the $|F_o(\mathbf{h})|_i$ after each cycle, so that each new cycle is started with the trial values of G_i all equal to unity. If $|F_c|$ denotes a calculated F without a scale factor, the necessary derivatives are

$$\frac{\partial(G_i|F_c|)}{\partial G_i} = |F_c| \text{ and}$$

$$\frac{\partial(G_i|F_c|)}{\partial G_j} = 0\,. \tag{10·1}$$

The LS equations have the form

$$\left.\begin{aligned}
&\sum_{h,1} wF^2 \,.\, \varepsilon_{G1} + 0 \ldots + 0 \\
&\qquad + \sum_{h,1} wF\frac{\partial F}{\partial p} \,.\, \varepsilon_p = \sum_{h,1} w\Delta_1 F \\
&0 + \sum_{h,2} wF^2 .\, \varepsilon_{G_2} + \ldots + 0 + \ldots \\
&\qquad\qquad = \ldots \\
&0 + 0 \ldots + \sum_{h,n} wF^2 .\, \varepsilon_{G_n} + \ldots \\
&\qquad\qquad = \sum_{h,n} w\Delta_n F \\
&\sum_{h,1} wF\frac{\partial F}{\partial p} \,.\, \varepsilon_{G_1} + \sum_{h,2} \ldots + \sum_{h,n} \ldots \\
&\quad + \sum_h (\sum_i w)\left(\frac{\partial F}{\partial p}\right)^2 .\, \varepsilon_p = \sum_h (\sum_i w\Delta_i)\frac{\partial F}{\partial p}
\end{aligned}\right\} \tag{10·2}$$

where $F = |F_c(\mathbf{h})|$, $\Delta_i = |F_o|_i - |F_c|$, $|F_o|_i$ includes the starting inverse scale factor, $\sum_{h,i}$ is a summation over all planes $\mathbf{h}$ observed in batch i, and p stands for all the other parameters of the structure.

We see that the normal equations are diagonal in the block concerned with the scale factors alone, and that the coupling between scale factors arises only through the off-diagonal terms involving other types of parameter. Correlation with the (possibly dummy) overall vibration parameters is particularly important.

11. Choice of weights

In the early stages of refinement artificial weights may be chosen to accelerate refinement (*e.g.* downweight high-order planes or those with small $|F_c|$).

In the final stages of refinement the weights must be related to the accuracy of the structure factors if the derived parameters are to be of maximum accuracy. There are two distinct ways, covering two ranges of error, in which this may be done.

A. The weights may reflect the precision of the $|F_o|$, and $w(hkl) = 1/\sigma^2(hkl)$ where σ^2 is the estimated variance of $|F_o|$ due to a specific class of random experimental errors. These weights are derived from an analysis of the experiment. Weights chosen in this way lead to estimates $\sigma^2(p_j)$ [Rollett (1970), Eq. (35)] which cover only the specific class of random experimental errors.

B. The weights may reflect the trends in the $|\Delta F|$. A weighting function with a small number of parameters is chosen so that the averages of $w\Delta^2$ are constant when the set of $w\Delta^2$ values is analysed in any pertinent fashion. Weights chosen in this way lead to estimates $\sigma^2(p_j)$ [Rollett (1970), Eq. (36)] which allow for all random experimental errors, such systematic experimental errors as cannot be paralleled in the calculated model, and such defects of the model as are not paralleled in the experimental data.

A typical weighting function of type B is

$$w = 1/(c_1 + |F_o| + c_2|F_o|^2)\,,$$

where $c_1 = 2F_{\min}$ and $c_2 = 2/F_{\max}$ are typical. It is often necessary that the weights should also be a function of sin θ and/or sin θ/λ. Thus K. W. Muir has used

$$w = w(1) \times w(2) ,$$

where $w(1)=1$ if sin $\theta/\lambda > P(1)$, otherwise $[(\sin\theta/\lambda)/P(1)]^2$; $w(2)=1$ if $|F_o| < P(2)$, otherwise $P(2)/|F_o|$.

There is debate (Dunning & Vand 1969) as to whether and how 'accidentally absent' reflections can be brought into the least-squares treatment.

A bad weighting scheme can yield misleading parameters, see *e.g.* Dunitz & Ryan (1966).

12. Concluding remarks

The extension of the SFLS refinement to include extinction parameters and group parameters is described elsewhere in this volume by Coppens, Larson & Doedens.

It is generally considered that the least-squares process converges to a unique minimum for any set of any parameters not too far distant from this minimum. However some false minima surprisingly close to the true minimum have been reported [especially by Donohue & Goodman (1967) for adamantane].

A lower minimum can always be found by reducing the symmetry since this introduces more parameters. Thus in thortveitite (Cruickshank, Lynton & Barclay, 1962) minima were found not only in the accepted space group $C2/m$, but also in $C2$ and Cm. The question whether additional parameters have significantly improved the residual may (sometimes) be settled by Hamilton's (1965) significance tests. However for a cautionary tale see Parthasarathy, Sime & Speakman (1969).

There are close relations between the least-squares and Fourier methods of refinement. These have been examined by Cochran (1948) and Cruickshank (1952, 1953).

Although published some years ago much useful information about the least-squares method of refinement and related topics can still be found in *International Tables* (1959), *Computing Methods and the Phase Problem in X-ray Crystal Analysis* (1961), and *Computing Methods in Crystallography* (1965).

References

Bloch, F. (1932). *Z. Phys.* **74**, 195.

Cerrini, S. (1970). *Acta Cryst.* In press.

Cochran, W. (1948). *Acta Cryst.* **1**, 138.

Computing Methods and the Phase Problem in X-ray Crystal Analysis. (1961). Oxford: Pergamon Press. *Computing Methods in Crystallography.* (1965). Oxford: Pergamon Press.

Cruickshank, D. W. J. (1952). *Acta Cryst.* **5**, 511.

Cruickshank, D. W. J. (1956). *Acta Cryst.* **9**, 747.

Cruickshank, D. W. J., Lynton, H. & Barclay, G. A. (1962). *Acta Cryst.* **15**, 491.

Cruickshank, D. W. J. & McDonald, W. S. (1967). *Acta Cryst.* **23**, 9.

Cruickshank, D. W. J., Pilling, D. E., Bujosa, A., Lovell, F. M. & Truter, M. R. (1961). *Computing Methods and the Phase Problem in X-ray Crystal Analysis*, p. 32. Oxford: Pergamon Press.

Cruickshank, D. W. J. & Robertson, A. P. (1953). *Acta Cryst.* **6**, 698.

Donohue, J. & Goodman, S. H. (1967). *Acta Cryst.* **22**, 352.

Dunitz, J. D. & Ryan, R. R. (1966). *Acta Cryst.* **21**, 617.

Dunning, J. D. & Vand, V. (1969). *Acta Cryst.* A**25**, 489.

Grison, E., Eriks, K. & Vries, J. L. de (1950). *Acta Cryst.* **3**, 290.

Hamilton, W. C. (1965). *Acta Cryst.* **18**, 502.

Johnson, C. (1970). This volume.

International Tables for X-ray Crystallography (1959). Vol. II. Birmingham: Kynoch Press.

Nelmes, R. J. (1969). *Acta Cryst.* A**25**, 523.

Parthasarathy, R., Sime, J. G. & Speakman, J. C. (1969). *Acta Cryst.* B**25**, 1201.

Patterson, A. L. (1959). In *International Tables for X-ray Crystallography*, Vol. II. Birmingham: Kynoch Press.

Peterse, W. J. A. M. & Palm, J. H. (1966). *Acta Cryst.* **20**, 147.

Rollett, J. S. (1970). This volume.

Ueki, T., Zalkin, A. & Templeton, D. H. (1966). *Acta Cryst.* **20**, 836.

Discussion

AHMED: Do you prefer to express the isotropic thermal vibration in terms of U_{iso} instead of B_{iso}?

CRUICKSHANK: Yes, because when one proceeds to anisotropic refinement, it is advantageous to use the U_{ij}'s.

JOHNSON: Since only 3 of the 6 U_{ij} parameters are in physical component form, it would be a good way to publish the principal axes of **U** referred to orthogonal axes.

CRUICKSHANK: I agree.

MATTHEWS: Is there any objection in calculating the structure factor derivatives numerically from the change in $|F_c|$ corresponding to changes in parameters?

CRUICKSHANK: It is not a very good way except possibly in group refinement.

BUSING: If you were going to do a rigid body refinement, you might like to use numerical derivatives of the coordinates with respect to the rigid body parameters, $\partial x_r/\partial p_i$, then combine these with the numerical derivatives, $\partial|F_c|/\partial x_r$, to obtain your final structure factor derivatives $\partial|F_c|/\partial p_i$.

DEWAR: In polar space groups, does the choice of a light atom to fix the origin result in ill-conditioned matrices?

ROLLETT: In cases like this, the ill-conditioning shows up as an apparent overestimate of the e.s.d.'s of the atomic coordinates in the polar direction, which is corrected only when you take into account the off-diagonal terms in the full matrix.

CRUICKSHANK: I agree completely, but the advantage of taking the heaviest atom at the origin is that when you invert the matrix to get the e.s.d.'s, you find the diagonal terms are closer to the true values than if you take a light atom at the origin. In fact, what one gets is the standard deviation of one atom with respect to the atom at the origin.

EISS: Is it permissible to double the number of reflections in the refinement by including $F(\mathbf{H})$ and $F(\bar{\mathbf{H}})$ since they are independent?

CRUICKSHANK: Yes, some people have done it.

JOHNSON: It is possible to include $\Delta f''$ as a parameter in the refinement, and this provides a very convenient method for getting the absolute configuration without doing two calculations.

CRUICKSHANK: This is true especially since $\Delta f''$ is essentially from the inner electrons and does not vary very much with $\sin\theta/\lambda$.

STEWART: Is it not best to bring all the data into action at the final stages of refinement?

CRUICKSHANK: If one can solve the structure from the 300 largest E's, it may also be possible to start refinement with only those reflections.

ROBINSON: Would you agree that in order not to bias the result in non-centrosymmetric cases, you should collect both forms, $F(\mathbf{H})$ and $F(\bar{\mathbf{H}})$?

CRUICKSHANK: What you say is true from common sense, but it does not seem to be necessary except in difficult cases such as where pseudosymmetry is present.

STEWART: What happens if you have weak interscaling correlations?

CRUICKSHANK: With full matrix, you should get appropriate correlation factors. With block diagonal you must include some temperature factor parameters with the structural parameters that you put into the block.

COPPENS: I would like to sound a warning in the use of weights based on $(|F_o|-|F_c|)$, because they would bias the outcome of the experiment.

CRUICKSHANK: The point is well taken.

G. BIRNBAUM: Is it safe to assume that if $\langle w\Delta^2\rangle$ are equal in ranges of $|F_o|$ and $\sin^2\theta$, that the weights have been correctly chosen?

CRUICKSHANK: These conditions should be satisfied, but this does not exclude the possibility that there might have been other significant parameters which are dependent on the Δ's.

JOHNSON: Do differential syntheses still have a place in crystallography?

CRUICKSHANK: The differential synthesis corresponds to a choice of artificial least-squares weights, and these weights are therefore not the best ones to produce estimates of highest precision.

Group Least-Squares Refinement

ROBERT J. DOEDENS

Department of Chemistry, University of California, Irvine, California, U.S.A.

In a group (or rigid-body) refinement of atomic parameters, certain portions of a crystal structure are constrained to a fixed geometry and refined as units. Group refinement is often motivated by a desire to minimize the number of parameters varied in a least-squares refinement. This results in reduced computer time and storage requirements as well as a more favorable data/parameter ratio. Many aspects of group least-squares refinement were first described by Scheringer (1963). This type of refinement has found its most frequent application in crystal structure analyses of complex compounds containing aromatic rings whose structures can be estimated with some degree of confidence. Many complex inorganic or organo-metallic compounds contain large numbers of phenyl or other rings, the detailed structures of which are not of great interest and cannot be determined with high precision. Representative early examples of such compounds to which group methods of full-matrix least-squares refinement have been applied are $RhH(CO)[P(C_6H_5)_3]_3$ (La Placa & Ibers, 1965) and $[(C_5H_5N)_6Fe][Fe_4(CO)_{13}]$ (Doedens & Dahl, 1966). It must be emphasized that systematic errors will be introduced into a group refinement if the actual configuration of the groups differs from that assumed. The best test of the validity of a group refinement (if merited by the quantity and quality of the intensity data) is the removal of the group constraints after convergence and continued refinement with individual atomic parameters. Even in cases in which the final refinement is to be one with individual atomic parameters, it may be advantageous to use group methods in the preliminary stages of refinement. It has been shown (Scheringer, 1963) that a group refinement will often have a larger range of convergence than an individual-atom refinement.

Computational aspects of group refinement

The position and orientation of a group may be described by 6 parameters – three positional parameters which define the origin of the group in the conventional unit cell coordinate system and three angular parameters which specify the orientation of the group. One possible way of defining these parameters is as follows: Let x^0 denote the coordinates of the origin of the group coordinate system with respect to the unit cell axes a.

Define as x' the coordinates (in Ångstroms) of an individual atom of a group with respect to a set of orthogonal internal axes a'. Each group will have its own internal coordinate system and will in general contain several atoms.

Define an orthogonal cell A of unit dimensions oriented as follows with respect to the crystal axes a:

$$A_2 \text{ parallel to } a_2$$
$$A_1 \text{ parallel to } a_2 \times a_3$$
$$A_3 \text{ parallel to } A_1 \times A_2$$

Let **U** be the transformation matrix which

relates cell A to the original cell a by:

$$A = \mathbf{U}a$$

If X and x are the coordinates of an atom referred to the A and a systems respectively, then it follows that:

$$x = \mathbf{U}^{tr}X$$

Define $\mathbf{R}$ as the rotation matrix which brings the orthogonal coordinate system A into alignment (except for translation) with the internal system a' of a group. This rotation may be expressed as the product of successive counterclockwise rotations as follows:

(1) Rotation by φ about A_3
(2) Rotation by θ about new A_1
(3) Rotation by ϱ about new A_2

The coordinates x of a group atom in the original coordinate system a are then given by:

$$x = x^0 + \mathbf{U}^{tr}\mathbf{R}^{tr}x' \qquad (1)$$

The matrix $\mathbf{U}$ may be written explicitly as:

$$\mathbf{U} = \begin{pmatrix} a^* & b^* \cos \gamma^* & c^* \cos \beta^* \\ 0 & 1/b & 0 \\ 0 & -\cot \alpha/b & 1/c \sin \alpha \end{pmatrix}$$

The rotation matrix $\mathbf{R}$ is simply the product of the matrices for the rotations as defined above:

$$\mathbf{R} = \begin{pmatrix} \cos \varrho & 0 & -\sin \varrho \\ 0 & 1 & 0 \\ \sin \varrho & 0 & \cos \varrho \end{pmatrix} \begin{pmatrix} 1 & 0 & 0 \\ 0 & \cos \theta & \sin \theta \\ 0 & -\sin \theta & \cos \theta \end{pmatrix} \begin{pmatrix} \cos \varphi & \sin \varphi & 0 \\ -\sin \varphi & \cos \varphi & 0 \\ 0 & 0 & 1 \end{pmatrix}$$

Once the coordinates of the group atoms have been obtained from (1), they may be employed in the usual manner in calculation of structure factors. The variable positional parameters for a group are the coordinates x^0 of the group origin and the angles φ, θ, and ϱ. The problem which remains is the formulation of appropriate contributions to the matrix of the normal equations for these parameters.

The derivatives of F_c with respect to the group positional parameters x_1^0, x_2^0, x_3^0, φ, θ, and ϱ are required for construction of the normal equations. These derivatives may be evaluated analytically or by the method of finite differences – both methods have been employed. The analytical expressions are based upon relations like the following which arise from the functional dependence of F_c upon the coordinates x and the dependence of these coordinates upon x^0, φ, θ, and ϱ:

$$\frac{\partial F_c}{\partial x_1^0} = \sum_{j=1}^{N} \frac{\partial F_c}{\partial x_{1j}}$$

$$\frac{\partial F_c}{\partial \varphi} = \sum_{j=1}^{N} \sum_{i=1}^{3} \frac{\partial F_c}{\partial x_{ij}} \frac{\partial x_{ij}}{\partial \varphi}$$

The j-summations in both cases are taken over the N atoms in the group under consideration. The terms $\dfrac{\partial F_c}{\partial x_{ij}}$ are simply individual-atom derivatives of F_c. Derivatives such as $\dfrac{\partial x_{ij}}{\partial \varphi}$ can be evaluated from Eq. (1) or its expanded versions. These derivatives will in general be functions of the unit cell parameters and the group orientation angles.

The final point which must be considered is the nature of the model to be adopted for the thermal motion of the group atoms. Since the use of group methods has commonly been motivated by a desire to minimize the number of parameters involved in a least-squares refinement, this model has usually been a simple one. The most frequently employed thermal models have included either a single isotropic temperature factor for each group or individual isotropic thermal parameters for group atoms. No new computational procedures are required if the latter model is adopted; if a single group temperature factor B_G is employed, the appropriate derivative may be evaluated simply by:

$$\frac{\partial F_c}{\partial B_G} = \sum_{j=1}^{N} \frac{\partial F_c}{\partial B_j}$$

where again the summation is taken over the N atoms in the group. More complex thermal models may also be used if a more detailed analysis of the group thermal motion is desired and warranted by the quality of the intensity data.

Additional points

Note from the definition of the group orientation angles that φ and ϱ become equivalent for $\theta=90°$. Thus, any attempted refinement from $\theta=90°$ in which both φ and ϱ are varied will result in a singular matrix. This problem may be avoided by defining the group axes x' in such a manner that θ is not close to 90°.

The origin of the group may be defined at any point, but for symmetric groups the center of gravity of the group is usually the best choice. Its coordinates depend upon all of the group atoms and are in general better known than those of many other possible choices.

A further application of group techniques arises in the refinement of disordered structures. Refinement of individual atomic parameters with partial multiplicity factors is a standard procedure in cases of disorder. Such a refinement may not be possible if atoms from different contributions to the disordered structure are in very close proximity. Some problems of this sort have been successfully dealt with by group refinement methods; an example is thiophene chromium tricarbonyl, in which the thiophene rings exhibit a threefold disorder (Bailey & Dahl, 1965).

References

Bailey, M. F. & Dahl, L. F. (1965). *Inorg. Chem.* **4**, 1306.

Doedens, R. J. & Dahl, L. F. (1966). *J. Amer. Chem. Soc.* **88**, 4847.

La Placa, S. J. & Ibers, J. A. (1965). *Acta Cryst.* **18**, 511.

Pawley, G. S. (1969). *Acta. Cryst.* **A25,** 531.

Scheringer, C. (1963). *Acta Cryst.* **16**, 546.

Discussion

BUSING: It is important to point out that you need to calculate the coefficients of the derivatives of the individual atom coordinates with respect to the rigid body coordinates only once and not for every reflection.

DOEDENS: Yes.

GASSMANN: It should be pointed out that in group refinement one should not refine groups with very different scattering powers. The process does not converge in such cases.

ROLLETT: I think there are also some complicaitons when the groups share one or more atoms.

DOEDENS: Yes, one should be very careful in those two cases.

PAWLEY: Beside the economic advantages of group refinement, the results of this method can also be more accurate.

DOEDENS: Yes, you are right.

Comments on Some Practical Aspects of Least-Squares Refinements

J. W. SCHILLING

Department of Chemistry, University of Michigan, Ann Arbor, Michigan, U.S.A.

Approximating the Sine and Cosine

A sin/cos routine using table lookup with parabolic interpolation has been found to be twice as fast as the system routine on the IBM 7090 and to reduce structure-factor running time by about 10% in an orthorhombic space group. Least-squares is used to determine tables of coefficients *A*, *B*, and *C* for the expression:

$$\sin x \simeq A_i + B_i\xi + C_i\xi^2$$

where x is the argument (in cycles); n is the number of points in one cycle for which values of *A*, *B*, and *C* are tabulated; i is the integer part of the product $n \,.\, x$; ξ is the fractional excess $nx - i$. The normal equations are derived from the condition:

$$\sum_{=0,k} \{\sin(x_n + j/k) - A - B(j/k) - C(j/k)^2\}^2 = \text{minimum}$$

which is imposed for each of the n abscissa points to be tabulated. The value $k = 10$ was arbitrarily chosen and different values of n were tried. A significant break-point in accuracy occurs between $n = 32$ and $n = 64$: the latter gives a maximum error of about 0·000006, the former about 0·0003. No improvement in accuracy was observed when $n = 128$ was used. The $n = 32$ routine was finally chosen to conserve space. $1\frac{1}{4}$ cycles (41 points in A, B, C) are stored; $\cos(x)$ is evaluated as $\sin(x+\frac{1}{4})$; $\sin(-x)$ as $\sin(x+\frac{1}{2})$. Entries SIN, COS accommodate arguments in radians; CSIN, CCOS, in cycles. The number of machine instructions executed is independent of n, being on the average about 25 on the IBM 7090 for CSIN, CCOS.

Short-floating-point numbers

The accuracy and range of data involved in crystallographic least-squares do not require a full word of bits, and seldom a sign. Compared to 'scaled integers', a floating-point representation generally gives a larger range of expressibility and a more nearly constant relative error over the whole range. A subroutine was written for the IBM 7090 to allow arbitrary specification of the number of bits in the characteristic (C) and mantissa (F), and the magnitude of the constant, K, added to the characteristic of numbers to be packed two or more per word. This routine has simple calling sequences for a setup entry which 'compiles' a sequence of packing and unpacking instructions for each type of packed numbers; and for 'store' and 'fetch' entries. For example CALL PACK (A, IALPHA, J) would have the effect of ALPHA(J) = A; A = UNPACK (IALPHA, J) has the effect of A = ALPHA(J). IALPHA contains an integer identifying the desired array, to be used by the subroutine in indexing its tables of parameters.

Such a scheme was experimentally introduced into a modified version of the Fortran II OR FLS least-squares program (Busing, Martin & Levy, 1962) with some success. Among the data stored for each reflection were:

Name	F	C	K	Largest number expressible	Smallest number expressible	Maximum relative error	Sign ?
F_{obs}	13	5	15	$0{\cdot}6554\times 10^{5}$	$0{\cdot}152\times 10^{-4}$	0·024%	no
$\varrho=(\sin\theta/\lambda)^2$	14	4	15	2·0	$0{\cdot}152\times 10^{-4}$	0·012%	no
f=atomic form factors	8	4	8	128·0	$0{\cdot}195\times 10^{-2}$	0·78%	no

These were stored in 36-bit words as follows:

1 … 18	19 … 36
F_{obs}	ϱ

1 … 12	13 … 24	25 … 36
f_1	f_2	f_3

There was no change in the accuracy of the program when this was done. Up to 17 quantities could be stored per reflection using 7 machine words (252 bits) per reflection. Blocks of these packed data were stored on tape, reducing the number of tape-read operations accordingly. A general discussion of this algorithm, suitable for adaptation to any machine, is given by Schilling (1968).

'Cascade' block-diagonal algorithm

A method of solving the normal equations based upon the Gauss-Seidel algorithm is discussed by Sparks (1961). Whereas the usual block-diagonal algorithms refine all atoms independently of the shifts being computed for other atoms on the same cycle, the 'Gauss-Seidel' approach is one in which atom (or group) 1 is refined, then atom (or group) 2, and so on; the shifts determined for atom 1 improve the estimate of the shifts for atom 2, and so on. In order to make this efficient, the least-squares routine must have a 'partial' structure factor capability; that is, on each cycle, the SF loop need include only the contribution(s) to A and B of the atom(s) most recently refined, the others (AP, BP) being constant. How this might be carried out is exemplified below for five atoms/groups $I \ldots M$:

Cycle	SF only or LS	Block or atom to be refined	SF/derivative loop includes	AP, BP input	AP, BP output
1	SF	—	I, J, K, L, M	0	$I+J+K+L+M$
2	LS	I	I	$I+J+K+L+M$	$J+K+L+M$
3	SF	—	I'	$J+K+L+M$	$I'+J+K+L+M$
4	LS	J	J	$I'+J+K+L+M$	$I'+K+L+M$
·					
·					
·					
11	SF	—	M'	$I'+J'+K'+L'$	$I'+J'+K'+L'+M'$

where the primes denote refined entities.

One pass through all the atoms is roughly equivalent to two structure-factor calculations; the optimum condition for this procedure is in an unequal-atom structure, with the atoms refined in order of decreasing atomic number ('cascade' algorithm). It might also be particularly advantageous in certain hardware configurations. That is, the storage required for the matrix is very small, allowing more memory to be used for data storage. Auxiliary storage devices, if necessary, must be efficient: sequential disk files, or tapes which may be read forward and backward.

'Segmented matrix' algorithm

Let 'c' be the amount of storage needed for resident data and the program which computes the normal equations; 'i', that required by the matrix inverter; and 'm', the total available fast memory. The size of the matrix is then limited to $m-c-i$. The quantity c is generally appreciable, but there is no simple way to utilize this space for the matrix. The following procedure has been found useful to permit a matrix as large as $m-i$:

(1) Compute the maximum number of rows of the matrix which can be accommodated in $m-c$ words; store these (a truncated triangular array) on an auxiliary device.

(2) Compute successive 'matrix segments' (each requiring a pass through the reflection data).

(3) Assemble the matrix segments in a separate overlay, in which only the matrix and the matrix inverter reside.

As distributed the OR FLS program accommodated 100 parameters with a 32K memory; a modification storing reflection data entirely on tape raised this to 185; the above algorithm permitted 230 parameters. On the IBM 7090, 218 parameters refined in three matrix segments required 55 minutes per cycle, with 1068 reflections. A more detailed discussion of this procedure is given by Schilling (1968).

The CLS program

This program is a drastic revision of OR FLS. The goals of this revision were:

(1) To make the program more useful in the early and intermediate stages of a structural problem, and for larger structures.

(2) To minimize as much as possible the cost of refinements. The following factors were included in this consideration:

A. minimizing the effects of user blunders by simplified input, extensive scrutiny of the input, and a 'restart' procedure;

B. accommodation of special algorithms for special cases;

C. minimizing memory requirements (especially in multiprogrammed environments);

D. minimizing frequency of extensive modifications;

E. saving a machine instruction here and there.

The major mathematical differences from OR FLS are:

(1) anomalous scattering correction (Ibers & Hamilton, 1964);

(2) secondary extinction correction (Zachariasen, 1968);

(3) atoms may be mixed isotropic and anisotropic;

(4) weights may be recomputed from cycle to cycle;

(5) overall anisotropic thermal parameters may be refined;

(6) input scale factors refer to observed structure factors;

(7) comprehensive reflection statistics may be compiled.

There are also major computational and structural differences:

(1) Modularity is emphasized for adaptability to different problems and different machines.

A. The calculation and solution of the normal equations occurs in one subroutine; this subroutine may be absent for a no-refinement cycle (SFC); it may be replaced for a full-matrix (FLS), block-diagonal (BLS), or diagonal (DLS) algorithm. Dynamic storage allocation is being used on the IBM 360/67 to obviate the necessity of recompilation for different problems (the matrix is redimensioned on each cycle).

B. Reflection data storage is relegated to one subroutine, so that dynamic storage allocation, word-packing, or other special techniques may be used.

C. Special features, such as the reflection statistics, are in optional subroutines, which may be replaced by dummies to conserve memory space.

(2) Extensive facilities exist for *ad hoc* modification of the program. For example, subroutine 'ADHOC' (normally a dummy) is called at carefully selected points. All quantities of any significance may be referenced through common by a special version of that routine.

(3) A 'partial' structure factor facility exists, including implementation of the 'cascade' algorithm above. This feature can also be used to refine structures in which some atoms are to be held constant for two or more cycles or in which it may be desired to repeat a refinement with, say, a different thermal representation for some atoms.

(4) BLS can accommodate a variable number of atoms per block.

(5) BLS and DLS contain a provision for refining an overall thermal parameter in a submatrix block with the scale factor (Cruickshank, 1961).

(6) Because of multiprogramming (and even more so, time-sharing), the elapsed time of a job on the machine may be several times the CPU time used. Thus, the *virtual* error-free time of a modern machine may be less than that of an older, intrinsically less reliable one. CLS allows a partial matrix to be dumped periodically (onto a tape, say) for later restart in case of a disaster. This can also be useful in case a refinement cycle is to be repeated with some reflections added, subtracted, or weighted differently, or because parameter oscillation has set in.

On the Michigan Terminal System for the IBM 360/67, CLS operates in three overlays. A parameter read by the second determines whether the BLS, DLS, FLS, or SFC version of the third overlay (SF loop and matrix inversion) is to be loaded. It is also structured to accommodate a segmented matrix approach, in which case a fourth overlay is required for the matrix inversion; this feature is not currently implemented, however.

Acknowledgement

The author wishes to acknowledge that this program incorporates some features inspired by programs of J. M. Stewart (1967), R. Shiono (private communication), C. K. Johnson (1967), and C. E. Nordman (private communication).

References

Busing, W. R., Martin, K. O. & Levy, H. A. (1962). *ORFLS: A Fortran Crystallographic Least-Squares Program.* Publication ORNL-TM-305 of the Oak Ridge National Laboratory.

Cruickshank, D. W. J. (1961). *Computing Methods and the Phase Problem in X-ray Crystal Analysis*, p. 43. Oxford: Pergamon Press.

Ibers, J. A. & Hamilton, W. C. (1964). *Acta Cryst.* **17**, 781.

Johnson, C. K. (1967). ORXLS/360, a revision to ORFLS. Uncataloged publication of the Oak Ridge National Laboratory.

Schilling, J. W. (1968). Ph. D. Thesis, University of Michigan.

Sparks, R. A. (1961). *Computing Methods and the Phase Problem in X-ray Crystal Analysis*, p. 175. Oxford: Pergamon Press.

Stewart, J. M. (1967). *X-ray* 67 *Program System for X-Ray Crystallography.* Technical Report 67–58, Computer Science Center, University of Maryland.

Zachariasen, W. H. (1968). *Acta Cryst.* A**24**, 212.

TOPIC F

ANALYSIS OF ATOMIC THERMAL VIBRATION

CHAIRMAN: C. K. JOHNSON

An Introduction to Thermal-Motion Analysis

CARROLL K. JOHNSON

Chemistry Division, Oak Ridge National Laboratory, Oak Ridge, Tennessee, U.S.A.

There are three distinct levels of thermal-motion analysis. In the first level, all atoms are considered as independent individuals and the actual correlations in the motions of different atoms are ignored. The structure-factor equation used in crystal-structure analysis incorporates this hypothesis. This approach leads to a statistical description of the displacements of individual atoms averaged over time and lattice.

The second level in thermal-motion analysis is concerned with the application of models, such as the rigid body with its fixed interatomic separations, to groups of atoms to analyze the major librational (*i.e.*, rotatory oscillational) and translational modes of motion for the groups of atoms. The major parts of the present paper and the following paper by Dr. Maslen are concerned with this problem and will illustrate methods for applying these model-fitting techniques.

The third level of thermal-motion analysis involves the dynamics of the entire crystal; hence, the correlations among the motions of a large number of atoms or groups must be considered. Dr. Pawley's paper is concerned with the lattice dynamics of molecular crystals and with intercomparisons of the results obtained from lattice-dynamics studies and crystal-structure studies.

Correlations between the motions of pairs of atoms can lead to an apparent shrinkage in the interatomic distances determined by a crystal-structure analysis. Methods for estimating the magnitude of this shrinkage are discussed in a separate paper by the author.

The following discussion is a general review of the mathematical properties associated with the thermal-vibration tensors determined by the anisotropic temperature-factor coefficients. Following this is a discussion of the kinematics of rigid-body motion.

Thermal motion of an atom

A. Statistical interpretations of the structure factor equation

One of the common formulations for the structure factor equations is

$$F(h_1,h_2,h_3)=\sum_{r=1}^{n} f_r(h_1,h_2,h_3)$$
$$\times \exp\,[2\pi i(x_r^1h_1+x_r^2h_2+x_r^3h_3)]$$
$$\times \exp\,[-(b_r^{11}h_1h_1+b_r^{22}h_2h_2+b_r^{33}h_3h_3+2b_r^{12}h_1h_2+2b_r^{13}h_1h_3+2b_r^{23}h_2h_3)]\,, \quad (1)$$

where x_r^i and b_r^{ij} $(i,j=1,2,3)$ are respectively the contravariant components for the mean position of atom r and for the anisotropic temperature factor of atom r. The term 'contravariant' means that the coefficients are based on the direct crystal-axis vectors $\mathbf{a}_1$, $\mathbf{a}_2$, $\mathbf{a}_3$. The covariant components h_i $(i=1,2,3)$ are the Miller indices, which are based on the reciprocal-axis vectors $\mathbf{b}^1$, $\mathbf{b}^2$, $\mathbf{b}^3$. The atomic scattering factor for atom r, $f_r(h_1,h_2,h_3)$, is a function of the reciprocal-lattice vector $h_1\mathbf{b}^1+h_2\mathbf{b}^2+h_3\mathbf{b}^3$ as is the complete structure factor

---

Research sponsored by the U.S. Atomic Energy Commission under contract with the Union Carbide Corporation.

$F(h_1,h_2,h_3)$. Eq. (1) is more manageable when written in the tensor form

$$F(\mathbf{h}) = \sum_{r=1}^{n} f_r(\mathbf{h}) \times \exp(2\pi i x_r^j h_j - b_r^{jk} h_j h_k), \quad (2)$$

where the repeating index summation convention is assumed ($j,k=1,2,3$). Eq. (1) may also be written in matrix notation as

$$F(\mathbf{h}) = \sum_{r=1}^{n} f_r(\mathbf{h}) \times \exp(2\pi i \mathbf{h}^t \mathbf{x}_r - \mathbf{h}^t \boldsymbol{b}_r \mathbf{h}), \quad (3)$$

where $\mathbf{h}^t$ is the transpose of the column vector $\mathbf{h}$. By making the change of variables $\mathbf{q}=2\pi\mathbf{h}$ and $\boldsymbol{\sigma}=\boldsymbol{b}/2\pi^2$, Eq. (3) becomes

$$F(\mathbf{q}) = \sum_{r=1}^{n} f_r(\mathbf{q}) \times \exp(i\mathbf{q}^t\mathbf{x}_r - \tfrac{1}{2}\mathbf{q}^t\boldsymbol{\sigma}_r\mathbf{q}). \quad (4)$$

The exponential function in Eq. (4)

$$\Phi_r(\mathbf{q}) \equiv \exp(i\mathbf{q}^t\mathbf{x}_r - \tfrac{1}{2}\mathbf{q}^t\,\boldsymbol{\sigma}_r\mathbf{q}) \quad (5)$$

is the characteristic function for the probability density function $\varrho_r(\mathbf{u})$ of a trivariate normal distribution (Wilks, 1962)

$$\varrho_r(\mathbf{u}) = \frac{[\det(\boldsymbol{\sigma}_r^{-1})]^{1/2}}{(2\pi)^{3/2}} \times \exp[-\tfrac{1}{2}(\mathbf{u}-\mathbf{x}_r)^t\boldsymbol{\sigma}_r^{-1}(\mathbf{u}-\mathbf{x}_r)]. \quad (6)$$

The matrix $\boldsymbol{\sigma}_n^{-1}$ is the inverse of the dispersion (variance-covariance) matrix $\boldsymbol{\sigma}_r$, which represents the second moments about the mean position $\mathbf{x}_r$ of atom r.

The scalar mean-square displacement (msd) of atom r from its mean position at $\mathbf{x}_r$ in the direction of any vector $\mathbf{w}$ is obtained by evaluating a quadratic form based on the dispersion matrix $\boldsymbol{\sigma}$

$$\text{msd}(\mathbf{w}) = \sigma^{jk} w_j w_k/(w^l w_l) \equiv \sigma_{jk} w^j w^k/(w^l w_l). \quad (7)$$

Since the components obtained for $\boldsymbol{\sigma}$ and $\mathbf{w}$ are contravariant, it is necessary to incorporate the metric tensor g_{ij}* to obtain

$$\text{msd}(\mathbf{w}) = \sigma^{jk} g_{jm} g_{kn} w^m w^n/(w^l w^p g_{lp}), \quad (8)$$

which may also be written in matrix form

$$\text{msd}(\mathbf{w}) = \mathbf{w}^t\mathbf{g}^t\boldsymbol{\sigma}\mathbf{g}\mathbf{w}/(\mathbf{w}^t\mathbf{g}\mathbf{w}). \quad (9)$$

B. Probability ellipsoids

The probability density for any displacement $\mathbf{u}-\mathbf{x}$ from the mean at $\mathbf{x}$ is given by Eq. (6). A surface of constant probability density is obtained by equating the quadratic form in the exponential factor of Eq. (6) to a constant,

$$(\mathbf{u}-\mathbf{x})^t\boldsymbol{\sigma}^{-1}(\mathbf{u}-\mathbf{x}) = C^2. \quad (10)$$

In order to write this equation in tensor form, a new symbol for $\boldsymbol{\sigma}^{-1}$ is defined as

$$\mathbf{p} \equiv \boldsymbol{\sigma}^{-1}, \quad (11)$$

where $\mathbf{p}$ is a matrix of the covariant components p_{jk}. Eq. (10) may now be formulated in tensor form as

$$p_{jk}(u^j - x^j)(u^k - x^k) = C^2. \quad (12)$$

Eqs. (10) and (12) describe an ellipsoid called the probability ellipsoid, which is centered on the mean position of the atom at $\mathbf{x}$. The square of the distance, $d^2(\mathbf{w})$, from the mean position to an equiprobability surface in the direction of any vector $\mathbf{w}$ is

$$d^2(\mathbf{w}) = C^2 g_{lm} w^l w^m/(p_{jk} w^j w^k), \quad (13)$$

or

$$d^2(\mathbf{w}) = C^2\mathbf{w}^t\mathbf{g}\mathbf{w}/(\mathbf{w}^t\mathbf{p}\mathbf{w}). \quad (14)$$

Note that Eqs. (9) and (14) *are not equivalent* and the equiprobability ellipsoid is *not* the surface generated by a radius vector with length proportional to the mean-square or root-mean-square displacement. The latter surface is peanut shaped rather than ellipsoidal. However, if C is unity, the

* The tensor g_{ij} is the covariant metric tensor whose components are $\mathbf{a}_i \cdot \mathbf{a}_j$; the scalar products of the unit-cell vectors.

distances to the ellipsoidal surface along the principal axes of $\boldsymbol{\sigma}$ (or $\boldsymbol{\sigma}^{-1}$) are equal to the rms displacements in those directions. This is only true along the principal axes.

The probability P that a displacement falls within the ellipsoid specified by the constant C is given by integrating the probability density function $\varrho(\mathbf{u})$ over the volume of the ellipsoid

$$P(C)=\iiint_{\text{ellipsoid}} \varrho(\mathbf{u})d\mathbf{u}\,, \qquad (15)$$

where $\varrho(\mathbf{u})$ is defined in Eq. (6). A change of variables first to the scaled principal axes of $\boldsymbol{\sigma}^{-1}$ and then to spherical coordinates produces an integral expression which is a function only of the constant C in Eq. (12)

$$P(C)=\left(\frac{2}{\pi}\right)^{1/2}\int_0^c r^2 \exp\left(-r^2/2\right)dr\,. \qquad (16)$$

When $C=1{\cdot}5382$, $P=0{\cdot}5$, and the corresponding ellipsoid is called the 50% probability ellipsoid; that is, the ellipsoid encloses 50% of the trivariate normal probability density. A tabulation of C values for given probability values P is given in the *Handbook of Statistical Tables* by Owen (1962) under the title 'Critical Values for the Spherical Normal Distribution'.

Several computer programs have been written to plot crystal structures with probability ellipsoids on the atomic sites. Examples of drawings made by computer-graphical methods appear occasionally in *Acta Crystallographica* and in other journals which publish crystallographic papers.

Thermal motion of a rigid body

The single most important model for interpreting the thermal motion of molecules in crystals is the rigid-body model. A rigid body is defined as a system of particles constrained to maintain constant interparticle separation within the rigid body. This model gives the simplest possible correlations among the motions of all the different atoms within a molecule.

The method for determining mean-square amplitude values for rigid-body motion from anisotropic temperature factor coefficients was first developed by Cruickshank (1956) and is known to be valid for the special case in which the molecule is situated on a site of $\bar{1}$, $\bar{3}$, $\bar{6}$ or $\bar{4}3m$ symmetry in the crystal. The general method for determining rigid-body motion in the quadratic approximation was developed by Schomaker & Trueblood (1968). The treatment given here follows that of Schomaker & Trueblood except that matrix notation is substituted for dyadic notation. A paper on the Brownian motion of rigid particles by Brenner (1967) contains a detailed discussion of many of the mathematical concepts used here.

A. Types of rigid motions

A rigid motion is any transformation of the rigid body which leaves the interparticle distances unchanged. One example of a rigid motion in a 3-dimensional Euclidean space is a rotation of every particle in the rigid body by an angle α about a directed rotation axis fixed in the rigid body. The most general rigid motion is a screw rotation. According to Chasles's theorem (Synge, 1960), any rigid motion is equivalent to a screw rotation; that is, it is always possible to find an axis for which the components of translation perpendicular to the axis are zero. If, on the other hand, the correctly oriented screw rotation axis is arbitrarily positioned, the motion is described by the same rotation correlated with a translation which includes both parallel and perpendicular components. The screw rotation needed to describe the rigid motion is invariant to the location of the screw axis, but the perpendicular translation is not. An example in two dimensions is given in the article on bond distance corrections.

B. Expansion of the rotation operator

Euler's theorem states that the resultant of any number of successive rotations about a fixed point can be accomplished by a single rotation about a fixed axis through the fixed point (Synge, 1960). The axis of rotation is specified by a three-dimensional unit vector **l** whose Cartesian components l_1, l_2, l_3 are the direction cosines. The rotation is in the sense of a right-handed screw propelled in the direction of the positive unit vector with rotational magnitude θ. If the coordinates of a point before and after a rotation are **r** and **r**′ in terms of a stationary Cartesian coordinate system with origin on the rotation axis, the coordinate transformation is written as

$$\mathbf{r}' = \boldsymbol{R}\mathbf{r} \tag{17}$$

where

$$\boldsymbol{R} = \cos\theta \begin{bmatrix} 1 & 0 & 0 \\ 0 & 1 & 0 \\ 0 & 0 & 1 \end{bmatrix} + \sin\theta \begin{bmatrix} 0 & -l_3 & l_2 \\ l_3 & 0 & -l_1 \\ -l_2 & l_1 & 0 \end{bmatrix} + (1-\cos\theta) \begin{bmatrix} l_1^2 & l_1l_2 & l_1l_3 \\ l_2l_1 & l_2^2 & l_2l_3 \\ l_3l_1 & l_3l_2 & l_3^2 \end{bmatrix} \tag{18}$$

(see Korn & Korn, 1968).

For small rotations, Eq. (18) can be expanded as

$$\boldsymbol{R} = \mathbf{I} + \theta\boldsymbol{L} + \frac{\theta^2}{2}\boldsymbol{L}^2 + \frac{\theta^3}{6}\boldsymbol{L}^3 + \frac{\theta^4}{24}\boldsymbol{L}^4 + \ldots,^* \tag{19}$$

where

$$\boldsymbol{L} = -\boldsymbol{L}^3 = \boldsymbol{L}^5 = -\boldsymbol{L}^7 \ldots = \begin{bmatrix} 0 & -l_3 & l_2 \\ l_3 & 0 & -l_1 \\ -l_2 & l_1 & 0 \end{bmatrix} \tag{20}$$

and

$$\boldsymbol{L}^2 = -\boldsymbol{L}^4 = \boldsymbol{L}^6 = -\boldsymbol{L}^8 \ldots = \begin{bmatrix} -l_2^2 - l_3^2 & l_1l_2 & l_1l_3 \\ l_2l_1 & -l_1^2 - l_3^2 & l_2l_3 \\ l_3l_1 & l_3l_2 & -l_1^2 - l_2^2 \end{bmatrix} \tag{21}$$

Equalities for all even and for all odd powers of $i\boldsymbol{L}$ occur because $l_1^2 + l_2^2 + l_3^2 = 1$.

A changes of variables is made by the introduction of an axial vector $\boldsymbol{\lambda} \equiv \theta\mathbf{l}$. The direction of the axis of rotation is given by the direction of $\boldsymbol{\lambda}$ and the angle of rotation by $|\boldsymbol{\lambda}|$. When the components of $\boldsymbol{\lambda}$, $(\lambda_1, \lambda_2, \lambda_3)$, approach zero, they are called infinitesimal rotation coordinates and correspond to rotations about the three axes of the coordinate system taken separately in any order. They are required to be infinitesimal coordinates for this description because successive rotations about different axes do not commute unless the rotations are infinitesimally small. The vector displacement **v** of a particle at **r** is described with the new coordinates as

$$\mathbf{v} = \boldsymbol{K}\mathbf{r} + \tfrac{1}{2}\boldsymbol{K}^2\mathbf{r} + \tfrac{1}{6}\boldsymbol{K}^3\mathbf{r} + \ldots, \tag{22}$$

where

$$\boldsymbol{K} = \begin{bmatrix} 0 & -\lambda_3 & \lambda_2 \\ \lambda_3 & 0 & -\lambda_1 \\ -\lambda_2 & \lambda_1 & 0 \end{bmatrix} \tag{23}$$

The same expression in vector notation is

$$\mathbf{v} = \boldsymbol{\lambda}\times\mathbf{r} + \tfrac{1}{2}\boldsymbol{\lambda}\times(\boldsymbol{\lambda}\times\mathbf{r}) + \tfrac{1}{6}\boldsymbol{\lambda}\times(\boldsymbol{\lambda}\times(\boldsymbol{\lambda}\times\mathbf{r})] + \ldots, \tag{24}$$

where the symbol $\times$ denotes the vector cross product.

C. The quadratic approximation for rigid-body motion

The simplest general treatment of rigid-body motion is obtained by using first-

* Eq. (19) is obtained either by expanding Eq. (18) or by expanding the matrix equation $\boldsymbol{R} = \exp(\Theta\boldsymbol{L})$ (see Moses, 1965).

order instantaneous displacement parameters in translation $\mathbf{t}$ and rotation $\boldsymbol{\lambda}$. In this approximation the displacement $\mathbf{u}$ of an atom from its equilibrium rest position $\mathbf{r}$ is

$$\mathbf{u}=\mathbf{t}+\mathbf{v}=\mathbf{t}+\boldsymbol{\lambda}\times\mathbf{r}=\mathbf{t}+A\boldsymbol{\lambda}\,, \qquad (25)$$

where

$$A\equiv\begin{bmatrix} 0 & r_3 & -r_2 \\ -r_3 & 0 & r_1 \\ r_2 & -r_1 & 0 \end{bmatrix}. \qquad (26)$$

The second moment for the time-lattice distribution of displacements from $\mathbf{r}$ (the dispersion or covariance matrix) is obtained by taking direct product of both sides of Eq. (25) and averaging each term to obtain

$$\langle\mathbf{u}*\mathbf{u}\rangle=\langle\mathbf{t}*\mathbf{t}\rangle+\langle A\boldsymbol{\lambda}*A\boldsymbol{\lambda}\rangle+\langle\mathbf{t}*A\boldsymbol{\lambda}\rangle+\langle A\boldsymbol{\lambda}*\mathbf{t}\rangle\,, \qquad (27)$$

or

$$U=T+ALA^t+AS+S^tA^t, \qquad (28)$$

The symbol $*$ in (27) designates a direct product (see Hall, 1963). The matrices S, T, and U are defined in detail below. Written out in full the direct products are

$$U\equiv\langle\mathbf{u}*\mathbf{u}\rangle=\begin{bmatrix} \langle u_1u_1\rangle & \langle u_1u_2\rangle & \langle u_1u_3\rangle \\ \langle u_2u_1\rangle & \langle u_2u_2\rangle & \langle u_2u_3\rangle \\ \langle u_3u_1\rangle & \langle u_3u_2\rangle & \langle u_3u_3\rangle \end{bmatrix} \qquad (29)$$

$$ALA^t\equiv\langle A\boldsymbol{\lambda}*A\boldsymbol{\lambda}\rangle$$

$$=A\begin{bmatrix} \langle \lambda_1\lambda_1\rangle & \langle \lambda_1\lambda_2\rangle & \langle \lambda_1\lambda_3\rangle \\ \langle \lambda_2\lambda_1\rangle & \langle \lambda_2\lambda_2\rangle & \langle \lambda_2\lambda_3\rangle \\ \langle \lambda_3\lambda_1\rangle & \langle \lambda_3\lambda_2\rangle & \langle \lambda_3\lambda_3\rangle \end{bmatrix}A^t \qquad (30)$$

$$AS\equiv\langle A\boldsymbol{\lambda}*\mathbf{t}\rangle$$

$$=A\begin{bmatrix} \langle \lambda_1t_1\rangle & \langle \lambda_1t_2\rangle & \langle \lambda_1t_3\rangle \\ \langle \lambda_2t_1\rangle & \langle \lambda_2t_2\rangle & \langle \lambda_2t_3\rangle \\ \langle \lambda_3t_1\rangle & \langle \lambda_3t_2\rangle & \langle \lambda_3t_3\rangle \end{bmatrix} \qquad (31)$$

$$S^tA^t\equiv\langle\mathbf{t}*A\boldsymbol{\lambda}\rangle \qquad (32)$$

$$T\equiv\langle\mathbf{t}*\mathbf{t}\rangle \qquad (33)$$

The T matrix has the same form as the U matrix. It is apparent that U, T, and L are symmetric, because terms such as $\langle u_iu_j\rangle$ and $\langle u_ju_i\rangle$ are equal since they are averages of the same two scalar quantities. On the other hand, $\langle\lambda_it_j\rangle\neq\langle\lambda_jt_i\rangle$ and S is not symmetric in general.

D. The 6-space of rotation and translation

Eq. (28) can be rearranged to the partitioned matrix form

$$U=[I|A]\left[\begin{array}{c|c} T & S^t \\ \hline S & L \end{array}\right]\left[\begin{array}{c} I \\ \hline A^t \end{array}\right]$$

$$\equiv E_{(3\times6)}C_{(6\times6)}E^t{}_{(6\times3)} \qquad (34)$$

The 6×6 symmetric matrix C can be interpreted as the averaged direct product of a 6-space vector $\mathbf{c}$ with itself. The 6-space vector $\mathbf{c}^t\equiv[c_1, c_2, c_3, c_4, c_5, c_6]=[t_1, t_2, t_3, \lambda_1, \lambda_2, \lambda_3]$ is a vector in the rigid-body configurational space of rotation and translation. At any instant the position of the rigid-body is uniquely specified by a point in the configurational 6-space. The S matrix is the mean-square displacement matrix for rigid-body configurational space.

The equation for transforming the configurational 6-space coordinates from one orthonormal base-vector triplet with origin at O in the physical rigid-body 3-space to a different orthonormal triplet with origin at O' is obtained in two steps. First the orthonormal triplet is rotated so that positional coordinates $\mathbf{x}$ in the physical space are transformed as $\mathbf{y}=R\mathbf{x}$. Then the origin is shifted from O to O' by the origin translation vector $\boldsymbol{\varrho}$ with components in the rotated system of coordinates such that the final positional coordinates become $\mathbf{y}'=\mathbf{y}-\boldsymbol{\varrho}$. The complete transformation may be succinctly written as

$$\left[\begin{array}{c} \mathbf{t}' \\ \hline \boldsymbol{\lambda}' \end{array}\right]=\left[\begin{array}{c|c} I & P \\ \hline O & I \end{array}\right]\left[\begin{array}{c|c} R & O \\ \hline O & Q \end{array}\right]\left[\begin{array}{c} \mathbf{t} \\ \hline \boldsymbol{\lambda} \end{array}\right]$$

$$=\left[\begin{array}{c|c} R & PQ \\ \hline O & Q \end{array}\right]\left[\begin{array}{c} \mathbf{t} \\ \hline \boldsymbol{\lambda} \end{array}\right] \qquad (35)$$

$$\mathbf{t}'=R\mathbf{t}+PQ\boldsymbol{\lambda}\,, \qquad (36)$$

$$\boldsymbol{\lambda}'=Q\boldsymbol{\lambda}\,, \qquad (37)$$

where $\boldsymbol{P}$ is a skew-symmetric matrix composed of the components of $\boldsymbol{\varrho}$, the origin transformation vector,

$$\boldsymbol{P} \equiv \begin{bmatrix} 0 & \varrho_3 & -\varrho_2 \\ -\varrho_3 & 0 & \varrho_1 \\ \varrho_2 & -\varrho_1 & 0 \end{bmatrix}, \quad (38)$$

$$\boldsymbol{Q} \equiv (\det \boldsymbol{R})\boldsymbol{R}, \quad (39)$$

$\boldsymbol{O}$ is the null matrix and $\boldsymbol{I}$ is the identity matrix. The $\boldsymbol{Q}$ matrix is needed because, in contrast to $\mathbf{t}$ which is a polar vector $\boldsymbol{\lambda}$ is axial (pseudo-vector) and changes sign for improper rotations with determinant $|\boldsymbol{R}| = -1$. The sign of a polar vector is invariant for all transformations.

The inverse of the transformation in (35) is

$$\begin{bmatrix} \mathbf{t} \\ \hline \boldsymbol{\lambda} \end{bmatrix} = \left[\begin{array}{c|c} \boldsymbol{R}^t & -\boldsymbol{R}^t\boldsymbol{P} \\ \hline \boldsymbol{O} & \boldsymbol{Q}^t \end{array}\right] \begin{bmatrix} \mathbf{t}' \\ \hline \boldsymbol{\lambda}' \end{bmatrix}. \quad (40)$$

The transformation properties of the rigid-body mean-square displacement matrix are given by

$$\left[\begin{array}{c|c} \boldsymbol{T}' & \boldsymbol{S}'^t \\ \hline \boldsymbol{S}' & \boldsymbol{L}' \end{array}\right] = \left[\begin{array}{c|c} \boldsymbol{R} & \boldsymbol{PQ} \\ \hline \boldsymbol{O} & \boldsymbol{Q} \end{array}\right] \times \left[\begin{array}{c|c} \boldsymbol{T} & \boldsymbol{S}^t \\ \hline \boldsymbol{S} & \boldsymbol{L} \end{array}\right] \left[\begin{array}{c|c} \boldsymbol{R}^t & \boldsymbol{O} \\ \hline -\boldsymbol{Q}^t\boldsymbol{P} & \boldsymbol{Q}^t \end{array}\right], \quad (41)$$

or

$$\boldsymbol{T}' = \boldsymbol{RTR}^t + \boldsymbol{PQSR}^t - \boldsymbol{RS}^t\boldsymbol{Q}^t\boldsymbol{P} - \boldsymbol{PQLQ}^t\boldsymbol{P} \quad (42)$$

$$\boldsymbol{S}' = \boldsymbol{QSR}^t - \boldsymbol{QLQ}^t\boldsymbol{P} \quad (43)$$

$$\boldsymbol{S}'^t = \boldsymbol{RS}^t\boldsymbol{Q}^t + \boldsymbol{PQLQ}^t \quad (44)$$

$$\boldsymbol{L}' = \boldsymbol{QLQ}^t = \boldsymbol{RLR}^t. \quad (45)$$

E. Constraints on rigid-body motion related to site symmetry

Note that if $\boldsymbol{P}$ in Eqs. (35)–(45) is replaced by a null matrix and $\boldsymbol{Q}$ is replaced by $\boldsymbol{R}$ det $\boldsymbol{R}$, the transformation properties of $\boldsymbol{T}$, $\boldsymbol{L}$, and $\boldsymbol{S}$ under general rotations become apparent. $\boldsymbol{T}$ and $\boldsymbol{L}$ are polar tensors since det $\boldsymbol{R}$, which can be ± 1, appears to an even power. However, $\boldsymbol{S}$ is an axial tensor, since det $\boldsymbol{R}$ appears to an odd power.

The symmetry of the site on which a rigid body is located can place restrictions on the coefficients of $\boldsymbol{T}$, $\boldsymbol{L}$, and $\boldsymbol{S}$. The symmetry restrictions, which are often different for polar and axial tensors, are tabulated by Birss (1964), Schomaker & Trueblood (1968), and Johnson (1970*b*).

F. The center of reaction

A unique shift of origin can always be effected which minimizes the trace of $\boldsymbol{T}'$ in (42) and simultaneously makes $\boldsymbol{S}$ symmetric (see Schomaker & Trueblood). To determine this unique origin which is termed the 'center of reaction' by Brenner, we set $\boldsymbol{R}$ and $\boldsymbol{Q}$ to $\boldsymbol{I}$, subtract (43) from (44), and equate the result to zero, to obtain

$$\boldsymbol{P}^0\boldsymbol{L} + \boldsymbol{LP}^0 = \boldsymbol{S} - \boldsymbol{S}^t. \quad (46)$$

where the superscript ^{0} on $\boldsymbol{P}$ indicates the origin shift to the center of reaction. After performing the matrix operations indicated in (46), we obtain a matrix equation containing two skew-symmetric matrices only. Since there are only 3 unique elements in a 3-dimensional skew-symmetric matrix, we obtain 3 linear equations in the three unknown components of $\boldsymbol{\varrho}^0$. The three linear equations may be arranged in the form of the matrix equation

$$\boldsymbol{\varrho}^0 = [(\text{tr } \boldsymbol{L})\boldsymbol{I} - \boldsymbol{L}]^{-1}\mathbf{s}, \quad (47a)$$

which written in component form becomes

$$\begin{bmatrix} \varrho_1^0 \\ \varrho_2^0 \\ \varrho_3^0 \end{bmatrix} = \begin{bmatrix} L_{22}+L_{33} & -L_{12} & -L_{13} \\ -L_{12} & L_{33}+L_{11} & -L_{23} \\ -L_{13} & -L_{23} & L_{11}+L_{22} \end{bmatrix}^{-1} \times \begin{bmatrix} S_{23}-S_{32} \\ S_{31}-S_{13} \\ S_{12}-S_{21} \end{bmatrix} \quad (47b)$$

Eq. (47) is simplified if all quantities are first transformed to a Cartesian system with axes parallel to the principal axes of $\boldsymbol{L}$. Denoting components in this coordinate system with carets we obtain

$$\hat{\varrho}_1^0 = (\hat{S}_{23} - \hat{S}_{32})/(\hat{L}_{22} + \hat{L}_{33}) \qquad (48a)$$

$$\hat{\varrho}_2^0 = (\hat{S}_{31} - \hat{S}_{13})/(\hat{L}_{33} + \hat{L}_{11}) \qquad (48b)$$

$$\hat{\varrho}_3^0 = (\hat{S}_{12} - \hat{S}_{21})/(\hat{L}_{11} + \hat{L}_{22}) . \qquad (48c)$$

G. The description in terms of non-intersecting screw-axes

A further reduction in the trace of $\boldsymbol{T}$ and an annihilation of the off-diagonal coefficients of $\boldsymbol{S}$ is possible if the rigid-body orientations are described in terms of subspaces of the 6-dimensional rigid-body configurational space. The particular subspaces to be used are not independent but instead display properties analogous to those found when a crystal structure is determined from the projections of the atoms onto two intersecting planes. The atomic positional components along the one-dimensional intersection are present in both projections. The 3-dimensional space is the union of the two 2-dimensional subspaces with a 1-dimensional subspace in common in the intersection.

The 6-dimensional configurational space can be described as the union of three 4-dimensional subspaces each of which contains the entire 3-dimensional translation space plus a one-dimensional rotation space defined by a vector coincident with one of the three principal axes of $\boldsymbol{L}$. Any vector in the 6-space can now be described as a sum of three vectors, one from each of the three 4-dimensional subspaces. The intersection of the subspaces is the 3-dimensional translation space.

The rigid-body mean-square displacement matrix $\boldsymbol{C}$ in Eq. (34) is now described as the sum of three matrices, one from each of the subspaces.

$$\left[\begin{array}{c|c} \hat{\boldsymbol{T}} & \hat{\boldsymbol{S}}^t \\ \hline \hat{\boldsymbol{S}} & \hat{\boldsymbol{L}} \end{array}\right] = \left[\begin{array}{ccc|ccc} & & & \hat{S}_{11} & 0 & 0 \\ & \hat{\boldsymbol{T}}_1 & & \hat{S}_{12} & 0 & 0 \\ & & & \hat{S}_{13} & 0 & 0 \\ \hline \hat{S}_{11} & \hat{S}_{12} & \hat{S}_{13} & \hat{L}_{11} & 0 & 0 \\ 0 & 0 & 0 & 0 & 0 & 0 \\ 0 & 0 & 0 & 0 & 0 & 0 \end{array}\right] + \left[\begin{array}{ccc|ccc} & & & 0 & \hat{S}_{21} & 0 \\ & \hat{\boldsymbol{T}}_2 & & 0 & \hat{S}_{22} & 0 \\ & & & 0 & \hat{S}_{23} & 0 \\ \hline 0 & 0 & 0 & 0 & 0 & 0 \\ \hat{S}_{21} & \hat{S}_{22} & \hat{S}_{23} & 0 & \hat{L}_{22} & 0 \\ 0 & 0 & 0 & 0 & 0 & 0 \end{array}\right] + \left[\begin{array}{ccc|ccc} & & & 0 & 0 & \hat{S}_{31} \\ & \hat{\boldsymbol{T}}_3 & & 0 & 0 & \hat{S}_{32} \\ & & & 0 & 0 & \hat{S}_{33} \\ \hline 0 & 0 & 0 & 0 & 0 & 0 \\ 0 & 0 & 0 & 0 & 0 & 0 \\ \hat{S}_{31} & \hat{S}_{32} & \hat{S}_{33} & 0 & 0 & \hat{L}_{33} \end{array}\right] \qquad (49)$$

where

$$\hat{\boldsymbol{T}} \equiv \hat{\boldsymbol{T}}_1 + \hat{\boldsymbol{T}}_2 + \hat{\boldsymbol{T}}_3 . \qquad (50)$$

The conceptual decomposition of $\boldsymbol{T}$ into $\boldsymbol{T}_1$, $\boldsymbol{T}_2$, and $\boldsymbol{T}_3$ is arbitrary and only a convenience for formulating Eq. (49). In the description of rigid-body motion, the origin of rotation for each of the three 4-dimensional configurational subspaces can be moved independently, since the intersection contains only the translation space, which by itself is simply a subspace of vector displacements requiring no origin. The $\boldsymbol{S}$-symmetrizing origin transformation, Eq. (48), is applied to each of the three terms in (49) to obtain shifts of the rotation axes in the subspaces. For subspace 1, which contains the λ_1 rotation axis, the shift in the position of the axis is

$$\begin{bmatrix} {}^1\hat{\varrho}_1^0 \\ {}^1\hat{\varrho}_2^0 \\ {}^1\hat{\varrho}_3^0 \end{bmatrix} = \begin{bmatrix} 0 \\ -\hat{S}_{13}/\hat{L}_{11} \\ \hat{S}_{12}/\hat{L}_{11} \end{bmatrix} \qquad (51)$$

where the presuperscript denotes which rotation axis is being shifted. This shift of axis 1 annihilates the S_{12} and S_{13} terms and reduces the trace of $\boldsymbol{T}_1$; therefore it reduces the trace of $\boldsymbol{T}$. The S_{11} term cannot be reduced since it corresponds to a correlation of rotation about axis 1 with translation along axis 1. Equations analogous to Eq. (51) also can be written for axes 2 and 3.

It should be noted that the positioning of the axis in a subspace to eliminate the off-diagonal elements of $\boldsymbol{S}$ and to minimize the trace of $\boldsymbol{T}$ is simply an application of Chasles's theorem in the sense of mean-square rigid motion; that is, a unique axis position is found where the average correlations between rotation and the perpendicular components of translation are zero; and consequently the perpendicular components of mean-square translations are minimized. A much more extensive discussion of the non-intersecting screw-axis description of mean-square displacement is given in the original paper by Schomaker and Trueblood.

It should be emphasized that the foregoing discussion is kinematic in character, elucidating essentially the geometrical aspects of the mean-square displacements. No claim is implied that the elementary motions so derived must have dynamic significance.

Determination of rigid-body motion

Eq. (34) is the fundamental equation relating the mean-square rigid-body motion to the mean-square displacements of atoms and can be restated in component form as

$$U_{ij} = \sum_{k,l=1}^{6} E_{ik}E_{jl}C_{kl} \quad (i,j=1,2,3) \quad (52)$$

where C_{kl} $(k,l=1,2,\ldots,6)$ are the components of the rigid-body-motion tensor to be determined. The E_{ik} $(i=1,2,3;\ k=1,2,\ldots,6)$ components are defined by Eqs. (34) and (26). Eq. (52) provides a convenient formulation for a program which generates observational equations for the linear least-squares determination of the C_{kl} coefficients from the U_{ij} coefficients.

A. Singularities related to the trace-of-S and to special configurations of atoms

Since $\boldsymbol{C}$ is symmetric, there are $6(6+1)/2=21$ unique elements in $\boldsymbol{C}$; however, there is one dependent linear combination in (52) which makes the trace of $\boldsymbol{S}$ (*i.e.*, $S_{11}+S_{22}+S_{33}=C_{41}+C_{42}+C_{43}$) indeterminate from the mean-square displacements U_{ij}. This singularity is most easily understood by writing out the coefficients $E_{ik}E_{jl}$ of the variables C_{kl} for each element of U_{ij} as shown in Table I with the components of $\mathbf{r}$ in $\boldsymbol{A}$ equal x,y,z. The sum of the coefficients for the three variables S_{11}, S_{22}, and S_{33} is zero for an arbitrary atom at x,y,z; hence this linear combination is zero for all atoms. This linear combination will always lead to a singularity in the least-squares normal-equations matrix. One possible corrective measure would be to set arbitrarily one element such as S_{11} to zero by not entering it into the least-squares matrix. A preferred corrective measure is to constrain the trace of $\boldsymbol{S}$ to the arbitrary value zero. An examination of Eqs. (43) and (47) shows that the magnitude of the trace of $\boldsymbol{S}$ remains invariant under all transformations as long as the coordinate systems are Cartesian.

In addition to the trace-of-$\boldsymbol{S}$ singularity, other numerical singularities may appear because of the geometrical arrangement of the atoms in the rigid body. For example, the L_{12} term is indeterminate if the molecule has 5 atoms located at $0,0,0$; $x,0,0$; $-x,0,0$; $0,y,0$; and $0,-y,0$ with a symmetry center at $0,0,0$. This occurs because all coefficients for L_{21} in Table I are zero for each atom.

For another example, suppose all the atoms of a rigid body lie on a conic section* (*i.e.*, ellipse, parabola or hyperbola)

* Schomaker and Trueblood use the term 'quadratic curve' instead of 'conic section'.

Table I Coefficients in observational equations for rigid-body motion in the quadratic approximation

	C_{11}	C_{21}	C_{22}	C_{31}	C_{32}	C_{33}	C_{41}	C_{42}	C_{43}	C_{44}	C_{51}	C_{52}	C_{53}	C_{54}	C_{55}	C_{61}	C_{62}	C_{63}	C_{64}	C_{65}	C_{66}
	T_{11}	T_{21}	T_{22}	T_{31}	T_{32}	T_{33}	S_{11}	S_{12}	S_{13}	L_{11}	S_{21}	S_{22}	S_{23}	L_{21}	L_{22}	S_{31}	S_{32}	S_{33}	L_{31}	L_{32}	L_{33}
U_{11}	1	0	0	0	0	0	0	0	0	0	$2z$	0	0	0	z^2	$-2y$	0	0	0	$-2yz$	y^2
U_{21}	0	1	0	0	0	0	$-z$	$-2y$	0	0	0	z	0	$-z^2$	0	x	$-y$	0	yz	xz	$-xy$
U_{22}	0	0	1	0	0	0	0	0	0	z^2	0	0	0	0	0	0	$2x$	0	$-2xz$	0	x^2
U_{31}	0	0	0	1	0	0	y	0	0	0	$-x$	0	z	yz	$-xz$	0	0	$-y$	$-y^2$	xy	0
U_{32}	0	0	0	0	1	0	0	y	$-z$	$-yz$	0	$-x$	0	xz	0	0	0	x	xy	$-x^2$	0
U_{33}	0	0	0	0	0	1	0	0	$2y$	y^2	0	0	$-2x$	$-2xy$	x^2	0	0	0	0	0	0

in the $x,y,0$ plane. This arrangement of atoms allows no unique separation of rotations about the two axes in the xy plane and translation among z normal to the plane; since when z is zero, the coefficients in the L_{11}, L_{22}, L_{12}, S_{23}, S_{13}, and T_{33} columns form the dependent linear combination $ax^2+by^2+cxy+dx+ey+f=0$. The a, b, c, d, e and f coefficients are constants defining the curve and are the same for each atom. Shifting the origin and rotating the coordinate axes makes the dependent linear combination have different coefficients and involves more columns from the Table but does not eliminate its existence.

The 5-atom example given above may be described geometrically as an improper hyperbola (*i.e.*, two intersecting lines) and the singularity may be described as the conic-section effect.

The possibility of a conic arrangement should be considered whenever all atoms are coplanar. In particular it should be noted that conic descriptions can always be given for five (or fewer) coplanar atoms (see Korn & Korn, 1968, p. 45). If no 4 of 5 given points in the xy plane are collinear, the equation of the conic through (x_1,y_1), (x_2,y_2), (x_3,y_3), (x_4,y_4), (x_5,y_5) is uniquely given by the determinant equation

$$\begin{vmatrix} x^2 & xy & y^2 & x & y & 1 \\ x_1^2 & x_1y_1 & y_1^2 & x_1 & y_1 & 1 \\ x_2^2 & x_2y_2 & y_2^2 & x_2 & y_2 & 1 \\ x_3^2 & x_3y_3 & y_3^2 & x_3 & y_3 & 1 \\ x_4^2 & x_4y_4 & y_4^2 & x_4 & y_4 & 1 \\ x_5^2 & x_5y_5 & y_5^2 & x_5 & y_5 & 1 \end{vmatrix} = 0\,. \quad (53)$$

This determinant equation has the same form as the previously mentioned columns in the matrix of observational equations. A determinant of zero denotes a singular matrix.

B. Regression on principal components

A practical procedure which permits all the singularities, including the trace of ***S*** to be eliminated simultaneously without having to predict the geometrical possibilities is to solve the linear least-squares problem by the statistical technique of regression on principal components.* This method requires a determination of all the eigenvalues and the corresponding orthonormal set of eigenvectors for the least-squares normal-equations matrix (product-moment matrix). Linear combinations of variables specified by the eigenvectors corresponding to non-positive eigenvectors are indeterminate and are set to zero. Actually a family of solutions exists for each eigenvalue whose numerical value is zero; but by setting the linear combination to zero, we specify the solution vector with minimum Euclidean length (Golub & Kahan, 1965). A disadvantage of the technique of regression on principal components is that reliable estimates of the standard errors for the variables are difficult to obtain when general linear combinations have been deleted.

Coupled rigid bodies

The rigid-body model requires that all intramolecular interatomic separations remain fixed. However, many molecules have large-amplitude modes of internal vibrations such as torsional oscillation about a single covalent bond. In such cases coupled rigid bodies can sometimes be used to allow for this internal motion. A 'segmented-body' model is described by Johnson (1970*a*) and similar models, devised by V. Schomaker, are described by Trueblood (1969). Current research in this field is being actively pursued but the details cannot be included conveniently in this introductory paper.

---

* Diamond (1958) and Scheringer (1968) describe a method called filtering which is identical to the long-established method of regression on principal components. For a comprehensive treatment of principal component analysis see Kendall & Stuart (1966). Massy (1965), Hanson & Lawson (1969), and Golub & Reinsch (1970).

Higher-order effects in rigid-body motion

The cumulant-expansion structure-factor equation (Johnson, 1969*a*, 1970*a*) provides a general mathematical model for the experimental determination of higher-order thermal-motion effects in crystals. The equation is

$$F(\mathbf{h}) = \sum_{\substack{\text{atoms}\\ \text{in cell}}} f(\mathbf{h}) \exp \Big[2\pi i \sum_{j=1}^{3} {}^{1}\kappa^{j} h_j + \frac{(2\pi i)^2}{2!} \sum_{j,k=1}^{3} {}^{2}\kappa^{jk} h_j h_k + \frac{(2\pi i)^3}{3!} \sum_{j,k,l=1}^{3} {}^{3}\kappa^{jkl} h_j h_k h_l + \ldots \Big] , \tag{54}$$

where $F(\mathbf{h})$ is the structure factor, $f(\mathbf{h})$ is the atomic scattering factor, $\mathbf{h}$ is the vector of Miller indices, and ${}^{r}\kappa$ $(r=1,2,3,\ldots)$ are symmetric cumulant tensors of order r. The first four cumulant tensors describe the mean, the dispersion, the skewness, and the kurtosis of the probability density function for the thermal motion of an atom. The coefficients of the cumulant tensors are fitted by a least-squares procedure to the observed structure-factor amplitudes $|F_o(\mathbf{h})|$. If higher-order thermal-motion effects are absent, the third and all other higher-order cumulant tensor coefficients are zero.

The two major physical causes of higher-order thermal-motion effects are curvelinear motion and anharmonic vibration. A curvilinear motion is a nonrectilinear motion such as bond-bending or rigid-body libration. Anharmonic vibration results from a potential well which is not parabolic in shape. Many interesting aspects of chemical bonds are related to the shape of the potential well; hence it is desirable to attempt to separate the higher-order effects arising from anharmonicity from those arising from curvilinear motion.

A Cartesian tensor expression (Johnson, 1969*b*) relating ${}^3\boldsymbol{\kappa}$ with $\boldsymbol{L}$ for a centrosymmetric rigid body undergoing harmonic libration is

$$ {}^3\kappa_{jkl}=2\sum_{m,n,p,q=1}^{3} L_{mp}L_{nq}[A_{jm}A_{kn}B_{lpq} + A_{lm}A_{jn}B_{kpq}+A_{km}A_{ln}B_{jpq}]\,, \quad (55) $$

where

$$ A_{jk}\equiv\sum_{l=1}^{3} e_{jkl}r_l\,, \quad (56) $$

$$ B_{jkl}\equiv B_{jlk} \equiv\tfrac{1}{4}\sum_{m,n=1}^{3} e_{jml}e_{kmn}r_n+e_{jmk}e_{lmn}r_n\,, \quad (57) $$

$\mathbf{r}$ = vector to the rest position of an atom from the center of symmetry in the molecule,

$e_{jkl}=1$ when jkl is a cyclic permutation of 1,2,3
$=-1$ when jkl is a cyclic permutation of 3,2,1
$=0$ for all other jkl.

Eq. (55) is derived by expanding a rotation operator about $\mathbf{r}$ in a power series in the curvilinear librational components $(\lambda_1,\lambda_2,\lambda_3)^t\equiv\boldsymbol{\lambda}$ to obtain the displacement $\mathbf{u}$ in rectilinear Cartesian components

$$ u_j=\sum_{k=1}^{3} A_{jk}\lambda_k+\sum_{k,l=1}^{3} B_{jkl}\lambda_k\lambda_l+\ldots\,. \quad (58) $$

Moments about the equilibrium position of an atom at r are obtained by raising u_j in Eq. (58) to the appropriate power and averaging each term in the expanded equation (see Table II). If the librational distribution has Gaussian form, the odd curvilinear moments $\langle\lambda_j\rangle$, $\langle\lambda_j\lambda_k\lambda_l\rangle$, etc. are zero and the even curvilinear moments are all completely specified by the second curvilinear moment $\langle\lambda_m\lambda_n\rangle\equiv L_{mn}$. The fourth curvilinear moment, for example, is

$$ L_{mnpq}=L_{mn}L_{pq} +L_{mp}L_{nq}+L_{mq}L_{np}\,. \quad (59) $$

The cumulants can be calculated from the moments about any point, and all cumulants except the first are invariant to the choice of origin. The first four cumulants are related to the first four rectilinear moments μ_j, μ_{jk}, μ_{jkl}, μ_{jklm} (Kubo, 1962) by

$$ {}^1\kappa_j=\mu_j\,, \quad (60) $$

$$ {}^2\kappa_{jk}=\mu_{jk}-\mu_j\mu_k\,, \quad (61) $$

Table II. Tensor moments about equilibrium position for rigid-body motion. The lowest order even and odd contributors to the first three rectilinear moments $\langle u\rangle$, $\langle uu\rangle$, and $\langle uuu\rangle$ of an atom are listed. A centrosymmetric site symmetry for the rigid body will allow only the even-polar and odd-axial tensors listed in the centrosymmetric column. If the site symmetry is noncentrosymmetric, the tensors listed in the 'Additional' column are also allowed in addition to those for the centrosymmetric site. The libration vector λ is axial and the translation vector t is polar.

Rectilinear moments for individual atoms		Moments for rigid body: Centrosymmetric, Even polar and odd axial tensors: Trans.	Coupling	Libr.	Additional, Odd polar and even axial tensors: Trans.	Coupling	Libr.
Mean $\langle u\rangle$	Even			$\langle\lambda\lambda\rangle$			
	Odd			$\langle\lambda\rangle$	$\langle t\rangle$		
Dispersion $\langle uu\rangle$	Even	$\langle tt\rangle\equiv T$		$\langle\lambda\lambda\rangle\equiv L$		$\langle\lambda t\rangle\equiv S$	
	Odd			$\langle\lambda\lambda\lambda\rangle$		$\langle\lambda\lambda t\rangle$	
Skewness $\langle uuu\rangle$	Even		$\langle\lambda\lambda tt\rangle$	$\langle\lambda\lambda\lambda\lambda\rangle$		$\langle\lambda\lambda\lambda t\rangle$	
	Odd		$\langle\lambda tt\rangle$	$\langle\lambda\lambda\lambda\rangle$	$\langle ttt\rangle$	$\langle\lambda tt\rangle$	

$$^3\kappa_{jkl}=\mu_{jkl}-\mu_j\mu_{kl}-\mu_k\mu_{jl}-\mu_l\mu_{jk}+2\mu_j\mu_k\mu_l\,, \quad (62)$$

and

$$^4\kappa_{jklm}=\mu_{jklm}-\{\mu_j\mu_{klm}+\mu_k\mu_{jlm}+\mu_l\mu_{jkm}+\mu_m\mu_{jkl}\}-\{\mu_{jk}\mu_{lm}+\mu_{jl}\mu_{km}+\mu_{jm}\mu_{kl}\}+2\{\mu_j\mu_k\mu_{lm}+\mu_j\mu_l\mu_{km}+\mu_j\mu_m\mu_{kl}+\mu_{jk}\mu_l\mu_m+\mu_{jl}\mu_k\mu_m+\mu_{jm}\mu_k\mu_l\}-6\mu_j\mu_k\mu_l\mu_m\,. \quad (63)$$

All terms of higher order than $L_{mn}L_{pq}$ are omitted from Eq. (55). This is a reasonable approximation for the libration amplitudes found in most molecular crystals. The fourth cumulant, Eq. (63), becomes zero in this approximation because all terms containing factors of the type $L_{mn}L_{pq}$ cancel when Eq. (59) is introduced.

An assumption implicit in the derivation of Eq. (55) is that no interaction occurs between rigid-body libration and rigid-body translation when the molecule is on a crystallographic center of symmetry. This is not strictly true. In particular the third-rank axial tensor for the interaction $\langle\lambda tt\rangle$ (*i.e.*, mean cubic libration-translation-translation) might be an important contributor to the skewness of thermal motion for an atom belonging to a centrosymmetric rigid molecule, since odd-rank axial tensors are not required to be zero by the center of symmetry. Table II lists the contributors to the first three moments (about the equilibrium positions) from which the cumulants are derived. Eq. (55) is not applicable to the noncentrosymmetric case.

Direct determination of rigid-body motion from diffraction data

Pawley (1964) showed that the parameters ***T*** and ***L*** in the Cruickshank (1956) model for the thermal motion of a centrosymmetric rigid molecule could be determined directly from the diffraction data. More recently, the Schomaker & Trueblood (1968) model has been incorporated into structure-factor least-squares programs by Hirshfeld (1969) and by Scheringer (1969, 1970) and has been used successfully for noncentrosymmetric rigid molecules.

A higher-order treatment for centrosymmetric rigid-body motion giving a result equivalent to Eqs. (54) and (55) is presented by Willis & Pawley (1970) and Pawley & Willis (1970). Their model gave fair improvement over the quadratic model when both were applied to the neutron diffraction data for hexamethylenetetramine (Duckworth, Willis & Pawley, 1970).

Acknowledgements

The author is indebted to Drs. V. Schomaker, H. A. Levy, G. M. Brown, and R. Desiderato for helpful discussions concerning this paper.

References

Birss, R. R. (1964). *Selected Topics in Solid State Physics*, Vol. III: *Symmetry and Magnetism.* Amsterdam: North-Holland.

Brenner, H. (1967). *J. Colloid Interface Sci.* **23**, 407.

Cruickshank, D. W. J. (1956). *Acta Cryst.* **9**, 754.

Diamond, R. (1958). *Acta Cryst.* **11**, 129.

Duckworth, J. A. K., Willis, B. T. M. & Pawley, G. S. (1970). *Acta Cryst.* A**26**, In press.

Golub, G. & Kahan, W. (1965). *J. SIAM Numer. Anal.* B**2**, 205.

Golub, G. & Reinsch, C. (1970). *Numer. Math.*, In press.

Hall, G. G. (1963). *Matrices and Tensors*, pp. 32-33 MacMillan, New York.

Hanson, R. J. & Lawson, C. L. (1969). *Math. of Comput.* **23**,787.

Hirshfeld, F. L. (1969). Private communication.

Johnson, C. K. (1969*a*). *Acta Cryst.* A**25**, 187.

Johnson, C. K. (1969*b*). Abstracts of American Crystallographic Association Winter Meeting. Seattle, Wash. p. 55.

Johnson, C. K. (1970*a*). In *Thermal Neutron Diffraction*, ed. Willis, B. T. M. London: Oxford.

Johnson, C. K. (1970*b*). *Acta Cryst.* A**26**, in preparation.
Kendall, M. G. & Stuart, A. (1966). *The Advanced Theory of Statistics*, Vol. III, Chap. 43. New York: Hafner.
Korn, G. A. & Korn, T. M. (1968). *Mathematical Handbook for Scientists and Engineers*, pp. 473, 478. New York: McGraw Hill.
Kubo, R. (1962). *J. Phys. Soc. Jap.* **17**, 1100.
Massy, W. F. (1965). *J. Amer. Statistical Assoc.* **60**,234.
Moses, H. E. (1965). *Nuovo Cimento* **40**A, 1120.
Owen, D. B. (1962). *Handbook of Statistical Tables*. Reading: Addison-Wesley.
Pawley, G. S. (1964). *Acta Cryst.* **17**, 457.
Pawley, G. S. & Willis, B. T. M. (1970). *Acta Cryst.* A**26**, in press.
Scheringer, (1969). C. Private communication.
Scheringer, C. (1970). *Acta Cryst.* A**26**, in press.
Schomaker, V. & Trueblood, K. N. (1968). *Acta Cryst.* B**24**, 63.
Synge, J. L. (1960). In *Encyclopedia of Physics*, Vol. III, No. 1, pp. 13, 14. Berlin: Springer-Verlag.
Trueblood, K. N. (1969). In *Molecular Dynamics and Structure of Solids*, ed. Carter, R. S. & Rush, J. J., p. 353, Nat. Bur. Standards Spec. Publ. 301. Washington: U.S. Gov. Print. Off.
Wilks, S. S. (1962). *Mathematical Statistics*. New York: Wiley.
Willis, B. T. M. & Pawley, G. S. (1970). *Acta Cryst.* A**26**, in press.

The Effect of Thermal Motion on Interatomic Distances and Angles

CARROLL K. JOHNSON

Chemistry Division, Oak Ridge National Laboratory, Oak Ridge, Tennessee, U.S.A.

The correction for thermal motion is in many cases the principal factor limiting the accuracy of the interatomic distances and angles determined through Bragg diffraction studies [Cruickshank, 1956, 1961; Busing & Levy, 1957, 1964]. The mean atomic positions and the distances between mean atomic positions are often determined to a high precision; however, the quantities of chemical significance are the mean separations of atoms which are usually longer (never shorter) than the observed intermean distances. The following discussion follows closely that given by Busing & Levy (1964). Similar discussions also are given by Marino, Nakamura & Moore (1962) and Cyvin (1968) for bond shrinkage effects in electron diffraction results.

Mean separation of two atoms

With no loss in generality, the mean separation of two atoms (A and B) is derived using a fixed Cartesian coordinate system with the z axis along the interatomic vector when both atoms are at their mean positions. The interatomic distance between the mean atomic positions is

$$d_0 = [(x_B - x_A)^2 + (y_B - y_A)^2 + (z_B - z_A)^2]^{1/2} . \quad (1)$$

At any instant the atoms are displaced from their mean positions by Δx_B, Δy_B, Δz_B, Δx_A, Δy_A, Δz_A. If we denote relative displacement by $\Delta x \equiv \Delta x_B - \Delta x_A$, $\Delta y \equiv \Delta y_B - \Delta y_A$, $\Delta z \equiv \Delta z_B - \Delta z_A$, the instantaneous interatomic distance may be written as

$$d = [(d_0 + \Delta z)^2 + \Delta x^2 + \Delta y^2]^{1/2} . \quad (2)$$

By using the binomial expansion, the instantaneous interatomic distance is approximated by

$$d \approx d_0 + \Delta z + (\Delta x^2 + \Delta y^2)/2(d_0 + \Delta z) - \dots . \quad (3)$$

In addition, the $(d_0 + \Delta z)^{-1}$ term is expanded as

$$(d_0 + \Delta z)^{-1} \approx d_0^{-1} - d_0^{-2}\Delta z + \dots \quad (4)$$

to give an approximation carried to third-order terms in relative displacement

$$d \approx d_0 + \Delta z + (\Delta x^2 + \Delta y^2)/2d_0 - (\Delta x^2 \Delta z + \Delta y^2 \Delta z)/2d_0^2 . \quad (5)$$

The nth moment of the distribution of interatomic distances about the distance between the mean positions is obtained by raising d in Eq. (5) to the nth power and averaging all terms in the expression. The first moment gives the mean or center of gravity of the probability density function for the interatomic distance. In the quadratic approximation, the mean interatomic distance d_g is

$$d_g \equiv \langle d \rangle \approx d_0 + (\langle \Delta x^2 \rangle + \langle \Delta y^2 \rangle)/2d_0 \quad (6)$$

where the angular brackets $\langle\ \rangle$ indicate time and lattice averaging. The term $\langle \Delta z \rangle$ is zero because the expansion is about the distance d_0 between the mean atomic posi-

Research sponsored by the U.S. Atomic Energy Commission under contract with the Union Carbide Corporation.

tions. Note that the factor $(\langle\Delta x^2\rangle+\langle\Delta y^2\rangle)$ is independent of the orientation of the x and y axes of the Cartesian system because of the invariance of a trace operation to rotation in Cartesian systems. The quantity $(\langle\Delta x^2\rangle+\langle\Delta y^2\rangle)/2$ is simply the average mean-square relative displacement in the xy plane.

If all terms in Eq. (6) could be determined from the Bragg diffraction data, the mean bond length could be calculated unambiguously. Unfortunately, this is not the case. Substituting the original Cartesian displacement variables Δx_A, Δy_A, Δz_A, Δx_B, Δy_B, Δz_B into the quadratic correction term $(\langle\Delta x^2\rangle+\langle\Delta y^2\rangle)$ of Eq. (6) we obtain

$$\langle\Delta x^2\rangle+\langle\Delta y^2\rangle=\alpha+\beta-2\gamma \qquad (7)$$

where

$$\alpha\equiv\langle\Delta x_A^2\rangle+\langle\Delta y_A^2\rangle$$
$$\beta\equiv\langle\Delta x_B^2\rangle+\langle\Delta y_B^2\rangle$$
$$\gamma\equiv\langle\Delta x_A\Delta x_B\rangle+\langle\Delta y_A\Delta y_B\rangle$$

The α and β terms of Eq. (7) are experimentally determinable but the γ term is not. The anisotropic temperature-factor coefficients for an atom define the mean-square displacements of the atom. The α term, for example, is the sum of the mean-square displacements of atom A along any two perpendicular vectors, each of which is also perpendicular to the mean interatomic vector. A convenient formulation for this calculation in Cartesian coordinate form is

$$\alpha=\mathrm{tr}\ \boldsymbol{U}_A-r^t\boldsymbol{U}_A\mathbf{r}/(\mathbf{r}^t\mathbf{r}) \qquad (8)$$

$$\beta=\mathrm{tr}\ \boldsymbol{U}_B-\mathbf{r}^t\boldsymbol{U}_B\mathbf{r}/(\mathbf{r}^t\mathbf{r}) \qquad (9)$$

where $\boldsymbol{U}_A$ and $\boldsymbol{U}_B$ are the mean-square displacement matrices for atoms A and B in any Cartesian coordinate system and $\mathbf{r}$ is the mean interatomic vector in the same coordinate system. The trace of $\boldsymbol{U}$ (*i.e.*, tr $\boldsymbol{U}\equiv U_{11}+U_{22}+U_{33}$) is invariant to the choice of the Cartesian system. Eqs. (8) and (9) can be verified by first considering the case $\mathbf{r}=(0,0,z)$.

Limits on bond distance correction

Limits can be placed on the γ term of (7) by noting that the mean-square displacement matrix

$$\boldsymbol{\Sigma}\equiv\begin{bmatrix}\alpha & \gamma\\ \gamma & \beta\end{bmatrix} \qquad (10)$$

must be positive semidefinite (*i.e.*, possess non-negative eigenvalues); consequently α, β, and the determinant $|\boldsymbol{\Sigma}|$ are non-negative. Multiplying the first row and column of $|\boldsymbol{\Sigma}|$ by $\alpha^{-1/2}$ and the second row and column by $\beta^{-1/2}$, we obtain

$$|\boldsymbol{\Sigma}|=\alpha\beta\begin{vmatrix}1 & \gamma(\alpha\beta)^{-1/2}\\ \gamma(\alpha\beta)^{-1/2} & 1\end{vmatrix}\geqq 0 \qquad (11)$$

Hence, it is clear that

$$-1\leqq\gamma(\alpha\beta)^{-1/2}\leqq 1\,. \qquad (12)$$

Table I. Mathematical models used for approximating the mean interatomic distance $d_g\approx d_0+(\langle\Delta x^2\rangle+\langle\Delta y^2\rangle)/2d_0$ where $\langle\Delta x^2\rangle+\langle\Delta y^2\rangle\equiv\alpha+\beta-2\gamma$. The α and β terms are derived from diffraction measurements. (The correction term $\langle\Delta x^2\rangle+\langle\Delta y^2\rangle$ decreases in magnitude from top to botton in the Table.)

Model	$\gamma(\alpha\beta)^{-1/2}$	γ	$\langle\Delta x^2\rangle+\langle\Delta y^2\rangle$
Upper limit	-1	$-(\alpha\beta)^{1/2}$	$\alpha+\beta+2(\alpha\beta)^{1/2}$
Independent motion	0	0	$\alpha+\beta$
Atom B riding on atom A	$0<(\alpha/\beta)^{1/2}<1$	α	$\beta-\alpha$
Lower limit	1	$(\alpha\beta)^{1/2}$	$\alpha+\beta-2(\alpha\beta)^{1/2}$
Rigid translation (*i.e.*, no correction)	1	$\alpha(\equiv\beta)$	0

Table I contains the correlation terms $\gamma(\alpha\beta)^{-1/2}$ and γ for various mathematical models used in calculating mean interatomic distances.

The upper-limit model corresponds to correlated displacements of the two atoms with displacement components perpendicular to the bond and 180 degrees out of phase. An example of this motion is torsional oscillation (*i.e.*, rotational oscillation) about an axis which is somewhere between the two atoms and perpendicular to the bond. The lower-limit model corresponds to in-phase perpendicular displacements such as those caused by a torsional oscillation about an axis perpendicular to an extension of the bond vector. As the torsion axis is moved along the bond extension and further from the atom pair, the motion approaches that of rigid translation where the perpendicular displacements of atoms A and B are equal and in phase.

If a torsion axis perpendicular to the bond goes through the center of atom A, the motion can be described as either independent motion or B-riding-on-A motion since the displacement of A is zero. However, if a translation independent of the torsional oscillation is simultaneously applied to both atoms, the riding model is the proper one. In a real crystal, there are many modes of vibration occurring simultaneously and the model to be used must be chosen with great care for each interatomic distance.

The lower limit model is of considerable practical importance. Note well that $\langle \Delta x^2 \rangle + \langle \Delta y^2 \rangle$ cannot be negative under any circumstance, hence the time average mean interatomic distance will never be shorter than the distance between the time average mean atomic positions.

The independent motion model is based on the assumption that there is no correlation between the motions of the two atoms. This description is a reasonable approximation for atoms in two different molecules if it is certain that there is no significant interaction between the molecules. It is never a reasonable approximation for bonded atoms.

The riding model

The riding motion model of Busing & Levy (1964) with atom B riding on atom A is based on the hypothesis that atom B has all of the translational motion of atom A plus an additional motion uncorrelated with the instantaneous position of atom A. The mean-square displacement of atom B in any direction is never less than the corresponding mean-square displacement of atom A. In other words, the probability density function for atom B is the convolution of that for atom A with an uncorrelated probability density function describing the instantaneous separation of the two atoms.

The riding model (atom B riding on atom A) provides a useful approximation if (1) atom B is strongly linked to atom A and to no other atoms and (2) the mass of atom B is considerably less than the combined mass of atom A plus its neighboring atoms (other than atom B) which are strongly linked to atom A. For example, if atom A belongs to a rigid group and if atom B is a relatively light atom bonded only to atom A, the riding model might be applicable. The riding model is often applied to the results from a neutron diffraction study when atom B is a hydrogen atom bonded to a heavier atom A.

The riding model is not strictly applicable when the motion transmitted to atom B through atom A has a curvilinear component such as that resulting from rigid-body libration about an effective axis removed from atom A. In the derivation of the riding model, the transmitted motion is assumed to be rectilinear (*i.e.*, translational) in nature. A model described in the next section is more appropriate to correct bond distances for the curvilinear librational component of rigid-body motion.

Correction for rigid-body libration

An important mode of motion which causes the distance between mean atomic positions to be less than the corresponding mean interatomic separation is the libration component of rigid-body motion in molecular crystals. Rigid-body motion is made up of libration (rotational oscillation) and translation (rectilinear oscillation).

A basic concept in rigid-body mechanics is illustrated in two dimensions in Fig. 1. Suppose a rigid body represented by the triangle *abc* is moved to a new position *a′b′c′*. There will be some point *d* in the plane which would have remained unmoved by the rigid motion had the point *d* been part of the rigid body. The point *d* would be at infinity if there were no rotational component to the rigid motion. The displacement of all points *a*, *b*, *c*, and *d* can be described as due to a rotation α about *d* with no accompanying translation. The same displacements of *a*, *b*, *c*, and *d* can also be produced by the same rotation α about any other point, such as *a*, plus a translation of the rigid body by the vector from *a* to *a′*. The important concept is that the rotational component of rigid motion is invariant to location of the axis of rotation but the translational component is not. Consequently, the rotation axis may be placed anywhere as long as the accompanying correlated rigid translation for the chosen axis position is known. Similar properties hold for the three-dimensional case, except that the rigid translation component along the rotation axis cannot be eliminated by positioning the axis; consequently the reduced motion is in general a screw rotation.

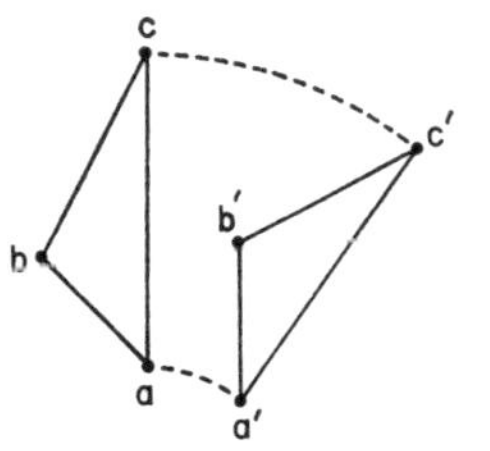

Fig. 1. A rigid motion in two-dimensions.

Eq. (6), which is an approximation for the mean separation of two atoms, may also be applied to any two atoms in the rigid body

$$d_g \approx d_0 + (\langle \Delta x^2 \rangle + \langle \Delta y^2 \rangle)/2d_0 \qquad (13)$$

[(6) restated]

The z axis of the coordinate system used for Eq. (13) is along the interatomic vector between the mean atom positions, and Δx, Δy, Δz are displacements of atom B relative to atom A (*i.e.*, $\Delta x = \Delta x_B - \Delta x_A$, etc.). Note well that rigid translations do not contribute to the relative displacement coordinates Δx, Δy, Δz. Placing all axes of rotation on the arbitrary origin atom A, we find

$$\langle \Delta x^2 \rangle + \langle \Delta y^2 \rangle = (L_{22} + L_{11})d_g^2 \qquad (14)$$

where L_{22} is the mean-square libration (angular oscillation) about the y axis and L_{11} is the mean-square libration about the x axis of the Cartesian coordinate system for the bond.

The equation for correcting interatomic distances for rigid-body libration is

$$d_g \approx d_0 + (L_{11} + L_{22})d_0/2\,, \qquad (15)$$

where the d_g factor in (14) which would appear on the right-hand side of (15) has been replaced by d_0 with negligible loss in accuracy. An equivalent form of Eq. (15) containing the vector $\mathbf{r}$ between the mean locations of two atoms in general positions is

$$d_g \approx \mathbf{r}^t \boldsymbol{M} \mathbf{r}/(\mathbf{r}^t \mathbf{r})^{1/2} \qquad (16)$$

where

$$\boldsymbol{M} = [1 + \tfrac{1}{2}\,\mathrm{tr}\,(\boldsymbol{L})]\boldsymbol{I} - \tfrac{1}{2}\boldsymbol{L} \qquad (17)$$

A more rigorous derivation of an equation equivalent to Eq. (16) is given by Schomaker & Trueblood (1968). They expand a rotation operator into a series about the rest position and form the first moment of the probability density function for displacements due to libration. Another result from the Schomaker & Trueblood paper is that an equilibrium interatomic vector **y** is related to the vector **r** by

$$\mathbf{y} \approx \boldsymbol{M}\mathbf{r} \tag{18}$$

where $\boldsymbol{M}$ is given by Eq. (17). The rest distance calculated from

$$d_r \approx (\mathbf{y}^t\mathbf{y})^{1/2} \equiv [(\boldsymbol{M}\mathbf{r})^t(\boldsymbol{M}\mathbf{r})]^{1/2} = [\mathbf{r}^t\boldsymbol{M}^2\mathbf{r}]^{1/2} \tag{19}$$

is numerically very close to d_g calculated from Eq. (16) because the matrix $\boldsymbol{M}$ is always nearly diagonal. When M is diagonal, the approximations (16) and (19) are equivalent. The mean-square libration tensor **L** is determined by methods discussed in the paper by Schomaker & Trueblood. It is interesting to note that the riding model gives the same corrected distance as the rigid-body model if and only if the origin atom A, on which atom B rides, is on a true center of libration with no correlated rigid translation.

Correction of bond angles

Eqs. (17) and (18) show that rigid-body libration makes the rigid molecule appear to shrink anisotropically if **L** is anisotropic and isotropically if **L** is isotropic. Anisotropic shrinkage will effect the calculated bond angles whereas isotropic shrinkage will not. Interatomic angles can be corrected for rigid-body libration because the rigid-body model with its invariant interatomic distances completely specifies all correlations of motion between the different atoms of the rigid body. In order to correct any bond angle, the correlations in motion among the three atoms involved must be specified.

The corrected angle α between two vectors **v** and **w** in a rigid body apparently distorted by libration is obtained by using Eq. (18) and the vector formulation for an angle between two vectors

$$\alpha = \arccos\{\mathbf{v}^t\boldsymbol{M}^2\mathbf{w}/[(\mathbf{v}^t\boldsymbol{M}^2\mathbf{v}) \times (\mathbf{w}^t\boldsymbol{M}^2\mathbf{w})]^{1/2}\}\,. \tag{20}$$

Note that Eq. (20) is in Cartesian coordinates. For general crystal coordinates we must appropriately include the metric tensor (see Appendix).

Rigid-body plus riding correction

The rigid-body correction for bond lengths may be combined with the riding model correction when a rigid body has an attached atom with large amplitude displacement resulting from internal molecular vibration. For example, in the case of neutron diffraction, hydrogen atoms attached to a ring system show large amplitude mean-square displacements greater than those calculated from the rigid-body motion of the ring system. Carbonyl oxygen atoms attached to ring systems often show this effect also. Such atoms should not be used directly in the determination of the rigid-body motion parameters. If enough other atoms are available to determine the parameters of the rigid-body motion, the expected mean-square displacements of some atom B due to rigid-body motion can be calculated and subtracted from the observed values for atom B to give a residual mean-square displacement matrix $\triangle U_B$ that is positive definite. Eq. (9) can then be used to calculate the residual mean-square displacement normal to the bond. The small residual motion of the atom on which B is riding can also be included in the riding calculation if desired. The bond length correction for riding is added to the bond length corrected for rigid-body motion.

Corrections incorporating spectroscopic results

The problem of correcting bond lengths when large amplitude modes of internal vibration are present can be approached in a much more rigorous manner if a complete spectroscopic normal coordinate analysis of the gaseous form of the molecule under study is available. To incorporate this additional information into the analysis, the assumption is made that there is negligible interaction between the internal molecular vibrations and the lattice vibrations which involve rigid-body motion. From the spectroscopic normal coordinate analysis, a mean-square amplitude matrix can be calculated which can be used for bond-distance correction for the internal molecular vibrations. In particular, a $3N$ by $3N$ (N=number of atoms in molecule) Cartesian mean-square amplitude matrix can be calculated in which the 3 by 3 submatrices along the diagonal correspond to the mean-square displacements of the individual atoms. If these 3 by 3 submatrices are subtracted from the corresponding 3 by 3 mean-square total displacement matrices obtained from Bragg diffraction data, the differences should represent the mean-square displacements due to rigid-body motion alone. This correction for internal molecular vibration should be made before attempting to determine the rigid-body motion parameters in order to obtain the most reliable estimates. This procedure often allows atoms to be used in the rigid-body analysis which would have to be excluded otherwise.

Suitable correction parameters for internal molecular vibration can often be obtained from the normal coordinate analyses of related molecules. For example, the mean-square displacements calculated for Benzene provide a reasonable approximation to use for a phenyl group (Johnson, 1970).

Appendix

Key equations reformulated in crystal coordinates

In the following equations we define

$$\boldsymbol{\sigma} \equiv \boldsymbol{b}/2\pi^2$$

where $\boldsymbol{b}$ is the matrix of anisotropic temperature-factor coefficients in the temperature factor

$$T = \exp(-\mathbf{h}^t \boldsymbol{b} \mathbf{h})$$

with $\mathbf{h}$ the vector of Miller indices.

1. The uncorrected interatomic distance between two atoms with interatomic vector $\mathbf{v}$ in crystal coordinates is

$$d_0 = (\mathbf{v}^t \boldsymbol{g} \mathbf{v})^{1/2} \qquad (A\ 1)$$

where $\boldsymbol{g}$ is the metric tensor with components $g_{ij} = \mathbf{a}_i \cdot \mathbf{a}_j$; the inner products of the unit-cell vectors.

2. The bond distance between atoms A and B corrected for 'atom B riding on atom A' is

$$d_g = d_0 + (\beta - \alpha)/(2d_0) \qquad (A\ 2)$$

where

$$\alpha = \mathrm{Tr}\,(\boldsymbol{g}\boldsymbol{\sigma}_A) - \mathbf{v}^t \boldsymbol{g} \boldsymbol{\sigma}_A \boldsymbol{g} \mathbf{v}/(\mathbf{v}^t \boldsymbol{g} \mathbf{v}) \,. \qquad (A\ 3)$$

where Tr denotes the trace operation. In order to calculate β, we substitute $\boldsymbol{\sigma}_B$ for $\boldsymbol{\sigma}_A$ in Eq. (A 3).

3. A rigid-body interatomic distance corrected for the effects of rigid-body libration may be calculated as

$$d_g = \mathbf{v}^t \boldsymbol{g} \boldsymbol{M} \mathbf{v}/d_0 \qquad (A\ 4)$$

where

$$\boldsymbol{M} \equiv [1 + \tfrac{1}{2}\,\mathrm{Tr}\,(\boldsymbol{L}\boldsymbol{g})]\boldsymbol{I} - \tfrac{1}{2}\boldsymbol{L}\boldsymbol{g} \qquad (A\ 5)$$

The matrix $\boldsymbol{L}$ contains the contravariant components of the rigid-body libration tensor. The matrix of covariant components is gLg. The symbol Tr denotes the trace operation and I is an identity matrix.

An alternate and equally valid formulation is derived by noting that libration in effect produces an anisotropic shrinkage of space so that a vector $\mathbf{v}_r$ between rest positions is related to the vector $\mathbf{v}$ between mean position by

$$\mathbf{v}_r = M\mathbf{v} \qquad (A\,6)$$

where M is defined by Eq. (A 5). The distance between rest positions is

$$d_r = (\mathbf{v}_r^t g \mathbf{v}_r)^{1/2} \qquad (A\,7)$$

Eqs. (A 4) and (A 7) produce similar numerical results.

4. A rigid-body interatomic angle corrected for rigid-body libration may be calculated from two uncorrected interatomic vectors $\mathbf{v}$ and $\mathbf{w}$ by

$$\alpha_r = \cos^{-1} \{\mathbf{v}_r^t g \mathbf{w}_r / [(\mathbf{v}_r^t g \mathbf{v}_r) \times (\mathbf{w}_r^t g \mathbf{w}_r)]^{1/2}\} \qquad (A\,8)$$

where $\mathbf{v}_r$ is defined by Eq. (A 6) and

$$\mathbf{w}_r = M\mathbf{w} \qquad (A\,9)$$

with M defined by Eq. (A 5).

References

Busing, W. R. & Levy, H. A. (1957). *J. Chem. Phys.* **26**, 563.
Busing, W. R. & Levy, H. A. (1964). *Acta Cryst.* **17**, 142.
Cruickshank, D. W. J. (1956). *Acta Cryst.* **9**, 757.
Cruickshank, D. W. J. (1961). *Acta Cryst.* **14**, 896.
Cyvin, S. J. (1968). *Molecular Vibrations and Mean Square Amplitudes*, Chap. 14. Amsterdam: Elsevier Publ. Co.
Johnson, C. K. (1970). In *Thermal Neutron Diffraction*, ed. Willis, B. T. M. London: Oxford.
Morino, Y., Nakamura, J. & Moore, P. W. (1962). *J. Chem. Phys.* **36**, 1050.
Schomaker, V. & Trueblood, K. N. (1968). *Acta Cryst.* B**24**, 63.

Discussion

MASLEN: It is important to distinguish between the various possible definitions of bond lengths and angles. One may use the mean values of these quantities, which are influenced by the thermal motion, or use the values computed from nuclear positions defined either as corresponding to the potential energy minima or the mean positions in the ground state. The definition in terms of the potential energy minimum is in accord with usage in quantum chemical calculations and seems to me to be preferable.

JOHNSON: I agree that definition in terms of the equilibrium position is to be preferred, but in work of this kind the extraction of such information is operationally difficult and involves theoretical assumptions which are not required in the treatment of the mean values of the bond lengths and angles.

CRUICKSHANK: If one wishes to accurately measure dimensions of structures, then I really think it must be done at the lowest possible temperatures, particularly in molecular crystals when the interest is not so much in stereochemical problems but in obtaining distances between potential energy minima. There is still the problem of the zero point energy, which is not negligible; but in any case for organic molecules where the bond corrections can be several hundredths of an Ångstrom, there is significant error in these corrections.

STEWART: Could you comment on the quality of structure to which these methods should be applied?

JOHNSON: If there is any approximation to a rigid body in a structure, I think it is legitimate to attempt to fit a rigid-body model. If there are errors in the temperature factors, they may tend to average out. However, if the temperature factors are not reliable, I would not recommend using a riding model as there is no averaging.

Drawing Crystal Structures by Computer

CARROLL K. JOHNSON

Chemistry Division, Oak Ridge National Laboratory, Oak Ridge, Tennessee, U.S.A.

The computer program ORTEP (Oak Ridge Thermal Ellipsoid Plot Program, Johnson, 1965) which produces crystal-structure illustrations has been used routinely in our laboratory for about five years. This article describes the operational procedures used with the program and gives some guidelines for making crystal-structure illustrations by computer-plotter graphical methods.

Graphics

Perhaps the most difficult problem of all in crystal-structure graphics is deciding what to include and what to leave out of the illustration. A crystal structure contains an unlimited number of atoms, and it is often tempting to include many unit cells so that the periodic properties are clearly visible. In general, this approach produces undesirable results. The goal of technical graphics is not to display everything, as an aerial photograph does, but rather to select the important and prominent features of the subject and to display these features in such a way that they are immediately obvious to the reader.

The artistic aspects of a crystal structure illustration are also important since it is highly desirable that an illustration have sufficient aesthetic appeal to tempt the reader to study the drawing. A reader often forms an initial impression about an article from the illustrations even before reading the abstract.

Research sponsored by the U.S. Atomic Energy Commission under contract with the Union Carbide Corporation.

Operational procedures

The illustrations produced by ORTEP are principally of the ball-and-stick type. The graphical representation of an atom can be a circle, a sphere, or a thermal ellipsoid.

ORTEP is controlled through instructions which are used to 'program' an illustration. The instructions are grouped into the following categories: (1) distance and angle calculation for preliminary analysis, (2) plotter control, (3) plot dimensions, (4) subject definition (atoms to be included), (5) orientation, (6) scale and position on the paper, (7) atom plotting, (8) bond plotting, (9) labeling, and (10) repetition of a series of instructions.

Individual atoms are denoted by a five-component code word for atom designation in which the last two digits signify the number of the symmetry operator, the preceding 3 digits give the lattice translation, and the leading digits denote the atom number. The translation components code is such that 555 means no lattice translation. The atom designator code word 3 654 02, for example, specifies that the third atom is transformed by symmetry operator number 2, then translated by $(1, 0, -1)$ along the unit cell vectors.

The subject-definition instructions are used to build a list of atoms by adding and eliminating any number of (1) 'runs' of atoms, (2) groups of atoms within a 'sphere of enclosure' or a 'box of enclosure', and (3) atoms around atoms already in the list found by 'convoluting spheres of enclosure'. For example, illustrations for metal coordination compounds can often be defined by first storing all the metal atoms in the unit cell, then using convoluting spheres of enclosure about the metal atoms to find all the atoms directly coordinated to the metal atoms.

Bonds between atoms are usually defined implicitly with 'vector search codes' which reflect the user's knowledge of the structural chemistry and bond distance ranges. By carefully arranging the input sequence of atoms, atom number runs can be used advantageously in the vector search codes to select bonds, *e.g.*, hydrogen bonds, from a population of similar interatomic distances.

Labels for atom symbols (and for bond lengths) can be plotted with a constant relative displacement from the center of each atom (or bond), or the labels may be positioned individually to obtain optimal results. Usually a combination of the two procedures is used with only the labels which tend to cause overlap being positioned individually. Labels and captions can be plotted in perspective in planes defined by the user. This feature is particularly useful with stereoscopic drawings.

Stereoscopic drawings

Stereoscopic crystal structure illustrations have a long history in crystallography. A two-volume atlas of stereoscopic drawings edited by M. von Laue & R. von Mises (1926, 1936) is still used as a visual aid in demonstrating simple structures. An outstanding paper on stereoscopic drawing techniques by Rule (1938) is a valuable reference and should be read by anyone interested in making stereoscopic drawings by computer. One of the out-standing advantages of computer-graphics techniques is that the precision obtainable through machine plotting makes feasible the routine production of detailed stereoscopic illustrations. This is a difficult and tedious task by conventional drafting methods (Watts & Rule, 1946).

Many computer programs for making stereoscopic drawings, including the distributed version of ORTEP, obtain the left-eye and right-eye projections by rotating the object about an axis perpendicular to an axis of projection. A recent paper by Saunders (1968) points out that this method leads to a slightly distorted stereoscopic image because the two projections are always placed in the same plane for viewing in a stereoscope. The correct procedure is to translate the axis of perspective projection from one eye to the other instead of rotating the axis (see Fig. 1), or equivalently, to translate rather than rotate the object. A comparison of stereoscopic drawings made by a modified version of ORTEP which incorporates both methods shows that the differences are indeed minor. However, if the maximum stereoscopic fidelity is desired, the translation approach should be followed.

Fig. 1 illustrates the geometrical consideration used to determine a set of operating parameters for a typical ORTEP stereo drawing. In order to produce any degree of fidelity, a stereoscope* must be used to view the stereo drawings. The optical characteristics of the stereoscope (*i.e.*, focal length and separation of optical centers) must be considered in making the drawing (Rule, 1938). When stereo slides are made for projection in a stereo projector,† the optics of the projector and the dimensions of the screen are the important

* A compact stereoscopic viewer of reasonable quality for printed stereo pairs is sold by T. M. Hubbard Scientific Company, P.O. Box 105, Northbrook, Illinois, for about two U.S. dollars. This viewer has a focal length of 12 cm and an adjustable interocular separation of 6 to 7 cm. Other stereo viewers have focal lengths ranging from 9 to 25 cm.

† The only stereo projector currently being manufactured in the USA is made by Compco Corp., 1808 N. Spaulding Ave., Chicago, Ill. Stereo viewing spectacles are available from Polaroid Corp., 549 Technology Sq., Cambridge, Mass. To prevent depolarization at the screen, a silver lenticular projection screen must be used. The common glass beaded screen is not usable.

variables which determine the drawing parameters (Rule, 1941).

Applications

Besides the most common application of making regular crystal structure drawings and stereoscopic pairs of such drawings for publication, several other uses of ORTEP should be mentioned.

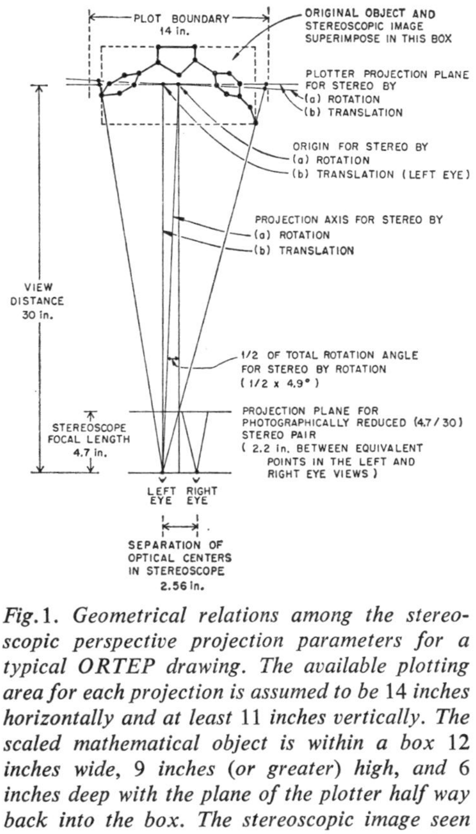

Fig. 1. *Geometrical relations among the stereoscopic perspective projection parameters for a typical ORTEP drawing. The available plotting area for each projection is assumed to be* 14 *inches horizontally and at least* 11 *inches vertically. The scaled mathematical object is within a box* 12 *inches wide,* 9 *inches (or greater) high, and* 6 *inches deep with the plane of the plotter half way back into the box. The stereoscopic image seen through a stereoscope with a* 12 *cm* (4·7 *in.) focal length and a* 6·5 *cm* (2·56 *in.) separation between optical centers should appear superimposed on the original object. The parameters for both 'translation stereo' and 'rotation stereo' are shown. The appropriate linear dimensions can be scaled to accommodate other plotting areas yet produce the same final image. The empirical parameters in common usage* [(*a*) 30-*inch view distance,* (*b*) 5 *to* 6 *degree stereo rotation, and* (*c*) *photographic reduction to achieve* 2·2 *to* 2·4 *inch separation of equivalent points in the stereo pair*] *also seem to produce acceptable results. Inches are used here only because most commercial plotting systems operate with this unit of measurement.*

ORTEP was originally written to aid in analyzing thermal-motion patterns. The major librational modes of displacement of a group of atoms often become apparent by studying thermal ellipsoid stereo drawings. Such drawings sometimes indicate that a model such as the rigid body (Schomaker & Trueblood, 1968) or the segmented body (Johnson, 1970) might be applicable. After the model has been fitted to the data, a plot of the calculated thermal ellipsoids for comparison with the observed thermal ellipsoids can sometimes give information about internal molecular motion. ORTEP has also been used to draw ellipsoids showing the mean-square displacement caused by internal molecular motion as indicated by a spectroscopic normal-coordinate analysis (Holmes & Deiters, 1969). Another interesting application is in plotting the differences between thermal ellipoids derived from X-ray and neutron data in order to investigate bonding electron effects (Coppens, 1969).

Stereoscopic drawings are also useful in screening *E*-maps when direct methods are used for solving crystal structures (*e.g.*, see Hamilton, 1968). The way this procedure is used at Oak Ridge with molecular crystals is to have a Fourier program* punch out cards containing the interpolated positions of the largest peaks in the *E*-map. Then ORTEP starts with a given peak (usually the largest) and does a reiterative sphere-of-enclosure convolution with a sphere radius slightly larger than the appropriate bond distances to isolate a molecule if one is present. A minor modification is incorporated which allows an atom (in any of its equivalent positions) to be used only once so that the reiterative

* A modified version of the program FORDAPER by A. Zalkin is used.

convolution process cannot continue endlessly if extraneous 'bridging peaks' between molecules are present.

References

Coppens, P. (1969). In *Molecular Dynamics and Structure of Solids*, ed. Carter, R. S. & Rush, J. J. Nat. Bur. Standards Special Publication 301, p. 385. Washington, D.C.: U.S. Gov. Print. Off.

Hamilton, W. (1968). *J. Comput. Phys.* **2**, 417.

Holmes, R. R. & Deiters, R. M. (1969). *J. Chem. Phys.* **51**,4043.

Johnson, C. K. (1965). *ORTEP: A Fortran Thermal-Ellipsoid Plot Program for Crystal Structure Illustrations*. ORNL-3794 Revised, Oak Ridge National Laboratory, Oak Ridge, Tennessee.

Johnson, C. K. (1970). In *Thermal Neutron Diffraction*, ed. Willis, B. T. M. London: Oxford University Press.

Rule, J. T. (1938). *J. Opt. Soc. Amer.* **28**, 313.

Rule, J. T. (1941). *J. Opt. Soc. Amer.* **31**, 325.

Saunders, B. G. (1968). *Appl. Opt.* **7**, 1499.

Schomaker, V. & Trueblood, K. N. (1968). *Acta Cryst.* B**24**, 63.

Von Laue, M. & Von Mises, R. (eds.) (1926, 1936). *Stereoscopic Drawing of Crystal Structures*, Vols. I and II. Berlin: Springer.

Watts, E. F. & Rule, J. T. (1946). *Descriptive Geometry*, pp. 268–287. New York: Prentice-Hall.

Higher Order Mechanistic Models for Thermal Motion in Crystal Structures

E. N. MASLEN

Department of Physics, University of Western Australia.

Crystal structure analysis consists of the determination of a set of parameters related to the density distribution of scattering material from a set of measured diffraction intensities. In the analysis of coherent elastic scattering of X-rays or thermal neutrons, it is normal to invoke the kinematic theory of scattering in which the intensity of diffraction from planes with the Miller index triple **H** (h,k,l) is proportional to the square of the amplitude or structure factor, $F_{\mathbf{H}}$, of the diffracted beam. The latter is just the Fourier transform of the average density of scattering material $\langle\varrho(\mathbf{r})\rangle$.

It is normal to subdivide this average scattering density into a set of scattering units, each with a scattering power given by a product fT, where the temperature factor T is the Fourier transform of the smearing due to the thermal motion. The scattering factor, f, is the Fourier transform of the distribution of scattering material in the limit of zero thermal motion.

The most common method of analysing crystal structures is to fit to a set of diffraction data a series of parameters, p_j, which define the nuclear positions, the thermal smearing functions, and the scattering factors in an iterative least-squares procedure. A residual of the form

$$\sum_{\mathbf{H}} w_{\mathbf{H}}(|F^o_{\mathbf{H}}| - |F^c_{\mathbf{H}}|)^2 ,$$

where $F^o_{\mathbf{H}}$ and $F^c_{\mathbf{H}}$ are the observed and calculated structure factors and $w_{\mathbf{H}}$ is a weighting factor, is commonly used. On statistical grounds, however, the residual

$$\sum_{\mathbf{H}} w_{\mathbf{H}}(|F^o_{\mathbf{H}}|^2 - |F^c_{\mathbf{H}}|^2)^2 ,$$

with $w_{\mathbf{H}}$ as the reciprocal of the variance of $|F^o_{\mathbf{H}}|^2$, is to be preferred.

Until recent years it has been common practice to assign to each scattering centre a temperature factor appropriate to a harmonic with three degrees of freedom. This has the form

$$T_3(V) = \exp\,(-2\pi^2 \sum_{i=x,y,z} \sum_{j=x,y,z} H_i H_j U_{ij}) \quad (1)$$

where the $\mathbf{U}_{ij}$ are the elements of a mean square amplitude tensor. The scattering factors f are usually those appropriate to atomic systems.

Both of these procedures impose serious limitations on work of high accuracy, especially for molecular systems. A substantial part of the thermal motion in such cases is due to rotational oscillations, or librations, for which the vibrational model is a poor approximation. Changes in the electron density distribution due to chemical bonding are substantial for the lighter nuclei and the use of atomic scattering factors involves appreciable errors. These interact quite strongly with the parameters defining the thermal motion, and conversely, and both may affect the nuclear positions. Errors produced by these interactions are far from negligible in an accurate structure analysis. For this reason improved models for thermal motion and scattering factors are required.

In broad terms there are two approaches to this problem. The first is to choose the models in as general a way as possible, applying the minimum constraint to the results which are obtained. An example of this technique applied to the analysis of thermal motion is the cumulant expansion method of Johnson, described in a preceding paper in this series. The advantage of the approach is that it makes few prejudgements of physical situations. Its disadvantage is that it may introduce an unnecessarily large number of variable parmeters into the analysis of the diffraction data.

The second method is to set up models for thermal motion and scattering power which have a minimum number of variable parameters *i.e.*, one in which the variation of a further parameter fails to produce a significant improvement in the agreement between sets of observed and calculated diffraction intensities (Hamilton, 1969). When properly used, this results in rapid refinement to results of high validity, but the technique is more prone to abuse than the general method. Models for applying the second approach, which are especially suitable for the analysis of accurate X-ray diffraction data from molecular crystals, are described below.

Models for scattering factors

The total electron density is regarded as the sum of component density functions. It is convenient to adopt a molecular orbital representation for these functions, which are then compatible with the usual framework for theoretical calculations. Following Stewart (1969) an electron density distribution may be written as a sum of orbital products thus

$$\varrho(\mathbf{r}) = \sum_m \sum_n P_{mn} \chi_m(\mathbf{r}) \chi_n(\mathbf{r}) \qquad (2)$$

where χ_m and χ_n are appropriately chosen orbitals and the P_{mn} are population coefficients. In an LCAO (linear combination of atomic orbitals) representation of the charge density in a molecule containing only first row atoms, the χ are the atomic $1s$, $2s$, and $2p$ orbitals.

For orbitals belonging to the same centre only products of types $1s^2$, $2s^2$, and $2p^2$ contribute appreciably to the density. The bulk of the remaining contribution to the density comes from nearest neighbour products and, to a reasonable level of approximation, all other contributions may be neglected. Theoretical calculations of electron density indicate that the core electron density is almost invariant to chemical bonding; this is especially true of the $1s^2$ cores for atoms heavier than helium. Thus the population parameter for this product may be set equal to 2 and two centre orbital products involving $1s$ orbitals for such atoms may be neglected. For first row atoms the core contribution dominates the scattering at high Bragg angles, which is useful both in normalizing the population coefficients and in determining the thermal parameters. The P_{mn} for the remaining products may be included as variable parameters in the least-squares refinement. Their values when determined give the valence electron distribution.

For heavier atoms the valence electron distribution is more complex than for first row atoms, but it makes a proportionally smaller contribution to the Bragg scattering. For second row atoms a reasonable approximation to the valence electron density may be obtained using only $3s$ and $3p$ functions, which are treated in the same manner as the valence orbitals for the lighter atoms.

Analytical formulae for the Fourier transformations of the relevant orbital products have been determined and are quoted by Stewart (1969). Numerical values for a set of SCF wavefunctions (Clementi, 1965) are also available (Stewart, 1969).

Mechanistic models for thermal motion

The only rigid constraint on the thermal

smearing function is that it conforms to the site symmetry. In setting up a mechanistic model for the thermal motion one must use constraints which are approximately valid in order to simplify the description of the smearing function. A useful starting-point is the harmonic approximation in which the restoring force produced by the displacement of a nucleus from its equilibrium position by an amount δ is proportional to that displacement. The smearing function is given by the well-known relation from statistical mechanics that

$$P(\delta) = (2\pi\langle\delta^2\rangle)^{-1/2} \exp\left(\frac{-\delta^2}{2\langle\delta^2\rangle}\right) \qquad (3)$$

At the level of accuracy of the harmonic approximation the thermal motion results from a number of independent harmonic motions, or modes. For a molecular crystal each mode may be regarded as a linear combination of periodic component distortions of the following types

(i) rigid body vibrations
(ii) rigid body librations
(iii) periodic rotations about bonds
(iv) periodic bond-angle distortions
(v) periodic bond-length distortions

There is ample evidence that there is very little coupling between the rigid body motions (i) and (ii) and the internal distortions (iii), (iv), and (v). The spectra of molecular solids contain all the features of the free molecule spectrum with little distortion in addition to lines due to intermolecular modes which in general are of low frequency. The major contribution to the mean square amplitude of thermal motion comes from the low frequency modes, and in most cases these are almost entirely simple motions of type (i), (ii), or (iii). For the purposes of calculating smearing functions a model involving only pure modes of types (i) to (v) is a good first approximation. The effects of coupling and anharmonicity may be treated as minor perturbations.

Calculation of temperature factors

Fortunately the temperature factor for a nucleus executing thermal motion due to several independent modes is just the product of the temperature factors for each mode. In order to represent the thermal motion at the level of accuracy of the harmonic approximation, only two basic modes are required, namely a linear harmonic motion, or vibration with one degree of freedom, and a harmonic oscillation about a rotation axis, *i.e.* a libration with one degree of freedom. All the modes referred to above may be represented as a combination of independent components of these basic types. Thus the vibrational and librational modes of a rigid body may be represented as combinations of three independent linear vibrations in mutually perpendicular directions and three independent librations about mutually perpendicular axes respectively. Bond-stretching and rotation about bonds are represented by a vibration and libration with one degree of freedom respectively. The bond-angle deformations at O resulting from harmonic motion of atom P normal to the bond OP may be represented as a libration with two degrees of freedom, *i.e.* with two independent components as shown in Fig. 1.

The calculation of temperature factors in the harmonic approximation at the limit of zero coupling thus requires temperature factors for the two basic mode types. It is helpful to refer these modes to an

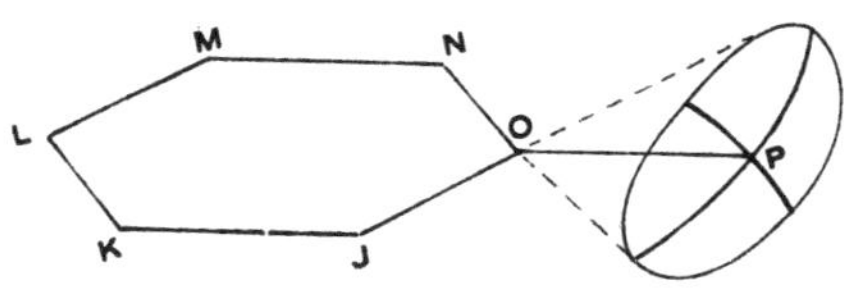

Fig. 1. Mode resulting from bond angle deformation due to motion of the nucleus at P in directions normal to the bond OP. The trajectory is confined to the surface of a sphere centred at O and describes a librational motion with two degrees of freedom.

appropriate orthogonal frame of reference. For a harmonic vibration with one degree of freedom we choose axis 3 parallel to the direction of vibration, with axes 1 and 2 perpendicular to axis 3 and to each other, but otherwise arbitrary [Fig. 2(*a*)]. The temperature factor is then given by

$$T_1(V) = \langle \exp(2\pi i H_3 r_3) \rangle$$
$$= \int_{-\infty}^{\infty} P(r_3) \exp(2\pi i H_3 r_3) dr_3$$
$$= \exp(-2\pi^2 H_3^2 \langle r_3^2 \rangle) \qquad (4)$$

The evaluation of temperature factors for harmonic vibrational modes with more than one component follows in an obvious manner. Thus for a harmonic vibration with three degrees of freedom with principal axes of vibration parallel to axes 1, 2, and 3

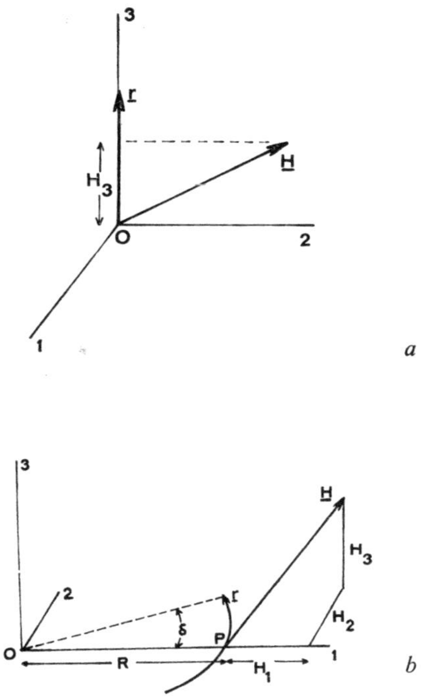

Fig. 2. Orthogonal frames of reference. (a) For a vibrational mode with one degree of freedom. (b) For a librational mode with one degree of freedom. **H** *is a vector normal to the set of Bragg planes with a magnitude equal to the reciprocal of the interplanar spacing.*

$$T_3(V) = \exp\{-2\pi^2(H_1^2\langle r_1^2\rangle + H_2^2\langle r_2^2\rangle + H_3^2\langle r_3^2\rangle)\}. \qquad (5)$$

This may be related to the more usual expression with the argument in the form of a second order tensor, which is given in Eq. (1), by a transformation from the orthogonal frame of reference to the crystallographic system.

An orthogonal frame of reference appropriate to a harmonic libration with one degree of freedom may be chosen as in Fig. 2(*b*). The temperature factor is given by the expression

$$T(L) = \int_{-\pi}^{+\pi} P(\delta) \exp(2\pi i \mathbf{H} \cdot \mathbf{r}) d\delta \qquad (6)$$

A variety of methods for evaluating this expression have been proposed (*e.g.*, Atoji, Watanabe & Lipscomb, 1953; Kay & Behrendt, 1963; Brown & Chidambaram, 1967; Maslen, 1968). The most exact method appears to be that of Atoji, Watanabe, and Lipscomb, but the resulting expression is cumbersome and is computationally prohibitive for large systems. Brown and Chidambaram make the approximation of extending the integration to infinity in either direction, using the expression for $P(\delta)$ given in Eq. (3). This yields

$$T_1(L) = \exp(-2\pi i R H_1) \sum_{p=0}^{\infty} \varepsilon_p i^p J_p(a) \times \exp(-\tfrac{1}{2} p^2 \langle \delta^2 \rangle) \cos(p\chi) \qquad (7)$$

where

$$a = 2\pi R (H_1^2 + H_2^2)^{1/2}$$
$$\chi = \tan^{-1}(H_2/H_1)$$
$$\varepsilon_p = 1 \text{ if } p = 0 \text{ and } 2 \text{ otherwise}$$

and J_p is the Bessel function with real argument of order p. This expression converges rapidly for motions with large $\langle \delta^2 \rangle$ and small R.

The expressions derived by Kay and Behrendt and by Maslen are simpler, but are applicable only in the range for which $\sin \delta \rightarrow \delta$ is a good approximation. It may be shown that the two expressions are identical in the limit of zero coupling between components.* For a single component

$$T(L)=(1+2\pi iRH_1\langle\delta^2\rangle)^{-1/2} \times \exp\{-2\pi^2R^2H_2^2\langle\delta^2\rangle \times(1+2\pi iRH_1\langle\delta^2\rangle)^{-1}\}. \quad (8)$$

Structure refinement using mechanistic models

The parameters defining a mode may be defined in a variety of ways and the optimum choice of the defining parameters is a central problem in using a mechanistic model. Thus a harmonic vibration with three degrees of freedom may be specified by the six independent elements of the mean square amplitude tensor or by three mean square amplitudes in the direction of the principal axes of vibration with three Eulerian angles relating the vibrational frame of reference to cell axes of the crystal in an orthogonal system. The former representation is generally preferred because of the simpler form of the derivatives of the temperature factor with respect to the defining parameters.

The situation for librations is complicated by the fact that the frames of reference for different nuclei affected by the same mode component have only their rotation axes in common. The parameters defining a librational mode are therefore specified in terms of a standard frame of reference for the mode and the librational frames for all nuclei affected by the mode are transformed to that frame.

For computational simplicity it is desirable that the parameters should be defined in the same manner for modes of all types, except the harmonic vibration with three degrees of freedom for the reasons given above. This may be accomplished by defining each mode in terms of a nine-parameter set as follows

- (i), (ii), and (iii) the coordinates of the centre of the standard frame for the mode with respect to the unit cell origin
- (iv), (v), and (vi) the orientation of the standard frame changes which are specified by rotation about the axes (i), (ii), and (iii)
- (vii), (viii), and (ix) three amplitude coefficients.

For some mode types one or more of these parameters will not be relevant. Thus only one amplitude coefficient is required for a harmonic motion with one degree of freedom. Parameters (i) to (iii) have no meaning when applied to vibrational modes.

The normal equations for the least-squares refinement may now be formulated in terms of the derivatives of the residual with respect to the mode parameters. In dealing with librational modes it is convenient first to evaluate the derivatives of the temperature factor for each nucleus with respect to its own frame. For the simplest form of librational temperature factor these are given in Table I. These derivatives may be used to evaluate those with respect to the standard frame, using the transformations given in Table II.

Programming

The parameters to be determined include those which define the structure, *i.e.* the coordinates of the nuclear positions, population coefficients, and thermal parameters. In general there will also be terms which are functions of the intensity data rather than the structure, such as the factor which relates the scales of the $|F_{\mathbf{H}}^c|^2$ and the $|F_{\mathbf{H}}^o|^2$ values, and the domain radius and

* I am indebted to Dr. M. I. Kay for pointing out that this is so.

Gaussian breadth parameter for an extinction correction if the kinematic theory of scattering is not completely valid (Zachariasen, 1967).

The normal equations for the non-linear least-squares refinement of the J parameters may be written as J equations of the type

$$\sum_{\mathbf{H}}\sum_{i=1}^{J} w_{\mathbf{H}} \frac{\partial |F_{\mathbf{H}}^{c}|^2}{\partial p_i} \frac{\partial |F_{\mathbf{H}}^{c}|^2}{\partial p_j} \delta p_i = \sum_{\mathbf{H}} w_{\mathbf{H}}(|F_{\mathbf{H}}^{0}|^2 - |F_{\mathbf{H}}^{c}|^2) \frac{\partial |F_{\mathbf{H}}^{c}|^2}{\partial p_j} \quad (9)$$

where δp_i is the change in the ith parameter. This may be written in shorthand notation

$$\sum_i a_{ij}\delta p_i = b_j \quad (9a)$$

In programming the calculations we evaluate the derivatives, sum products of pairs to obtain the elements of the matrix $\boldsymbol{a}$, and products of residuals and derivatives to obtain the elements of the column vector **b**. The solutions to the equations are then given by

$$\delta p_j = \sum_i (a^{-1})_{ji} b_i \quad (10)$$

where $\boldsymbol{a}^{-1}$ is the matrix inverse of $\boldsymbol{a}$.

Table I. Derivatives of $T_1(L)$ with respect to $r_1, r_2, r_3, \varphi_1, \varphi_2, \varphi_3$ and $\langle\delta^2\rangle$. $\varphi_1, \varphi_2, \varphi_3$ are rotations about the axes 1, 2, and 3 respectively.

$$\frac{\partial T_1(L)}{\partial r_1} = \pi B^{-1/2}\left[4\pi R H_2^2\langle\delta^2\rangle \exp(-i\gamma) + H_1\langle\delta^2\rangle \exp\left\{i\left(\frac{\pi}{2}-\gamma\right)\right\} - 4A^2B^{-1/2}H_R^{-1}\exp\left\{i\left(\frac{\pi}{2}-2\gamma\right)\right\}\right]T_1(L)$$

$$\frac{\partial T_1(L)}{\partial r_2} = -\pi B^{-1/2}\left[4A\exp(-i\gamma) - H_2\langle\delta^2\rangle \exp\left\{i\left(\frac{\pi}{2}-\gamma\right)\right\} + 4A^2H_1^{-2}H_2B^{-1/2}\exp\left\{i\left(\frac{\pi}{2}-2\gamma\right)\right\}\right]T_1(L)$$

$$\frac{\partial T_1(L)}{\partial r_3} = 0$$

$$\frac{\partial T_1(L)}{\partial \varphi_1} = -4\pi^2R^2H_2H_3\langle\delta^2\rangle B^{-1/2}\exp(-i\gamma)T_1(L)$$

$$\frac{\partial T_1(L)}{\partial \varphi_2} = \pi R H_3 B^{-1/2}\left[\langle\delta^2\rangle \exp\left\{i\left(\frac{\pi}{2}-\gamma\right)\right\} - 4A^2H_1^{-2}B^{-1/2}\exp\left\{i\left(\frac{\pi}{2}-2\gamma\right)\right\}\right]T_1(L)$$

$$\frac{\partial T_1(L)}{\partial \varphi_3} = 0$$

$$\frac{\partial T_1(L)}{\partial\langle\delta^2\rangle} = -\pi R B^{-1/2}\left[2\pi R H_2^2\exp(-i\gamma) + H_1\exp\left\{i\left(\frac{\pi}{2}-\gamma\right)\right\} - 4A^2H_1^{-1}(\langle\delta^2\rangle)^{-1}B^{-1/2}\exp\left\{i\left(\frac{\pi}{2}-2\gamma\right)\right\}\right]T_1(L)$$

where

$$A = \pi R H_1 H_2\langle\delta^2\rangle$$

$$B = 1 + 4A^2H_2^{-2}$$

$$\gamma = \tan^{-1}(2\pi R H_1 A^2),\quad -\pi/2 < \gamma < \pi/2$$

In many cases, however, it is impractical or unnecessary to evaluate all the off-diagonal elements of the matrix $\boldsymbol{a}$. Many pairs of parameters in a structure analysis affect the intensity data in an uncorrelated manner. The corresponding cross-product in the normal equations matrix will not differ appreciably from zero and may be neglected with very little loss in accuracy. If the parameters can be arranged in groups between which there is very little correlation, the normal equations may be divided into corresponding blocks. Only those blocks containing diagonal elements are evaluated and may be inverted separately. Care is required in using this procedure, since the wrongful neglect of a cross-product will affect the covariance matrix $\boldsymbol{a}^{-1}$ and thence both the shifts in parameters and standard deviations. Properly used, however, the block diagonal techniques is computationally more efficient than full matrix methods. Thus in an ideal program for structure refinement, the normal equations matrix may be partitioned at will. It is also an advantage to be able to apply acceleration factors to the parameter shifts to ensure convergence. The optimum factors may vary considerably for different types of parameter.

A program for the refinement of atomic coordinates, population coefficients, and the parameters of a mechanistic model is conveniently organized in a series of four routines which are described in the Appendix.

Selection and grouping of parameters

The preparation of a program for carrying out the computations does not guarantee

Table II. Derivatives of $T_1(L)$ for a nucleus with cylindrical polar coordinates R, θ, z in the standard frame for the mode. Derivatives with respect to its own frame are denoted prime.

$$\frac{\partial T_1(L)}{\partial r_1} = \frac{\partial T_1(L)}{\partial r_1'} \cos\theta - \frac{\partial T_1(L)}{\partial r_2'} \sin\theta$$

$$\frac{\partial T_1(L)}{\partial r_2} = \frac{\partial T_1(L)}{\partial r_1'} \sin\theta + \frac{\partial T_1(L)}{\partial r_2'} \cos\theta$$

$$\frac{\partial T_1(L)}{\partial r_3} = 0$$

$$\frac{\partial T_1(L)}{\partial \varphi_1} = -z \left(\frac{\partial T_1(L)}{\partial r_1'} \sin\theta + \frac{\partial T_1(L)}{\partial r_2'} \cos\theta \right) + \frac{\partial T_1(L)}{\partial \varphi_1'} \cos\theta - \frac{\partial T_1(L)}{\partial \varphi_2'} \sin\theta$$

$$\frac{\partial T_1(L)}{\partial \varphi_2} = z \left(\frac{\partial T_1(L)}{\partial r_1'} \cos\theta - \frac{\partial T_1(L)}{\partial r_2} \sin\theta \right) + \frac{\partial T_1(L)}{\partial \varphi_1'} \sin\theta + \frac{\partial T_1(L)}{\partial \varphi_2'} \cos\theta$$

$$\frac{\partial T_1(L)}{\partial \varphi_3} = 0$$

$$\frac{\partial T_1(L)}{\partial \langle \delta^2 \rangle} = \frac{\partial T_1(L)}{\partial \langle \delta^2 \rangle'}$$

a successful refinement using the mechanistic model which depends both on the choice of parameters and the arrangement of the computation for the reasons given above. When introducing variable parameters both the magnitudes of the resulting changes in the intensity data and the range of data affected must be considered. Analyses of the structure factor agreement as a function of $|F^o_{\mathbf{H}}|^2$ and of Bragg angle are frequently helpful in optimizing parameter selection. Thus errors confined to large $|F^o_{\mathbf{H}}|^2$ may result from extinction terms; uniformly poor agreement in the low angle region is more likely to require variation of a valence shell population coefficient; discrepancies in the high angle region suggest a need for an improved thermal motion model, and so on. There is, however, an enormous range of structural problems and, since the optimum choice of parameters is both structure and data dependent, no precise rules for parameter selection can be given.

The subdivision of the least-squares matrix is more readily described since a quantitative estimate of the interaction for any pair of parameters may be obtained. It is helpful to discuss this in terms of the elements of a correlation matrix **c**, defined by the equation

$$c_{ij} = (a^{-1})_{ij}\{(a^{-1})_{ii}(a^{-1})_{jj}\}^{-1/2} \qquad (11)$$

As a rough general rule it is safe to neglect the interaction entirely if the corresponding matrix element is less then 0·1 and the interaction may be neglected in the intermediate cycles of refinement if the matrix element is less than 0·5. Some typical values of the c_{ij} are as follows:

(i) two coordinates of the same scattering centre 0·0 to 0·5

(ii) scale factor and the diagonal elements of the rigid body vibrational motion tensor 0.3 to 0.6

(iii) scale factor and valence shell population parameter 0.2 to 0.6

(iv) two population parameters for orbital products with the same centre and similar shapes 0·99

Interactions which normally may be neglected are those between

(i) coordinates of different scattering centres

(ii) coordinates and thermal modes which affect different scattering centres

(iii) the scale factor and the parameters for librational modes.

Nevertheless, there are cases which do not conform to the general rules and where possible it is advisable to evaluate the correlation coefficients in the initial and final cycles of refinement.

Coupling and anharmonicity

The method of structure refinement outlined above may be extended to include the effects of coupling between modes and anharmonicity, provided these are small. Consider first of all the cases of coupling between vibrational and librational modes which are illustrated in Fig. 3. In Fig. 3(*a*) the trajectories of the motions are confined to the same plane. An oscillation of the point P through an angle δ about the

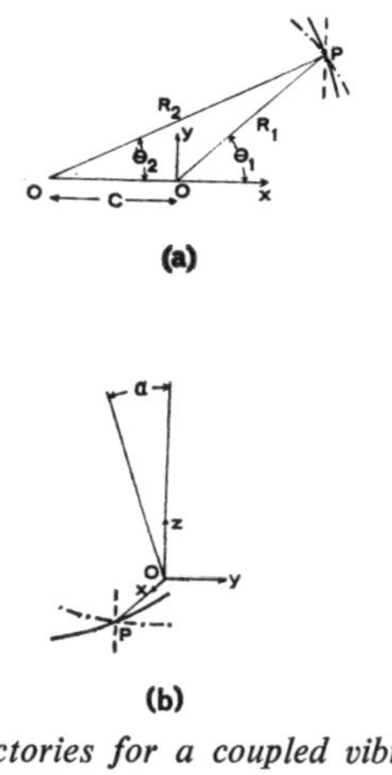

Fig (3) Trajectories for a coupled vibration and libration. (a) With both components in the same plane. (b) With the vibration parallel to the rotation axis. The displacements due to vibration, libration, and resultant motions are shown dotted, chain-dotted, and as full lines respectively.

centre at O is accompanied by a translation of $C\delta$ in the y direction. It may be readily shown that in the limit of small δ, this is identical to a libration about the point O', displaced by a distance C from O at right angles to the direction of the vibration. This origin displacement is independent of R_1 and θ, and is thus identical for all points undergoing the same libration. Thus coupled motion of this type may be treated using the procedures given above without further extension.

Consider now the effect of coupling where the vibration is normal to the plane of the libration, introducing a screw component to the motion [Fig. 3(*b*)]. For small coupling and angles of libration the effect is to leave the radius vector to the rest position unchanged, but to rotate the axis of libration about the radius vector by an angle α, where

$$\alpha = \tan^{-1}(C/R) \qquad (12)$$

Coupling of this kind does not change the form of the temperature factors, but C must be introduced as an additional parameter, which may be included in the least-squares refinement by evaluating the derivative of $T(L)$ with respect to C in the usual manner. Coupling between librations may be handled by a similar procedure.

For anharmonic motions the temperature factor expressions themselves must be altered, since $P(\delta)$ no longer has a Gaussian form. In general the smearing function will have the form

$$P(\delta) \propto \exp\left(-\sum_{m=2}^{\infty} a_m \delta^m\right) \qquad (13)$$

and in principle the coefficients a_m may be included as variable parameters in the refinement. There are no simple analytical forms for the resulting temperature factor expressions in the general case. For small distorting terms of simple form, approximate expreisson may be derived, but the optimum choice of function varies considerably according to the problem and no general rules can be given.

References

Atoji, M., Watanabe, T. & Lipscomb, W. N. (1953). *Acta Cryst.* **6**, 62.

Brown, G. M. & Chidambaram, R. (1967). Program and Abstracts. American Crystallographic Association Meeting, Atlanta, Georgia, 25–28 Jan. 1967.

Clementi, E. (1965). *Tables of Atomic Functions.* Suppl. IBM *J. Res. Devel.* **9**, 2.

Hamilton, W. C. (1969). *Acta Cryst.* A**25**, 194.

Kay, M. I. & Behrendt, D. R. (1963). *Acta Cryst.* **16**, 157.

Maslen, E. N. (1968). *Acta Cryst.* A**24**, 434.

Stewart, R. F. (1969). Program and Abstracts. American Crystallographic Association Meeting, Seattle, Washington, 23–28 March 1969.

Zachariasen, W. H. (1967). *Acta Cryst.* **23**, 558.

Appendix

ROUTINE 1

Calculation of scattering factors

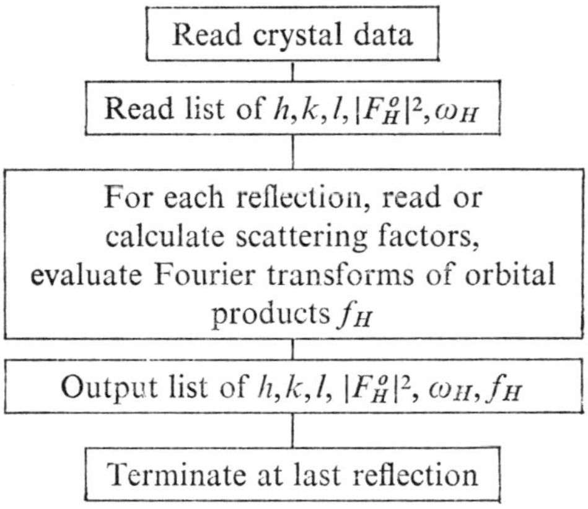

ROUTINE 2

Input data for least-squares refinement

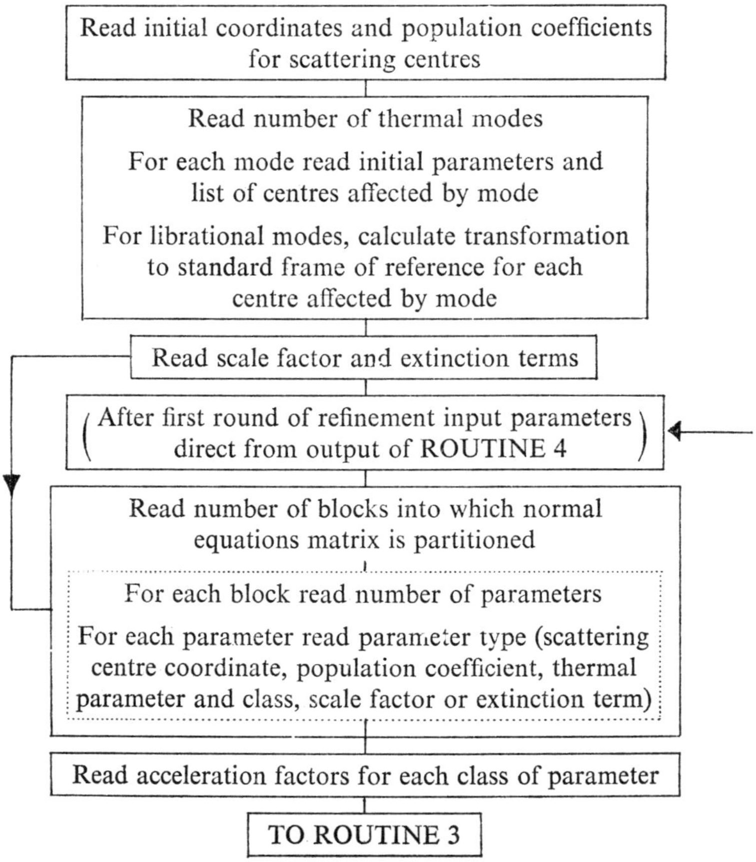

ROUTINE 3

Structure factor and derivative calculation

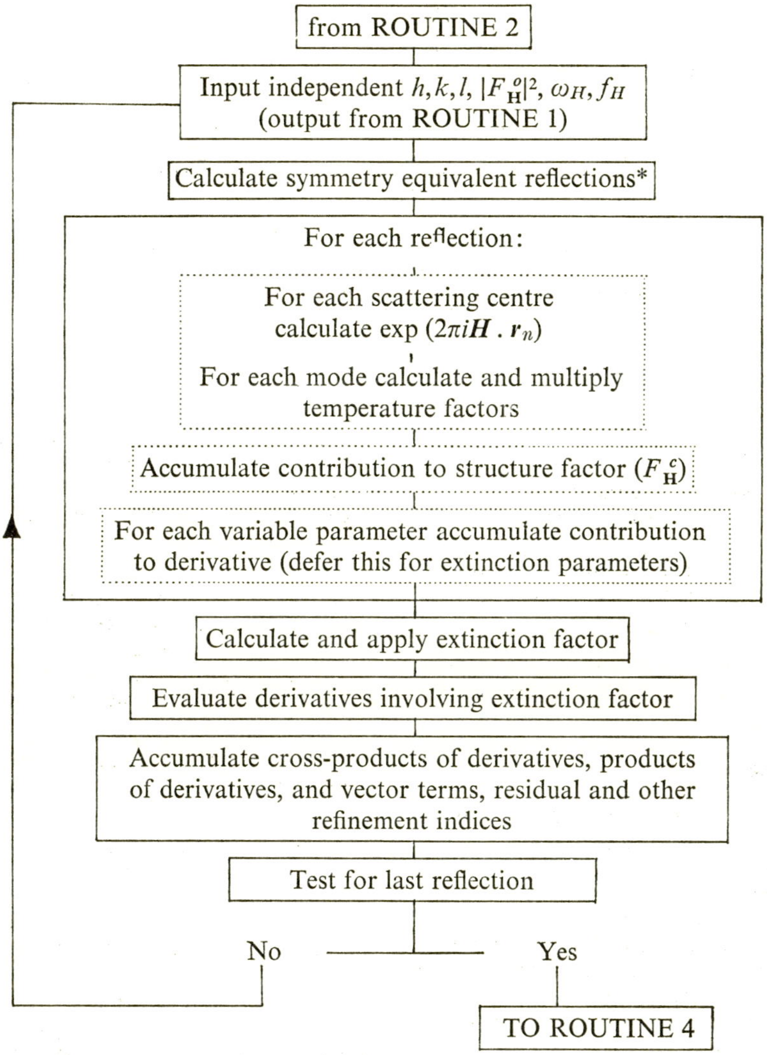

* When evaluating scalar products of the type $\boldsymbol{H} \cdot \boldsymbol{r}_n$ for equivalent positions, it is preferable to apply the rotational part of the symmetry transformation to $\boldsymbol{H}$ rather than $\boldsymbol{r}_n$. This technique is known as the equivalent indices method and is equally valid when applied to the calculation of temperature factors.

ROUTINE 4

Matrix inversion and results

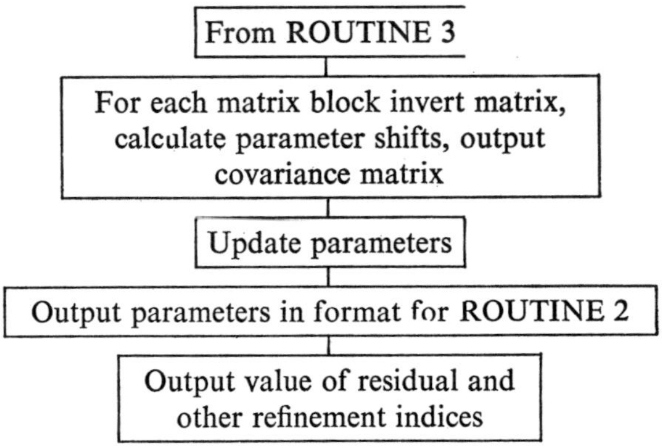

Discussion

SILVERMAN: Were the internal modes of vibration considered in the example you gave?

MASLEN: Internal modes of vibration have not been considered whereas the individual librational components were.

PAWLEY: It is difficult to separate the internal and external modes in the least-squares procedure.

MASLEN: I agree.

COPPENS: I would like to comment on the population coefficients as defined by Dr. R. F. Stewart and described by Dr. Maslen. We have a full matrix program which refines these and the conventional temperature parameters. However, we take the thermal and positional parameters from a neutron analysis and use them to refine the population of valence electron orbital distributions. Agreement factors of the X-ray data typically drop by about 50% using this procedure.

ROLLETT: How significant are these population parameters?

COPPENS: This is difficult to answer because correlation between the parameters is large. However, by choosing a minimum set of population parameters, one can find significant deviations from the atomic model.

MASLEN: We have analysed the population coefficients using X-ray data only. The valence shell populations are almost in all cases significantly less than in the atomic model. With good data our precision is to within 0·05 e.Å^{-3}.

ROLLETT: Are these electrons appearing somewhere else?

MASLEN: Yes, some are appearing in the bonds but not as many as we would like. However, I think this is an interaction problem with the thermal model involving librational temperature factors for atoms in the outer bonds.

ROLLETT: Does this enable you to distinguish between the different theoretical models such as valence bond and molecular orbital?

MASLEN: We do not have sufficient confidence in our model to try and differentiate this as yet.

The Use of Molecular Lattice Dynamical Models

G. S. PAWLEY

Chemistry Department, Århus University, Denmark; on leave from: *Physics Department, Edinburgh University, Scotland.*

The crystallographic process yields information on the average motion of atoms or molecules in a crystal. In even a very small crystal the number of atoms present, N, is enormous, and we can say that there are $3N$ normal modes of vibration of this crystal. The average motion of the atoms therefore depends on this large number of thermal excitations. To give us an understanding of the dynamics of the crystal, studies other than crystallographic are necessary. These other studies, however, do lead to conclusions about the average atomic behaviour that can be tested by accurate crystallography.

First let us consider the dynamical problem, the way in which we calculate the frequencies of the lattice modes of vibration or phonons when we know the potential function Φ for the crystal. Then, after section 2 on experimental methods, we will discuss the problem of fitting theoretical models to the experimental results, which is in essence a search for the best Φ. Finally, in section 4, we will see how these results can be compared with Bragg diffraction studies. We will not consider anharmonic effects.

1. The dynamical problem

We must first write down the equations of motion for all the atoms in the crystal. Let n be the number of atoms in a primitive unit cell. Let $r_x(lk)$ be the x-coordinate of the atom k in the unit cell l and $u_x(lk)$ a small displacement from equilibrium, then, knowing the crystal potential Φ, we can find

$$\Phi_{xy}(lk,l'k') = \frac{\partial^2\Phi}{\partial u_x(lk)\partial u_y(l'k')} \quad (1\cdot1)$$

the force constants relating all interacting atoms. The equation of motion in the x-direction for the atom of mass $m(k)$ at $\mathbf{r}(lk)$ is then

$$m(k)\ddot{u}_x(lk) = -\sum_{l'k'y} \Phi_{xy}(lk,l'k') \times u_y(l'k') \quad (1\cdot2)$$

and there are $3n$ such equations for the atoms in one cell. We now try for a travelling wave solution with angular frequency ω and wave-vector $\mathbf{q}$ and write

$$u_x(lk) = U_x(k\mathbf{q}) \times \exp i[\mathbf{q}\,.\,\mathbf{r}(lk) - \omega(\mathbf{q})t] \quad (1\cdot3)$$

and similar equations for all the other indices k and for all three Cartesian coordinates, giving $3n$ in all. Substituting (1·3) in (1·2), we get

$$\begin{aligned} m(k)\omega^2(\mathbf{q})U_x(k\mathbf{q}) &= \sum_{l'k'y} \Phi_{xy}(lk,l'k') \times U_y(k'\mathbf{q}) \times \exp\{i\mathbf{q}\,.\,[\mathbf{r}(l'k')-\mathbf{r}(lk)]\} \\ &= \sum_{k'y} U_y(k'\mathbf{q}) \sum_{l'} \Phi_{xy}(lk,l'k') \times \exp\{i\mathbf{q}\,.\,[\mathbf{r}(l'k')-\mathbf{r}(lk)]\} \\ &= \sum_{k'y} U_y(k'\mathbf{q}) M_{xy}(k,k',\mathbf{q}) \quad (1\cdot4) \end{aligned}$$

thereby defining $M_{xy}(k,k',\mathbf{q})$, the dynamical matrix elements. Note that the elements $M_{xy}(k,k',\mathbf{q})$ are complex in character and this causes the secular matrix to be hermitean. The secular matrix is derived from $\boldsymbol{M}$ by dividing each row and each column by the appropriate $\sqrt{m(k)}$, whereupon the eigenvalues can be found by standard techniques (see Appendix).

For molecular crystals it is reasonable to assume to a first approximation that the molecules move as rigid bodies. This movement is governed by the external modes of vibration which are determined by the crystal forces. The appropriate coordinates are not now Cartesian atomic coordinates, but translational and rotational coordinates for the molecules. Instead of the label k for the n atoms, we now need two labels, kj, for the jth atom in the kth molecule. We then need to let $\mathbf{u}(lk)$ be a vector with six coordinates, three translational and three rotational, and we must allow x and y of Eq. (1·1) etc. to run over these six coordinates. To Eq. (1·2) must be added Newton's equations for angular motion, and here we see that the problem is simplified by choosing a Cartesian coordinate system for the molecule with axes parallel to the molecular inertia axes. The equations extra to (1·2) contain the moments of inertia $I_z(k)$, $z=1,2,3$ in place of the molecule mass $m(k)$, and with these additions in mind Eqs. (1·3) and (1·4) remain unaltered.

The secular equation obtained from (1·4) then gives six roots for each molecule in the unit cell. Each root belongs to a different phonon branch which can be plotted out by doing the calculation for various values of $\mathbf{q}$.

2. Experimental results

Of the methods for measuring the frequencies of phonons, that of coherent inelastic neutron scattering has proved the most successful. The method is described by Cochran & Cowley (1967) § B where references to earlier work are given. They also describe other techniques which, for

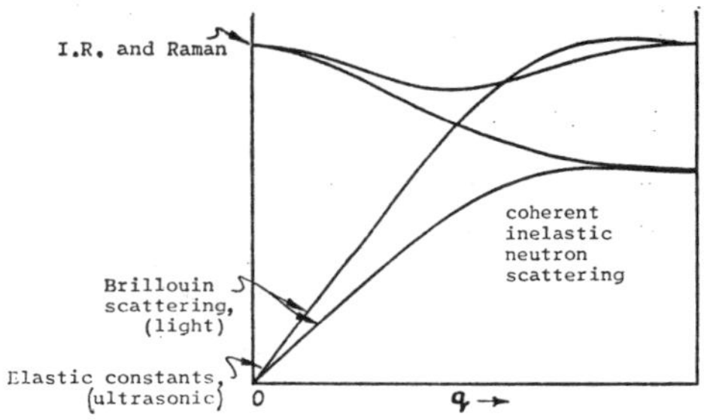

Fig. 1. Dispersion diagram showing the ranges of usefulness of methods for measuring the frequencies of phonons.

the present purpose, can best be reviewed by a dispersion diagram in Fig. 1 showing their ranges of usefulness.

With none of the methods besides the neutron technique can wave-vectors far from $\mathbf{q}=0$ be studied.

The diagram in Fig. 1 omits mention of X-ray thermal diffuse scattering. With the neutron technique the phonon frequencies are found by doing an energy analysis of the scattered neutrons, but this is impossible in the X-ray case as thermal energies ($\sim$0·025 eV) are so much smaller than X-ray energies. To make use of thermal X-ray scattering accurate intensity measurements must be made. These measurements taken throughout reciprocal space do have $\mathbf{q}$ dependence, but at each point the effects of all phonon branches are superimposed. Only for crystals where high symmetry reduces the number of phonon branches can the X-ray technique be used effectively, and then extra care must be taken to allow for two phonon scattering.

The equations governing one- and two-phonon X-ray scattering are given by Cochran & Cowley (1967). A detailed description of how these equations are adapted for computation for the molecular crystal hexamethylenetetramine (HMT) and of how the computation can be done is given by Pawley (1970). The result, which emphasizes the difficulty of the experimental procedure, is that at room temperature at large scattering angles for

Cu $K\alpha$ radiation the two-phonon scattering sometimes exceeds the one-phonon scattering.

In the study of molecular crystals the measurements with neutrons of the external modes of vibration is not sufficient to give a good understanding of the dynamics. Gas-phase spectroscopic measurements of internal mode frequencies are also required. These latter results allow the fitting of a force field for an individual molecule which can then be taken over in a lattice dynamical calculation. The internal force constants should not change on crystallization, but the internal mode frequencies will be altered by the crystal forces. This is discussed at the end of the next section.

3. Model fitting

We have tacitly assumed that the external modes of vibration are governed entirely by the crystal forces, that there is no distortion of the molecules in these modes, and that the internal modes are unaffected by the crystal forces and have no **q** dependence. These inter-related assumptions we will call the separation assumption; and if valid, we would expect a large separation of the frequencies of the internal and external modes. Let us now investigate models which use this assumption.

In section 1 it was pointed out that the fundamental problem is in finding a suitable crystal potential function Φ. The simplest model we can propose is one in which all the symmetry independent $\Phi_{xy}(lk, l'k')$ of Eq. (1·1) are treated as parameters. Remember that the suffices x and y now cover rotations as well as translations. This model was proposed for HMT by Cochran & Pawley (1964) and has been used for fitting phonon measurements by Powell (1968), see also Pawley (1969). The number of parameters required is small in comparison with the results available, and the method of least squares can be used.

The procedure is not as straightforward as in crystallographic SFLS calculations. For each value of **q** a set of eigenvalues must be calculated. These must be assigned to the appropriate frequency measurement by analysing the eigenvectors, from which can be found the symmetry of the modes. Calculations and measurements alike must be grouped according to their irreducible space group representations. It is possible to write the program so that the dynamical matrix is block diagonalized into blocks of different representation; but as this diagonalization is different for **q** along different symmetry directions, this presents a considerable programming task. The ordering of frequencies in one representation is sometimes not sufficient for assigning a calculated result to a measurement, and often the physical aspect of the eigenvector has then to be investigated. As an example, let us suppose that there is a mode predominantly translational and another predominantly librational of higher frequency in one representation. There is no guarantee that a model calculation will give the translational mode as the lowest frequency, but it is essential that its eigenvalue should be fitted to the lowest measurement. Another complication is that differentials needed for the least-squares matrix are often obtained numerically, so the calculation has to be done with considerable accuracy.

When the crystal symmetry is low, the number of independent force constants $\Phi_{xy}(lk, l'k')$ becomes so large that they cannot be used as independent model parameters. A smaller, more basic parameter set is required, which is furnished by the Buckingham potential function

$$V_i(r) = -A_i r^{-6} + B_i \exp(-\alpha_i r) \qquad (3\cdot1)$$

$V_i(r)$ is the potential for a pair of atoms separated by a distance r interacting only by van der Waals attraction and short range repulsion. The index i refers to a specific chemical pairing. By summing over all atom pairs (out to $r = 5{\cdot}5$ Å is sufficient), we obtain Φ as a function of the parameters A_i, B_i, α_i and the positional coordinates of the molecules. Thus we can obtain $\Phi_{xy}(lk, l'k')$ and calculate phonon frequencies and cigenvectors. This has been done by Pawley (1967) for naphthalene and anthracene using the parameters of Kitaigorodskii (1966). The results of phonon fitting with this type of model for HMT will soon be reported (Dolling, Powell & Pawley, 1970).

The separation approximation used so far leads to dispersion curves such as Fig. 2(*a*). However, during motion the molecule must distort, and this distortion must be such as to lower the frequencies of the external modes as in Fig. 2(*b*). The distortion decreases the restoring forces. At the same time the internal mode frequencies are raised and become **q**-dependent. These shifts can be much larger than the experimental errors in frequency measurements and must be considered (Pawley & Cyvin, 1970).

In their normal coordinate analysis of the 18-atom naphthalene molecule, Hagen & Cyvin (1968) set up 48 internal coordinates in the manner of Wilson, Decius & Cross (1955). They then fitted a molecular force field to all the internal mode frequencies. This was then used as follows for the crystal calculations: The 48×48 force field matrix was transformed by the *B*-matrix (see Wilson *et al.*, 1955, Eq. (1), p. 55) to give the 54×54 force constant matrix $\boldsymbol{\varphi}$, which is in Cartesian coordinates,

$$\begin{array}{l}\text{Force on atom}\\ i \text{ in direction } j\end{array} = \sum_{kl} \varphi_{ij,kl} \times \begin{array}{l}\text{displacement}\\ \text{of atom } k \text{ in}\\ \text{direction } l\end{array} \qquad (3{\cdot}2)$$

The matrix $\boldsymbol{\varphi}$ has six zero roots corresponding to unimpeded translation and

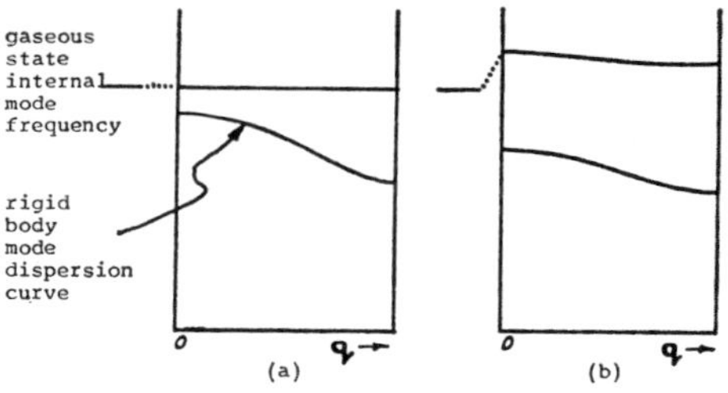

Fig. 2. The dispersion curves showing internal and external mode phonon frequencies for separation approximation (a) good, (b) poor.

rotation of the molecule. The dynamical problem is then done by including the $\Phi_{xy}(lk, l'k')$ of Eq. (1·1) where x and y refer as originally to atomic coordinates but now in neighbouring molecules. These are furnished by the Buckingham potential. As there are two naphthalene molecules in the primitive unit cell ($k = 1$ or 2), the dynamical matrix becomes 108×108 hermitean – hence the need for the Appendix.

The problem can be reduced to a 54×54 hermitean matrix by using the symmetry of the problem, obtaining answers only for the diad symmetry direction [010]. In this direction the wave solutions (1·3) are either symmetric or antisymmetric. If $\mathbf{D}$ represents the diad symmetry operation, the wave amplitudes have the relationships $\mathbf{U}(1\mathbf{q}) = \mathbf{DU}(2\mathbf{q})$ for the symmetric modes and $\mathbf{U}(1\mathbf{q}) = -\mathbf{DU}(2\mathbf{q})$ for the antisymmetric modes. Substitution of either of these relationships in (1·3) and (1·4) reduces the order of the dynamical matrix by half.

In order to find the frequencies that this program would give if the separation approximation were good, the following procedure was used. All the components of $\boldsymbol{\varphi}$ were increased by a factor of 100, the calculations were repeated, and the internal mode eigenvalues (squares of frequencies) were reduced by 100. These then showed negligible differences from the free molecule results, so the external modes could be assumed to correspond to rigid-body motion.

Rough values for the differences in these calculations are:

Depression of external modes $\sim$ 5 cm^{-1}

Raising of internal modes $\sim$ 20 cm^{-1}

Symmetric-antisymmetric splitting $\sim$ 5 cm^{-1}

These values show the importance of considering molecular distortion when doing model fitting, for which a suggested procedure is as follows. Knowing $\boldsymbol{\varphi}$, calculate the frequency differences between the rigid-body and the non-rigid-body models for all $\mathbf{q}$ where measurements have been made. Increase all the external mode frequency measurements by the appropriate amounts and use these as observations in a rigid-body model fit. An iteration for self-consistency should not prove necessary.

4. Relation to structural studies

If we know the full phonon dispersion curves, internal and external, we can calculate the average motion of all the atoms or molecules in the crystal at any given temperature. The mean energy in a mode is given by

$$\varepsilon(\omega)=\hbar\omega\{\tfrac{1}{2}+[\exp(\hbar\omega/kT)-1]^{-1}\} \quad (4{\cdot}1)$$

The mean atomic displacements in this mode then follow as the eigenvector is known. When all the coordinates are Cartesian atomic coordinates, the sum of the mean square displacements is ellipsoidal. Thus if $\mathbf{X}$ is the vector displacement of appropriate magnitude for one atom in one mode, then the mean square displacement in the direction given by the unit vector $\mathbf{l}$ is

$$\tfrac{1}{2}X_iX_jl_il_j \quad (4{\cdot}2)$$

Summing over all the modes we get

$$T_{ij}l_il_j = \tfrac{1}{2}\left\{\sum_{\text{modes}} X_iX_j\right\}l_il_j \quad (4{\cdot}3)$$

defining the mean square displacement tensor $\mathbf{T}$. This gives directly the anisotropic temperature factors when transformed into the crystal fractional coordinates, and a comparison with the crystallographic result can be made.

In this article we are concerned with molecular crystals where the coordinates of the system are molecular translations and rotations. The eigenvector for a mode can be written as four vectors for each molecule

$$(\mathbf{u}, i\mathbf{v}, \boldsymbol{\theta}, i\boldsymbol{\varphi})$$

where the magnitudes are appropriate for the mean energy in the mode. $\mathbf{u}+i\mathbf{v}$ gives the translational component, $\boldsymbol{\theta}+i\boldsymbol{\varphi}$ the rotational component. That these are complex results from the hermitean nature of the dynamical matrix, and a simplification only occurs when the molecules are on centrosymmetric sites, whereupon $\mathbf{v}$ and $\boldsymbol{\theta}$ are zero. As we wish to find the average motion of one molecule, we do not need to consider the phase relationships between different molecules in the unit cell. For the atom at $\mathbf{r}$ in this molecule the motion is

$$(\mathbf{u}+\boldsymbol{\theta}\times\mathbf{r})\cos\omega t+(\mathbf{v}+\boldsymbol{\varphi}\times\mathbf{r})\sin\omega t \quad (4{\cdot}4)$$

giving a mean square displacement

$$\tfrac{1}{2}(\mathbf{u}+\boldsymbol{\theta}\times\mathbf{r})^2+[\mathbf{v},\boldsymbol{\varphi}] \quad (4{\cdot}5)$$

where $[\mathbf{v},\boldsymbol{\varphi}]$ means 'similar term in $\mathbf{v}$ and $\boldsymbol{\varphi}$'.

$\mathbf{u}+\boldsymbol{\theta}\times\mathbf{r}$ is the vector displacement analogous to $\mathbf{X}$ in Eq. (4·2), and so we can sum over all modes as in Eq. (4·3) and obtain the mean square displacement parallel to $\mathbf{l}$, namely

$$(U_{\mathbf{r}})_{ij}l_il_j=\tfrac{1}{2}\left\{\sum_{\text{modes}}(\mathbf{u}+\boldsymbol{\theta}\times\mathbf{r})_i(\mathbf{u}+\boldsymbol{\theta}\times\mathbf{r})_j+[\mathbf{v},\boldsymbol{\varphi}]\right\}l_il_j \quad (4{\cdot}6)$$

Expanding this for typical components we

have

$$(U_r)_{11} = \tfrac{1}{2} \sum_{\text{modes}} (u_1 + \theta_2 r_3 - \theta_3 r_2)^2 + [\mathbf{v}, \boldsymbol{\varphi}]$$
$$= T_{11} + L_{22} r_3^2 + L_{33} r_2^2 - 2L_{23} r_2 r_3$$
$$+ 2S_{21} r_3 - 2S_{31} r_2$$
$$(U_r)_{23} = T_{23} + L_{31} r_1 r_2 + L_{12} r_3 r_1 - L_{23} r_1^2$$
$$- L_{11} r_3 r_2 + S_{33} r_1 - S_{13} r_3$$
$$+ S_{12} r_2 - S_{22} r_1 \qquad (4{\cdot}7)$$

where we have defined

$$T_{ij} = \tfrac{1}{2} \sum_{\text{modes}} (u_i u_j + v_i v_j)$$
$$L_{ij} = \tfrac{1}{2} \sum_{\text{modes}} (\theta_i \theta_j + \varphi_i \varphi_j)$$
$$S_{ij} = \tfrac{1}{2} \sum_{\text{modes}} (\theta_i u_j + \varphi_i v_j) \qquad (4{\cdot}8)$$

These are the equations of Schomaker & Trueblood (1968) as developed by Pawley (1968), the numerical differences between the latter and the present equations being a matter of convenience in the eigenvector definition.

Pawley (1968) goes on to show how these equations can be included in the SFLS refinement. However, work on this has shown that data of very high accuracy are essential for success. Any attempt to fit the tensors ***T***, ***L*** and ***S*** to the results of conventional SFLS refinements cannot give the best estimates of these tensors that the data could provide, and therefore the SFLS must be constrained so that the anisotropic temperature factors agree always with Eq. (4·7) throughout refinement. ***S*** is only non-zero when the molecules are not on centres of symmetry, and a lattice dynamical calculation for such a case still remains to be done.

One difficult point has been evaded until now. In the harmonic approximation using molecular coordinates it is clear that the probability distribution functions for the atoms will not be ellipsoidal but, in colloquial terms, banana-like. This is in contrast to the ellipsoidal result using atomic Cartesian coordinates. This dilemma will be left for contemplation. Rigid-body lattice dynamics yields the tensor ***L*** which governs the distortions from ellipsoidal shape, and these manifest themselves as higher cumulant terms in the structure factor equation (see Johnson, 1969). The exact form of these terms has been derived by Willis & Pawley (1970) and applied to HMT by Duckworth, Willis & Pawley (1970) and for the general centrosymmetric case by Pawley & Willis (1970). This leaves undone the noncentrosymmetric case, which is also in need of some lattice dynamical calculations and some good experimental data, both inelastic and crystallographic.

Appendix

Eigenvalues of a hermitean matrix

The positive definite hermitean $n \times n$ matrix ***M*** can be written

$$\boldsymbol{M} = \boldsymbol{A} + i\boldsymbol{B}$$

where ***A*** is symmetric positive definite and ***B*** is anti-symmetric. Matrices such as ***A*** can be solved by Householder's method or by Jacobi's method of successive rotations. Householder's method sometimes fails with repeated zero roots whereas Jacobi's method does not, otherwise Householder's method is favoured for speed and accuracy, especially on large matrices.

We can always form a $2n \times 2n$ matrix

$$\boldsymbol{C} = \begin{pmatrix} \boldsymbol{A} & \boldsymbol{B} \\ -\boldsymbol{B} & \boldsymbol{A} \end{pmatrix}$$

where ***C*** is symmetric positive definite with roots in degenerate pairs which are the roots of ***M***. This can then be solved by the methods mentioned above, but a variant on the Jacobi method can be used which does not seem to be generally well known.

In the Jacobi method a search is made

for an off-diagonal element, (i,j), which exceeds a threshhold value. When one is found, the matrix is pivoted to bring the (i,j) element to zero, altering all the elements in the ith row and jth column. When no elements are found above the threshhold, the threshhold is reduced until all off-diagonal elements are made negligible. This requires working with just the diagonal and, say, lower triangle of the original matrix.

For the matrix $\boldsymbol{C}$ the same procedure is used; but when pivoting about (i,j) by a certain amount, we must pivot by the same amount about $(i+n \bmod 2n, j+n \bmod 2n)$ as these elements must have the same value as (i,j). In this way the symmetry of the pivoted matrix remains the same as $\boldsymbol{C}$ throughout the process and we need only retain the diagonal and lower triangle of $\boldsymbol{A}$ and the upper triangle of $\boldsymbol{B}$. The whole hermitean matrix is then stored as $n \times n$ giving a saving in computer storage, but more important is the fact that fewer elements have to be calculated at each pivot than in a standard solution of $\boldsymbol{C}$ and this gives a large saving in computer time.

References

Cochran, W. & Cowley, R. A. (1967). *Handbuch der Physik*, **25/2a**, 59.
Cochran, W. & Pawley, G. S. (1964). *Proc. Roy. Soc.* A**280**, 1.
Dolling, G., Powell, B. M. & Pawley, G. S. (1970). In preparation.
Duckworth, J. A. K., Willis, B. T. M. & Pawley, G. S. (1970). Submitted to *Acta Cryst.* A.
Hagen, G. & Cyvin, S. J. (1968). *J. Phys. Chem.* **72**, 1446.
Johnson, C. K. (1969). *Acta Cryst.* A**25**, 187.
Kitaigorodskii, A. I. (1966). *J. chim. phys.* **63**, 6.
Pawley, G. S. (1967). *Phys. Stat. Sol.* **20**, 347.
Pawley, G. S. (1968). *Acta Cryst.* B**24**, 485.
Pawley, G. S. (1969). *J. Chem. Phys.* **50**, 2785.
Pawley, G. S. (1970). *Acta Cryst.* A, In press.
Pawley, G. S. & Cyvin, S. J. (1970). *J. Chem. Phys.* In press.
Pawley, G. S. & Willis, B. T. M. (1970). Submitted to *Acta Cryst.* A.
Powell, B. M. (1968). *Neutron Inelastic Scattering* 4th I.A.E.A. Symposium, Copenhagen. II, 185.
Schomaker, V. & Trueblood, K. N. (1968). *Acta Cryst.* B**24**, 63.
Willis, B. T. M. & Pawley, G. S. (1970). Submitted to *Acta Cryst.* A.
Wilson, E. B. Jr., Decius, J. C. & Cross, P. C. (1955). *Molecular Vibrations.* New York: McGraw-Hill.

Discussion

JOHNSON: Exactly what do you mean by dynamical matrix?

PAWLEY: The dynamical matrix is that matrix which when solved will give the eigenvalues and eigenvectors for the modes of vibration. Basically, it is the matrix used to obtain phonon frequencies.

JOHNSON: Why are your eigenvalues complex?

PAWLEY: Because of the phase relationship in Eq. (1.4). In general, we are not going to have a $\pi/2$ phase difference between the librational and translational components and this gives rise to a complex eigenvector.

ROLLETT: Can you discuss the various ways the frequency wave numbers operate to control the amplitudes of various modes in X-ray analysis?

PAWLEY: The acoustic modes of vibration are most important for correction purposes. Any mode which is flat at $q=0$ in the dispersion diagram is going to scatter intensity evenly, while a mode which is steep at $q=0$ will give an intensity which falls off. Thus, the intensity change we obtain about a Bragg peak will be determined almost entirely by the acoustic modes of vibration. It is these that need to be considered when correcting for thermal diffuse scattering. However, the motion due to the other modes is still significant due to their large number.

JOHNSON: Do you think that crystallographers should try to calculate frequencies from their libration tensors?

PAWLEY: I do not think so, because the librational modes sometimes fall so dramatically across the Brillouin zone that the average often depends on the larger regions of this zone. I do not think this average has much physical significance at the moment.

Panel Discussion on Thermal Motion

JOHNSON: I would like to summarize what I consider to be the advantages of the statistical approach to thermal motion. Firstly, results from this type of analysis are unbiased by preconceived notions of what the molecule is doing. This way you are less likely to miss unexpected physical phenomena. Secondly, it is simple since we handle all problems in the same way, thus allowing general computer programs and relative ease of operation. On the other hand, a disadvantage is that a large number of parameters are required to describe a certain level of approximation of the thermal motion. Also, the precision of fit to a specific model, such as a rigid body, would be better if the model were tailored to the structure factor equation rather than first fitting the anisotropic temperature factors and then fitting the rigid body to these. However, we have the problem of internal vibration which is not always negligible. Sometimes you may wish to subtract it from spectroscopic data, or you may want to try the segmented body model, putting in some internal modes. This may be done at the structure factor level but it is also convenient to do it with the anisotropic temperature factor. My experience is that it is easier to interpret the residuals in a physical way when fitting the rigid body to the temperature factors rather than to the structure factors.

What about the validity of the **S**-tensor in the Schomaker–Trueblood analysis? Dr. Pawley has not been able to fit the **TLS** model to the diffraction data and yet several of us have been able to fit this model to the anisotropic temperature factors with some physical meaning. This obviously bears further investigation. The **S**-tensor can be determined with reasonable precision. Work by Cromer & Kay on an alum compound indicates that the sulfate group on a 3-fold axis has a large torsional oscillation and translation at the same time. They fitted a model directly to their diffraction data and we fitted the Schomaker–Trueblood analysis to their anisotropic temperature factors; there is reasonable agreement in terms of the pitch and libration along the 3-fold axis.

LARSON: Cromer & Kay did not obtain quite so good a statistical fit as you did with the anisotropic temperature factor model. This may be because the rigid body model has a smaller number of parameters and thus it was possible to take into account many of the small variations.

JOHNSON: I think this is generally the case.

PAWLEY: I would like to comment on some of the points Dr. Johnson has made. I consider the best way of obtaining results from thermal analyses is to do both constrained and unconstrained refinements; and if the difference between them is not significant, one may assume the validity of the constrained model. If, on the other hand, there is a significant difference, this indicates something of physical interest. I do not think, therefore, that this is an argument for not doing the constrained refinement. The refinements that I have been doing, trying to fit **TLS** to the results, have been concentrated on the few examples in the literature where 'significant' results have already been obtained. Looking at Schomaker & Trueblood's original paper there is just one result in it which purports to show that there is a significant improvement in fitting **TLS** over fitting **TL** and a centre of libration, **X**, and that example was benzo-*c*-phenanthrene. They obtained a highly significant difference between the β's, fitting the anisotropic temperature factors for the **TLS** and **TLX** models. When I tried doing a constrained refinement on the original results I found that this model was completely divergent and would not be stable. I then tried the unconstrained model and that was unstable. I think that in this particular example the molecule is rigid but there are a fair number of internal modes of quite low frequency. The internal motion makes the whole problem impossible to put into a least-squares refinement.

I consider one of the problems in doing the **TLS** on the anisotropic temperature factors is that it is a linear process. You put the numbers into the computer and if you have enough accuracy you get out a result. You will probably get a good fit to your anisotropic temperature factors, but how you interpret this I do not know. Dr. Johnson says that some results have been interpreted physically and this is what I question, considering that they are averages over so many modes of vibration. My question to Dr. Johnson is, then, how can he give a physical interpretation to the fit of **TLS**?

JOHNSON: The sort of physical interpretation that I have made is not in a dynamical sense, but comes rather from looking at the arrangements of atoms where you expect to have, say, a screw motion. If the neighboring atoms are undergoing a vibration such that they would sterically interfere in one direction and not in the other, then this is an empirical type of reasoning which seems worthwhile in some analyses. This is what occurs in the Cromer & Kay alum case, where the motion of the oxygens around the 3-fold axis is as you would have expected from steric considerations. Now this is quite empirical reasoning because those other atoms are doing something at the same time, and to handle this properly would require a dynamical treatment. On the other hand, the empirical approach does provide a reasonable type of physical interpretation of the effect.

PAWLEY: It would seem really worthwhile then to do some dynamical calculation to see whether there really are modes of that type.

JOHNSON: I did not realize there was a strong objection to our method. The theory is, of course, correct for the quadratic approximation; it is just that we now need to prove experimentally that this is really what happens in nature.

MASLEN: The point which I would like to make right at the outset is that I do not think that the statistical and the mechanistic models are essentially competitive; rather there is a transiion between the two. The mechanistic models which I have described undoubtedly have their prime use in the interpretation of X-ray data. However, I do not think that anybody would prefer to interpret thermal motion using X-ray data when neutron diffraction was a reasonable alternative. For instance, in the molecule *s*-trinitro triamino benzene, if you use the X-ray data and in one analysis refine only the mean square amplitudes of vibration and then in another refine for each atom the best isotropic form factor, you find that the mean square amplitudes change in a rather unexpected way. The changes are in fact markedly anisotropic. For the oxygens there is a pronounced decrease of the thermal motion in plane and virtually no change normal to the plane of the molecule. The carbons show this effect, but to a lesser degree. Another example is orthanilic acid, which was analysed by Dr. Hall, using both X-ray and neutron diffraction. This structure contains a benzene ring with a SO_3^- and NH_3^+ attached in the ortho position. When one does the analysis of thermal motion with neutron diffraction data, the centre of libration is at the centre of mass of the molecule, whereas for the X-ray data it is somewhere between the NH_3^+ and SO_3^- groups. The difference between the two is highly significant and undoubtedly arises from the errors associated with using atomic form factors for these models. I think, therefore, that in investigating thermal motion one should obtain neutron data, preferably using a statistical approach because of the prejudgements inherent in the mechanistic model. On the other hand, if you are doing an X-ray structure analysis to look for movements of charge, the thermal motion is rather a nuisance and the mechanistic models are very effective because constraints can be applied in a very efficient way. Although the temperature factor formulas are much more complicated in the case of mechanistic models, the application of constraints is extremely easy. In the case of the statistical models the temperature factor formulas are very simple but the programming of constraints is exceedingly difficult. Therefore, if you have a problem where there is a large number of constraints, the mechanistic model is effective. When you are looking at thermal motion and wanting a general model, there are not many constraints and then the statistical model is undoubtedly best.

JOHNSON: I think that none of us is really in disagreement about these things. We have been trying to present the various pros and cons in the approach to thermal motion analysis.

HANSON: Is there any problem in attaching significance to the higher-order cumulants when they have such a small effect? Do they have a much smaller effect than, say, a shift in a coordinate? What is the effect on the *R* factor ratio?

JOHNSON: I have done this on structures where it makes a great improvement and on other structures where it does not matter much at all. There have also been significant changes in the *R* factor ratio but more important effects are seen in some difference-Fourier results.

TOPIC G

ABSORPTION AND EXTINCTION CORRECTIONS

CHAIRMAN: P. COPPENS

The Evaluation of Absorption and Extinction in Single–Crystal Structure Analysis

PHILIP COPPENS

Chemistry Department, State University of New York at Buffalo, Buffalo, New York, U.S.A.

In the last few years the emphasis in crystal-structure analysis has increasingly shifted from the mere determination of molecular conformation and molecular packing to a more detailed analysis of the diffraction data. Thus, with accurate data one may detect small changes in geometry resulting from interactions such as hydrogen bonding or charge transfer. Also in some cases it may be desirable to measure the charge density in a molecule or ion. Another subject receiving renewed attention is the analysis of molecular dynamics through the effect of thermal displacements on the intensities of the Bragg peaks.

Such aims require increasingly accurate data as well as an improved understanding of the many effects which influence the intensities of the diffracted beams. As was evident at the Cambridge Conference on Accurate Determination of X-ray Intensities and Structure Factors (Arndt & Mathieson, 1969) a great many such effects at present limit the accuracy of our single-crystal structure factors to perhaps 2–3%. While no single effect is the main culprit, it is fair to say that allowance for absorption and extinction in the crystals is a *conditio sine qua non* for the advanced diffraction studies which are now undertaken at a number of crystallographic laboratories.

In this article we describe some more recent procedures for evaluating absorption and extinction. Both can be calculated with the diffraction geometry and the crystal dimensions. But while calculation of absorption requires knowledge of the linear absorption coefficient, a quantity which can be tabulated with fair accuracy, extinction depends on the mosaic spread or on the size of the perfect crystallites (particles) in the specimen crystal. As these quantities may be anisotropic, it is much more difficult to obtain the *a priori* information necessary for the numerical computation of the extinction correction. It has therefore become the practice to calculate and correct for absorption in the crystal, but to determine the effect of extinction by least-squares refinement of appropriate parameters.

A third potential source of error, multiple reflection, is closely related to extinction. While extinction is caused by transfer of energy between the incident and the diffracted beam, multiple reflection implies energy transfer among the incident beam and several reflected beams. Although it could be treated like extinction in the least-squares refinement, it is possible on four-circle diffractometers to select the experimental conditions such as to avoid multiple reflection completely (Coppens, 1968). This is always preferable to computational corrections, no matter how sophisticated the treatment may be.

1. Basic operations

The use of vector and matrix algebra

Very complicated and unwieldly expressions have appeared in the past in some articles describing absorption corrections. Often, a number of different coordinate systems are used. It is the purpose of this section to demonstrate the relative simplicity of the algorithms when matrix and vector algebra are used.

The use of arbitrarily defined crystal-based orthogonal coordinate systems becomes superfluous when both direct and reciprocal crystal axes are utilized. If N_i and m_i are respectively the direct and reciprocal space components of the vectors **N** and **M** the scalar product **N** . **M** becomes

$$\mathbf{N}\,.\,\mathbf{M}=N_1m_1+N_2m_2+N_3m_3$$

Direct space and reciprocal space components are interconverted by the relationships

$$(n_1n_2n_3)=\mathbf{G}\{N_1N_2N_3\} \qquad (1a)$$

and

$$(N_1N_2N_3)=\mathbf{g}\{n_1n_2n_3\}^* \qquad (1b)$$

in which **G** and **g** are the direct space and reciprocal space metrics (*International Tables for Crystallography*, Vol. II, p. 56).

It is efficient programming to calculate the matrices **G** and **g** early in the routine and to perform the interconversions (1*a*) and (1*b*) in a simple matrix-vector multiplication subroutine.

The only step in the calculations which depends on the diffraction technique is the evaluation of the components of the unit vectors $\mathbf{s}_0$ and **s**, which are parallel to the incident and diffracted beams, respectively.

Other calculations, such as evaluation of the vertices and edges of the crystal or the path lengths of the beams in the crystal are performed within the crystal coordinate system. They are therefore independent of the experimental technique once **s** and $\mathbf{s}_0$ have been determined.

* Here and elsewhere in this article the curled brackets indicate a column vector, while direct and reciprocal space components are represented by capital and lower-case characters respectively.

1·1. *Calculation of the components of* $\mathbf{s}_0$ *and* **s**

The procedure described here for photographic data was outlined by Coppens, Leiserowitz & Rabinovich (1965).

1·1·1. The Weissenberg equi-inclination technique. If we assume that the crystal is mounted with the *a*-axis along the spindle of the Weissenberg camera, then, for the equi-inclination geometry:

$$\mathbf{s}_0\,.\,\mathbf{a}=-\mathbf{s}\,.\,\mathbf{a}$$

and

$$(\mathbf{s}+\mathbf{s}_0)\,.\,\mathbf{a}=0$$

If [Fig. 1(*a*)] $\mathbf{I}=\mathbf{s}+\mathbf{s}_0$ and

and

$$\mathbf{D}=\mathbf{s}-\mathbf{s}_0=\lambda\mathbf{d}^*=\lambda(h\mathbf{a}^*+k\mathbf{b}^*+l\mathbf{c}^*)$$

then

$$\mathbf{I}\,.\,\mathbf{a}=i_1=0 \qquad (2)$$

and

$$\mathbf{I}\,.\,\mathbf{D}=(i_2\mathbf{b}^*+i_3\mathbf{c}^*)\times(h\mathbf{a}^*+k\mathbf{b}^*+l\mathbf{c}^*)\,.\,\lambda=0 \qquad (3)$$

The length of **I** is also known

$$(\mathbf{I}\,.\,\mathbf{I})=(i_2^2b^{*2}+i_3^2c^{*2}+2i_2i_3b^*c^*\cos\alpha^*)=4\cos^2\theta \qquad (4)$$

Since (4) is a quadratic expression, it will still be satisfied after a change in sign of both i_2 and i_3.

This corresponds to reversal of **I**, and therefore an interchange of the incident and diffracted beams. These two cases correspond to two different spots on the film and a choice can be made when the directions of the positive *b* or *c* axes with respect to the incident beam have been recorded. But, as absorption is invariant to the interchange of incident and diffracted beams, we may arbitrarily assign the positive sign to either i_2 or i_3 and obtain **s** and $\mathbf{s}_0$ from

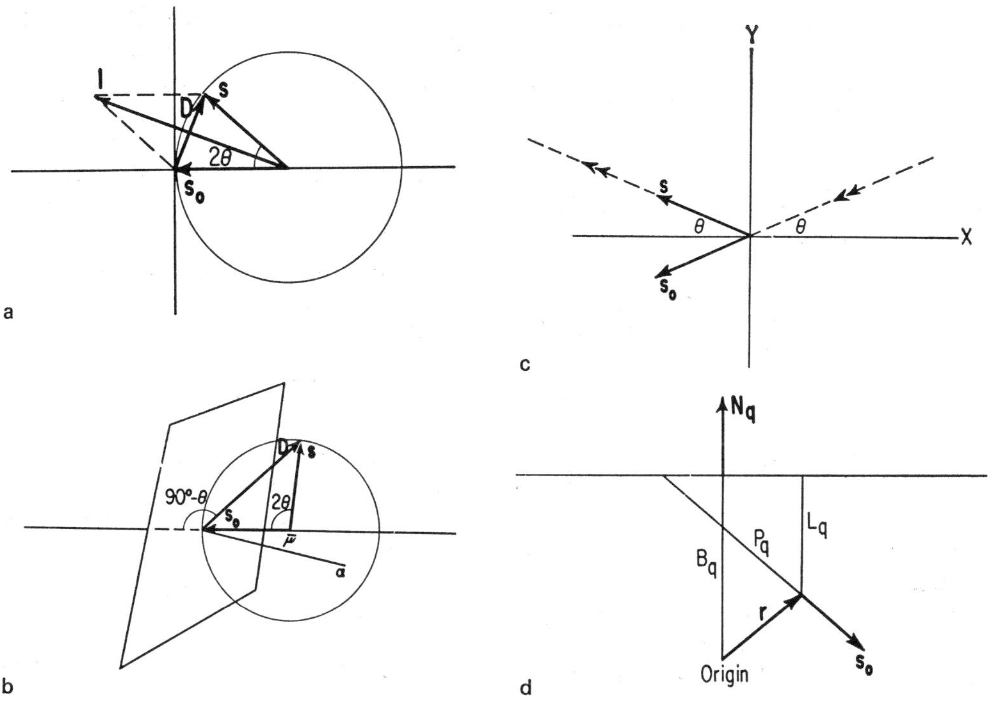

Fig. 1. (a) The vectors **I** *and* **D** *and the sphere of reflection. (b) The geometry of the precession camera. (c) Definition of the laboratory-based XYZ system. The Z axis is perpendicular to the plane of the paper, such as to give a right-handed coordinate system. (d) Distance of a point* **r** *from a face q, and the path length of the incident beam.*

$$\mathbf{s}_0=(\mathbf{I}-\mathbf{D})/2 \qquad (5)$$

$$\mathbf{s}=(\mathbf{I}+\mathbf{D})/2\,. \qquad (6)$$

1·1·2. The precession camera. If the precession axis is parallel to **a** and the precession angle is $\mu°$

$$\mathbf{s}_0\,.\,\mathbf{a}=-a\cos\mu$$

therefore

$$\mathbf{s}_0\,.\,\mathbf{a}=s_1^0=-a\cos\mu \qquad (7)$$

Since our basic equation does not involve $\mathbf{s}+\mathbf{s}_0$ we do not introduce the vector **I**, but use the general result [Fig. 1(*b*)]

$$\mathbf{s}_0\,.\,\mathbf{D}=|s_0|\;|D|\cos(90+\theta)$$
$$=-|D|\sin\theta=-2\sin^2\theta$$

or

$$s_1^0D_1+s_2^0D_2+s_3^0D_3=-2\sin^2\theta \qquad (8)$$

Since $\mathbf{s}_0$ is also a unit vector,

$$\mathbf{s}_0\,.\,\mathbf{s}_0=1 \qquad (9)$$

(8), like (4), is a quadratic equation. So there are two sets of values for s_2^0 and s_3^0 satisfying (8) and (9). They do not correspond to an interchange of incident and diffracted beams [the former is unambiguously defined by (7)], but rather to the reciprocal lattice point entering and leaving the sphere of reflection during the precession motion. As is well-known the two diffracted beams are recorded on the same spot of the film if the crystal is properly oriented.

The absorption in the crystal is generally different for the two beams, as is the Lorentz factor (Waser, 1951). As was pointed out by Wuensch & Prewitt (1965), each absorption factor is to be multiplied by the appropriate Lorentz factor before averaging over the two beams.

Eqs. (7), (8), and (9) lead to the components of $\mathbf{s}_0$; and $\mathbf{s}$ is calculated from

$$\mathbf{s} = \mathbf{s}_0 + \mathbf{D}$$

1·1·3. Four-circle diffractometer. It is possible to obtain the components of $\mathbf{s}_0$ and $\mathbf{s}$ for the four-circle diffractometer without knowing all four setting angles. Such a procedure, in which the value of φ is required for four-circle diffractometer data or for three-circle diffractometer at $\chi = 90°$, has been described by Coppens *et al.* (1965) and it has, indeed, been applied successfully in a large number of calculations.

However, this method has two disadvantages:

A. It is easier to specify ω as the angle that is arbitrarily fixed than φ, because ω is usually kept zero or constant during data collection, while φ varies widely. Errors in φ values (or for example in the sense of the φ circle) may easily escape attention.

B. The solution does not distinguish between the incident and diffracted beams. This is irrelevant for the absorption correction but for anisotropic extinction refinement of type II crystals the incident beam has to be identified (see section 3 below). Since diffractometer data are often sufficiently accurate to warrant anisotropic extinction correction this is a serious drawback.

A different procedure, given here, eliminates these disadvantages. It follows the programs in use of Brookhaven National Laboratory (Bednowitz, Coppens & Hamilton, unpublished) and is similar to the method described by Busing & Levy (1967) in their comprehensive treatment of diffractometer-angle calculations.

A laboratory system of unit axes is defined as follows: at zero setting angles, $\mathbf{X}$ points towards the radiation source, $\mathbf{Y}$ is horizontal in the plane of χ-circle, such that the incident beam arrives at the origin (crystal) from the positive $\mathbf{X}$ positive $\mathbf{Y}$ direction, $\mathbf{Z}$ is parallel to the φ axis such that the coordinate system is right handed.

The orientation of the crystal on the goniometer head is described by the matrix $\mathbf{W}$, such that

$$(\mathbf{XYZ})\mathbf{W} = \{\mathbf{a}^*\mathbf{b}^*\mathbf{c}^*\} \tag{10}$$

The matrix $\mathbf{W}$ can be determined from the known setting angles of two reflections (provided the cell constants are known).

When the crystal is brought into the reflection position the reciprocal axes are rotated such that

$$(\mathbf{a}^*\mathbf{b}^*\mathbf{c}^*)_{\varphi,\chi,\omega} = \mathbf{R}\{\mathbf{a}^*\mathbf{b}^*\mathbf{c}^*\}_{0,0,0} \tag{11}$$

When the diffractometer angles increase with anti-clockwise rotation when looking down the X or Z axes, $\mathbf{R}$ is defined as

$$\begin{pmatrix} \cos\varphi\cos\omega - \sin\varphi\cos\chi\sin\omega & \cos\varphi\sin\omega + \sin\varphi\cos\chi\cos\omega & \sin\varphi\sin\chi \\ -\sin\varphi\cos\omega - \cos\varphi\cos\chi\sin\omega & -\sin\varphi\sin\omega + \cos\varphi\cos\chi\cos\omega & \cos\varphi\sin\chi \\ \sin\chi\sin\omega & -\sin\chi\cos\omega & \cos\chi \end{pmatrix}$$

Thus, after the crystal has been set

$$(\mathbf{XYZ})\mathbf{W} = \mathbf{R}^{-1}\{\mathbf{a}^*\mathbf{b}^*\mathbf{c}^*\}$$

or

$$\mathbf{RW}'\{\mathbf{XYZ}\} = \{\mathbf{a}^*\mathbf{b}^*\mathbf{c}^*\} \tag{12}$$

Before rotation the components of $\mathbf{s}_0$ are given by [Fig. 1(c)]

$$(-\cos\theta \quad -\sin\theta \; 0)$$

Since the inverse of (12) holds for the components along the two axial systems, we get for the reciprocal-axes components s_i^0 and s_i^t:

$$(s_1^0 s_2^0 s_2^0) = \{-\cos\theta \;\; -\sin\theta \;\; 0\}\,(\mathbf{RW}')^{-1} \tag{13}$$

Similarly, for the diffracted beam,

$$(s_1 s_2 s_3) = \{-\cos\theta \quad \sin\theta \quad 0\}\,(\mathbf{RW}')^{-1} \quad (14)$$

With these equations the components can be calculated from the diffractometer setting angles and the orientation matrix.

1·2. *Is a point inside the crystal?*

The unit vector $\mathbf{N}_q$ from a point inside the crystal normal to a boundary plane q with Miller indices h_q, k_q, l_q, and lattice spacing d_q is given by

$$\mathbf{N}_q = \frac{h_q\mathbf{a}^* + k_q\mathbf{b}^* + l_q\mathbf{c}^*}{d_q}$$

If the distance from an origin within the crystal to the boundary plane is B_q [Fig. 1(*d*)], then a point with direct space coordinates r_1, r_2, r_3 is within the crystal or on the surface, if for all boundary planes

$$\mathbf{N}_q \cdot \mathbf{r} = \frac{h_q\mathbf{a}^* + k_q\mathbf{b}^* + l_q\mathbf{c}^*}{d_q} \cdot \times (r_1\mathbf{a} + r_2\mathbf{b} + r_3\mathbf{c}) = \frac{h_q r_1 + k_q r_2 + l_q r_3}{d_q} \leq B_q \quad (15)$$

Analogous expressions for quadratic boundary surfaces have been given by Wuensch & Prewitt (1965).

1·3. *The path length of the incident and diffracted beams for a point in the crystal*

The distance $L_q(\mathbf{r})$ [Fig. 1(*d*)] from the point $\mathbf{r}$ to the face q is given by

$$L_q(\mathbf{r}) = B_q - \mathbf{N}_q \cdot \mathbf{r} \quad (16)$$

The distance traveled by the incident beam to the point $\mathbf{r}$ after crossing the plane of face q is

$$P_q(\mathbf{r}) = L_q(\mathbf{r})/(-\mathbf{s}_0 \cdot \mathbf{N}_q) \quad (17)$$

The face for which $P_q(\mathbf{r})$ has the smallest positive value is the one through which the beam enters the crystal. The path length of the diffracted ray is found by replacing $\mathbf{s}_0$ by $-\mathbf{s}$ in (17).

1·4. *The vertices of a crystal*

For a point on a boundary plane, equation (15) becomes

$$h_q r_1 + k_q r_2 + l_q r_3 = B_q d_q$$

For the intersection of three planes we find three such conditions:

$$h_1 r_1 + k_1 r_2 + l_1 r_3 = B_1 d_1$$
$$h_2 r_1 + k_2 r_2 + l_2 r_3 = B_2 d_2$$
$$h_3 r_1 + k_3 r_2 + l_3 r_3 = B_3 d_3$$

or

$$\mathbf{H} \cdot \mathbf{r} = \boldsymbol{Bd} \quad \text{or} \quad \mathbf{r} = \mathbf{H}^{-1}\boldsymbol{Bd} \quad (18)$$

If the point is not outside the crystal [*i.e.* (15) is satisfied for all faces], it is a vertex of the crystal.

1·5. *The intersection of a line and a plane*

A line may be described by a vector

$$x\mathbf{a} + y\mathbf{b} + z\mathbf{c} \text{ through a point } r_1, r_2, r_3\,.$$

If the equation of the plane is given by

$$H_1\mathbf{a}^* + H_2\mathbf{b}^* + H_3\mathbf{c}^* = B$$

then the point of intersection has coordinates r_1+px, r_2+py, r_3+pz such that

$$H_1(r_1+px) + H_2(r_2+py) + H_3(r_3+pz) = B.$$

The multiplier p is found from

$$p = \frac{B - (H_1 r_1 + H_2 r_2 + H_3 r_3)}{H_1 x + H_2 y + H_3 z} \quad (19)$$

1·6. *The normal to a plane defined by two vectors*

A plane defined by two vectors

$$\mathbf{L} = L_1\mathbf{a} + L_2\mathbf{b} + L_3\mathbf{c}$$

and

$$\mathbf{M} = M_1\mathbf{a} + M_2\mathbf{b} + M_3\mathbf{c}$$

has a normal $\mathbf{N}$ which is parallel to $\mathbf{L} \times \mathbf{M}$. Since $\mathbf{a} \times \mathbf{a} = 0$ and $\mathbf{a} \times \mathbf{b} = \frac{|a|\,|b| \sin\gamma\, \mathbf{c}^*}{|c^*|}$ etc., the reciprocal-space components of **N** are:

$$n_1 = L_2M_3 - M_2L_3$$
$$n_2 = M_1L_3 - L_1M_3$$
$$n_3 = L_1M_2 - M_1L_2 \qquad (20)$$

in which the constant V is omitted. **N** can be normalized by dividing each component by $(\mathbf{NgN})^{1/2}$ (**g** is the reciprocal space metric).

If a point r_1, r_2, r_3 is known to lie in the plane, the equation of the plane is:

$$N_1X + N_2Y + N_3Z = N_1r_1 + N_2r_2 + N_3r_3 \qquad (21)$$

where X, Y, Z are the coordinates of any point in the plane, and the term on the right-hand side is a constant equal to the distance of the plane from the origin.

2. Absorption correction of single-crystal intensity data

Fast and relatively accurate numerical evaluation of the absorption integral

$$A = \frac{1}{V}\int_{\text{crystal}} \exp(-\mu T)dV \qquad (22)$$

was first introduced by Busing & Levy (1957), who used the Gaussian integration method (Margenau & Murphy, 1943; Lanczos, 1956) for the absorption correction of zero-layer reflections. The method was extended to Weissenberg, precession, and three- and four-circle diffractometer data by Wuensch & Prewitt (1965) and Coppens, Leiserowitz & Rabinovich (1965). If the crystal has re-entrant angles, the Gaussian method cannot be used and recourse has to be taken to the far less accurate summation over an isometric grid of equidistant points.

A totally different approach has been proposed by De Meulenaer & Tompa (1965), who extended Howell's concept of dividing the crystal into a number of polyhedra, in each of which the path length is a linear function of the coordinates of the point considered. The method leads to an analytical evaluation of the integral (22) and has a definite edge over the numerical integration when absorption is severe (see Section 2·1·3).

Huber & Kopfman (1966) have described an experimental determination of (22) through the measurement of a transmission surface. Their method has advantages when the shape of the crystal is irregular or when other bodies, such as drops of mother liquor, participate in the absorption process. For a crystal of well-defined shape the method is probably less accurate than computational methods. The approximation

$$\frac{1}{V}\int \exp(-\mu T)dV = \frac{1}{V} \times \int \exp(-\mu T_{\text{incident}})\,dV \times \int \exp(-\mu T_{\text{diffracted}})\,dV$$

is used, which is poor when absorption is severe. The transmission surface is determined from the variation in intensities when reflections of intermediate strength are recorded at various values of the angle of rotation around the scattering vector. Since intermediate reflections are often weakened by extinction the experimental transmission surface would be sensitive to anisotropy of the extinction.

2·1·1. The Gaussian integration method. In the Gaussian quadrature method a non-isometric grid is set up in the crystal along a set of axes which may be the crystallographic axes a, b, and c. The grid points are located such as to minimize the difference between the exact value of the absorption integral and the summation over all the gird points. According to Lanczos (1956) a three-dimensional Gaussian integration with n sampling points is about equivalent to an integration with an isometric grid containing 8 times as many points.

Suppose the a axis is selected first. The maximum and the minimum x coordinates of the set of vertices of the crystal are determined, and points along the a axis are selected such that

$$x_i = x_{\min} + (x_{\max} - x_{\min})\upsilon_i$$

in which υ_i are the Gaussian constants.

For each x_i value a line is drawn parallel to the b axis; the maximum and the minimum values of the $y(x_i)$ coordinate are calculated. Points are then selected such that

$$y_{ij} = y_{\min}(x_i) + [y_{\max}(x_i) - y_{\min}(x_i)]\upsilon_j$$

Finally, the procedure is repeated in the third dimension to give

$$z_{ijk} = z_{\min}(x_i, y_j) + [z_{\max}(x_i, y_j) - z_{\min}(x_i, y_j)]\upsilon_k$$

The example of Fig. 2 illustrates how the grid points accumulate near the surface where the change of absorption with path length is largest. The grid points are always within the crystal. For an isometric grid, on the other hand, the calculated transmission factor can vary considerably if some points are just outside the crystal and therefore excluded, or just inside the crystal and included. The absence of this source of error is one of the reasons for the accuracy of the Gaussian integration.

A weight is associated with each grid point which is proportional to the volume element represented by the point. The tables of Gaussian constants [Margenau & Murphy (1943) for $i \leq 6$; Davis & Rabinovich (1956); Davis & Polonsky (1964)] contain values for υ_i and the associated weight R_i. The weight of a grid point ijk is $R_iR_jR_k$.

Once the grid has been set up, the vectors **s** and $\mathbf{s}_0$ have to be calculated for each reflection (Section 1·1). Then, for each grid point the path lengths of the incident

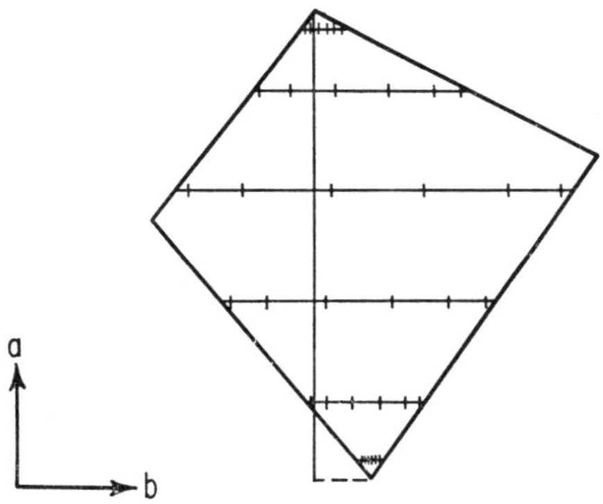

Fig. 2. A 6 × 6 Gaussian grid in a two-dimensional crystal.

and diffracted beams and the corresponding transmission factor have to be evaluated. The integral A is finally obtained from the polynomial

$$A = 1/V \sum_i \sum_j \sum_k R_iR_jR_k \exp(-\mu T_{ijk}) \qquad (26)$$

For strong absorption the number of grid points required to obtain good accuracy (say $< 1\%$) becomes large and the summation over the grid points becomes very time consuming. This point is further discussed in Section 2·1·3.

2·1·2. The analytical method of De Meulenaer and Tompa. The analytical method will be treated in detail by Alcock (this volume) and only a brief discussion is given here. First, it is worthwhile to point out that calculation of the vertices of the crystal and calculation of the direction cosines, are independent of the integration technique employed.

The most complicated calculations in the analytical method concern the evaluation of the Howell's polyhedra. These calculations are simplified considerably, however, when the vector algebra outlined in Section 1 is employed and when all calculations are performed in the crystallographic axial systems. This is illustrated by the following operations by which the vertices of the Howell's polyhedra are calculated (De Meulenaer & Tompa, 1965).

A. A face is in the light of the incident beam when $\mathbf{s}_0 . \mathbf{N} < 0$.

B. A ray through a vertex in the light is projected onto a face of the crystal or onto a plane of light through an edge (Eq. 19).

C. A plane of light passes through the vector $\mathbf{s}_0$ (or $\mathbf{s}$) and an edge. (Eq. 20).

D. A plane of light through an edge cuts an opposite edge (Eq. 19).

E. To find the intersection of the projections of two edges l_1 and l_2 onto a face q: bring planes through l_1 and the normal to q, and through l_2 and the normal to q (Eq. 20); the intersection of these two planes with the face q (Eq. 18) is the required point.

The Howells' polyhedra, once evaluated, are divided into tetrahedra and the total absorption is calculated as a function of the path lengths at the vertices of the polyhedra.

2·1·3. Comparison of the numerical and analytical methods. Ultimately both the numerical and the analytical methods should lead to the same value for the absorption correction. But there are differences in computer-time and computer-storage requirements. The accuracy of the numerical method depends on the number of grid points. Table I, which is from Coppens, De Meulenaer & Tompa (1967), shows the convergence of the numerically evaluated values towards the analytical results for the highly absorbing crystal of $YFeO_3$ ($\mu = 878{\cdot}8$ cm^{-1}). But in this case, the number of grid points is such that the calculation becomes prohibitively expensive in comparison with the analytical method. This is partly a result of increased storage requirements: for each grid point we must store the three coordinates and the weight.

The case of $YFeO_3$, with transmissions of 3–7%, is an extreme one, which can generally be avoided by selecting radiation of a different wavelength. Calculations on a crystal of γ, γ'-dipyridyl dihydrate (Coppens, unpublished results) are given in Table II for varying grids and two values of μ differing by a factor 2. The crystal was used for the collection of neutron-diffraction data and was pillarshaped, its length and width being 5·4 and 0·9 mm respectively. The grids were selected to give more points along the long dimension of the specimen. Analytical calculations were performed with a program kindly supplied by Cullen and Watenpaugh of the University of Washington at Seattle. For $\mu = 2{\cdot}28$ cm^{-1} a grid of $4 \times 8 \times 4$ gives sufficient accuracy (to within 1% of the analytical value) and the numerical method is faster than analytical evaluation. When μ is increased by a factor 2, the spacing of the grid points should be *decreased* by the same factor, in order to keep the ratio of the function $\exp(-\mu \bar{T})$ for two adjacent points constant. Therefore a grid of $8 \times 16 \times 8$ points is required to obtain the same accuracy for $\mu = 4{\cdot}56$ cm^{-1} and the same crystal dimensions.

Table I. Transmission of some reflections of a crystal of $YFeO_3$ (Coppens & Eibschütz, 1965) computed by Gaussian integration on the grids indicated and by the analytical method of De Meulenaer & Tompa (1965)

hkl	$12 \times 12 \times 32$	$24 \times 24 \times 32$	$32 \times 32 \times 32$	Analytical
025	0·0463	0·0378	0·0387	0·0390
$0\bar{2}5$	0·0463	0·0378	0·0387	0·0390
125	0·0375	0·0335	0·0339	0·0340
$1\bar{2}5$	0·0377	0·0332	0·0337	0·0339
225	0·0613	0·0583		0·0587
$2\bar{2}5$	0·0541	0·0509		0·0514
325	0·0695	0·0681		0·0684

The computer time per reflection is proportional to the number of grid points, and at a transmission of 75%, the Gaussian quadrature method becomes more time-consuming than the analytical method (Fig. 3). This assumes that an accuracy of better than 1% is desired; the Gaussian integration remains competitive for accuracies better than 2% down to a transmission of 0·56.

These accuracy estimates are quite conservative and, as can be seen in Table II, the average errors are much smaller. It is evident that at low absorption the numerical method is faster; it leads to good accuracy with a surprisingly small number of points. The analytical method is to be strongly preferred for heavily absorbing crystals. These conclusions are summarized in Table III.

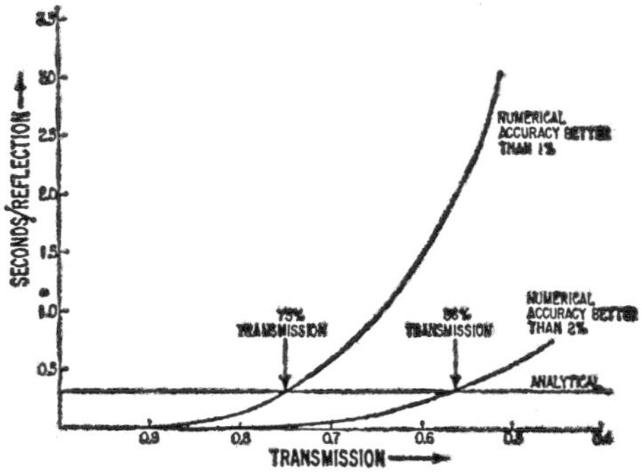

Fig. 3. CDC 6400 time in seconds/reflection against transmission for the analytical and numerical methods. Numbers are based on calculations for needle-shaped crystal of γ,γ′-dipyridyl dihydrate with six boundary planes.

Table II. Transmission of some reflections of a crystal of γ, γ′ dipyridyl dihydrate computed by Gaussian integration on the grids indicated and analytically for $\mu=2{\cdot}28$ cm^{-1} and $\mu=4{\cdot}56$ cm^{-1}. The last column lists CDC 6400 central-processor time in seconds/reflection.

	$2\,9\,\bar{2}$	$4\,0\,0$	$0\,4\,0$	$9\,\bar{2}\,\bar{1}\bar{1}$	$3\,8\,\bar{8}$	$3\,9\,3$	$0\,1\,\bar{2}$	Seconds/ reflection
$\mu=2{\cdot}28$ cm^{-1}								
numerical								
2×4×2	0·8090	0·7989	0·7833	0·8257	0·8040	0·7486	0·7782	0·04
8×4×8	0·8112	0·8143	0·7970	0·8289	0·8087	0·7486	0·7805	0·13
8×16×8	0·8112	0·8134	0·7969	0·8301	0·8091	0·7506	0·7859	0·86
analytical	0·8110	0·8137	0·7960	0·8278	0·8088	0·7474	0·7862	0·26
$\mu=4{\cdot}56$ cm^{-1}								
numerical								
2×4×2	0·6560	0·6384	0·6135	0·6843	0·6491	0·5659	0·6056	0·04
4×8×4	0·6615	0·6647	0·6361	0·6897	0·6582	0·5679	0·6093	0·13
8×16×8	0·6615	0·6635	0·6364	0·6919	0·6589	0·5711	0·6186	0·86
analytical	0·6612	0·6639	0·6358	0·6899	0·6583	0·5688	0·6190	0·26

Table III. Comparison of various methods for absorption correction

	Handles non-planar faces	Handles re-entrant angles	Accuracy when absorption is strong	Relative computing costs
Numerical integration with isometric grid	yes	yes	no	poor
Numerical integration with Gaussian grid	yes	no	at expense of computer time and storage	good unless absorption is severe
Analytical method	no	no	yes	independent of magnitude of the absorption
Experimental determination of transmission surface	yes	yes	no	—

2·1·4. The effect of errors in the absorption coefficient and the crystal dimensions. The absorption coefficient for a crystal of mixed atomic composition is usually derived as the weighted average of the atomic absorption coefficients, such that the absorption is taken to be equal to that of a stoichiometric mixture of the same atomic composition. This 'mixture rule' has been criticized (Deslattes, 1969) because it neglects any effect of the state of aggregation on the wave functions.

The question arises: how accurate do we have to know the absorption coefficient to attain an accuracy of about 1% in the transmission factor. Normally in structure analysis only the *relative* corrections are of importance because the calculations are not performed on an absolute scale. The errors in the relative transmission factors are smaller than the absolute errors: if the transmission is calculated too low by n% for some and by m% for other reflections, the error in the relative transmission factors will only be $n-m$%. This can be illustrated for a spherical crystal. The two extremes are forward reflection ($\theta=0°$) and back reflection ($\theta=90°$). For a spherical crystal Figs. 4 and 5 show, respectively, $\frac{\Delta A}{A}$ and $\left(\frac{\Delta A}{A}\right)_{0°}-\left(\frac{\Delta A}{A}\right)_{90°}$ divided by the percentage error in μr, as a function of μr (r is the radius of the spherical crystal).

The curve in Fig. 5 indicates that:

$$\left|\left(\frac{\Delta A}{A}\right)_{0°}-\left(\frac{\Delta A}{A}\right)_{90°}\right| < 0{\cdot}15\,\frac{\Delta\mu r}{\mu r}$$

for transmission factors larger than 50%. Therefore for all but severe absorption an accuracy of about 7% in μ is sufficient to achieve a 1% accuracy in the relative correction.

When the transmission is larger, the accuracy of μ becomes even less important. Fig. 5 indicates that for a spherical crystal with $A\approx0{\cdot}80$, a 10% error in μ would lead to an error of about 0·3% in the relative absorption correction.

For a non-spherical crystal the absorption correction varies more from reflection to reflection than for a spherical crystal. It follows that the relative errors will be somewhat larger. This is illustrated in Table IV which relates the results of a model calculation on a crystal of γ, γ'-dipyridyl dihydrate.

The dimensions of the crystal are as described in paragraph 2·1·3: it is a long needle with a length more than five times the width. The calculations were done for the value of μ obtained from the 'mixture rule' and for a second value 10% larger. The reflections represented in the Table contain the extremes of lowest and highest absorption and include low- as well as high-order data. The 10% increase in μ causes a relative change in the calculated transmission of almost 1%. This is significantly larger than for a spherical crystal, but still much smaller than the error in the absorption coefficient.

The μ dependence for a very strongly absorbing crystal of BiOF has been investigated by Werner (1964). At transmissions of around 10%, he finds that a 25% change in μ leads to a 40% change in the relative absorption corrections. This is in good, qualitative agreement with Fig.5, because 10% transmission corresponds to $\mu r\sim2{\cdot}4$.

Table IV. The change in calculated transmission for some reflections of a crystal of γ, γ'-dipyridyl dihydrate with a 10% increase in the absorption coefficient

μ (cm^{-1})	$2\,9\,\bar{2}$	$4\,0\,0$	$0\,4\,0$	$9\,2\,\overline{11}$	$3\,8\,\bar{8}$	$3\,3\,13$	$3\,9\,3$	$0\,1\,\bar{2}$
2·28	0·8112	0·8134	0·7969	0·8301	0·8091	0·8108	0·7506	0·7859
2·51	0·7945	0·7967	0·7790	0·8149	0·7923	0·7943	0·7298	0·7671
$\frac{A_{2\cdot28}-A_{2\cdot51}}{A_{2\cdot28}}$	0·021	0·021	0·022	0·018	0·021	0·020	0·028	0·024

Errors in the crystal dimensions, however, affect the results more strongly, because the different reflections depend to a different extent on the individual dimensions. If the width of the specimen has been measured incorrectly, the correction for the ray almost parallel to the long axis will not change, unlike the absorption for the other reflection in the figure.

The curve of Fig. 4, in which $(\Delta A/A)/(\Delta\mu r/\mu r)$ is plotted, is now appropriate. The ratio rapidly increases with μr, indicating that great care should be taken to obtain accurate crystal dimensions. Werner (1964) has pointed out that it is especially important to measure accurately the smallest dimension of a strongly absorbing crystal. For strongly absorbing materials well-defined crystals with smooth faces and sharp edges are to be preferred.

3. The refinement of anisotropic secondary extinction

Absorption correction of accurate crystal data is essential. But in many cases the treatment of secondary extinction is equally important. A breakthrough, which allowed rapid subsequent development, was reached with a number of articles by Zachariasen (1963, 1967). The earlier (1963) approximative expressions were used successfully in least-squares refinement of the extinction coefficient by Åsbrink & Werner (1966) and by Larson (1967). The more recent expressions were generalized by Coppens & Hamilton (1969) to include anisotropy of extinction. The latter generalization is discussed below.

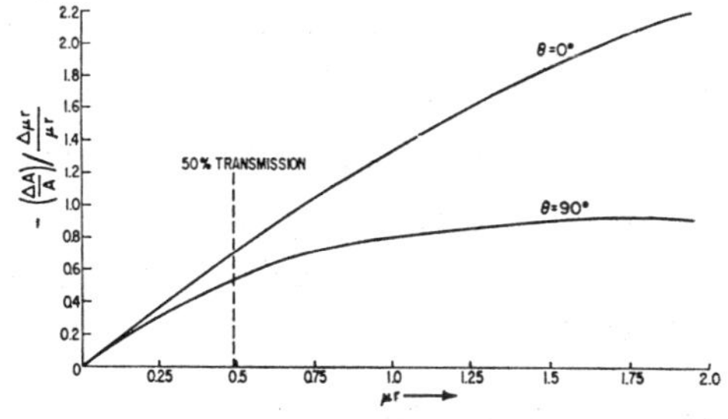

Fig. 4. Percentual error in A divided by percentual error in μr for a spherical crystal at various values of μr.

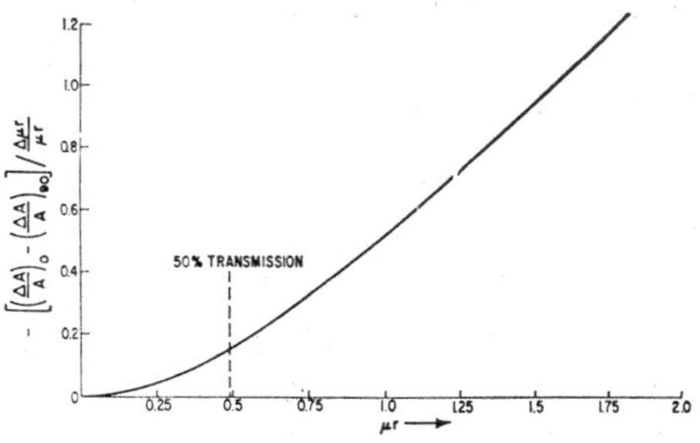

Fig. 5. Difference between the function at Bragg angles of 0° and 90°.

3·1. *Introduction*

The equations for the power of the incident and diffracted rays are given by Zachariasen as:

$$\frac{\partial I_0}{\partial t_1} = -\sigma I_0 + \sigma I \qquad (27a)$$

$$\frac{\partial I}{\partial t_2} = -\sigma I + \sigma I_0 \qquad (27b)$$

Zachariasen makes a number of approximations to solve these equations and obtains an expression for the extinction-affected integrated intensity.

These expressions will be described in detail by Larson (this volume).

The expressions for spherical particles are, when primary extinction is neglected,

$$P = P_k y$$

$$y = \{1 + (2p_2/p_1)\bar{x}_0\}^{-1/2} \qquad (28)$$

and

$$\bar{x}_0 = \beta Q_0 \bar{T}\{1 + (\beta/g)^2\}^{-1/2} \qquad (29)$$

in which

$$\beta = 2\bar{t}_\perp/3\lambda$$

$\bar{t}_\perp$ is the average maximum dimension of the spherical particle in planes parallel to the plane of the incident and diffracted beams. Thus, $\bar{t}_\perp$ equals the area of a circle divided by its diameter, or

$$\bar{t}_\perp = \pi r/2 \tag{30}$$

for

$$\beta \gg g: \quad \bar{x}_0 = gQ_0\bar{T} \tag{31}$$

This means that the broadening of the beam due to the limited size of the particles is negligible compared with the mosaic-spread broadening (type I crystals). In the other extreme

$$g \gg \beta \text{ and } \bar{x}_0 = \beta Q_0 \bar{T} \tag{32}$$

(type II crystals) in which case the beam profile is dependent on the particle size.

3·2. *Anisotropy of mosaic spread*

In the isotropic treatment it is assumed that the misalignment of the lattice-plane normals obeys a Gaussian distribution law. The expression for the distribution as a function of the angular deviation Δ is then:

$$\psi(\Delta) = \sqrt{2}\, g \exp(-2\pi g^2 \Delta^2) \tag{33}$$

The mosaic spread influences the coupling constants in the power equations (27). In the anisotropic case they depend on the value of $\psi(\Delta)$ in the plane of the incident and diffracted beams. Suppose this quantity is represented by the length of a vector parallel to the unit vector **D** normal to the plane of the incident and diffracted rays. The surface of the end points of these vectors, which describes mosaic spread for the whole crystal, may be assumed to be ellipsoidal and represented by

$$\eta(\mathbf{D}) = \frac{1}{\sqrt{2\pi}} (\mathbf{D}'\mathbf{Z}\mathbf{D})^{-1/2}$$

The anisotropic expression for the distribution becomes a three-dimensional Gaussian, in which $g^2(\mathbf{D})$ equals $\mathbf{D}'\mathbf{Z}\mathbf{D}$. After proper normalization:

$$\psi(\Delta, \mathbf{D}) = |Z|^{1/2} 2^{3/2} \exp(-2\pi \mathbf{D}'\mathbf{Z}\mathbf{D}\Delta^2)$$

For type I crystal we obtain:

$$\bar{x}_0 = g(\mathbf{D}) Q_0 \bar{T} \tag{34}$$

3·3. *Anisotropy of particle size*

In the anisotropic treatment the particle size is assumed to be ellipsoidal, such that the length of the radius vector in the direction of a unit vector **N** is given by

$$r(\mathbf{N}) = (\mathbf{N}'\mathbf{W}\mathbf{N})^{-1/2}$$

Since an ellipse is a projection of a circle, it can be shown easily that (30) is to be replaced by:

$$\bar{t}_\perp = \frac{\pi}{2} r(\mathbf{N})$$

The coupling constants σ for type II crystals depend on the dimension of the particle in the diffraction plane perpendicular to the ray. Since the particles are no longer spherical, the coupling constants multiplying I and I_0 are no longer equal and (27) becomes

$$\frac{\partial I_0}{\partial t} = \sigma(I \to I_0)I - \sigma(I_0 \to I)I_0 \tag{35a}$$

$$\frac{\partial I}{\partial t} = \sigma(I_0 \to I)I_0 - \sigma(I \to I_0)I \tag{35b}$$

It follows that the extinction factor $\bar{x}_0$ is dependent on the particle dimensions perpendicular to the incident and perpendicular to the diffracted beam, but it is much less dependent on the latter because in general I is smaller than I_0. Hence, a good approximate solution can be obtained by neglecting the 'feedback' term $\sigma(I \to I_0)I$ in expression (35) which leads to

$$\bar{x}_0 = r(\mathbf{N})\lambda^{-1} Q_0 \bar{T} \tag{36}$$

in which **N** is the unit vector in the diffraction plane perpendicular to the incident beam.

3·4. *Anisotropic refinement*

Expressions (34) and (36) are substituted in (28) to obtain the equations for anisotropic refinement. The explicit formulas have been derived by Coppens & Hamilton (1969).

For each reflection the components of the vectors **D**, **N** and the mean path length $\bar{T}$ have to be known. These quantities can be derived in the data-processing program in which the absorption correction is evaluated. **D** is obtained by normalizing $\mathbf{s} \times \mathbf{s}_0$, and **N** similarly from $\mathbf{s}_0 \times \mathbf{D}$. $\bar{T}$ is calculated from

$$\bar{T} = -\frac{\ln A}{\mu}$$

For each reflection the six components and the mean path length are transferred to the least-squares program.

Since the crystal-shape and the mosaic spread and particle size do not necessarily conform to crystal symmetry, symmetry-related reflections other than Friedel pairs are no longer averaged before least-squares refinement.

It is usually simplest to start with refining the isotropic extinction parameter g. The anisotropic extinction refinement requires six parameters: one for each independent element of the symmetrical tensors **Z** or **W**. Introduction of five additional extinction parameters is not justified unless extinction is severe. But experience with the method indicates that in this case extinction is often significantly anisotropic.

3·5. *Examples of extinction refinement*

A number of cases have been discussed by Coppens & Hamilton (1969). As an example, we give here the discussion for α-deutero and α-proto oxalic acid dihydrate (α-DOX and α-POX respectively) in which pronounced anisotropy of extinction was observed.

The X-ray data on the regular and the deuterated forms of oxalic acid dihydrate were collected by Delaplane & Ibers (1969). For both structures the results of the initial refinements were disappointing, as agreement factors remained at about 15%. Introduction of an isotropic extinction parameter in the refinements led to a marked improvement. Nevertheless, the agreement factors were still larger than anticipated from a comparison of the intensities of symmetry-related reflections. Examination of the structure factor list revealed that reflections with $k=0$ for both compounds were generally undercorrected, whereas other reflections appeared to be overcorrected. An attempt to explain this anisotropy through the anisotropy of the crystal shape was made by using the numerical integration procedure developed by Hamilton (1963). This attempt led to a slight improvement, although a number of strong reflections remained poorly corrected. The strong reflections for both compounds were then recollected using very small crystals with volumes of 2×10^{-4} mm^3 and 13×10^{-4} mm^3 for α-DOX and α-POX respectively. Although extinction effects were effectively suppressed by this procedure, the counting statistics were poor even for the stronger reflections, and little was gained from the additional experimental effort.

Introduction of the anisotropic extinction finally led to a more acceptable refinement of the initial set of data. All elements of the **Z** tensor (type I) were essentially zero except for Z_{22}, indicating that according to this interpretation the mosaic spread was very much smaller in planes with normals perpendicular to the b-axis. The anisotropy of the extinction was slightly less pronounced for the neutron data on α-DOX, which were collected on an appreciably larger crystal (Coppens & Sabine, 1969). The results showed that the larger crystal was more nearly perfect. Thus particle size varied between 1·4 and 3·0μ, compared with 0·4 and 0·9μ for the X-ray data on the same compound. This situation is analogous to that found for hydrazinium sulfate (Coppens & Hamilton 1969). This effect is not surprising, because the larger crystals needed for neutron diffraction are usually grown more slowly and with more care than the small X-ray crystals. It follows that extinction will often be more severe for neutron data for this reason, not only because the path lengths inside the crystal are longer.

In most of the compounds studied up till now, the R factors are very similar for type I and type II treatments, the sole exception being the X-ray data on α-POX. It is therefore usually not possible to distinguish between type I and type II crystals on the basis of the R factor alone. Other information, such as a second set of data at a different wavelength, is required if the distinction is to be made. But for the compounds investigated the structural parameters are equal to within one standard deviation for the two treatments. It is therefore of no consequence for the structural analyses which of the extreme cases is closer to reality. Only when the textural parameters of the specimen crystals are to be studied is further experimental work indicated.

Zachariasen (1968) has shown that for type II crystals for which the particle size is larger than 10^{-4} cm and $\mu T > 1$ the Borrmann effect may become important. Under these conditions (28) and (29) have to be modified as described by Zachariasen. Since the modified expressions are similar in form to (28) and (29), our use of (28) and (29) will still lead to a reasonable correction for extinction in the least-squares refinement, but the resulting textural parameters no longer provide a reliable estimate of the particle size.

4. Errors in the parameters resulting from incomplete allowances for absorption and extinction

Since transmission factors are generally lower for low-order reflections, for which the beams have to traverse most of the crystal, neglect of absorption causes temperature factors to be systematically low. In extreme cases negative temperature factors may occur. Some systematic studies have been done by Srivastava & Lingafelter (1966), who concluded that for the test compound bisethylenediamine palladium (II) chloride, the isotropic temperature factors change by about the same amount when absorption was not allowed for. On the other hand, for the more heavily absorbing crystal of BiOF, studied by Werner (1964), mainly the oxygen temperature parameters were affected.

The changes in the positional parameters may be quite large, but they are usually within the standard deviations. The most dramatic change here is a large reduction of the standard deviations on correction for absorption, rather than a shift in atomic positions. Significant shifts have however been reported by Huber & Kopfmann (1969) after application of the experimental method for absorption correction. Apparently, such changes have not been found for computationally corrected intensity data.

The errors introduced by neglect of extinction are very similar, but often not recognized. This is because an excellent fit between F_o and F_c can usually be obtained by excluding a number of strong reflections. It has been found by Åsbrink & Werner (1966) and by Coppens & Sabine (1969) that the intermediate reflections are affected sufficiently to produce a systematic lowering of the diagonal elements of the thermal motion tensors, which becomes apparent when extinction is allowed for. For example, the neutron-diffraction temperature factors for α-deutero oxalic acid dihydrate (α-DOX) with extinction refinement and without extinction refinement, but excluding the 53 strongest reflections (out of 540), show that all the diagonal tensor elements increase. The change is typically about 7%, or roughly three standard deviations. At the same time, the R factors decreased from 6·2 to 4·3%. No significant changes were observed in the positional parameters. Similarly, bond lengths and angles derived from the X-ray data of α-POX did not change by more than two standard deviations when the R factor was reduced from 19·5 to 2·8% through the application of a number different extinction procedures, but the standard deviations of the positional parameters generally decreased by a factor 3.

5. Conclusion

Though absorption and extinction can be properly allowed for, the accuracy of the correction is less satisfactory when the corrections are large. In the case of absorption this is mainly a result of the difficulty of obtaining very precise measurements of the crystal dimensions. The treatment of extinction in the Coppens and Hamilton extension of the Zachariasen theory contains many approximations which become less reliable when extinction is strong. It is therefore good experimental practice to avoid, if possible, situations for which absorption and extinction are severe.

6. Acknowledgement

The author would like to thank Drs. Cullen, Jensen, and Watenpaugh for providing a version of their analytical absorption correction program.

Various methods described in this article were developed in collaboration with Drs. A. L. Bednowitz, W. C. Hamilton, L. Leiserowitz, and D. Rabinovich. Their contribution is gratefully acknowledged.

Appendix

Symbols not defined specifically in the text.

$\mathbf{d}^*$	reciprocal lattice vector
d_q	spacing of lattice planes q
F_c	the absolute calculated structure factor
F_k	the scaled structure factor $F_k = kF_c$
g	isotropic extinction parameter
$\mathbf{g}$	reciprocal space metric
$\mathbf{G}$	direct space metric
I_0	the intensity in the direct beam inside the crystal in cm^{-2} sec^{-1}
I	the intensity in the scattered beam inside the crystal in cm^{-2} sec^{-1}
P	integrated intensity of a Bragg reflection in sec^{-1}
P_k	integrated intensity in the kinematic approximation of a Bragg reflection in sec^{-1}
$p_2 = 1 + K^2$ $p_1 = 1 + K$	in which K is the polarization factor for the parallel component of polarization in the incident beam. 1 for neutrons, $\cos^2 2\theta$ for X-rays
Q_0	average scattering cross-section per unit volume of crystal (cm^{-1}) excluding the polarization factor
R	$\Sigma \lvert F_o - \lvert F_k \rvert \rvert / \Sigma F_o$
R_w	$\left\{ \dfrac{\Sigma w \lvert F_o - \lvert F_c \rvert \rvert^2}{\Sigma w F_0^2} \right\}^{1/2}$
$\mathbf{s}_0$	unit vector parallel to incident beam
$\mathbf{s}$	unit vector parallel to diffracted beam
T	total path length in crystal for rays diffracted by a volume element in the crystal.

t_1	distance in the incident beam of a single domain, cm
t_2	distance in the diffracted beam in a single domain, cm
V	volume of crystal
δ_{ij}	$=1$ for $i=j$; 0 for $i\neq j$
φ, χ, ω	diffractometer setting angles
λ	wavelength
σ	coupling constant for transfer of energy between two rays
θ	Bragg angle

References

Arndt, V. W. & Mathieson, A. McL. (1969). *Acta Cryst.* A**25**, 1276.

Åsbrink, S. & Werner, P. E. (1966). *Acta Cryst.* **20**, 407.

Busing, W. R. & Levy, H. A. (1957). *Acta Cryst.* **10**, 180.

Coppens, P. (1968). *Acta Cryst.* A**24**, 253.

Coppens, P., De Meulenaer, J. & Tompa, H. (1967). *Acta Cryst.* **22**, 601.

Coppens, P. & Hamilton, W. C. (1969). *Acta Cryst.*, in press.

Coppens, P., Leiserowitz, L. & Rabinovich, D. (1965). *Acta Cryst.* **18**, 1035.

Coppens, P. & Sabine, T. M. (1969). *Acta Cryst.* **B25**, 2442.

Davis, P. J. & Polonsky, I. (1964). *Handbook of Mathematical Functions,* ed. M. Abramowitz & I. A. Stegun. Nat. Bur. Standards Appl. Math. Series **55**, 916. See also C. H. Love, NBS Monograph No. 98.

Davis, P. & Rabinovich, P. (1956). *J. Res. Nat. Bur. Standards* **56**, 35.

Delaplane, R. & Ibers, J. A. (1969). *Acta Cryst.* **B25**, 2423.

De Meulenaer, J. & Tompa, H. (1965). *Acta Cryst.* **19**, 1014.

Deslattes, R. D. (1969). *Acta Cryst.* A**25**, 89.

Hamilton, W. C. (1963). *Acta Cryst.* **16**, 609.

Huber, R. & Kopfmann, G. (1969). *Acta Cryst.* **A25**, 143.

Lanczos, C. (1956). *Applied Analysis.* Englewood Cliffs, N. J.: Prentice-Hall Inc.

Larson, A. C. (1967). *Acta Cryst.* **23**, 664.

Margenau, H. & Murphy, G. M. (1943). *The Mathematics of Physics and Chemistry.* New York: Van Nostrand.

Srivastava, R. C. & Lingafelter, E. C. (1966). *Acta Cryst.* **20**, 703.

Waser, J. (1951). *Rev. Sci. Instr.* **22**, 563.

Werner, P. (1964). *Acta Chem. Scand.* **18**, 1851.

Wuensch, B. J. & Prewitt, C. T. (1965). *Z. Krist.* **122**, 24.

Zachariasen, W. H. (1963). *Acta Cryst.* **16**, 1139.

Zachariasen, W. H. (1967). *Acta Cryst.* **23**, 558.

Zachariasen, W. H. (1968). *Acta Cryst.* A**24**, 421.

Discussion

: Do you always use the indices to define the crystal faces even when they are at non-crystallographic planes?

COPPENS: Yes we do; in that case we approximate to very high Miller indices.

MCCAULEY: Do you measure the crystal dimensions on the diffractometer or do you take it off?

COPPENS: We do it both ways, but it is best done on the diffractometer. I find it most convenient to have the telescope mounted along the χ axis; then you can rotate the crystal until it is in a diffracting position for a particular plane. You are then certain that the axis of the telescope is parallel to the plane in which you are interested.

CRUICKSHANK: Why has not extinction theory been redone in terms of spiral dislocations?

COPPENS: I do not know; certainly the mosaic block model is crude, as electron micrographs have shown.

HAYDEN: At what point in the structure determination do you apply the extinction correction?

COPPENS: After we have located all the atoms. In cases where extinction was severe, we could not find the hydrogen atoms until after the correction was applied.

The Analytical Method for Absorption Correction

N. W. ALCOCK

School of Molecular Sciences, University of Warwick, Great Britain.

The discovery that it was possible to calculate absorption corrections for polyhedral crystals exactly, by an analytical method, rather than by a numerical approximation, was made by Howells (1950). His solution applied to two dimensions and was relatively straightforward compared to the extension to three dimensions by Dr. H. Tompa and co-workers (de Meulenaer & Tompa, **MT**, 1965). This method has been incorporated in a complete absorption correction program in Fortran; it allows many variations in type of data, recording method, etc., and so the control routine is complicated. After initialization, angle calculations are performed for each reflection, to produce direction cosines of the incident and diffracted rays. The diffracted ray is then reversed, so that both rays enter the crystal and can be simply labelled 1 and 2 (cosines u_1, v_1, w_1, u_2, v_2, w_2). The calculations are standard and follow Wells (1960) with certain improvements (Alcock, 1969) for photographic recording and Busing & Levy (1967) for four-circle geometry. These cosines are calculated relative to a standard set of orthogonal axes, on which the crystal is defined by giving the equations of the faces ($a_i x + b_i y + c_i z = d_i$). The axes have $x = \mathbf{a}^*$, y in the $\mathbf{a}^* - \mathbf{b}^*$ plane, and z perpendicular ($=\mathbf{c}$). (The origin must be within the crystal, and re-entrant angles are prohibited.) The use of orthogonal axes simplifies the equations if they are presented, as here, in explicit rather than vector form, and it may also speed up computation.

During initialization, the shape of the

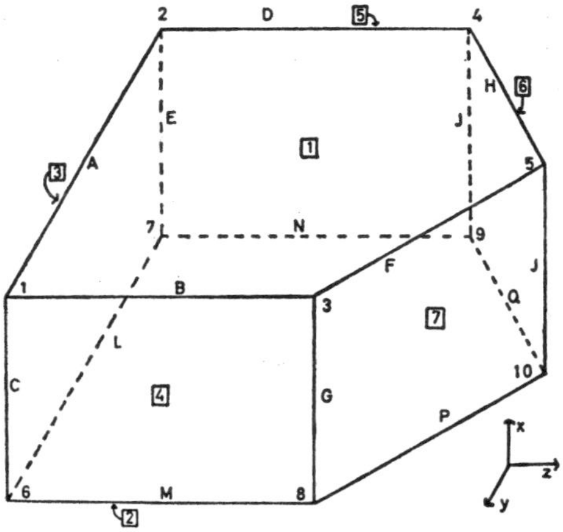

Fig. 1 *Crystal faces, vertices, and edges. Faces are numbered in squares; vertices are numbered; edges are lettered.*

crystal is further defined by finding the coordinates of vertices (X_i, Y_i, Z_i) and by identifying edges. The identification of vertices relies on the fact that no face lies between a real vertex and the origin; there must be an edge between each pair of vertices with two faces in common. All these are labelled in sequence. Fig. 1 shows the application to an illustrative crystal, and Flow Sheet 1 shows the method of calculation (cf. Coppens, 1970, §2·4). The coordinates of the vertices are listed in Table III. The results are checked against Euler's equation (Faces + Vertices − Edges = 2) to detect non-finite crystals. In order to reduce the effect of rounding errors, the whole crystal is scaled so that the maximum value of any coordinate is 1·0; μ, the absorption coefficient, is reduced in proportion. Then, two points are considered coincident if none of their coordinates differ by more than 3×10^{-5}.

The actual calculation of absorption correction by the analytical method is now performed for each reflection. The rigorous statement is given by **MT** and here it is illustrated by a particular example.

The essence of the method is the division of the crystal into polyhedra, within which the rays are entering or leaving through one face only. Within these polyhedra (Howells, polyhedra) the *loci* of constant absorption are planes. This is most easily seen in two dimensions – Fig. 2, a crystal divided into four Howells, polygons. The *locus* planes may lie in any direction, but their positions can be determined from the absorption of rays diffracted at the corners of the polyhedron.

Flow Sheet 1 – Crystal Components

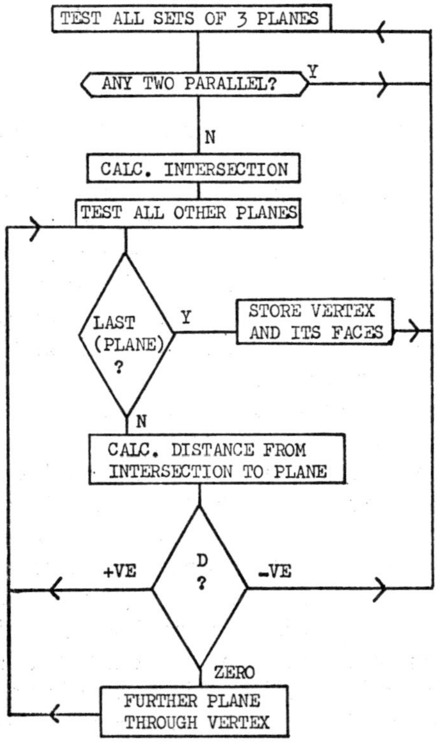

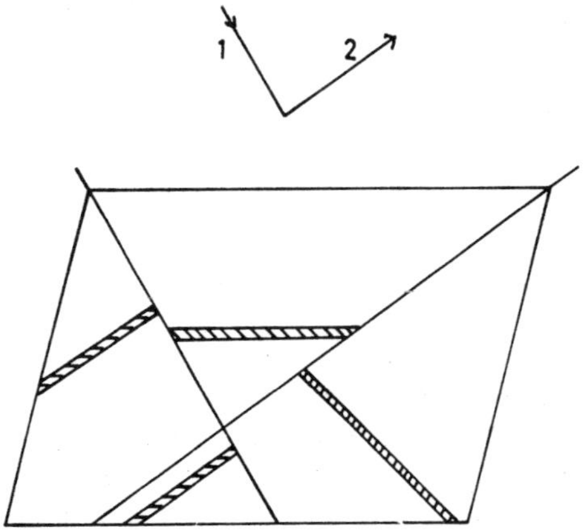

Fig. 2. A two-dimensional example of Howells' polygons, showing the loci of constant absorption.

Table I. Auxiliary points

Name	Type
S	Vertex
S 10	Where a ray (1) through a lit (1) vertex intersects the opposite face.
S 20	Similar.
S 12	Where the ray from a vertex to its S 10 is intersected by the plane of rays (2) through a lit (2) edge.
S 21	Similar.
S 01	Where a plane rays (1) through a lit (1) edge intersects an opposite edge.
S 02	Similar.
S 00	Intersection on a shaded face of planes of rays (1) and (2) through lit (1) and (2) edges.

Auxiliary points

For each reflection, a series of 'auxiliary' points on and in the crystal are found, which will be the vertices of the polyhedra. The first step is to decide whether the faces are lit or shaded by rays 1 (incident) or 2 (reversed diffracted), by calculating the angle between the face normal and the ray direction (*i.e.* whether $a_j u_i + b_j v_i + c_j w_i$ is positive or negative). Then an edge is lit, borderline, or shaded for direction 1 if both, one, or neither of its faces are lit by ray 1. Similarly a vertex is lit if all its faces are lit, or it may be borderline or shaded.

The auxiliary points are of five types, listed in Table I, and are shown in Fig. 3 for a rather symmetrical case with a cube crystal. Every time a point is found, the sum of the path lengths to the surface in each direction is computed, and multiplied by the absorption factor to give the optical path length. The distance from the point (x,y,z) to a face (j) in direction (i) is:

$$r_j = (d_j - a_j x - b_j y - c_j z) / (a_j u_i + b_j v_i + c_j w_i)\,.$$

The minimum value of this for any face is the distance to the surface [Ahmed, 1970, Eq. (9)].

$S10$ and $S20$ points are found first, produced by each *lit* vertex $(X_0 Y_0 Z_0)$ on an opposite face. (The face-distance calculation with the direction reversed gives the distance r and then the coordinates are $X = X_0 - r\,.\,u_i$, and similarly for Y and Z). A *borderline* vertex is ignored; whereas for a *shaded* one, just the path length to the surface is calculated, and the face shading it identified. In the example (Figs. 4 and 5) point 11 is an $S10$, the shadow of point 1,

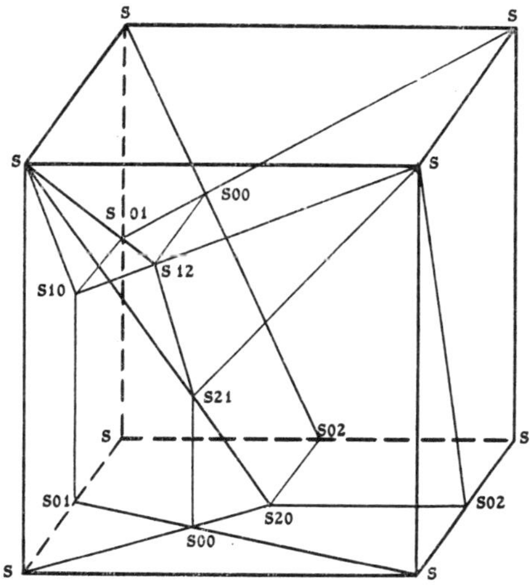

Fig. 3. The various auxiliary points illustrated on a cube crystal (from MT).

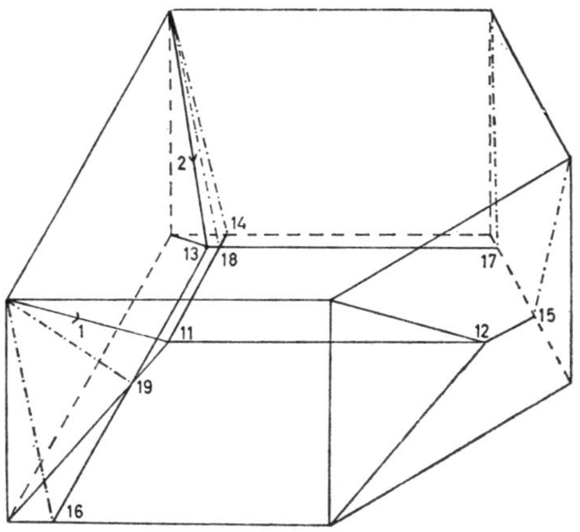

Fig. 4. Example showing the auxiliary points (11–19). *The visible crystal edges, the shadow lines of points and the shadows of edges are shown solid; the intersections of shadow planes have broken* (–·–·–) *lines. Directions* 1 *and* 2 *are indicated.*

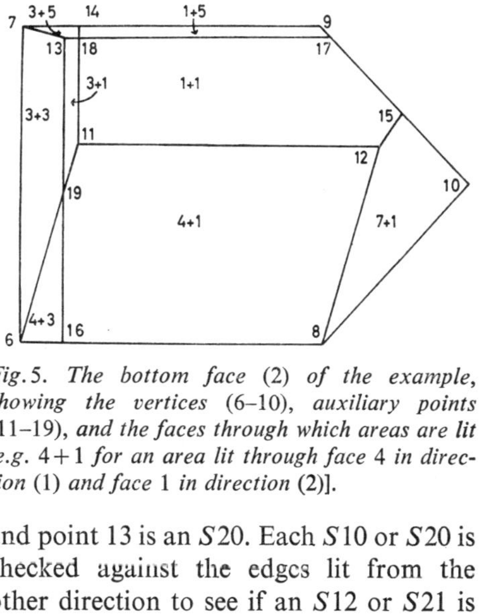

Fig. 5. The bottom face (2) *of the example, showing the vertices* (6–10), *auxiliary points* (11–19), *and the faces through which areas are lit* [*e.g.* 4+1 *for an area lit through face* 4 *in direction* (1) *and face* 1 *in direction* (2)].

and point 13 is an $S20$. Each $S10$ or $S20$ is checked against the edges lit from the other direction to see if an $S12$ or $S21$ is formed. (These are the only points not on faces and for clarity the example has been chosen to avoid them.)

For an $S12$ to be real it must lie between the vertex $S(X_A, Y_A, Z_A)$ and the $S10$; the ray through it in direction 2 must pass between the end points (X_1, Y_1, Z_1 and X_2, Y_2, Z_2) of the edge (a fractional distance α along the edge). Then the coordinates (X, Y, Z) of the point can be specified in terms of the distances to the surface, r_1 and r_2, in directions 1 and 2 by

$$X = X_A + u_1 \,.\, r_1$$
$$= [X_1 + \alpha(X_2 - X_1)] + u_2 \,.\, r_2$$

or:

$$u_1 \,.\, r_1 - u_2 \,.\, r_2 = \alpha(X_2 - X_1) - (X_A - X_1)$$

with similar equations in Y and Z. If these are solved for r_1, r_2, and α, the conditions for a real $S12$ are: $r_1 < (r_1$ for the $S10)$; $0 < \alpha < 1$. The sum of r_1 and r_2 gives the total path length.

Next, each lit edge is considered to see if it produces a point of type $S01$ or $S02$ – where its shadow cuts a shaded edge. In the example, point 14 ($S01$) is produced by the shadow of edge A cutting edge N. The

Flow Sheet 2 – S 00 Points

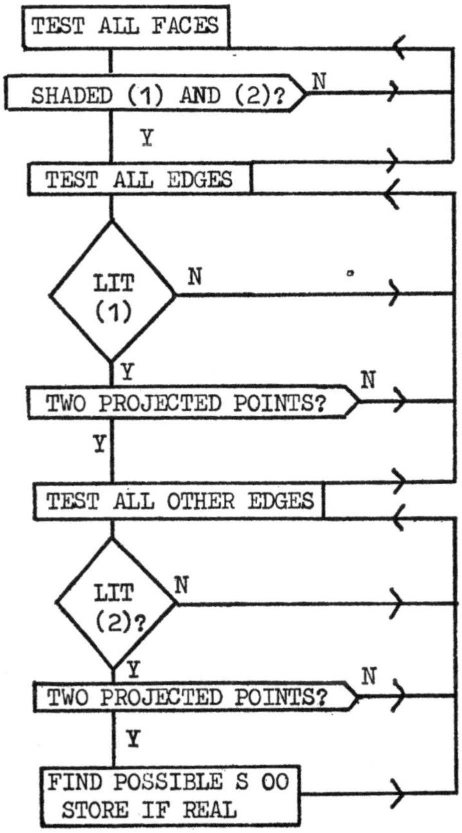

conditions for reality are that the point lies between the vertices (1 and 2) of the shaded edge and that the ray through it passes between the vertices (3 and 4) of the lit edge. Then its coordinates can be specified in terms of its fractional distance along the shaded edge (α_1), of the ray along the lit edge (α_2), and of the distance, r, from the point to the surface by:

$$X = X_1 + \alpha_1(X_2 - X_1)$$
$$= [X_3 + \alpha_2(X_4 - X_3)] + r \,.\, u_1$$

or:

$$\alpha_1(X_1 - X_2) + \alpha_2(X_4 - X_3) - r \,.\, u_1 = X_3 - X_4$$

with similar equations in Y and Z. If these are solved for α_1, α_2, and r, the conditions for a real $S01$ are: $0 < \alpha_1, \alpha_2 < 1$. Finally, as shown in Flow Sheet 2, points $S00$ are found. They occur only on faces shaded from both directions, and are produced by one edge lit from direction 1 and one from direction 2. Point 18 is an $S00$, from the intersection of the shadows of edges D and A. The conditions for reality are that each edge must have two projected points on the face (which may be vertices, $Sn0$ or $S0n$ points), and that the lines joining the points (1-2 and 3-4) must cross between them (at fractional distances α_1 and α_2). Then the coordinates of the point are

$$X = X_1 + \alpha_1(X_2 - X_1) = X_3 + \alpha_2(X_4 - X_3)$$

or:

$$\alpha_1 \Delta_1 + \alpha_2 \Delta_2 = \Delta_3$$

with similar equations in Y and Z. These can be solved for α_1 and α_2, and for reality $0 < \alpha_1$, $\alpha_2 < 1$. One of the three equations is redundant because the point must lie in the face concerned and if the face is perpendicular to an axis this equation is trivial. To avoid difficulty, the equations are solved by least squares, being converted to the form:

$$\alpha_1 \Sigma(\Delta_1)^2 + \alpha_2 \Sigma(\Delta_1 \Delta_2) = \Sigma(\Delta_1 \Delta_3)$$
$$\alpha_1 \Sigma(\Delta_1 \Delta_2) + \alpha_2 \Sigma(\Delta_2)^2 = \Sigma(\Delta_1 \Delta_3) \,.$$

Division into polyhedra

The crystal can now be dissected into Howells, polyhedra. This can be seen on Fig. 5, a drawing of face 2, where the regions are numbered according to the faces they are lit by. Fig. 6 shows the polyhedra individually. The dissection is done by

Table II. Selection of polyhedron vertices

Face *A* [lit (1)] with Face *B* [lit (2)]

Point	Condition (1)	Condition (2)
S	In or shaded by *A*	In or shaded by *B*
S 10	Generated by a vertex in *A*	In or shaded by *B*
S 12	From an *S* 10 generated by a vertex in *A*	From an edge in *B*
S 01	From an edge in *A*	In or shaded by *B*
S 00	From an edge in *A*	From an edge in *B*

S 20, *S* 21, and *S* 02 are similar, with the exchange of *A* and *B* in the conditions.

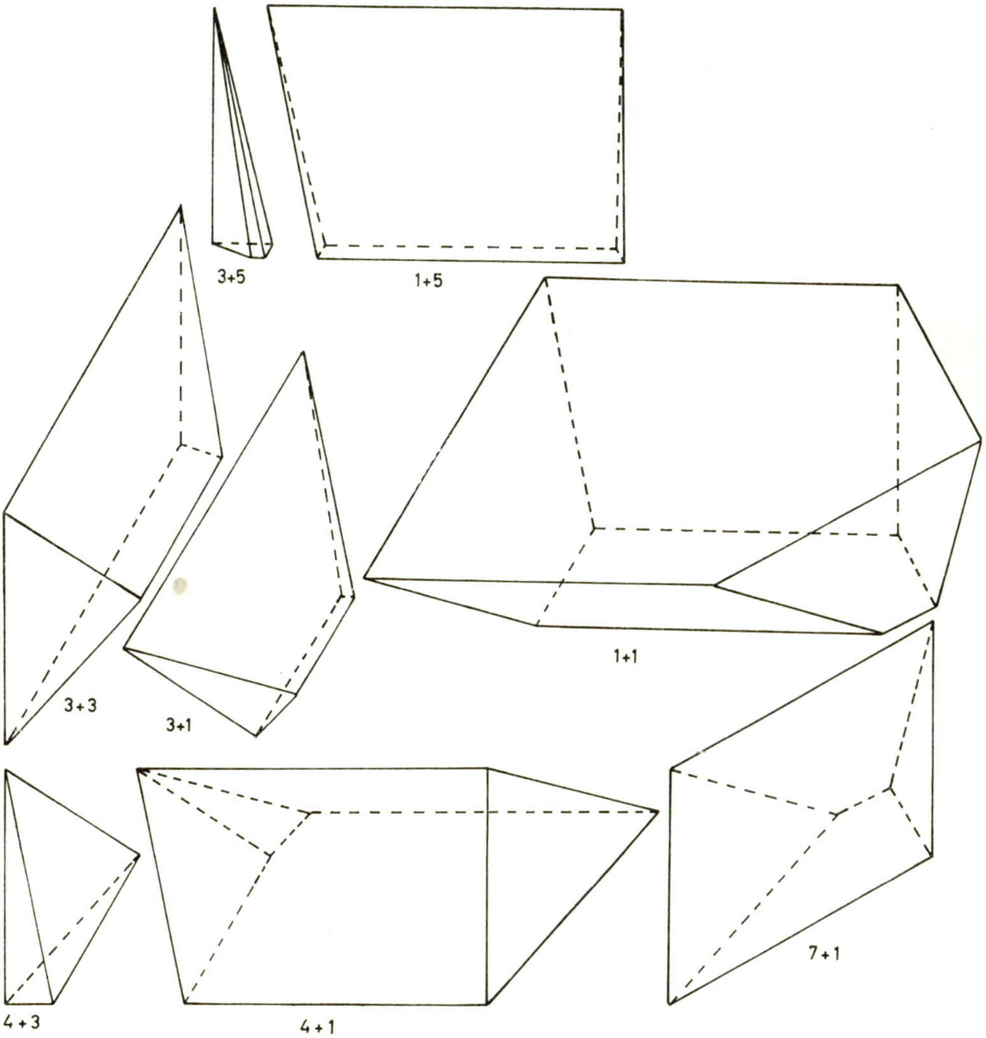

Fig. 6. The individual polyhedra of the example, with the faces through which they are lit.

selecting the points according to a series of logical decisions, as in Table II. The list of points for each polyhedron is searched to remove any that are accidentally coincident with others.

Consider for example, ray (1) lighting

Table III. Statistics of the example

A: Face planes

	a	b	c	d
1	1	0	0	10
2	−1	0	0	10
3	0	0	−1	40
4	0	1	0	15
5	0	−1	0	15
6	0	−1	1	5
7	0	1	1	5

B: Direction cosines

$u_1 = -0{\cdot}7164$	$v_1 = -0{\cdot}6705$	$w_1 = 0{\cdot}1928$
$u_2 = -0{\cdot}9798$	$v_2 = 0{\cdot}0537$	$w_2 = 0{\cdot}1928$

C: Points

The optical path lengths (the sum for both directions) are calculated with $\mu = 0{\cdot}1$

Vertices	No.	X	Y	Z	μ^* Path	
	1	2·5	3·75	−10·0	0	
	2	2·5	−3·75	−10·0	0	
	3	2·5	3·75	− 2·5	0	
	4	2·5	−3·75	−2·5	0	
	5	2·5	0·0	1·25	0	
	6	2·5	3·75	−10·0	0	
	7	−2·5	−3·75	−10·0	0	
	8	−2·5	3·75	−2·5	2·04	
	9	−2·5	−3·75	−2·5	2·79	
	10	−2·5	0·0	1·25	2·04	
S 10	11	−2·5	−0·92	−8·65	4·83	shaded by 1
	12	−2·5	−0·92	−1·15	4·83	shaded by 3
S 20	13	−2·5	−3·48	−9·01	4·08	shaded by 2
S 01	14	−2·5	−3·75	−8·65	2·79	from edge *A*, in edge *N*
	15	−2·5	−1·67	−0·42	4·83	from edge *D*, in edge *Q*
S 02	16	−2·5	3·75	−9·01	2·04	from edge *A*, in edge *M*
	17	−2·5	−3·48	−2·22	4·83	from edge *D*, in edge *Q*
S 00	18	−2·5	−3·48	−8·65	4·83	by edges *A* and *D*
	19	−2·5	0·33	−9·01	4·08	by edges *C* and *A*

D: Individual polyhedra

Face lit (1)	Face lit (2)	Vol.	Diff. intens.	Points
1	1	216·3	60·61	1 2 3 4 5 11 12 15 17 18
1	5	4·6	0·61	2 4 9 14 17 18
3	1	4·2	0·50	1 2 11 13 18 19
3	3	15·4	6·25	1 2 6 7 13 19
3	5	0·4	0·08	2 7 13 14 18
4	1	84·9	13·44	1 3 8 11 12 16 19
4	3	2·8	0·78	1 6 16 19
7	1	22·8	4·25	3 5 8 10 12 15
	totals	351·4	86·52	

Transmission = 86·52/351·4 = 0·2462

face 4 and ray (2) lighting face 1: The vertices 1 and 3 are in both faces, while vertex 8 is in face 4 and shaded by 1; 11 and 12 are $S10$'s generated by vertices in face 4 and again are shaded by face 1; 16 is an $S02$ from edge A (in face 1) and is itself in face 4; finally 19 is generated by edges of faces 4 and 1. Table III lists the points which form each polyhedron.

Summation over polyhedra

The crux of the method is in the next step. For each of the polyhedra the contribution to the total diffracted intensity is found (A_T); these are summed, and the transmission is given by ΣA_T, proportional to the actual intensity, divided by V, the crystal volume, proportional to the ideal intensity. The polyhedra are themselves divided into irregular tetrahedra (whose corners are auxiliary points) and their contributions to volume and transmission are summed. The transmission is given by **(MT)**:

$$A_t = 6V_T \exp(-g) \times \left\{ \frac{\dfrac{h(a)-h(a+b)}{b} - \dfrac{h(a+b)-h(a+b+c)}{c}}{b+c} \right\} \qquad (1)$$

where $g=g_1$, $a=g_2-g_1$, $b=g_3-g_2$, $c=g_4-g_3$, and the g_i are the optical path lengths: $\mu\times$ [distance to surface (1) + distance to surface (2)], for the corner (i), ordered so that

$$g_1<g_2<g_3<g_4\,,$$

V_T = tetrahedron volume,

$$h(x)=[1-\exp(-x)]/x\,.$$

Eq. (1) can normally be evaluated directly, but if any of a, b, c is small, then series expansions are used, ignoring cubic terms. Then, if the computer accuracy is 10^{-10}, the same loss of accuracy occurs in direct evaluation or in expansion when a, b, or c is about 3×10^{-3}.

The appropriate parts of the complete expression can be evaluated as follows:

{$\varepsilon=3\times10^{-3}$. $H(1)=A_T/[6V_T\exp(-g)]$.
The nth derivative of $h(x)$,
$h_n(x)=(-1)^n\,.\,[h(x)-(1+n\,.\,(-1)^n\,.\,h_{n-1}(x))/x]$.}

If one of a, b, $c<\varepsilon$

1a. $a<\varepsilon$ (1) $h(a)=1-a/2+a^2/6$
(2) $h(b+a)=h(b)+a\,.\,h_1(b)+a^2/2\,.\,h_2(b)$

1b. $b<\varepsilon$ (1) $h(a+b)$, as 1a(2)
(2) $[h(a)-h(a+b)]/b=-h_1(a)-b/2\,.\,h_2(a)-b^2/6\,.\,h_3(a)$

1c. $c<\varepsilon$ $[h(a+b)-h(a+b+c)]/c$, as 1b (2)

If two of a, b, $c<\varepsilon$

2a. $a>\varepsilon$ $H(1)=h_2(a)/2+(2b+c)\times h_3(a)/6+(3b^2+3bc+c^2)\,.\,h_4(a)/24$

2b. $b>\varepsilon$ (1) $h(a)$, as 1a (1)
(2) $[h(a+b)-h(a+b+c)]/c$, as 1b (2)

2c. $c>\varepsilon$ (1) $h(a+b)$, as 1a (1)
(2) $[h(a)-h(a+b)]/b=\frac{1}{2}-a/3-b/6+a^2/8+ab/8+b^2/24$

If all of a, b, $c<\varepsilon$

3. $H(1)=\frac{1}{6}-(a+b)/8+(b-c)/24+[(a+b+c)\,.\,(4a+3b)+2a^2+ab+c^2]/120\,.$

The selection of the individual tetrahedra to make up a polyhedron is fairly simple (Flow Sheet 3). Four vertices are chosen and the volume they enclose calculated. When a set has been found that is not coplanar, it is re-arranged so that the volume as calculated is negative. The list of points that make up each face is then stored. Then the remaining vertices are successively considered in conjunction with each free face. If the volume of this new tetrahedron is positive, the point lies 'outside' the free face, the tetrahedron is genuine, and its contribution to intensity is included; the previous free face and others now covered are deleted and new faces are added to the list. If the volume is negative, the new tetrahedron intersects the polyhedron and is ignored. The volume of the tetrahedron formed by a face with vertices 1,2,3 and a vertex 4 is given by

$$6V = \begin{vmatrix} X_1-X_2 & X_1-X_3 & X_1-X_4 \\ Y_1-Y_2 & Y_1-Y_3 & Y_1-Y_4 \\ Z_1-Z_2 & Z_1-Z_3 & Z_1-Z_4 \end{vmatrix}$$

Flow Sheet 3 – Tetrahedra in Polyhedron

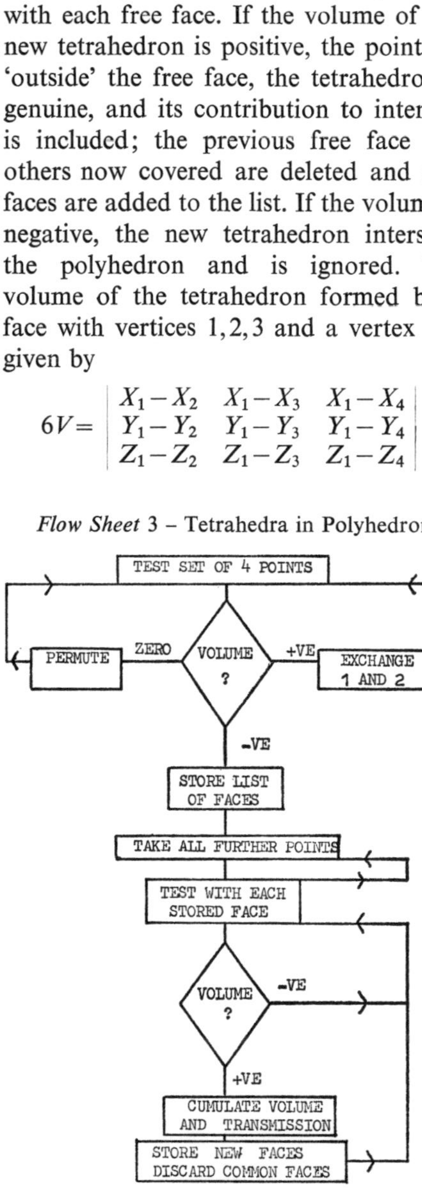

and columns 1 and 2 can be stored for each face.

If the volume is zero, the new point is coplanar with the face and no action is needed. Care must be taken here, because a coplanar point can, through rounding errors, give a small positive or negative volume, while a genuine tetrahedron may have a very small volume indeed but, if neglected, will cause a very large error in the total volume. The test limit will depend on the minimum separation of non-coincident points; and with 3×10^{-5} for this distance, a minimum volume of 1×10^{-9} has been found satisfactory.

As an overall check, the volumes of the tetrahedra are summed for comparison with the expected value. The final results for the example are shown in Table III.

Finally, in the control routine, the computed transmission factor is applied to the observed intensity. It is quite easy to calculate the coefficient $dA/d\mu$, where A is the transmission factor, during the main calculation, and a numerical differentiation, by varying the path lengths used in Eq. (1), takes only 10% more time. This coefficient is needed for secondary extinction calculations (Larson, 1970), and gives $\bar{T}$, the mean path length, by the expression:

$$\bar{T} = -A \,.\, dA/d\mu \,.$$

References

Ahmed, F. R. (1970). This volume.
Alcock, N. W. (1969). *Acta Cryst.*, **A25,** 518.
Busing, W. R. & Levy, H. A. (1967). *Acta Cryst.* **22,** 457.
Coppens, P. (1970). This volume.
De Meulenaer, J. & Tompa, H. (1965). *Acta Cryst.* **19,** 1014.
Howells, R. G. (1950). *Acta Cryst.* **3,** 366.
Larson, A. C. (1970). This volume.
Wells, M. (1960). *Acta Cryst.* **13,** 722.

Discussion

COPPENS: How do you define a tetrahedron of negative volume?

ALCOCK: This emerges in the calculation, depending on whether the vertex of the tetrahedron is on one or the other side of the face plane.

Organization of an Absorption Correction Program for Three Circle Diffractometer Data

F. R. AHMED

Biochemistry Laboratory, National Research Council of Canada, Ottawa 7, Ontario, Canada.

Mathematical expressions

Absorption by the crystal results in reducing the X-ray intensity I_0, by the transmission factor T, to the measured value I_m, so that

$$I_m = I_0 \, . \, T \tag{1}$$

As described by Busing & Levy (1957), the transmission factor through a crystal with planar faces and no re-entrant angles can be calculated from the m-point Gaussian integration formula (see Margenau & Murphy, 1943; Hartree, 1955)

$$T = \frac{1}{V} \sum_{i=1}^{m_1} \sum_{j=1}^{m_2} \sum_{k=1}^{m_3} (B-A)\,[D(x_i)-C(x_i)] \times [F(x_i,y_j)-E(x_i,y_j)] \times R_i R_j R_k \exp(-\mu r_{ijk}) \tag{2}$$

where

V	crystal volume
μ	linear absorption coefficient
r_{ijk}	path length of the primary and diffracted beams through the crystal to and from a point (x_i, y_j, z_k) within the crystal
A, B	lower and upper limits of the crystal in the x direction
C, $D(x_i)$	lower and upper limits in the y direction at $x = x_i$
E, $F(x_i, y_j)$	lower and upper limits in the z direction at $x = x_i$ and $y = y_j$

$$x_i = A + (B-A)u_i \quad i = 1 \ldots m_1 \tag{3}$$

$$y_j = C(x_i) + [D(x_i) - C(x_i)]u_j \quad j = 1 \ldots m_2 \tag{4}$$

$$z_k = D(x_i, y_j) - [F(x_i, y_j) - E(x_i, y_j)]u_k \quad k = 1 \ldots m_3 \tag{5}$$

R, u	Gaussian integration constants as calculated by Lowan, Davids & Levenson (1942).

A program written by Ahmed & Singh (1967) in FORTRAN IV for the IBM/360 system calculates the absorption corrections for data measured on three-circle diffractometers and applies them to the intensities or structure amplitudes. It accepts equal or different values of m_1, m_2, m_3 within the limits $4 \leq m \leq 16$, and executes all the necessary operations in a single run on the computer. In order to select suitable subdivisions that requier reasonable computer time, a sample calculation can be performed for every Nth reflexion in a preliminary run, employing different trial values of m to examine the convergence of the calculations and the computer time as m is varied. The program applies the correction to $|F_o|$ or $|F_o|^2$.

The crystal is described by the coordinates of its corners relative to any suitable Cartesian system x', y', z' having its origin within the crystal. Each face and edge is defined by stating on the input cards the serial numbers of 3 or 2 corners as appropriate. This eliminates from the program any search for faces or edges. Each of the x', y', z' axes is defined by the angles $\alpha_i, \beta_i, \gamma_i$ that it makes with the x, y, z axes respectively; transformation to the x, y, z system is obtained from equations of the form

$$x = x' \cos \alpha_1 + y' \cos \alpha_2 + z' \cos \alpha_3 \quad (6)$$

and similarly for y and z.

The Gaussian integrations are computed in the x, y, z system, which is defined as a right-handed Cartesian system fixed to the φ circle and moves with it as the angles χ, φ, 2θ are changed. When $\chi = \varphi = 2\theta = 0$, the x and y axes are assumed to be in the equatorial (horizontal) plane with x in the bisecting plane at $\varphi = 0°$, y on the line between source and detector, and z pointing vertically upwards. In this system, when a reciprocal lattice point *hkl* with cylindrical coordinates (φ, χ, $2 \sin \theta$ in r.l.u.) is in a diffracting position, the reverse of the primary beam (*OP* in Fig. 1) has direction cosines

$$\begin{aligned} l_p &= \sin \theta \cos \chi \cos \varphi - \cos \theta \sin \varphi \\ m_p &= \sin \theta \cos \chi \sin \varphi + \cos \theta \cos \varphi \quad (7) \\ n_p &= \sin \theta \sin \chi \end{aligned}$$

and the diffracted beam (*OD* in Fig. 1) has direction cosines

$$\begin{aligned} l_d &= \sin \theta \cos \chi \cos \varphi + \cos \theta \sin \varphi \\ m_d &= \sin \theta \cos \chi \sin \varphi - \cos \theta \cos \varphi \quad (8) \\ n_d &= \sin \theta \sin \chi \end{aligned}$$

or vice versa, depending on whether the diffractometer is a General Electric XRD 5 or a Picker, respectively.

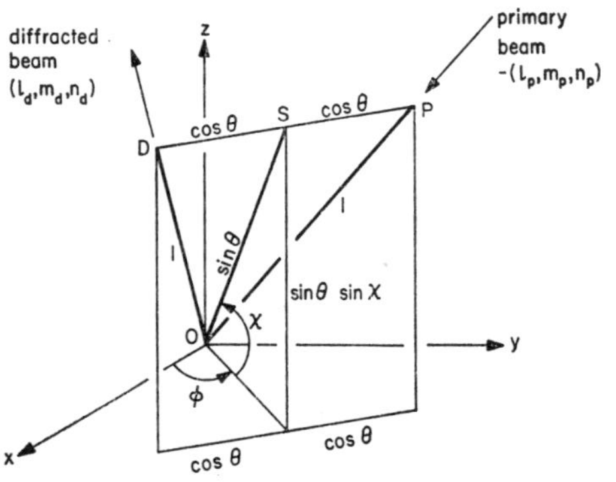

Fig. 1. Geometry of diffraction for a general reflexion. OS is half the scattering vector.

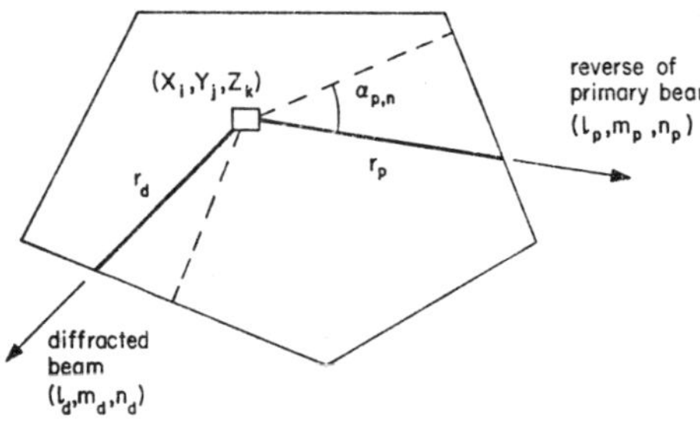

Fig. 2. Path lengths through a crystal.

Assuming the crystal faces have the equations $a_n x + b_n y + c_n z + d_n = 0$, then the path length of the primary beam through the crystal to a point (x_i, y_j, z_k) inside it is derived by calculating the expression

$$r_{p,n} = \frac{-a_n x_i - b_n y_j - c_n z_k - d_n}{a_n l_p + b_n m_p + c_n n_p} \quad (9)$$

for all faces which have the denominator in Eq. (9) as -ve, and selecting the smallest of these values. As shown in Fig. 2, the intersection of the primary beam with face (n) lies towards the source from the point (x_i, y_j, z_k) if and only if $\alpha_{p,n}$ is $< 90°$, *i.e.* if the denominator of the $r_{p,n}$ expression is -ve. Similarly, the path length of the diffracted beam is derived by calculating the expression

$$r_{d,n} = \frac{-a_n x_i - b_n y_j - c_n z_k - d_n}{a_n l_d + b_n m_d + c_n n_d} \quad (10)$$

for all faces which have a -ve denominator for Eq. (10), and selecting the smallest of these values (d_n is assumed to be +ve).

In calculating the y limits $C(x_i)$, $D(x_i)$, the program makes use of the fact that the plane $x = x_i$ does not intersect an edge PQ unless x_i is within the x values of the two corners P and Q. The y coordinate of the point of intersection is evaluated from the expression

$$y_i = \frac{y_Q - y_P}{x_Q - x_P}(x_i - x_P) + y_P \quad (11)$$

Summary of procedure

The program performs the calculations in steps as follows:

A. Preliminary operations

Read the 3 angles which each of the x', y', z' axes makes with x, y, z.

Read serial number and coordinates (x', y', z') of the crystal corners, transform them to the x, y, z system, and set $A = x_{\min}$ and $B = x_{\max}$.

Read the serial numbers of the corners defining the crystal faces and edges, and derive the coefficients of the crystal faces.

Read the Gaussian integration constants (R, u) for the selected values of m_1, m_2, m_3.

Calculate the coordinates (x_i, y_j, z_k) of the Gaussian points from Eqs. (3), (4), (5) for m_1 values of x_i, m_1m_2 values of y_j, and $m_1m_2m_3$ values of z_k.

Calculate the crystal volume from the expression

$$V = (B-A) \sum_i \sum_j \sum_k [D(x_i) - C(x_i)] \times [F(x_i, y_j) - E(x_i, y_j)] R_i R_j R_k \quad (6)$$

Read a, b, c, α, β, γ, λ, and calculate a^*, b^*, c^*, cos, and sin $(\alpha^*, \beta^*, \gamma^*)$.

Derive the constants for calculating χ, φ, 2θ of the lattice points.

Copy the lists preceding the planes' list from the input tape onto the output tape.

B. Operations for each reflexion

Read the record of one reflexion from the input tape.

Calculate cos and sin $(\chi, \varphi, \theta)_{hkl}$.

Calculate the direction cosines of the reverse primary and the diffracted beams from Eqs. (7), (8).

Calculate from Eq. (2) the transmission factor, using three nested loops, summing over k in the innermost loop, over j in the second loop, and over i in the outer loop.

Apply the absorption correction to the measured intensities or structure amplitudes.

Store the modified record of the reflexion on the output tape.

Desirable features

Added to the features already described by the author, the following have been proposed by N.W. Alcock for consideration in a general absorption program:

to handle the different data recording techniques and geometries. It should be remembered that in all the camera methods except the equi-inclination Weissenberg, a single reflexion (hkl) has two absorption factors.

to modify the statistical weights by the absorption factor.

to be able to skip to a certain reflexion on a magnetic or paper tape, or to change the conditions after a certain reflexion.

to calculate the derivative $dT/d\mu$, for use in secondary extinction corrections.

to calculate for a crystal setting about a non-standard axis, *e.g.* [110].

References

Ahmed, F. R. & Singh, P. (1967). Absorption Correction Program for the 3-Circle Goniostat Geometry. Unpublished.

Busing, W. R. & Levy, H. A. (1957). *Acta Cryst.* **10**, 180.

Hartree, D. R. (1955). *Numerical Analysis.* Oxford: Clarendon Press.

Lowan, A. N., Davids, N. & Levenson, A. (1942). *Bull. Amer. Math. Soc.* **48**, 739.

Margenau, H. & Murphy, G. M. (1943). *The Mathematics of Physics and Chemistry.* New York: Van Nostrand.

Discussion

BUSING: Your procedure of calculating the equation of a crystal face from the measured coordinates of three corners would leave the other corners off that plane. Also, if you fit the plane to more than three corners by least-squares, the corners will no longer be consistent with the planes. You should start with the equations of the faces and then solve for all the corners.

AHMED: The errors in my approach are small if the three corners defining each face are selected as far away from each other as possible. Except in cases of highly absorbing crystals, the errors should not be too serious. This approach eliminates the problem of finding the coefficients of the faces experimentally and simplifies the programming.

ZELWER: How do you measure the crystal cor ners?

AHMED: With the aid of a Vickers image splitting eye piece.

BUSING: We used to measure crystals in this way. We now do it on the diffractometer by orienting the crystal until a face is in the reflecting position (now face is parallel to telescope) and measure the distance of the face from the instrument centre, then use this in the program with χ and φ to calculate the plane.

AHMED: I think that is well worth trying.

Absorption Correction for Diffraction Measurements in Complex Cases

A. SANTORO

Center for Radiation Research, National Bureau of Standards, Washington, D.C., U.S.A.

When other objects in addition to the crystal itself take part in the absorption, the transmission factor for a reflection *hkl* is given by:

$$T(hkl) = (1/V) \int_V \exp[-\Sigma_i \mu_i l_i] dV \qquad (1)$$

In the above expression the integration is over the volume V of the crystal and $\exp[-\Sigma_i \mu_i l_i]$ is the transmission factor associated with a volume element dV of the crystal. The μ_i's and l_i's are, respectively, the linear absorption coefficients and the paths of incident and diffracted beams in the various materials traversed by the beams.

The numerical evaluation of integral (1) requires the determination of the l_i's for any desired sampling point of the crystal. The problem then is to find the direction cosines of incident and diffracted beams in the reference systems in which the shapes of the objects involved in the absorption are described.

1. Direction cosines of incident and diffracted beams

In order to calculate the direction cosines of interest it is convenient to define the following right-handed reference systems:

$(O\Xi_i)$ $(i=1,2,3)$: reference system attached to the reciprocal lattice of the crystal and formed by three non-coplanar reciprocal vectors $\boldsymbol{r}_i^*$.

(OX_i): Cartesian reference system attached in some way to the reciprocal lattice.

(OK_i): Cartesian reference system attached to the crystal in which the crystal shape is described.

(OK_i'): Cartesian reference system attached to some object taking part in the absorption in which the shape of the object is described.

(OY_i) and (OZ_i): Cartesian reference systems attached to the diffraction apparatus and to the laboratory, respectively, and defined so that they are coincident when the diffractometer settings are zero.

We may define a set of transformation matrices relating these reference systems as follows: let $\mathbf{v}$ be some physical vector whose components in the respective systems are the elements of column vectors $\boldsymbol{\xi}$, $\mathbf{x}$, $\mathbf{k}$, $\mathbf{k}'$, $\mathbf{y}$, and $\mathbf{z}$. These column vectors are then related by:

$$\mathbf{x} = \mathbf{R}\boldsymbol{\xi} \qquad (2)$$

$$\mathbf{x} = \mathbf{U}\mathbf{k} \qquad (3)$$

$$\mathbf{x} = \mathbf{U}'\mathbf{k}' \qquad (4)$$

$$\mathbf{y} = \mathbf{M}\mathbf{x} \qquad (5)$$

$$\mathbf{z} = \mathbf{D}\mathbf{y} \qquad (6)$$

It can be shown that all the matrices, except $\mathbf{R}$, are orthogonal. In the next section we will describe in detail the meaning of the transformation matrices and give methods for their evaluation.

The incident beam is a vector directed from the center of the reflection sphere* of coordinates z_{ic} to the origin, *i.e.* a vector described by $-\mathbf{z}_c$ in the system (OZ_i). On the other hand, the diffracted beam is a

* Here the radius of the reflection sphere is taken equal to unity.

vector directed from the center of the reflection sphere to the point of contact between the diffracting reciprocal lattice point and the reflection sphere. If z_{im} are the coordinates of the point of contact, this vector is described by:

$$\boldsymbol{\varepsilon}=\mathbf{z}_m-\mathbf{z}_c \qquad (7)$$

The direction cosines of incident and diffracted beams in any other reference system are simply given by the column vectors describing these two vectors in that reference system. For example, the direction cosines of the two beams in the system (OK_i) are given by:

$$-\boldsymbol{\gamma}=\tilde{\mathbf{U}}\tilde{\mathbf{M}}\tilde{\mathbf{D}}\mathbf{z}_c \qquad (8)$$

and

$$\boldsymbol{\delta}=\tilde{\mathbf{U}}\tilde{\mathbf{M}}\tilde{\mathbf{D}}\boldsymbol{\varepsilon} \qquad (9)$$

where $\tilde{\mathbf{U}}\tilde{\mathbf{M}}\tilde{\mathbf{D}}$ is the transpose of $\mathbf{DMU}$.

2. Evaluation of the transformation matrices

The vectors $\boldsymbol{r}_i^*$ defining the system $(O\Xi_i)$ can be chosen arbitrarily. However, for describing the precession and Weissenberg methods it is convenient to choose $\boldsymbol{r}_1^*$ and $\boldsymbol{r}_2^*$ so that

$$\boldsymbol{r}_1^*\times\boldsymbol{r}_2^*=\boldsymbol{V} \qquad (10)$$

where $\boldsymbol{V}$ is a vector pointing in the direction of the rotating or precessing zone axis.

The expression for matrix $\mathbf{R}$ depends on how the system (OX_i) is attached to the reciprocal lattice. In the general case we have:

$$\boldsymbol{r}_i^*=\Sigma_j R_{ji}\,(\boldsymbol{OX}_j) \qquad (11)$$

and, as (OX_i) is a Cartesian system:

$$(\boldsymbol{OX}_i)\,.\,(\boldsymbol{OX}_j)=\delta \qquad (12)$$

where $\delta=0$ if $i\neq j$ and $\delta=1$ if $i=j$.

From Eqs. (11) and (12) we obtain

$$R_{ij}=r_j^*\cos\angle r_j^*X_i\dagger$$

† $\angle r_j^*X_i$ is the angle between the vectors $\mathbf{r}_j^*$ and $\mathbf{OX}_i$.

It is possible to show (Santoro & Zocchi, 1964) that if OX_1 is taken coincident with $\boldsymbol{r}_1^*$ and OX_3 coincident with $\boldsymbol{V}$, the expression for matrix $\mathbf{R}$ becomes:

$$\mathbf{R}=\begin{bmatrix} r_1^* & r_2^*\cos\angle r_1^*r_2^* & r_3^*\cos\angle r_1^*r_3^* \\ 0 & r_2^*\sin\angle r_1^*r_2^* & -r_3^*\cos\angle r_2r_3\sin\angle r_1^*r_3^* \\ 0 & 0 & r_3^*\sin\angle r_2r_3\sin\angle r_1^*r_3^* \end{bmatrix} \qquad (14)$$

where the $\boldsymbol{r}_i$ are the direct vectors corresponding to the $\boldsymbol{r}_i^*$.

The relationship between the Cartesian system (OX_i) and (OY_i) depends on how the crystal has been attached and on the settings of the arcs of the goniometer head. It is therefore convenient to define $\mathbf{M}$ as the matrix relating the two systems after the crystal has been oriented as required by the particular diffraction technique. We shall see later that this definition permits a simple evaluation of matrix $\mathbf{M}$.

Matrix $\mathbf{U}$ relates the reference systems (OX_i) and (OK_i) and is defined by transformation (3). This matrix can be evaluated with a method which has been found useful in high-pressure diffraction work. Here we will describe the method briefly. Let us define as a standard setting of the goniometer head that setting for which the axes of the rotations on the arcs are coincident with the axes of the reference system (OY_i). When the goniometer head is in its standard setting, we have:

$$\mathbf{x}=\mathbf{L}_x\mathbf{y}'$$
$$\mathbf{k}=\mathbf{L}_k\mathbf{y}' \qquad (15)$$

In the general case $\mathbf{L}\neq\tilde{\mathbf{M}}$.

In order to orient the crystal we will have to perform rotations about the axes of the arcs and about the dial to which the head is directly attached. If $\mathbf{C}_x$ is the matrix representing this operation, we have:

$$\mathbf{L}_x=\tilde{\mathbf{M}}\mathbf{C}_x \qquad (16)$$

From Eqs. (15) and (16) we obtain:

$$\mathbf{x}=\mathbf{L}_x\tilde{\mathbf{L}}_k\mathbf{K}=\tilde{\mathbf{M}}\mathbf{C}_x\tilde{\mathbf{L}}_k\mathbf{k} \qquad (17)$$

i.e.

$$\mathbf{U}=\tilde{\mathbf{M}}\mathbf{C}_x\tilde{\mathbf{L}}_k \qquad (18)$$

Matrices $\mathbf{C}_x$ and $\mathbf{L}_k$ are evaluated from the known angles of the rotations performed for orienting the crystal and for bringing (OK_i) into coincidence with (OY_i). With a similar procedure we obtain:

$$\mathbf{U}'=\tilde{\mathbf{M}}\mathbf{C}_x\tilde{\mathbf{L}}'_k \qquad (19)$$

Matrix $\mathbf{D}$ relates the systems (OY_i) and (OZ_i). By definition, when the settings of the diffraction apparatus are zero, $\mathbf{D}=\mathbf{I}$, where $\mathbf{I}$ is the identity.

The system (OY_i) is rotated with respect to the system (OZ_i) in order to bring the reciprocal lattice points into reflecting position. For this reason $\mathbf{D}$ is called the 'diffraction' matrix and its expression depends on the particular diffraction technique.

2·1. *The precession method*

In the precession method the system (OY_i) is chosen so that OY_3 is coincident with the precession axis and OY_1 coincident with the axis of the dial and pointing from the dial to the crystal. The system (OZ_i) is chosen coincident with (OY_i) when the precession angle $\bar{\mu}=0°$. This method requires the crystal to be oriented with a zone axis coincident with the precession axis. If $\mathbf{R}$ has the expression given by (14), this means that OX_3 and OY_3 have to be coincident and matrix $\mathbf{M}$ has the form

$$\mathbf{M}=\begin{bmatrix}\cos\sigma & -\sin\sigma & 0\\ \sin\sigma & \cos\sigma & 0\\ 0 & 0 & 1\end{bmatrix} \qquad (20)$$

where σ can be easily measured on the photographic film.

If the reference system (OZ_i) is defined as previously indicated and if Mark II suspensions are used (Buerger, 1964), the precession motion can be described by means of a rotation β about OZ_1 followed by a rotation α about OZ_2 (Santoro, Weir, Block & Piermarini, 1968). Therefore matrix $\mathbf{D}$ has the form:

$$\mathbf{D}_p=\begin{bmatrix}\cos\alpha & -\sin\alpha\sin\beta & -\sin\alpha\cos\beta\\ 0 & \cos\beta & -\sin\beta\\ \sin\alpha & \cos\alpha\sin\beta & \cos\alpha\cos\beta\end{bmatrix} \qquad (21)$$

where the angles α and β are related to the precession angle $\bar{\mu}$ and to the angle ω of the rotation provided by the motor by the equations (Waser, 1951):

$$\begin{aligned}\sin\alpha\cos\beta&=\sin\bar{\mu}\cos\omega\\ \cos\alpha\cos\beta&=\cos\bar{\mu}\\ \sin\beta&=\sin\bar{\mu}\sin\omega\end{aligned} \qquad (22)$$

The angle ω for which a reciprocal lattice point is in reflecting position can be calculated with the following procedure. If lattice parameters and orientation of the crystal are known, we can calculate

$$\mathbf{y}=\mathbf{MR}\boldsymbol{\xi}$$

for any given reciprocal lattice point $\boldsymbol{\xi}$. From Eqs. (21) and (22) it can be shown that the point is in reflecting position when:

$$\tan\omega=[\tan(\tau\pm\eta)]/\cos\bar{\mu} \qquad (23)$$

where

$$\tan\tau=y_2/y_1$$

and

$$\cos\eta=[\sin^2\bar{\mu}+(y_1^2+y_2^2)-\sin^2\bar{\nu}]/[2(y_1^2+y_2^2)^{1/2}\sin\bar{\mu}] \qquad (24)$$

with $\cos\bar{\nu}=\cos\bar{\mu}-y_3$.

2·2. *The Weissenberg method*

In the Weissenberg method the system (OY_i) is chosen with OY_3 coincident with the axis of rotation of the crystal (rotation ω) and pointing from the dial to the crystal. The axis OY_2 is chosen so that it is coincident with the primary beam when the camera settings are zero. The rotation μ (Buerger, 1942), therefore, takes place about OY_1.

The crystal has to be oriented with a

zone axis coincident with the axis of the rotation ω. If $\mathbf{R}$ has the form given by (14), with the proper choice of the origin for the rotation ω we have $\mathbf{M}=\mathbf{I}$.

The Weissenberg motion is described by a rotation ω about OY_3 followed by a rotation μ about OY_1, *i.e.*

$$\mathbf{D}_W=\begin{bmatrix} \cos\omega & -\sin\omega & 0 \\ \cos\mu\sin\omega & \cos\mu\cos\omega & \sin\mu \\ -\sin\mu\sin\omega & -\sin\mu\cos\omega & \cos\mu \end{bmatrix} \tag{25}$$

The rotation μ depends on the Weissenberg method and on the geometry of the reciprocal lattice. It can be shown (Santoro & Zocchi, 1966) that a convenient expression for μ is:

$$\sin\mu=R_{33}(n/2-\xi_3) \tag{26}$$

where ξ_3 is the index of the level under study, R_{33} is the period along the rotation axis and it is given by matrix (14), and n is a parameter which defines the Weissenberg method. The values of n for the various methods are:

$n=2\xi_3$: Normal-beam

$n=\xi_3$: Equi-inclination

$n=0$: Flat-cone

n arbitrary: General case

A reciprocal lattice point of coordinates y_i is in reflecting position when

$$2[(y_1\sin\omega+\cos\omega)\times\cos\mu+y_3\sin\mu]+d^{*2}=0 \tag{27}$$

where

$$d^{*2}=y_1^2+y_2^2+y_3^2$$

From (27) we obtain:

$$\tan\frac{\omega}{2}=\frac{-b\pm(b^2-4ac)^{1/2}}{2a} \tag{28}$$

where:

$$a=2(y_3\sin\mu-y_2\cos\mu)+d^{*2}$$
$$b=4y_1\cos\mu$$
$$c=2(y_3\sin\mu+y_2\cos\mu)+d^{*2}$$

In general, a reciprocal point reflects twice. Of the two coordinates of each spot on the photographic film one, ω, is given by (28), and the other, called usually Υ, is given by:

$$\tan\Upsilon=\frac{y_1\cos\omega-y_2\sin\omega}{y_1\sin\omega+y_2\cos\omega+\cos\mu} \tag{29}$$

2·3. 3- *and* 4-*circle diffractometers*

In this method OZ_3 is taken coincident with the axis of the θ circle and directed from the plane of the detector circle to the crystal, OZ_2 is taken along the direction defined by the primary beam, and OZ_1 is directed so that the z_1-component of the scattering vector is positive. The system (OY_i) is attached to the shaft of the Φ circle and it is coincident with (OZ_i) when: (I) the counter is in position to pick-up the primary beam; (II) the plane of the χ circle is perpendicular to the primary beam; and (III) the axis of the Φ circle is coincident with OZ_3.

The operation performed by the diffractometer is a rotation φ about OZ_3, a rotation χ about OZ_2, a rotation ω about OZ_3, and a rotation θ about OZ_3. We have then:

$$\mathbf{D}=\mathbf{\Theta\Omega\chi\Phi} \tag{30}$$

A reciprocal lattice point of coordinates y_i can be brought into reflecting position in an infinite number of ways. It is possible to show (Santoro & Zocchi, 1964; Busing & Levy, 1967) that:

$$\mathbf{\Omega\chi\Phi}=\mathbf{\Psi\chi_0\Phi_0} \tag{31}$$

where $\mathbf{\Psi}$ represents a rotation (essentially arbitrary) about OZ_1, and

$$\sin\varphi_0=y_1/\varrho \qquad \cos\varphi_0=y_2/\varrho$$
$$\sin\chi_0=y_3/d^* \qquad \cos\chi_0=\varrho/d^*$$

where

$$\varrho=(y_1^2+y_2^2)^{1/2} \text{ and } d^*=(\varrho^2+y_3^2)^{1/2}$$

With this technique there are no requirements regarding the orientation of the crystal. If the lattice parameters are known, matrix **M** can be obtained from the observation of two reflections from non-parallel planes of known indices (Busing & Levy, 1967). The two reflections define two reciprocal lattice vectors $\boldsymbol{h}'$ and $\boldsymbol{h}''$ whose components $\mathbf{h}'_x$ and $\mathbf{h}''_x$ in (OX_i) and $\mathbf{h}'_y$ and $\mathbf{h}''_y$ in (OY_i) are known. It is possible to show (Busing & Levy, 1967) that matrix **M** is orthogonal if it is computed according to the expression:

$$\mathbf{M} = \mathbf{T}_y \mathbf{T}_x \qquad (32)$$

In this equation the columns of $\mathbf{T}_x$ are the components in (OX_i) of the vectors $\boldsymbol{t}_{1x}$, $\boldsymbol{t}_{2x}$, and $\boldsymbol{t}_{3x}$ of an orthogonal unit-vector triple defined so that $\boldsymbol{t}_{1x}$, is parallel to $\boldsymbol{h}'_x$, $\boldsymbol{t}_{2x}$ lies in the plane of $\boldsymbol{h}'_x$ and $\boldsymbol{h}''_x$, and $\boldsymbol{t}_{3x}$ is perpendicular to this plane. The columns of $\mathbf{T}_y$ are the components of the vectors $\boldsymbol{t}_{1y}$, $\boldsymbol{t}_{2y}$, and $\boldsymbol{t}_{3y}$ defined in the same way in the system (OY_i).

3. Transmission factors

The evaluation of integral (1) requires the knowledge of the transmission factor associated with any given sampling point located in the volume or on the surface of the crystal. The problem is then to find the intersections of incident and diffracted beams through the sampling point with the crystal and the objects taking part in the absorption.

If the crystal and the other objects have the shape of convex polyhedra, they can be described by the equations of the bounding planes obtained from the measured coordinates of the corners (Busing & Levy, 1957). A treatment for more general cases, including crystals with re-entrant angles, has been given by Wuensch & Prewitt (1965).

As an example, let us consider here a crystal having the shape of a circular right cylinder with the cylindrical surface surrounded by a highly absorbing metal gasket of thickness equal to that of the crystal (Fig. 1). This case is encountered in high-pressure diffraction studies (Weir, Block & Piermarini, 1965).

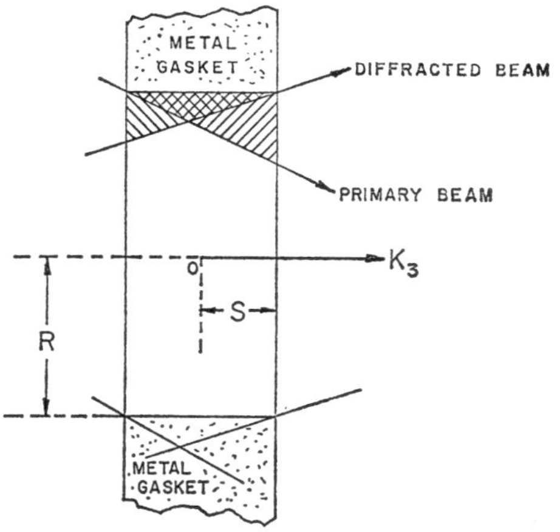

Fig. 1. Cylindrical crystal of radius R and height 2S surrounded by a metal gasket. The absorption associated with points located in the shaded region of the figure is affected by the metal gasket.

The cylinder is described in a reference system (OK_i) chosen so that OK_3 is coincident with the axis of the cylinder. Under these conditions a sampling point $k_1k_2k_3$ will be located in the volume or on the surface of the crystal if:

$$\begin{aligned} -R &\leq k_1 \leq R \\ -(R^2-k_1^2)^{1/2} &\leq k_2 \leq +(R^2-k_1^2)^{1/2} \\ -S &\leq k_3 \leq +S \end{aligned} \qquad (33)$$

where R is the radius and S the half-height of the cylinder. The equations of a straight line of direction cosines γ (or $\boldsymbol{\delta}$) through a sampling point $k_1k_2k_3$ are:

$$(k'_1-k_1)/\gamma_1 = (k'_2-k_2)/\gamma_2 = (k'_3-k_3)/\gamma_3$$

where $k'_1k'_2k'_3$ are the coordinates of a generic point. The intersections of the line with the cylindrical surface $k_{1c}^2+k_{2c}^2=R^2$ and with the planes $k_{3p}=\pm S$ can be easily calculated according to standard procedures of analytic geometry. Of the four intersections obtained in this way, only two are of interest for the problem at hand. For example, if $\gamma_3>0$, the intersections with

$k_{3c}-k_3$ and $k_{3p}-k_3$ negative are retained and the other two rejected (if $\gamma_3=0$, the test is performed on γ_2, and if also $\gamma_2=0$, on γ_1). The intersection of the incident beam with the crystal is the one with the shortest distance from the sampling point $k_1k_2k_3$. If this intersection happens to be with the cylindrical surface, the beam goes through the metal gasket and the path in the gasket is equal to the distance between the intersection with the cylindrical surface and that with one of the planes $k_{3p}=\pm S$.

A similar procedure can be applied to more complex cases if the geometry of the sample is known and if the objects can be described with sufficient accuracy.

4. Sampling points

In most cases integral (1) is evaluated by taking the weighted average of the transmission factors associated with a number of sampling points located in the irradiated volume of the crystal. The points needed for the calculation can be selected at regular intervals (Wuensch & Prewitt, 1965), at random (Alberti & Gottardi, 1966), or according to the Gauss method of numerical integration (Busing & Levy, 1957).

Calculations on spherical crystals ($\mu R=1{\cdot}0$) analyzed with the precession method have indicated that the method of spacing the points at regular intervals and weighting them equally should be used with care. In Fig. 2 are reported the results obtained by establishing in the spherical crystal a grid of points located on a cubic lattice. From the Figure it can be seen that increasing the number of sampling points does not necessarily mean obtaining better values of the transmission factors and it is difficult to predict how fine the grid should be in a practical problem. For example, if relative values of the intensities are desired, a grid with 1419 sampling points gives better results than a grid with 4159. It can be shown that the results of Fig. 2 depend on the lack of uniformity in the distribution of the points, especially those near and on the surface of the crystal.

We have tested a Monte Carlo method with the same problem. A sequence of pseudo-random numbers $U_1, U_2, \ldots, U_n$ has been generated according to the following recursive scheme (Taussky & Todd, 1956; Berger & Cooper, 1959):

$$x_0=c$$
$$x_{n+1}=5^k x_n \bmod 2^{35}$$
$$U_n=2^{-35}x_n$$

with $k=11$ and $c=16{,}743{,}785{,}845$.* The numbers U_n, in turn, have been used to generate the coordinates of the sampling points according to the expression

$$k_i=a+(b-a)U_i$$

where (a,b) is the interval of k_i. For a reflection with $\sin\theta=0{\cdot}3$ the transmission factors relative to the passages of the reciprocal lattice point through the reflection sphere obtained with 3057 points are $T_1=0{\cdot}289$ and $T_2=0{\cdot}278$ (the value reported in *International Tables for X-ray Crystallography*, Vol. II, is 0·245). This result shows that the distribution of the sampling points is not sufficiently uniform for the number of points used in the calculations.

The above considerations strongly suggest that the Gauss method should be used whenever possible. Abscissas and weights for Gaussian quadrature are available for $N\leq 100$ and for $N=125$, 150, 175, 200 (Love, 1966). If objects other than the crystal take part in the absorption and if their effect is not uniform throughout the volume of the crystal, one should expect that for a given required accuracy a finer grid of points is needed than if such objects are not present. Depending on the

* The program for generating pseudo-random numbers has been written by W. Gilbert, NBS, 1967.

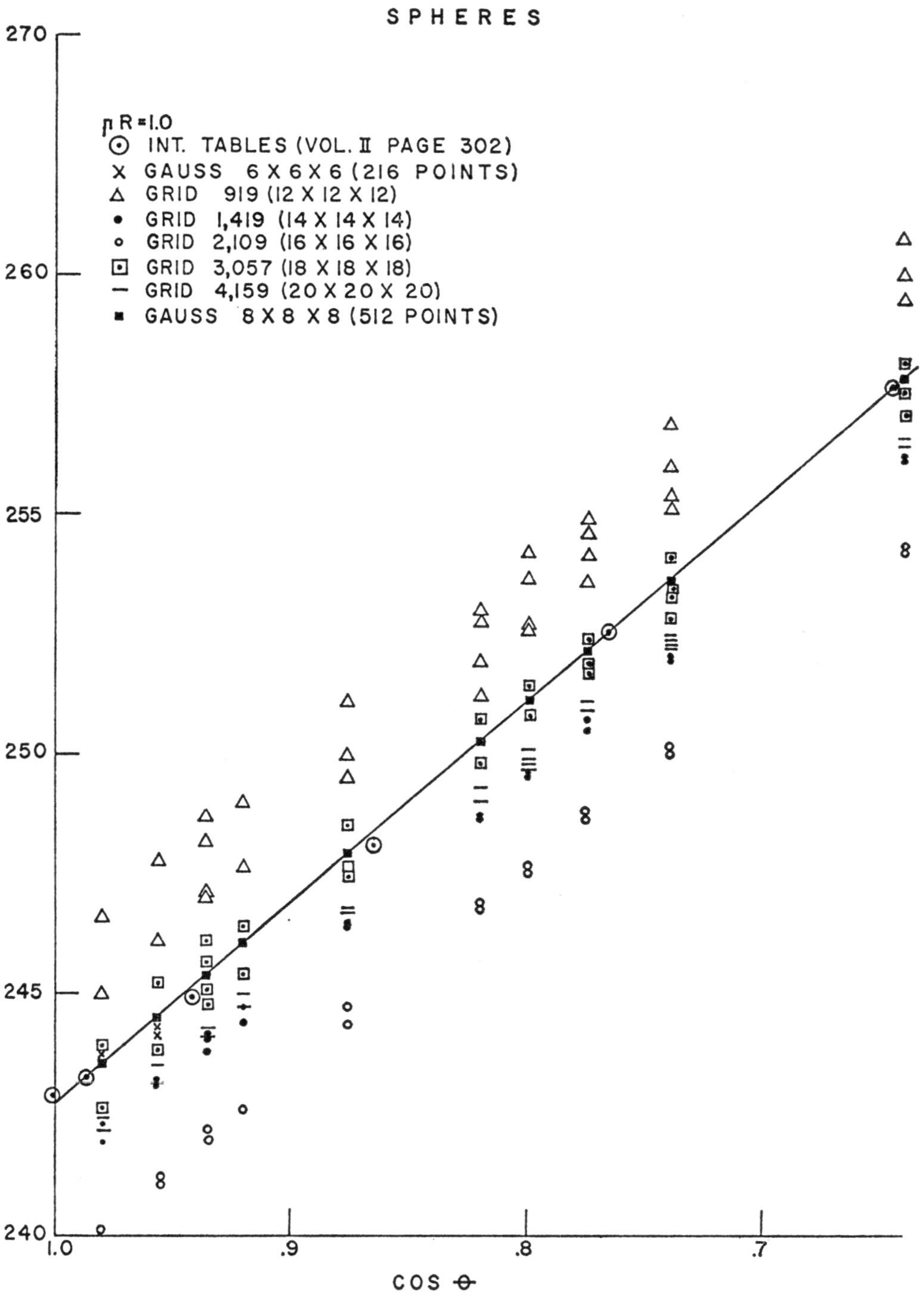

Fig. 2. Transmission factors for a sphere ($\mu R = 1{\cdot}0$) calculated with the Gauss method and with a grid of points regularly spaced.

problem at hand, one should also consider the possibility of evaluating the integral separately in different regions of the crystal.

5. Conclusions

The procedure discussed in the previous sections can be successfully applied if it is possible to describe the geometry of the sample with sufficient accuracy. This is the case of diffraction studies at high pressure.

There are cases in which the crystal morphology and the crystal mounting are not suitable for an accurate description. For example, in the study of proteins, crystals are enclosed in glass capillaries with some mother liquor trapped between the crystal and the inner wall of the capillary.

Empirical methods of absorption corrections for such samples have been proposed by Kopfman & Huber (1968) and by North *et al.* (1968). These methods do not require any knowledge of the geometrical shape of the sample, as they make use of the fact that the variation of absorption is related to the variations of the diffracted intensity from a reflection while the crystal is rotated about the scattering vector.

The validity of the empirical absorption corrections is based on the assumption that the variation of intensity in an azimuthal scanning is entirely due to absorption, that the effect of simultaneous diffraction can be disregarded, and that the calculated corrections are invariant with the diffraction angle θ. Another limitation is that such corrections can be evaluated only for those techniques and those samples for which it is possible to measure the intensity of a number of reflections while the crystal is rotated about the corresponding scattering vectors.

References

Alberti, A. & Gottardi, G. (1966). *Acta Cryst.* **21**, 833.

Berger, M. J. & Cooper, J. W. (1959). *J. Res. Nat. Bur. Standards* **63**A, 101.

Buerger, M. J. (1942). *X-Ray Crystallography.* New York: J. Wiley.

Buerger, M. J. (1964). *The Precession Method.* New York: J. Wiley.

Busing, W. R. & Levy, H. A. (1957). *Acta Cryst.* **10**, 180.

Busing, W. R. & Levy, H. A. (1967). *Acta Cryst.* **22**, 457.

Kopfman, G. & Huber, R. (1968). *Acta Cryst.* A**24**, 348.

Love, C. H. (1966). *Nat. Bur. Standards Handbook*, 98.

North, A. C. T., Phillips, D. C. & Mathews, F. S. (1968). *Acta Cryst.* A**24**, 351.

Santoro, A. & Zocchi, M. (1964). *Acta Cryst.* **17**, 597.

Santoro, A. & Zocchi, M. (1966). *Acta Cryst.* **22**, 918.

Santoro, A., Weir, C., Block, S. & Piermarini, G. (1968). *J. Appl. Cryst.* **1**, 101.

Taussky, O. & Todd, J. (1956). *Symposium on Monte Carlo Methods*, ed. H. A. Meyer. New York: J. Wiley.

Waser, J. (1951). *Rev. Sci. Instr.* **22**, 563.

Weir, C., Block, S. & Piermarini, G. J. (1965). *J. Res. Nat. Bur. Standards*, **69**C, 275.

Wuensch, B. J. & Prewitt, C. T. (1965). *Z. Krist.* **122**, 24.

The Inclusion of Secondary Extinction in Least-Squares Refinement of Crystal Structures

ALLEN C. LARSON

University of California, Los Alamos Scientific Laboratory, Los Alamos, New Mexico 87544, U.S.A.

The effects of both primary and secondary extinction on diffraction intensities have been recognized for a long time. The causes of extinction were outlined by early workers (*e.g.* Darwin, 1922) and methods of correcting observed data were suggested. The effect of primary extinction was generally assumed to be negligible for single-crystal studies and the effect of secondary extinction was usually corrected for by means of Eq. (1),

$$I_{o(\mathrm{cor})} = I_o(1 - gI_o)^{-1}\,, \qquad (1)$$

or else the most strongly affected reflections were omitted in the calculation. Several years ago at Los Alamos we decided that it would be better to include the extinction in our calculations and to refine g, the extinction parameter, along with the other structural parameters. Eq. (1), since it corrects an *observed* quantity, should not be used in a least-squares observation equation. Hence, Eq. (2),

$$I_{c(\mathrm{cor})} = I_c(1 + gI_c)^{-1}\,, \qquad (2)$$

which is derivable from Eq. (1), was inserted in our least-squares program. The observation equations were then in the form

$$\Delta F = |F_o| - F^*\,, \qquad (3)$$

where $F_c^* = k|F_c|\,(1 + gLp|F_c|^2)^{-1/2}$, k is a scale factor, Lp is the Lorentz and polarization factor, and $|F_c|$ is the usual calculated structure factor.

Shortly after this modification had been made, Zachariasen's 1963 paper on extinction appeared. Careful examination of his paper revealed that our procedure was basically correct and that the only change needed was to take account of his corrected polarization terms, p_2/p_1^2, where $p_n = 1 + \cos^{2n} 2\theta$. Eq. (4) resulted and was inserted into our program.

$$F_c^* = k|F_c|(1 + g|F_c|^2 Lp_2/p_1)^{-1/2} \qquad (4)$$

Zachariasen (1967) has given a general theory of diffraction which, through the introduction of some useful approximations, provides a much firmer basis for calculations in crystal structure refinements than have earlier formulations. His final equations are summarized below.

The integrated intensity of a diffracted beam is given by

$$\mathbf{P} = \mathbf{P}_k Y \qquad (5)$$

where

$$\mathbf{P}_k = \mathbf{I}_0 V A(\mu) Q_0\,, \qquad (6)$$

$\mathbf{I}$ is the intensity of the incident beam, V is the volume of the crystal, $A(\mu)$ is the transmission factor, and

$$Q_0 = \left|\frac{e^2FK}{mc^2V}\right|^2 \frac{\lambda^3}{\sin 2\theta} \qquad (7)$$

for X-rays and

$$Q_0 = \left|\frac{FK}{V}\right|^2 \frac{\lambda^2}{\sin 2\theta} \qquad (8)$$

for neutrons. K, the polarization factor, is 1 for neutrons and is, respectively, 1 or $\cos 2\theta$ for the normal or parallel components of polarization of the X-ray beam. He has shown that

$$Y = [1 + (2p_2/p_1)\bar{X}_0]^{-1/2} \qquad (9)$$

with

$$\bar{X}_0=\beta Q_0\{\bar{t}+[\bar{T}-\bar{t}][1+(\beta/g)^2]^{-1/2}\} \quad (10)$$

is a very good approximation to the extinction corrections for most crystals. The mean path in a single domain is $\bar{t}$, and the mean path length in the crystal, $\bar{T}$, is given by

$$\bar{T}=\frac{1}{A(\mu)}\int (T_1+T_2) \times \exp[-\mu(T_1+T_2)]d\tau\,, \quad (11)$$

where $\int d\tau$ implies integration over all diffraction paths in the crystal. The quantity β is defined as

$$\beta=2\bar{t}_\perp/3\lambda\,, \quad (12)$$

and g is related to the mosaic spread distribution. Assuming an isotropic Gaussian distribution function for the mosaic blocks,

$$g=\Delta_{1/2}^{-1}(2\pi/\log 2)^{1/2}=3{\cdot}01\ \Delta_{1/2}^{-1}\,, \quad (13)$$

where $\Delta_{1/2}$ is the half-width of the peak.

If (as is usually the case) primary extinction can be neglected ($\bar{t}\ll\bar{T}$),

$$\bar{X}_o=\beta Q_0\bar{T}[1+(\beta/g)^2]^{-1/2}\,. \quad (14)$$

For spherical domains of radius r

$$\bar{t}_\perp=\bar{t}=3r/2\,, \quad (15)$$

and hence

$$\beta=r/\lambda\,. \quad (16)$$

If $\beta\gg g$

$$\bar{X}_0=gQ_0\bar{T}\,, \quad (17)$$

and extinction is dependent only on the mosaic spread parameter, $\Delta_{1/2}$. This type of crystal has been designated by Zachariasen as Type I and its mosaic spread is much wider than the diffraction pattern of a single domain.

If $\beta\ll g$

$$\bar{X}_0=\beta Q_0\bar{T}=Q_0\bar{T}r/\lambda\,. \quad (18)$$

For this type of crystal, designated as Type II by Zachariasen, the natural width of a reflection from a single domain is much greater than its mosaic spread parameter.

For our purposes we shall remold some of these equations slightly. First we shall define the quantity r^* as

$$r^*=\beta[1+(\beta/g)^2]^{-1/2}\,. \quad (19)$$

Thus, Eq. (14) becomes

$$\bar{X}_0=r^*Q_0\bar{T}\,, \quad (20)$$

and Eq. (4), using Eq. (9) for Y, becomes

$$F_c^*=k|F_c|(1+2r^*Q_0\bar{T}p_2/p_1)^{-1/4}\,, \quad (21)$$

or with a rearrangement and regathering together of terms

$$F_c^*=k|F_c|(1+2r^*|F_c|^2\delta)^{-1/4}\,, \quad (22)$$

where

$$\delta=\left(\frac{e^2}{mc^2V}\right)^2\frac{\lambda^3}{\sin 2\theta}\frac{p_2}{p_1}\bar{T}\,.$$

The partial derivatives become

$$\frac{\partial F_c^*}{\partial k}=|F_c|(1+2r^*|F_c|^2\delta)^{-1/4}\,, \quad (23)$$

$$\frac{\partial F_c^*}{\partial P_i}=\frac{k(1+r^*|F_c|^2\delta)}{(1+2r^*|F_c|^2\delta)^{5/4}}\frac{\partial|F_c|}{\partial P_i}\,, \quad (24)$$

where the P_i are the parameters of the structure, and

$$\frac{\partial F_c^*}{\partial r^*}=-\frac{1}{2}k|F_c|^3\delta(1+2r^*|F_c|^2\delta)^{-5/4}\,. \quad (25)$$

Let us now take up the matter of deciding when it may be necessary to include extinction in our calculations. If we assume that we can measure our strongest intensities to an accuracy of about 1%, we should include extinction if $\bar{X}_0$ is greater than $\sim 0{\cdot}02$. To make an educated guess as to what conditions can cause extinction, we shall consider a reflection at $\theta=15°$ with $\lambda=1$ Å and compute the value of

$\left|\frac{F}{V}\right|^2 r^*$ needed to make $\bar{X}_0 = 0{\cdot}02$. From Eqs. (7) and (20)

$$\bar{X}_0 = r^* Q_0 \bar{T} = r^* \left|\frac{F}{V}\right|^2 \left(\frac{e^2}{mc^2}\right)^2 \frac{\lambda^3}{\sin 2\theta} \bar{T},$$

or

$$r^* = \bar{X}_0 \left|\frac{F}{V}\right|^{-2} \left(\frac{e^2}{mc^2}\right)^{-2} \frac{\sin 2\theta}{\lambda^3 \bar{T}}$$

$$= \bar{X}_0 \left|\frac{F}{V}\right|^{-2} \frac{0{\cdot}5\, \bar{T}^{-1}}{7{\cdot}9 \times 10^{-26} \times 10^{-24}}$$

$$r^* = \bar{X}_0 \left|\frac{F}{V}\right|^{-2} 6{\cdot}3 \times 10^{48} \bar{T}^{-1}.$$

With

$$\bar{X}_0 = 0{\cdot}02 \text{ and } \bar{T} = 0{\cdot}01 \text{ cm},$$

$$r^* = 12{\cdot}6 \times 10^{48} \left|\frac{F}{V}\right|^{-2}.$$

Thus the value of the quantity $\left|\frac{F}{V}\right|$ can be used as a measure of the likelihood that extinction will be a problem. Typical values for this quantity taken from a recent issue of *Acta Crystallographica* and from work performed at Los Alamos in recent years are given in Table I. If we take $\left|\frac{F}{V}\right| = 1 \times 10^{-24}$ cm^{-3}, a reasonably typical maximum value for an intermetallic compound or a heavy metal salt, we find that the intensity of the strongest reflection will be affected by extinction even if the mosaic blocks are extremely small, 13 Å, or the mosaic spreads are very large, $\Delta_{1/2} \simeq 14°$. However, if we take $\left|\frac{F}{V}\right| = 0{\cdot}1 \times 10^{-24}$ cm^{-3}, a reasonably typical maximum value for an organic compound, then the mosaic blocks may be as large as 1300 Å or the mosaic spread as small as $\Delta_{1/2} = 0{\cdot}14°$ before extinction would have a measurable effect on the intensity.

These figures show that, in general, extinction is likely to have a strong effect on the intensity data for heavy-metal and

Table I. Some typical values of $\left|\frac{F}{V}\right|$

Compound	Primitive cell contents	Largest $\left\|\frac{F}{V}\right\| \times 10^{-24}$
Trimesic acid	$24[C_6H_3(CO_2H)_3]$	0·12
	$2[C_{21}H_{31}O_4Br]$	0·10
	$2[Ni(C_6H_6N_4)_2(H_2O)_2(NO_3)_2]$	0·16
	$4[Cu(NH_3)_4SO_4 . H_2O]$	0·13
Cs Alum	$4[CsAl(SO_4)_2 . 12H_2O]$	0·16
Benzotrifuroxane	$4[C_6N_6O_6]$	0·1
	$4[CuCN . N_2H_4]$	0·26
	$4[PuGa_3]$	1·2
	$4[UMoC_2]$	1·6
	$4[UF_4]$	0·75
	$4[UO_2(OH)_2]$	0·64
	$4[(NH_3)_2CeF_6]$	0·5

Table II. Values of $\bar{T}$ as a function of μR and θ for a sphere of unit radius

$\theta/\mu R$	0	1	2	4	8
0 °	1·50	1·32	1·12	0·74	0·43
22·5	1·50	1·22	0·90	0·45	0·19
45	1·50	1·02	0·65	0·32	0·15
67·5	1·50	0·85	0·52	0·27	0·13
90	1·50	0·79	0·47	0·25	0·13

intermetallic compounds, but little, if any, effect on that for complex organic compounds. For the heavy-metal compounds, absorption reduces the mean path lengths somehwat, but with all crystals of reasonable size this will never be sufficient to warrant excluding extinction from the refinement. Values of $\bar{T}$ as a function of μR and θ for a sphere of unit radius are given in Table II.

References

Darwin, C. G. (1922). *Phil. Mag.* **43**, 800.
Larson, A. C. (1967). *Acta Cryst.* **23**, 664.
Zachariasen, W. H. (1963). *Acta Cryst.* **16**, 1139.
Zachariasen, W. H. (1967). *Acta Cryst.* **23**, 558.

TOPIC H

ORGANIZATION OF CRYSTALLOGRAPHIC COMPUTER SYSTEMS

CHAIRMAN: J. M. STEWART

Organization of Crystallographic Computer Systems

JAMES M. STEWART

Computer Science Center, University of Maryland, College Park, Maryland 20742, *U.S.A.*

The organization of programs to carry out scientific calculations is dependent upon a number of factors. Chief among these factors are the nature of the calculations, the type of computer available, and the characteristics of the data.

If the calculations to be carried out are exceedingly problem specific, then organization at the level that we are here to consider may be wasted. It is only when the nature of the problem is such that the calculations are to be made many times and the programs are to be used for many different data sets that the problem of organization is worthy of the effort it requires. Crystallographic calculations by and large are of this second kind. The International Tables bear witness to the highly stylized nature of many of the basic computations which must be performed.

There are many approaches to the problem of how to write codes for solving diffraction problems. All find their roots in the general principles laid down during the development of crystallography by Bragg, Ewald, Laue, Patterson, James, and many others. Considered in detail, the calculations necessary for the successful elucidation of the detailed structure of crystalline material are readily programmable. Many of the methods established when things were calculated 'by hand' are very useful when carrying out calculations by means of digital computers. One must, however, be careful about certain hand methods since computers are based at a very high level on simple on-off logic. This means that what is easy for a brilliant man to 'see' is very difficult for a programmer to break down into its simplest yes-no steps. Therefore, with each problem to be programmed, the confining limitations of the particular machine must be taken into account. Over the years, since the advent of computers, many refinements to the early 'hand' methods have been developed which play on the raw speed and simple logic of the computer. Many people over the years have contributed to this effort. In England, Dr. Rollett (1965) has written *Computing Methods in Crystallography*. In it are contributions by Prout, Cruickshank, Rogers, the Karles, Woolfson, Holmes, Vand, and others. In America, many codes have been prepared at various times and places by many individuals. Shoemaker-Sly, Busing & Levy, Carroll Johnson, Sparks & Trueblood, Marsh, Hamilton, Larson, Jensen, and many others have developed codes and techniques for utilizing various computers. Japanese, German, Scandinavian, Italian, Canadian, Australian, and French groups have also produced general crystallographic programs. It is our purpose here to summarize the organization of, and application of, these programs or ones like them, on various computers likely to be available to structure-solving groups.

In the case of crystallographic data, we find a stylization which suggests the possibility of organization. I believe, in fact, that the organization of the data is at the heart of any crystallographic computing system. Dr. Rollett, with others, has worked out details of lists of crystallographic data. We have made 'files' of data in XRAY63 and again more carefully in XRAY67 which mimic the FORTRAN file system. Each person who has written programs has given considerable attention to the organization of data. Busing & Levy's ORXLS for the 704 was an interesting, early example. It not only saved the data on a given magnetic tape but copies of the programs as well. It has taken many years and two 'generations' of computers to get back to this level of sophistication.

Of course, there are different styles or approaches to organization necessary, depending upon the type of computer, its language, and the peripheral equipment available. Dr. Rollett will speak about organization for medium-sized machines, with ALGOL as a programming language. Dr. Ahmed will speak on the organization for medium to small-scale machines utilizing FORTRAN as the programming language. Dr. Shiono will speak on the organization of small-scale computers in machine language and their merging to large or medium machines and FORTRAN. And I will try to give an indication of organization for large-scale machines on an 'all' FORTRAN basis.

It is, of course, true that many, many programs have now been written and checked out by various individuals and groups throughout the world. These people have used many different computers and computer systems. We cannot, in this short time, show the myriad, clever methods which have been employed to accomplish the objective of providing crystallographic computing. We can, however, outline some of them and give specific examples within our experience for organizing a system of programs.

The World List of Crystallographic Computer Programs (1966) is too long to quote here, but it serves as the best current source on crystallographic computer programs available. Currently, the ACA is setting up a library of computer programs which will serve the crystallographic community. In addition, there is, has been, and will continue to be an informal sharing of programs among crystallographers.

Considerations of computer languages for crystallography

The general factors which govern all programming for large-scale, repetitive problems are efficiency, generality, and ease of use. If the first two factors are well executed, the third may not follow automatically unless the user is considered as part of the overall process.

The ultimate in efficiency and generality may only be obtained by machine language coding. These ultimates, however, are bought too dear for most crystallographic groups to tolerate. First, with machine language coding, all the programs become machine specific; second, the cost of preparation and check-out is too high; and third, modification and updating is too demanding a task. Thus, for all but the most elegant and esoteric details, a 'higher' language is best. FORTRAN and ALGOL are the languages most used at this time. They are not free of the vagaries of the machine or system with which they are used. They are, however, somewhat standardized and do offer a variety of features which give programs written in them a degree of interchangeability and stability. These languages provide many subroutines and functions which are required to carry out large calculations. Finally, they are easily mastered and are more widely known by programmers in general. A word of caution is in order – if used 'correctly' the codes which are generated by a 'higher' language can be nearly as good (efficient) as machine language codes. This depends upon the the care and skill of the user of the 'higher' language and the skill of the person who prepared its compiler. There are ways of using compilers which are so powerful that the general meaning of the statements require that huge machine codes be generated. For crystallographic service programs, it is necessary to avoid these operations.

Notice that commenting and a glossary of the symbols used are necessary conditions for any program to be useful. Busing & Levy have, over the years, done an outstanding job in this regard. One might just as well throw away any program, no matter how good, if the author is unwilling to annotate it or give a glossary of symbols and write a careful, detailed set of directions for its use. The last step is by far the most demanding and difficult step in the preparation of any program. Many fine codes have been lost for want of documentation. In essence, organization of computer systems is tied directly to good documentation.

Organization of crystallographic data

At the heart of an organized system is the structure of the data sets and the method of presentation and filing of the data. Therefore, the first consideration should be the general nature of the experimental data and the final results to be obtained from it. In order to organize the data in XRAY67, we think in terms of records and files, each of which is of two types – the logical and the physical.

In the session on 'novel' computing techniques more details are given of the file management utilized in XRAY67 but let me give here the definitions necessary to our organization. Physical records and files are defined by the hardware in use and are of concern as far as efficiency is to be served. They are defined as a parameter of the system and a set of rules arbitrarily laid down for their structure and identification. Logical records and files are defined on the basis of grouping of data that are related 'logically'. It is on this basis that the working scientist wishes to think of his data. What we have done then is to set up a FORTRAN-compatible scheme for maintaining a logical file as a group of logical records written as physical records in a physical file.

The system of crystallographic programs is then designed to create and maintain files of this structure. Every logical file will be utilized by the system to carry out the calculations on the crystal for which it contains data. The programs are all specific to the defined file and in general are 'driven' by the structure of the logical file itself. As an example, consider an F_o Fourier calculation.

The program is 'loaded' into immediate access storage when the XRAY67 system encounters a card with FOURR in cols. 1–5. In this card are punched the scales of the map and the type of coefficients to be utilized. The input data file specified is read and all the cell parameters and other data previously stored are entered from the file.

Once started, the Fourier program is coded to read the reflections from the compound specific file. In the 'logical' record of the reflection data, there is the magnitude of F_o, the phase information, and the list of symmetry-related reflections (highly encoded). Each reflection encountered causes the Fourier program to contribute to the appropriate 'first' sums. At the end of the logical record of reflection information, the second and third sum calculation is initiated and carried out. The whole process, the number of passes required as a function of compound size and core availability, is programmed to be a function of the logical file size. We say the calculation is 'driven' by the properties of the file. These properties are the crystal-specific data.

Data and control cards

A few programs serve to create the file. These read the original data of the crystal under study and insert it into the data file.

In our card layouts for the system, we have made another set of logical distinctions about the preparation of data. Every card has a format in which the first six columns are reserved for an identification of the card type. These cards fall into three categories: control, function, and data. The control cards cause the calling of the specific calculation to be carried out (*e.g.* FC for structure factors) and contain the keys for the particular operation desired. The function cards provide for operations like file reassignment (*e.g.* FILES 7 9). The data cards contain actual crystallographic data (*e.g.* ATOM Cl 0·7 0·3 0·2). The major program of the system of programs screens out the first two kinds of cards. The latter are read by the appropriate crystallographic programs.

Because of the structuring of the deck, we have created a 'submonitor' of any FORTRAN monitor which provides 'chain' or 'overlay' as a programming feature.

At this point, it should be emphasized that the main distinction between a large-scale computer 'system' like XRAY67 and smaller computer systems is due mainly to the availability of immediate access storage to dedicate to the submonitor and to the large arrays of data. In general, efficiency is enhanced with a large core storage because multiple passes for least-squares, Fouriers, etc. are mitigated. All of our programs could be run on a submission by submission basis. Moreover, the principles of fixed data input and fixed file format can still be used to advantage with any size machine. Our use of the technique finds its roots in the University of Washington IBM 650 programs, but it is a common characteristic of all computer monitors, executives, or operating systems.

Generation of a set of crystallographic programs

The overall programming procedure using the approach taken in XRAY67 was begun by writing the submonitor in FORTRAN. This submonitor consists of a set of programs for handling the screening of the control cards and data files mentioned before. That is to say, we simply redefined our data cards and binary data files (which were initially on tape but later on drum and disc) from our IBM 650 through IBM 709–7094 X-ray system and then wrote a set of subroutines to recognize and handle these cards and files. As we began to write the codes, we tried to study the various dialects of FORTRAN that were available from computer manufactures. From this study we derived a set of rules for what we call 'PIDGIN' FORTRAN to be a sort of lingua franca among the commonly available large-scale computers. We next began creating on a task-by-task basis, the programs to do the usual crystallographic calculations. These programs were designed to make use of our defined card formats and the defined binary data file. We already had many FORTRAN II codes from XRAY63 and these were adapted wherever possible. We found, however, that we could only use these codes for their logic and had to recode them in order to keep within our self-imposed rule of 'PIDGIN' FORTRAN. Concurrent with the production of the crystallographic codes, we produced write-ups on punched cards for each of the calculations. A special link was created to make possible the printing of the write-up directly from the punched card data.

Finally, all routines and their write-ups were given catalog mnemonics and a set of test decks were generated. All this material could then be packed on a tape and the whole distributed to other interested users as a library of programs.

Other approaches

Most of the programs of interest created so far are specific crystallographic calculations of general usefulness. That is, structure factor, least-squares, Fourier, bond length and angle, or what have you. There are other possible approaches to the problem. One is to write compound-specific programs. This approach is a guaranteed loser; check-out costs are high, and each program generated is suspect because of the possibility of error. In addition, it means that all members of a crystallographic group must become very familiar with the computer and computer language of the local shop. Another approach which has been suggested is the creation of a 'crystallographic' compiler. A language like FORTRAN would be defined which would have a crystallographic vocabulary and from a series of statements, a machine code could be generated to carry out the specified set of crystallographic steps. Peter Pauling has actually generated such a program but it has, as far as I know, not received widespread use. Yet another approach is the use of interpretive routines – here one writes a series of 'sentences' directing an interpretive program to operate on a defined data set. This is actually what is generally done but at one level removed. That is to say, each step is decoded every time the operation is carried out. It is tantamount to compiling the FORTRAN statement over again every time it is to be executed. This method of using a computer for usual crystallographic calculations must be avoided. It is so wasteful that it is usually immediately apparent that something is wrong.

Summary

In summary, what we have done is to look at the nature of crystallographic data, design a file structured to account for the limitations imposed by FORTRAN and machine hardware, define a series of data input cards, write a set of FORTRAN subroutines to maintain the data file, screen the input cards, and, finally, write crystallographic calculations as FORTRAN programs to update the file and produce a refined crystal structure stored in it. All along the way, there is sufficient printed output to give a full record of the progress of the solution and refinement.

The XRAY67 system is available in a number of forms from the Computer Science Center at the University of Maryland, College Park, Maryland, U.S.A. 20742.

References

Rollett, J. S. (1965). *Computing Methods in Crystallography*. Oxford: Pergamon Press.

Shoemaker, D. P. (1966). *The World List of Crystallographic Computer Programs*. 2nd ed. Cambridge, U.S.A.: IUCr. Commission on Crystallographic Computing.

Discussion

AHMED: To run your system, what would be the minimum computer hardware requirements, and how much storage does the nucleus take?

STEWART: IBM 7094 with 5 tape units, or IBM 360/50 well equipped with auxiliary devices. The XRAY67 system requires 32K words for data +10K words for working links. There is no doubt that this is quite wasteful.

AHMED: When testing a particular routine do you have to generate the full system?

STEWART: No, in cases like this we have a routine which strips out from the nucleus all material having to do with overlay, and just has the call to the particular routine we are checking. We then generate a binary file and proceed with the checks just handling the source deck.

Crystallographic Data Structures

J. S. ROLLETT

Oxford University Computing Laboratory, Oxford OX1 3PL, England.

Need for Standardization

A typical crystal structure analysis involves a long sequence of diverse calculations. They include data reduction, intensity statistics, maps, convolutions in reciprocal space, structure factors and least-squares refinements, distances and angles, analysis of structure factor agreement and of atomic vibrations, packing energies, and so on. Many of these processes depend on the structure factor data, and others share and generate common data such as unit cell dimensions, symmetry operators, form-factors, weighting schemes, atomic parameters, and least-squares normal matrices. It is beyond the ability of a single programmer to produce routines for all these things quickly after the arrival of a new computing machine. If a large number of people work on the problem at once, there is a danger of waste of programming effort on conversions between various data formats. What is worse, users of the programs may need to prepare the same data by hand repeatedly in a variety of formats, which can mean confusion and mistakes. Extra machine operations may have to be run for format conversions. All this wastes human and machine time and slows down the progress of an analysis by introducing artificial interruptions.

When it became clear that a number of universities would install similar KDF9 computers, a working party from Glasgow (D. W. J. Cruickshank, J. G. Sime and J. G. F. Smith), Leeds (M. R. Truter and M. Wells), Oxford (J. S. Rollett), and Sydney (H. C. Freeman) set out to solve the problem by laying down a fixed set of data formats. No restrictions would be placed on any program except that it should accept its data and give its results in these forms. It was hoped that they would be convenient but standardization would be more important. It was felt that if the designs were published in a compact and convenient form before the installation of the computers, it would be easier to conform than to think up idiosyncratic designs. This was true.

The standard forms referred to data held inside the computer, input and output being seen as separate from the main computing. This has several advantages. The most machine-dependent operations are kept compact, and the transfer to a new computer is simplified. External data forms can be changed for new media or for data from unusual sources. The standard forms can contain addresses and counts to help programmers, though these would be tiresome to prepare by hand.

Difficulties

The whole scheme had to be prepared in ignorance of plans for the KDF9 operating system. This was as well, since eventually the parties to the scheme found themselves forced to use incompatible competing operating systems. As a result questions of the indexing and storage of the information during and between runs had to be left open and different implementations of the scheme solved these problems in quite different ways. They were not, for us in Oxford, a major obstacle.

It was not easy to ensure that the formats produced contained all the necessary items. In a symmetry list it is easy to overlook requirements for specifying lattice types and for allowing subsets of equivalent positions to be used when distances between special positions are needed. Complex formfactors can be forgotten. It is all too easy to restrict the variety of atomic and overall parameters unduly, for example by omitting provision for a 'dummy U iso' to be used in block diagonal treatments of scale factors. There are many items which may or may not be needed in planes lists. A map list must take account of the need for sections which are variously oriented and need not have parallel sides. Normal matrices need to be stored in original and inverted forms and at intermediate stages in the process. They are also conveniently associated with sums of squares of residuals and numbers of degrees of freedom. If various block diagonal schemes are used, the matrix list must take account of multiple blocks. Links must be provided from it to the lists (parameters, refinement instructions, planes, weights, and symmetry) that generated it. Flexibility of relationship between atomic and least-squares parameters implies an extensive refinement instructions list, allowing links from each atomic to any least-squares parameter, and for varying the sensitivity of the least-squares parameter to the atomic one, as well as leaving the latter unrefined. The solution to these problems, reached in advance of the arrival of our KDF9 machines, is indicated in the 5 May 64 edition, a sample of which is given as an appendix to these notes. One of the devices which helped to avoid rigidity of format and allow for future extensions was to refer to each type of item in a list by a key which served the same purpose as a column heading in a conventional printed table. A certain range of values for any key, scheme number, list number, and so on was reserved for publicly agreed meanings and higher values were to be free for temporary use by individuals.

Other problems became apparent when the scheme was implemented on KDF9 at Oxford. Certain lists, notably those for parameters and refinement instructions, were effectively frozen in a more restricted format than the scheme provided because the refinement programs could not deal with the full generality. The form of the list was, however, kept compatible with the scheme so that old lists would remain valid if the programs were improved later. The planes list was so large, for our problems, that we were forced to abandon the provisions of the scheme completely in our first implementation, which was in a programming system giving very limited machine space.

The address arithmetic for some lists of the 5 May 64 edition becomes complicated, because each list is a collection of sublists, each of which is of variable length. Space and time are needed in any program using the list to derive the addresses of the sublist heads. This wastes programming labour and machine time; and in our second set of programs for KDF9, based on Algol, we removed the trouble by placing a set of address pointers in a block of fixed positions at the head of the list, each giving a sublist start address. The address column of the list definition became extremely simple and the bulk of the address arithmetic was transferred to the input routines where it was done just once for each list type. A bonus from this organization was that new sections could be added to lists without invalidating any programs except input routines based on earlier forms.

Index system

At first we arranged to specify the serial number of the version needed whenever a list was fetched for a computing process. This was cumbersome and error prone.

We resorted to a strategy based on the realization that the usual need is for the latest version input or generated for any given list type. This was called the active version and our index contained a pointer to the entry for it. The entry gave the serial number, the address in store, and enough information on list size to enable the program to transfer the list from backing store. Now and then it may be required to cycle between two or more versions of a list type, to vary a weighting scheme, or to use two different block diagonal least-squares approximations, for example. For this reason the index was constructed to hold entries for the last three, and later four, versions of each list type. A 'wind-back' could then be arranged by simply adjusting an active pointer. This operation is relatively rare and we have not found four versions insufficient in practice.

Algol implementation

Advantages of Algol

The Algol version of our system of crystallographic programs was made by Dr G. C. Ford in the course of his research for a D. Phil. degree on constrained least-squares refinement and automatic heavy atom analysis. It is sufficient comment on the convenience of Algol as a programming language that he was able, single-handed, to produce a comprehensive set of programs compiling into ten million bits of machine code at the same time as solving the crystallographic problems involved and extending the structure of data formats. The freedom of Algol from dialect variations, except in the input-output area, is a great advantage in terms of the ultimate need to transfer to a newer machine. The power of the language also implies that it is in principle very easy to allow any user to write a control routine of his own which uses the subroutines of the system in just the way he desires.

Drawbacks of Algol

The drawbacks we encountered are, strictly speaking, more to do with the Algol system for KDF9 than with the Algol language itself. We suspect, however, that the nature of Algol is such that systems for other machines are liable to suffer from similar shortcomings. On KDF9 the running speed of Algol programs has proved to be low in comparison with that for machine-coded programs. The inner loops of all our least-squares and Fourier map routines have been re-coded in machine language. Because of the overheads of entering and leaving code body procedures, the tendency has been for large parts of any routine to turn into machine code rather than just the vital inner loops. The Algol system has rather slow input and output, so that it takes more central processor time (not elapsed time) to output a Fourier map than it does to compute and machine scan it for peaks, which is absurd.

The compilers available when the system was built were also slow and it was necessary to approach the machine three times to establish a text, compile it, and then test the binary version. A long queue for the machine, coupled with this, implied that it took several days to try out a program after amending it. This severe discipline forced us to adopt a standard control program which effectively operated as a load-and-go interpreter for a control tape of directives written in a much more restricted language than Algol.

There was a considerable delay in providing KDF9 Algol with segmentation facilities and this contributed to a long time lag between the arrival of our machine and the emergence of our Algol crystallographic programs. It must be said, however, that the segmentation system is now quite readily usable. There is still some difficulty over the restrictions of the magnetic tape handling routines which do not allow the user to continue with a tape on which a parity error has been found.

Extensions of the system

We added to the system of lists, more or less immediately, a list 14 to provide for Fourier map symmetry and other details such as ranges of real space to be covered. It is possible in principle to derive the required combinations of sines and cosines from the symmetry equivalent positions in list type 2, but we did not have time to do the analysis in the context of the need to establish our first system very quickly. We also required freedom to choose from the variety of alternative real space regions possible in, say, space group $P2_12_12_1$ and to do more or less than the full unique volume. We later added a list 15, analogous to symmetry list 2, but designed to deal with false symmetries arising from heavy atom phasing of Fourier maps. List 16 provided for constraints on atomic separations and vibrations. It had to be redesigned to allow for bond-stretching constraints and to allow the symmetry equivalent of a parameter list position to be specified. List 17 gave the atomic positions and molecular vibrations of a rigid body to be fitted to a set of refined sites. These extensions are all of general use and we feel that they should be included in any later system.

We again found it prudent to restrict the format of the planes, list 6, so that each reflection was represented by just seven words. This restriction on the original design of the scheme simplified storage on and transfer from magnetic tape, and it has not proved inconvenient so far.

Data storage

Between runs on the machine the data for any particular analysis are held on a magnetic tape. It is a limitation of our present scheme that we cannot put more than one analysis on each tape. The data are called from the tape by standard Algol binary read procedures and they are located by means of an index with entries for the four most recent versions of each type of list. This index is written to the tape as the last action whenever lists are written so that it is always available as the last block of information. Marker blocks are also kept indicating the number of blocks since the most recent index, in case of failure before the index is written up.

During a run all lists required are held on the disc backing store. They are put there from the magnetic tape by the initial routine and any which are changed or generated are returned at the end of the run. Any failure detected by a routine of the system leads to the terminal routine to write back to tape. The index for the lists has 10 words for general disc and tape addresses, 20 words for pointers to active lists of types 1–20, 400 words for 4 entries of 5 words each for 20 types of list, and finally 20 words which hold control parameters. These serve the same communication purpose as COMMON storage in a Fortran system and give information from master program to subroutines and vice versa. This index is held as a real array in the core store.

Program structure

The system runs under the control of a master program of compiled size 300 words (of 48 bits each). This communicates with a large number of segments each no more than 3000 words long. The executive routine of the operating system and the Algol control routine need about 6000 words, and we have 23000 of the 32768 words of core store for data. The index occupies 450 of these and it is the only parameter of each of the subprogram segments, so that the calling statement for a segment is simple and standard. Segments are known to the master as SEGMENT001 and so on, and these names are associated with the actual names and versions of routines by a 'call tape' which selects the routines required from those held on the program magnetic tape. The first segment entered reads from paper tape to disc a list of command directives, each of which refers to a particular segment number and is followed by a set of parameters. Each segment number is associated with a fixed number of parameters, which is suitable for the operation normally used under that number. Unusual operations are used under 'spare' numbers which have a full 20 parameters. This system provides economical means of calling for the commonly used routines, while retaining plenty of flexibility for adding routines of greater complexity.

If a routine fails in certain ways, the system will skip all operations until the next one which is specially marked. This may be the terminating routine, or may be the next map routine if the 'failure' is the convergence of a least-squares process to a negligible shift size. More complicated loops and jumps are not provided by the present master program.

Conclusions

A well-defined, even if sub-optimal, scheme of data formats which is available before the arrival of a new computer is a great aid in the rapid construction of an integrated set of programs by a cooperating group of people. The Algol language is a practical vehicle for implementing such a scheme on a computer with a 32K word store, a 4M word disc store, and magnetic tape units. Some concessions to efficiency have to be made at the expense of strict adherence to the preliminary scheme, but they are not numerous nor troublesome. The lack of a load-and-go compiler allowing pre-compiled overlaid subroutines is a drawback which we have not completely overcome.

Discussion

AHMED: One of the ideas for standardizing the data files was to be able to exchange programs among the different centres. How effective has this been?

ROLLETT: Not very, mainly because the operating systems were different, and this prevented us from using the KDF9 programs that were available in Glasgow when we got our computer.

BASSI: I tried to do the same thing in France 4 or 5 years ago but it did not work either.

Appendix

KDF9 *Programs for crystal structure analysis*

Edition 1 (5 May 1964)

Index of lists

List type	Title
1	Unit-cell
2	Symmetry
3	Formfactors
4	Weighting function
5	Parameters
6	Planes
7	Data reduction
8	Map
9	Distance and angle
10	Peaks
11	Matrix
12	Refinement instructions
13	Modification function

List Type 1: Unit cell

Word		Contents	Word		Contents
0		Job number	28	g^{11}	
1		List type	29	g^{12}	
2		Serial number	30	g^{13}	
3		List length	31	g^{21}	Inverse of previous
4			32	g^{22}	matrix. Metric ten-
5	a		33	g^{23}	sor for covariant vec-
6	b	Å	34	g^{31}	tors such as (h,k,l)
7	c		35	g^{32}	giving (length) in Å^{-2}
8	α		36	g^{33}	as $s^2=\sum_{ij} g^{ij}h_ih_j$
9	β	radians	37	N_{11}	
10	γ		38	N_{12}	
11	V	Å^3	39	N_{13}	$N\mathbf{x}=\mathbf{X}$ where $\mathbf{x}$ is in
12	a^*		40	N_{21}	fractional coordi-
13	b^*	Å^{-1}	41	N_{22}	nates and $\mathbf{X}$ in Å
14	c^*		42	N_{23}	units referred to or-
15	α^*		43	N_{31}	thogonal axes –
16	β^*	radians	44	N_{32}	axis 1 parallel to a^*,
17	γ^*		45	N_{33}	axis 3 parallel to c
18	V^*	Å^{-3}	46	M_{11}	
19	g_{11}		47	M_{12}	
20	g_{12}	Metric tensor for	48	M_{13}	$M=(N^T)^{-1}$
21	g_{13}	contravariant vectors,	49	M_{21}	$M\mathbf{h}=\mathbf{H}$, where $\mathbf{h}$ are
22	g_{21}	*i.e.* $d^2=\sum_{ij} g_{ij}x^ix^j$ gives	50	M_{22}	Miller indices and $\mathbf{H}$
23	g_{22}	the square of the dis-	51	M_{23}	are referred to or-
24	g_{23}	tance in Å between	52	M_{31}	thogonal reciprocal
25	g_{31}	two points separated	53	M_{32}	axes in Å^{-1} (axes
26	g_{32}	by the vector with	54	M_{33}	parallel to those for
27	g_{33}	components x^i in fractional coordinates.			N in words 37 to 45).

List Type 2: Symmetry

Word		Contents
0		Job number
1		List type
2		Serial number
3		List length
4		
5		Lattice type number
6	c	centre of symmetry ($c=2$), or not ($c=1$)
7	p	number of equivalent positions for which $\mathbf{R}_s$, $\mathbf{t}_s$ are supplied below. $1 \leq p \leq 24$
8	m	number of different symmetries
9	q	$= \sum_i^m n_i$
10	n_1	number of equivalent positions in symmetry 1
11	n_2	number of equivalent positions in symmetry 2
—		—
—		—
—		—
$9+m$	n_m	number of equivalent positions in symmetry m
$10+m$ — — —	s	the n_1 index numbers of $\mathbf{R}_s$, $\mathbf{t}_s$ to be used for symmetry 1
$9+m+n_1$ — — — $9+m+q$		Similar entries for symmetries 2, -----, m
$10+m+q$	$R_{11,1}$	$\mathbf{R}_1$ (for equivalent position 1)
—	$R_{12,1}$	
—	$R_{13,1}$	
—	$R_{21,1}$	
—	$R_{22,1}$	
—	$R_{23,1}$	
—	$R_{31,1}$	
—	$R_{32,1}$	
—	$R_{33,1}$	
—	$t_{1,1}$	$\mathbf{t}_1$ (for equivalent position 1)
—	$t_{2,1}$	
$21+m+q$	$t_{3,1}$	
$22+m+q$ — — $9+m+q+12p$		Similar entries for $\mathbf{R}_2$, $\mathbf{t}_2$, $\mathbf{R}_3$, --- $\mathbf{t}_p$

A Crystallographic System in FORTRAN IV for Medium-Scale Computers

F. R. AHMED

Biochemistry Laboratory, National Research Council of Canada, Ottawa 7, Ontario, Canada.

The crystallographic system of programs at the National Research Council of Canada was written in FORTRAN IV by Ahmed, Hall, Pippy & Huber (1966) for use on the IBM/360 computer. The system at the present time includes 14 basic programs for structure determination and refinement, and more programs are being added. The overall plan for the data lists of this system is basically a modified version of that for the KDF9 crystallographic system by Cruickshank, Sime, Smith, Truter, Wells, Rollett & Freeman (1964).

Each program in the NRC system is one separate entity containing all the subprograms it needs. Formats for the input/output data on punch cards and magnetic tapes are standardized for the whole system, so that the programs can be executed in succession with minimal effort on the user's part. Each standard magnetic tape contains the relevant data and tables for one crystal structure, arranged in lists (similar to those for the KDF9 system) each containing several unformatted records. For simplicity in programming, it is not necessary for tapes generated solely for internal use within a program to conform to the standard tape format.

The IBM/360 system is available in different models with different speeds and different memory sizes. Thus, in order to make the NRC programs applicable to the medium and small computer installations, certain restrictions were laid down at the planning stage, as follows:

1. All statements should be in the basic FORTRAN IV specifications, and should not include any of the extended FORTRAN statements such as logical IF statements. This restriction was relaxed for the more recent programs.

2. All programs should include options to produce typewritten instructions for the operator to mount and dismount the input/output tapes, followed by PAUSE statements. These instructions are bypassed automatically when the computer is under control of the more versatile operating systems such as OS/360.

3. None of the programs is allowed to call more than 3 magnetic tape units or to occupy more than 32K words of core memory.

Running these programs on disk-oriented systems instead of magnetic tapes presents no problems, since the necessary changes are made on control cards rather than in the programs.

The NRC crystallographic system has been adapted with some minor modifications by J. D. Lee to the ICT 1905 computer.

The standard lists' tape

Each of the standard lists' tapes is assumed to contain the following:

1. Identification number that is checked by the calling programs;
2. directory of the lists stored on the tape, and their serial numbers;

3. the lists in ascending order of the list number and its serial number;

4. end of file mark.

The lists which have been used thus far are:

1. unit cell
2. symmetry
3. form factors
4. atomic parameters
5. cos, sin table
6. planes.

The leading record of each list contains four words giving the list number, its serial number, the number of records in the list following this leading record, and the number of words in each record. Thus, with the exception of the leading record, all other records of the same list have an equal number of words; however, records in different lists may be of different lengths. All records are assumed to be in unformatted binary. Each of the records in the planes list contains the following information for one reflexion:

h, k, l

obs/unobs indicator

sign indicator (to distinguish between hkl, $\bar{h}kl$, $h\bar{k}l$, $hk\bar{l}$)

order indicator (to specify which index has changed)

parity

multiplicity

$\sin^2 \theta$

$\sqrt{w}$ (based on the statistics of measurements*)

structure amplitudes, such as F_o, F_{c1}, F_{c2}, F_{c3}, F_o^2, A_o, B_o, A_c, B_c

interpolated f-values for all the given f-curves.

* The least-squares program allows the weights to be recalculated with one of six available weighting functions.

Special program features

1. In general, the programs can be applied to any space group utilizing the symmetries and simplified electron-density expressions given in the *International Tables for X-ray Crystallography* (1952).

2. Each program is carefully and adequately documented with details about the mathematical procedure, the order and format of the input and output data and directive cards, an estimate of the time whenever possible, a list of the arrays whose limits have to be changed to reduce or expand the memory requirements, and well-planned data sheets for the input information.

3. The programs include options, whenever necessary, to exclude from the calculations certain specified reflexions, or to modify their structure amplitudes, their obs/unobs status, and weights. Omission of reflexions can also be conditional on the agreement between the $|F_o|$ and $|F_c|$ structure amplitudes.

4. The atomic parameters can always be read in either from the lists' tape or from punch cards.

5. Invariant quantities such as $\sin^2 \theta$, $f(hkl)$ are calculated only once and stored on the data tape. The scattering factor curves are accepted at the same intervals of $\sin \theta/\lambda$ as given in the *International Tables for X-ray Crystallography* (1962), and

accurate interpolation is achieved by utilizing different interpolation functions depending on the range of sin θ/λ.

References

Ahmed, F. R., Hall, S. R., Pippy, M. E. & Huber, C. P. (1966). *NRC Crystallographic Programs for the IBM* 360 *System, World List of Crystallographic Computer Programs.* 2nd ed. Appendix, p. 52.

Cruickshank, D. W. J., Sime, J. G., Smith, J. G. F., Truter, M. R., Wells, M., Rollett, J. S. & Freeman, H. C. (1964). KDF9 Programs for Crystal Structure Analysis. Private communication.

International Tables for X-ray Crystallography (1952). Vol. I. Birmingham: Kynoch Press.

International Tables for X-ray Crystallography (1962). Vol. III. Birmingham: Kynoch Press.

Discussion

MASLEN: On a machine that does not have suitable input/output buffer, it may be more convenient to do the interpolation on the scattering factor curves each time you calculate the structure factors, rather than reading them from magnetic tape.

AHMED: No, personally I am against that. I think the interpolation should be done to high accuracy at the data reduction stage where one can afford to do it.

STEWART: If you do not have buffered I/O now, you should make every effort to obtain it. I tend to agree very strongly with Dr. Ahmed.

JOHNSON: I wonder if you would like to comment on the least-squares versus differential syntheses. When would you use one over the other, and how do they compare in the rate of convergence?

AHMED: At the present time, differential syntheses would be used mainly as a check on the final parameters. The discrepancies in the results of the two methods should not be significant. The refinement with differential syntheses converges fairly well, but the thermal parameters are not easily refined with this method.

ALCOCK: In structures with a very heavy atom and several light atoms, the differential synthesis is more effective than other methods, especially if you want to find out which of the peaks belong to genuine atoms.

Use of a Small Computer in Crystallographic Research

RYONOSUKE SHIONO

Crystallography Department, University of Pittsburgh, Pittsburgh, Pa. 15213, *U.S.A.*

In recent years, the number of small computers used in scientific and engineering calculations has increased tremendously. In many instances, they are used as on-line control devices of certain instruments, but they are also used as standalone computers. Because of their relatively low cost, a small computer can be installed in a department or laboratory for its own use. With this easy and ready access to a small computer, the net turnaround time of many calculations is much less as compared to that of using a large centrally located computer.

This is particularly useful for crystallographic calculations, where the amount of data to be handled is usually too large to utilize a low-cost typewriter terminal for a time-shared large computer, and one could often make decisions to determine the course of calculation by intermediate results.

There are, of course, general advantages not restricted to crystallography such as the ease of program testing and the ease in training of programmers with this hands-on approach.

The way a small computer can be favorably used in a crystallographic laboratory depends on the availability of a larger and faster computer. In general, however, most calculations up to the final full-scale refinement of a structure can be done on a small computer. Many preliminary calculations and data reductions and early stages of sign determination are done more conveniently on a small computer on hand. Only a full-matrix least-squares refinement seems to require a bigger computer. Fourier summations and block-diagonal least-squares refinements may be used on both computers, depending on the number of parameters and reflexions.

Many small computers on the market now are equipped with disk storage and this relieves the limitation on core storage size. With a disk, only the part of the program needed in execution has to be in core at any particular time, and the rest of the program can be stored on disk. Either multiple-core load mode (LINK) or subroutines loaded on call (LOCAL) mode may be utilized to accommodate a big program.

When more than one computer is available, programs on both computers should be set up in a coordinated way so that calculations can be switched from one computer to the other and vice versa in different stages of structure analysis. This requires that the formats of reflexion data cards and parameter cards in particular should be the same for all programs that need them. In this way, for example, coordinate cards used in some preliminary structure factor calculations on a small computer may be used for a full-scale refinement on a big computer. Whenever any program at any stage produces a set of new parameters, such as cell constants or positional and thermal parameters, cards should be punched by the program with these in the right format so that they can be used in further calculations. This reduces the necessity of hand-punching of cards to a minimum and thereby decreases the chance of an error.

Another possible use of a small computer is to connect it to a large computer and utilize it as a terminal as well as a stand-alone computer. The author has no direct experience in this mode of operation, but it is a reasonable assumption that in order to get an equivalent efficiency from this dual use of a small computer, the computers on both sides have to be upgraded in hardware considerably compared to the stand-alone operation.

Programming language and core storage limitation

The technique of compiler writing has made much progress and many small computers have FORTRAN or similar higher level language compiler available. However, in many instances, it is very limited in its capability, in the size of allowable programs as well as in the speed of execution. Most small computers lack the floating point arithmetic hardware and therefore fixed point calculations are done much faster than the floating point calculations. However, most FORTRAN languages are not designed to do fixed point calculations. For many programs that are repeatedly used and have general applications, such as structure factor calculations and Fourier summations, it will pay to spend extra time in writing a machine language program and thus gain in shorter execution time and higher capability of the program.

In many instances, FORTRAN cannot be used because of the core storage limit or because it does not allow the operation to utilize certain unique features of the computer. Contrary to the popular misconception that a program written in FORTRAN can be universally used in different installations, more often than not it is compiler-dependent and considerable effort has to be made before it can be compiled for another computer of the same type. A program written in symbolic machine language, on the other hand, can be used on the same type of computer in other installations without change because it is not a system-dependent.

Apart from the difference in compilers, for a small computer with a particular set of input-output devices, the logical structure of a program to do certain calculations would be different depending on the characteristics of input/output devices and also on the size of core storage. Therefore, even a program written in FORTRAN may be configuration-dependent.

For some types of program such as Fourier or differential-Fourier synthesis, the first part of the program may be needed to set up the main calculation part to fit a given space group expression. Once this has been done, this part is no longer needed. With machine language, the program may be easily set up to use this program space as a part of the storage space for intermediate sums, etc., by overwriting it. A FORTRAN program generally cannot use the same space for program and data. A machine-language program can also modify the program itself to custom-fit the flow of calculation by eliminating tests required, whereas a FORTRAN program can only change the flow in the tests, but cannot eliminate the tests themselves.

One disk storage drive system

Many small computers are now equipped with random access disk storage. Usually a monitor system with compiler, assembler, and utility subroutines reside on the disk and the user's program and data may also be stored on it. This is a convenient arrangement, except very often the disk is not big enough to store all programs and a set of data to be processed along with working storage space for the program to be executed. There are several steps to solve this problem. First, an object program deck can be used every time and thus storage of the programs on the disk is avoided. In this way programs do not accumulate on the disk and there are always fixed amounts of storage space available for execution of a program. If this is not enough, a separate disk with only the required parts of monitor and subroutines on it may be used. Finally, a self-loading object deck complete with all necessary subroutines may be prepared and an entire disk may be used for the calculation without depending on the presence of a monitor.

When more than one disk pack is needed to store the data, the program can be written to switch to a second disk easily by a machine language, but generally this is harder to do with FORTRAN.

Sorting with single disk storage

To sort a fairly large number of data with a small core storage and a single disk drive requires some planning. For example, in sorting reflexion data with h, k, l indices, the core is not big enough to store all data at once. Therefore the data have to be divided into smaller groups and sorted one by one.

The first approach is to assign fixed areas on the disk for each group and to store the data presorted on the disk. Then each group can be brought into the core and sorted successively. This works as long as the maximum size of each group can be predetermined and enough space can be allocated for each of them. Since the distribution of data with regard to one index is not uniform, especially as the index may become negative, it is sometimes difficult to preassign enough space for a sufficient number of groups on the disk.

If this first approach cannot be used, then the second approach is to store all data continuously on the disk, then let the program scan through all data and pick up and store into the core those data within a given group. After sorting one group, the program would go back to bring in the next group of data. This method is not so efficient as the first one, but it can accommodate much more data. The definite advantage of this method is that the size of a group may be made variable to fit a particular set of data. The program may start with a pre-fixed group size, such as reflexions of index value from h to $h+n$, if h is the slowest running index, and n may be some suitable value including zero. If in the process of reading-in reflexions for a block, the program finds that there are more data to be read, then it can go back to re-read the data with a reduced group size by reducing the number n in $h+n$ or, if n is already zero, by subdividing the second index, for example, from k to $k+n'$ instead of all k with a given h index. If the information about the total number of data, the lowest and the highest values of two slower-running indices are available before the start of the sorting, a possible optimum size of the groups may be calculated to minimize the necessity of re-try and to maximize the sorting efficiency. In this way, for instance, 18,000 reflexions can be sorted on IBM 1130 with both sorted and unsorted data files on disk.

Mathematical expressions

Matrix method is generally neat and concise in describing some computational steps and often takes less storage space than the same calculation in analytical form. However, in a small computer particularly, the analytical expression yields a faster execution time.

For example, in order to calculate the distance between two points, given a set of fractional coordinates, the part that requires 12 multiplications and 12 additions by matrix method may be done by 12 multiplications and 6 additions by analytical expression for triclinic symmetry. For a higher symmetry, for instance monoclinic, the calculation will be reduced to 9 multiplications and 3 additions in analytical expression easily, but the same cannot be done in matrix approach.

Furthermore, if the nature of calculation is such that the calculation is repeatedly done with only a part of the parameters changing, the analytical expression usually makes it possible to calculate only the changing part every time and keep the rest of the unchanging part intact. This will save a great deal of execution time.

Summary

In summary, a small computer in a crystallographic laboratory is a very useful tool and it provides better over-all efficiency in many calculations, than a big computer elsewhere. In order to utilize fully a small computer, it may require a little bit more programming effort, but that is amply compensated by the faster execution time or the added convenience. With the advance of technology, computer speed is increasing and the cost of storage decreasing – there will be more use for a small computer in the future.

Discussion

HARDGROVE: A. Cooper at Buffalo has written a package of programs for the 8K IBM 1130 computer which includes block diagonal least-squares, Fourier, and some other programs which may be of interest.

STEWART: I would like to make two points here. Firstly, I consider the sort of optimization within loops that Shiono has discussed is important also with faster computers. Secondly, I think that larger machines should be used if possible to compile programs for the smaller machines. In this way sophisticated compiler routines can be used to produce an efficient code for the small machine.

TOPIC I

PROGRAMS FOR COMPUTER-CONTROLLED DIFFRACTOMETERS

CHAIRMAN: W. R. BUSING

Least-Squares Refinement of Lattice and Orientation Parameters for Use in Automatic Diffractometry

WILLIAM R. BUSING

Chemistry Division, Oak Ridge National Laboratory, Oak Ridge, Tennessee 37830, *U.S.A.*

The accurate determination of lattice parameters is a necessary part of any crystal structure determination, and both lattice and orientation parameters are usually established as a prerequisite to automatic single-crystal data collection. With a computer-controlled diffractometer these parameters can be found in a way which is almost automatic.

The purpose of this report is to describe in detail the procedures which we have found useful for measuring the diffractometer angles associated with individual X-ray reflections, for using these angles to obtain preliminary estimates of the lattice and orientation parameters, and, finally, for making a least-squares refinement of these parameters. The procedures have been programmed for use with the Oak Ridge computer-controlled X-ray diffractometer (Busing, Ellison, Levy, King & Roseberry, 1968), but they should be applicable to any four-circle instrument. This report will not contain mathematical details or derivations since these have been reported elsewhere (Busing & Levy, 1967*a*,*b*).

The Oak Ridge computer-controlled X-ray diffractometer programs (1968) consist of individual procedures. Some of these interact directly with the diffractometer and others are purely computational. Each procedure is called into operation by the user who types on the teletype keyboard a slash (/) followed by the appropriate two-letter symbol. It will be convenient for us to describe our methods in terms of these discrete procedures, a few of which are summarized in Table I.

Research sponsored by the U.S. Atomic Energy Commission under contract with the Union Carbide Corporation.

General procedure

If approximate lattice parameters are available from the literature or from precession photographs, they are read into the computer by means of procedure /CB (see Table I). If the orientation is known approximately, the /CU procedure may be executed to provide this information to the computer program. At this point a tentative matrix **UB** (Busing & Levy, 1967*a*,*b*) has been computed.

The procedure /IA can now be used to calculate the approximate settings for a low-angle reflection strong enough to be found easily. After some intensity has been located (see Busing *et al.*, 1968, for further details), the procedure /CR is used to maximize the intensity of the Bragg reflection and to center it in the counter aperture. This automatic procedure will be described in detail below. It is usually executed first with the aperture wide open and then repeated with reduced aperture.

A second low-angle reflection is similarly located and centered. The measured angles of these two reflections are then input by means of the /CU routine together with their indices, which are presumed to be known. The /IA procedure is then used to calculate angles for other low-angle reflections, which are then observed as a

Table I. Some of the individual procedures available in the program for the Oak Ridge computer-controlled X-ray diffractometer. See Busing *et al.* (1968) for more details. Equation numbers refer to Busing & Levy (1967*a,b*).

Procedure: /CB, Calculate **B**
Input: λ, a, b, c, cos α, cos β, cos γ
Result: Computes matrix **B**
Reference: Eq. 3

Procedure: /CU, Calculate **U**
Input: h_i, k_i, l_i, ω_i, χ_i, φ_i; $i=1,2$. Assumes /CB executed
Result: Computes matrix **U B**
Reference: Eqs. 22, 25–27

Procedure: /MT, Matrix from three reflections
Input: h_i, k_i, l_i, $2\theta_i$, ω_i, χ_i, φ_i; $i=1,2,3$
Result: Computes matrix **U B**
Reference: Eqs. 22, 28–31

Procedure: /CA, Calculate lattice parameters
Input: Assumes **U B** available from /MT
Result: Prints a, b, c, cos α, cos β, cos γ
Reference: Eqs. 32–36

Procedure: /IA, Given indices, compute angles
Input: h, k, l. Assumes **U B** available
Result: Prints 2θ, ω, χ, φ
Reference: Eqs. 37–38, 41

Procedure: /DA, Drive all shafts
Input: Assumes angles have been calculated by /IA
Result: Drives diffractometer to calculated angles

Procedure: /CR, Center reflection
Input: Option indicators
Result: Maximizes intensity and centers reflection in counter aperture; then prints 2θ, ω, χ, φ
Reference: Busing *et al.* (1968, p. 17). Also see text and Figs. 1–4

Procedure: /PO, Punch observation
Input: TYPEW, h, k, l
Result: Punches TYPEW, h, k, l, 2θ, ω, χ, φ on paper tape for use as least-squares observations
Reference: Busing *et al.* (1968, pp. 17 and 21)

Procedure: /RO, Read observations
Input: Paper tape punched by /PO for up to twelve reflections
Result: Stores observations for use by least-squares program
Reference: See text

Procedure: /RH, Read indices of orienting reflections
Input: h_1, k_1, l_1, h_2, k_2, l_2
Result: Stores indices for use by least-squares program
Reference: See text

Procedure: /RP, Read trial parameters
Input: Wavelength, λ; lattice constants, a, b, c, cos α, cos β, cos γ; orientation angles, ω_1, χ_1, φ_1, ω_2, χ_2, φ_2
Result: Stores parameters for use by least-squares program
Reference: See text

Procedure: /RK, Read variable-selection integers
Input: An indicator for each of the 13 parameters: 0 to hold it constant; 1 to vary it
Result: Stores them for use by least-squares program

Table I. (*Continued*)

Procedure:	/LS, Perform cycle of least-squares refinement
Input:	Assumes /RO, /RH, /RP, and /RK have been executed
Result:	Prints comparison of observations with calculations based on trial parameters. Calculates and prints adjusted parameters with standard errors
Reference:	Eqs. 59–82; Busing & Levy (1967*a*, pp. 51–53); Busing *et al.* (1968, p. 22). Also see text and Figs. 5 and 6
Procedure:	/CM, Calculate matrix
Input:	Assumes least-squares parameters are available
Result:	Uses parameters to calculate **U B**
Reference:	Eqs. 3, 22, 25–27

check on the identification of the two orienting reflections.

If approximate lattice parameters are not known or if the preliminary measurements indicate that they may be in considerable error, then better results can be obtained by using the measured angles of three reflections as input to procedure /MT, which computes the matrix **UB** directly. Program /CA can then be used to obtain the lattice parameters based on these three reflections. In general these parameters will not show the lattice symmetry exactly. The agreement with the requirements of symmetry provides a partial check on the identification of the reflections. Other reflections are observed as a further check.

The results obtained from low-angle reflections are still regarded as preliminary, because these reflections are insensitive to errors in the parameters, and because the X-ray α-doublet is unresolved at these angles. Sometimes it is necessary to repeat the initial procedures using reflections with intermediate Bragg angles. Eventually we must locate reflections at as high a counter angle as possible. Often a rapid automatic data collection is run overnight to identify the most intense high-angle reflections.

Each high-angle reflection is measured with the /CR procedure using a large counter aperture initially, and a theta-two-theta scan is recorded to identify the α_1 peak with certainty. The /CR operation is repeated with reduced aperture and with counting time increased to improve statistics. The final angles are punched on paper tape together with the indices, using the /PO procedure. Up to 12 reflections are measured in this way, care being taken to insure that the reflections included are distributed in all directions in reciprocal space.

These measurements are used for the least-squares refinement of lattice and orientation parameters as described in detail in a separate section below. Input of the observations, trial parameters, and other data is made by means of procedures /RO, /RH, /RP, and /RK. Any constraints imposed by symmetry are introduced as described below. A least-squares cycle performed by routine /LS yields a listing of the calculated angles and a comparison with those observed. At this point any blunders in input or in the indexing of reflections will become evident.

The program prints the adjusted parameters together with their standard errors. Two or three cycles of refinement are usually sufficient for convergence, which is indicated by the fact that no further parameter changes occur.

If data collection is to follow, the user will execute procedure /CM, which causes matrix **UB** to be calculated from the adjusted parameters.

Procedure for measuring angles

An ideal reciprocal-lattice vector is de-

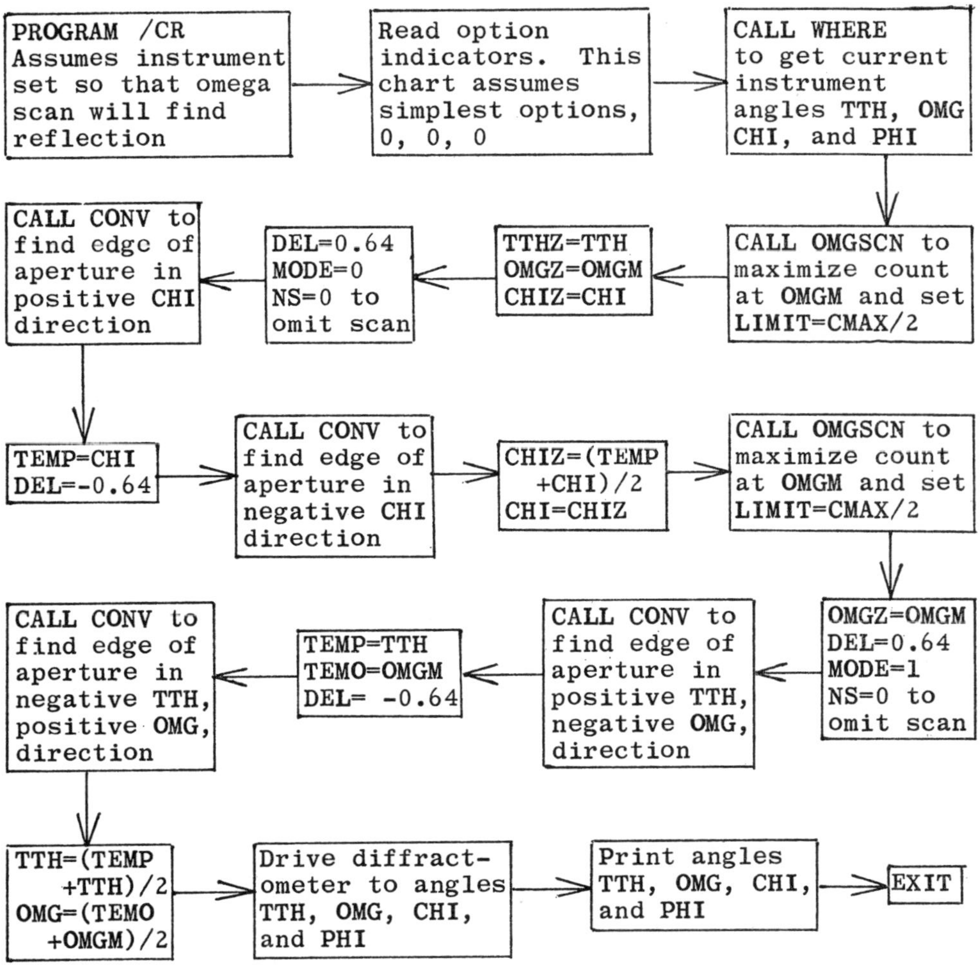

Fig. 1. *Program* /CR *to maximize the intensity of a Bragg reflection and center it in the counter aperture. The program calls subroutines shown in Figs.* 2–4.

scribed by three parameters; two are associated with its direction and one, the Bragg angle, with its length. Although four instrument angles are available for adjustment, only three need to be varied to achieve the ideal reflecting position. When operating near the bisecting position (Busing & Levy, 1967*a,b*), we choose to adjust 2θ, ω, and χ, leaving φ unchanged. In the parallel position φ is varied and χ is held constant.

Reflection will occur whenever the reciprocal lattice vector makes an angle with the primary beam which satisfies the Bragg condition. If we start with this plane normal near the ideal position, the Bragg condition can always be established by adjusting ω. If the counter aperture is large enough, the intensity of this reflection can be maximized even though it is not centered in the counter.

Next, we center the reflection vertically in the counter aperture, that is, we make the scattering vector horizontal. In the bisecting position this is best accomplished by adjusting χ. In the parallel position it is better to adjust φ. (This option, which is omitted from the discussion below, is described in the references.) In general, the preceding adjustment may interact with and alter to some extent the Bragg condition which was established previously. It may therefore be necessary to readjust ω when χ or φ is changed.

Finally, the reflection is centered horizontally by changing the counter angle without moving the crystal. On the Picker four-circle instrument this is accomplished by compensating for each change in 2θ by one in ω, so that $\Delta\omega = -\Delta 2\theta/2$.

A detailed description of the /CR procedure is available in the flow charts of Figs. 1 to 4, which show the main program and the subroutines which it calls. Table II provides a glossary of the symbols used in these charts, which are the same as those used in the PDP-5/8 programs (Busing *et al.*, 1968, pp. 130–134). For simplicity three options included in the programs are omitted from these flow charts.

After making certain initial settings, program /CR (Fig. 1) calls subroutine OMGSCN (Fig. 2) to maximize the intensity by adjusting ω. OMGSCN does this in two stages by calling SCAN (Fig. 3), first for a coarse search ($\Delta\omega = 0{\cdot}04°$) and then for a fine search ($\Delta\omega = 0{\cdot}005°$).

Next the program centers the reflection vertically by calling subroutine CONV

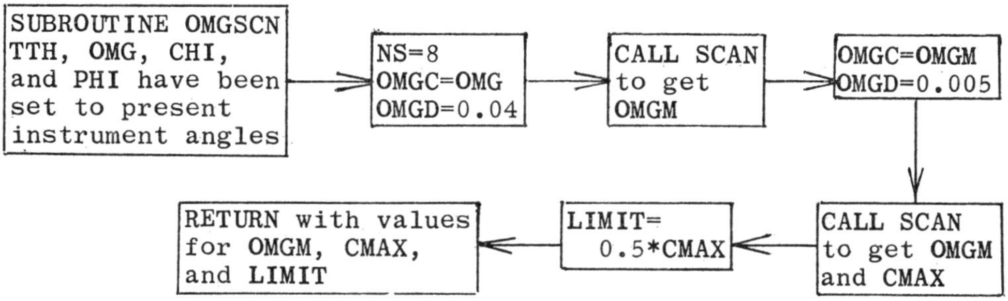

Fig. 2. Subroutine OMGSCN *to maximize the intensity of a Bragg reflection by making coarse and fine* ω *scans.*

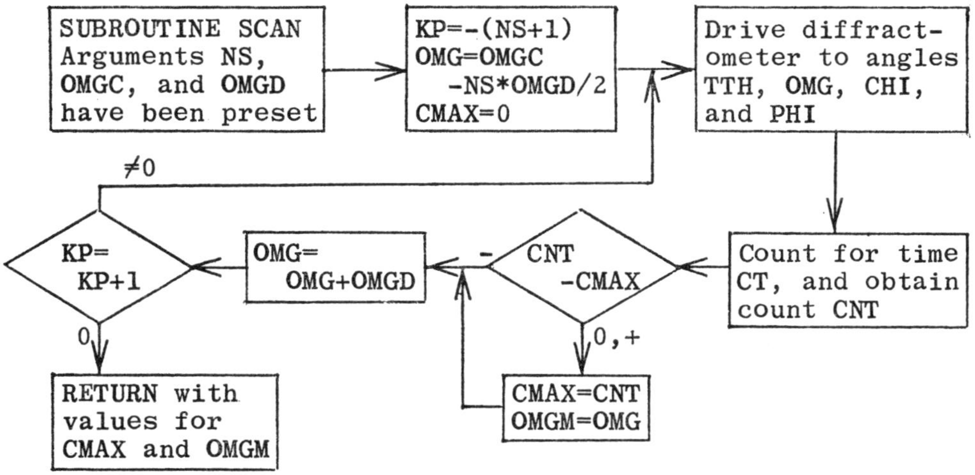

Fig. 3. Subroutine SCAN *to make one step scan in* ω, *finding the maximum count and the angle at which it occurs.*

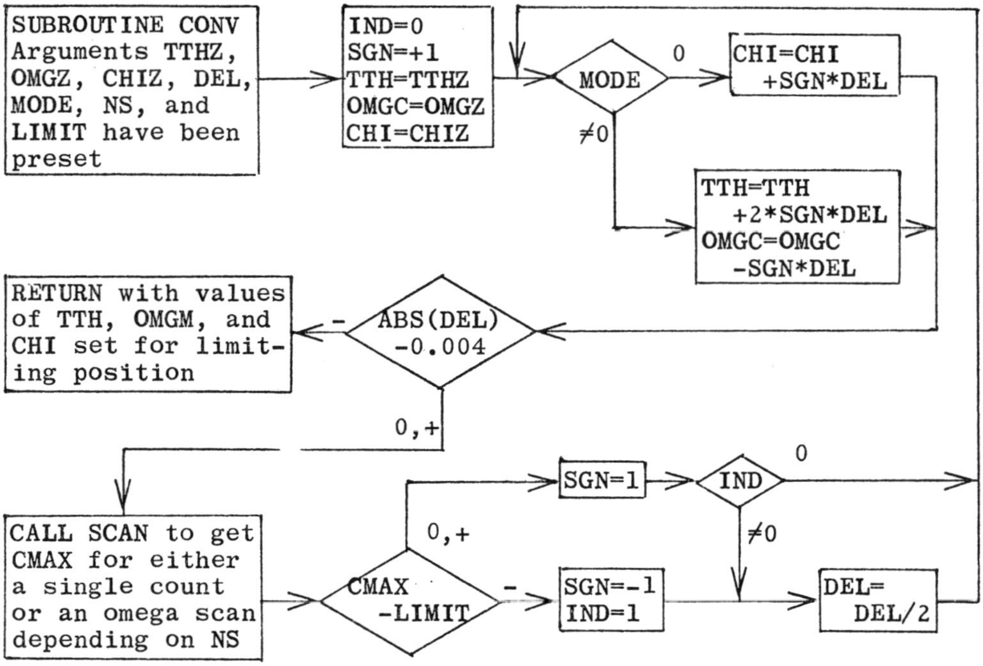

Fig.4. Subroutine CONV *to adjust either χ or 2θ and ω until the observed count is close to a desired value,* LIMIT, *which has been set at some fixed fraction of the maximum count.*

(Fig. 4). CONV adjusts χ until the count is close to LIMIT=CMAX/2, thus splitting the reflection intensity across the edge of the counter aperture. Two calls of CONV determine values of χ for the reflection at lower and upper edges of the aperture, respectively. These are averaged to obtain the desired χ setting. OMGSCN is called again to insure that the Bragg condition is maintained.

Then CONV is called twice more to adjust 2θ while making compensating changes in ω. The angles which reduce the intensity to half at each side of the aperture are averaged to obtain the desired 2θ and ω. The diffractometer is driven to the position thus located, and the final angles are printed.

Observations for least-squares refinement

Each reflection measured provides three observations which are independent, since they are actually measured separately. Thus the measurement of ω answers the question: At what value of ω does the Bragg condition occur when 2θ, χ, and φ are set at some arbitrary but known values? The corresponding calculated quantity is obtained using equations 79–81 (Busing & Levy, 1967*a*, *b*), assuming temporarily that the trial parameters (see below) are correct. The settings for 2θ, χ, and φ are used in this calculation without making any assumption that they are 'correct' or equal to those which would be calculated.

Similarly, the observation of χ answers the question: At what value of χ will the scattering vector be horizontal when 2θ, ω, and φ are set at arbitrary known values. The corresponding calculated quantity is obtained from Eqs. 68–70 (Busing & Levy, 1967*a*, *b*).

In the parallel position an observation of

φ would correspond to the quantity calculated from Eq. 73 (Busing & Levy, 1967*a,b*).

Strictly, the observation of 2θ should be compared with a value calculated from Eqs. 64 and 65, but we find that satisfactory results are obtained with the limiting Eqs. 17 and 18 (Busing & Levy, 1967*a,b*).

Fig. 5 is a schematic flow chart of subroutine CALC which evaluates the quantities corresponding to these observations. The algorithms used are available (Busing & Levy, 1967*a*, pp. 51–56), as is the PDP-5/8 program (Busing *et al.*, 1968, pp. 176–180).

Parameters for least-squares refinement

Thirteen parameters are included in the list of possible variables, but not more than nine can be varied at one time and these nine must be properly chosen.

The wavelength λ is considered as a possible variable, but in X-ray work it is

Table II. Glossary of symbols used in the flow charts, Figs. 1 to 4, for the program /CR to measure a reflection

CHI	Current diffractometer angle χ or position to which χ shaft will be driven when motors are next started
CHIZ	Initial value of χ from which subroutine CONV starts its search
CMAX	Maximum count found during an ω scan
CNT	X-ray count observed during time CT
CT	Counting time in tenths of seconds
DEL	Angular increment used in subroutine CONV. Starting at $\pm 0{\cdot}64°$ it is repeatedly halved until it becomes $\pm 0{\cdot}005°$
IND	Indicator used in subroutine CONV; IND=0 if LIMIT has not yet been found
KP	Negative step counter for ω scan
LIMIT	Value of count on which subroutine CONV attempts to converge
MODE	MODE=0 if CONV is searching with χ; MODE=1 if CONV is searching with 2θ and ω
NS	Number of steps in an ω scan. Number of counts is NS+1. NS is set to zero to omit scan
OMG	Current diffractometer angle ω or position to which ω shaft will be driven when motors are next started
OMGC	Central value of ω for an ω scan
OMGD	Increment of ω for an ω scan
OMGM	Value of ω at which maximum count is found in an ω scan
OMGZ	Initial value of ω from which subroutine CONV starts its search
PHI	Current diffractometer angle φ or position to which φ shaft will be driven when motors are next started
SGN	Plus or minus one to determine direction of step in search by subroutine CONV
TEMP	Temporary storage for limiting CHI or TTH
TEMO	Temporary storage for limiting OMG
TTH	Current diffractometer angle 2θ or position to which 2θ shaft will be driven when motors are next started
TTHZ	Initial value of 2θ from which subroutine CONV starts its search

usually assumed to be known and therefore held constant.

Six lattice parameters are included in the list. These are the cell translations *a*, *b*, and *c*, and the interaxial angles α, β, and γ. In practice we use the cosines of the angles as the parameters in the Oak Ridge programs. These lattice parameters are used to calculate the Cartesian transformation matrix **B**, using the same method as procedure /CB.

For triclinic crystals all six lattice constants are adjusted in the refinement, but for higher symmetry there are fewer variables. The application of constraints to the lattice parameters will be discussed below.

The final six parameters in our list are the angles ω_1, χ_1, φ_1 and ω_2, χ_2, φ_2 of the primary and secondary orienting reflections respectively. The indices of these reflections, h_1,k_1,l_1, and h_2,k_2,l_2, are supplied to the program by procedure /RH. These indices and angles are used to calculate the orientation matrix **U** using the same method as procedure /CU. In fact these angles simply define the orthogonal matrix **U** and need not correspond to any reflections actually observed.

Because an orthogonal rotation matrix such as **U** is specified completely by three parameters, we know that only three of the six angles can be varied in the orientation refinement. The choice of orienting reflections and the selection of three variables is discussed by Busing & Levy (1967*a,b*) and summarized in a table (1967*a*, p. 34).

The primary reflection defines a direction in space, and the secondary reflection determines only the angle of rotation about this direction (1967*a,b*). It is clear therefore that we may vary two angles of the former and only one of the latter. A simple situation which is readily visualized is the following: Let both reflections have ω's and χ's near zero and φ's which differ by about 90°. Then the appropriate variables are χ_1 and χ_2, which correspond approximately to adjustments of the arcs of a goniometer head, and φ_1, which provides a reference for rotation about the phi axis.

After these parameters have been varied the angles ω_2, χ_2, and φ_2 no longer correspond exactly to a reciprocal lattice vector, but this is of no importance. The angles are used only to define the orthogonal matrix **U**.

Least-squares refinement procedure

Fig. 6 is a schematic flow chart of the least-squares procedure /LS. The symbols used are defined in the glossary, Table III, and correspond as nearly as possible to those used in the PDP-5/8 program (Busing *et al.*, 1968, pp. 168–174). The method used is similar to that of the OR GLS program (Busing & Levy, 1962*a*).

In setting up the least-squares normal equations, the program loops through the measured reflections, using each for up to three observations specified by TYPEW. (This is the $\Sigma T''$ of Busing *et al.*, 1968, p. 21.) For each observation the program calls CALC (Fig. 5) to obtain YC, the corresponding calculated angle. Subroutine CALC also sets YO equal to the observed angle TTH, OMG, CHI, or PHI depending on the type of observation.

The difference WDY = YO − YC is also calculated. The sum of the squares of WDY is the quantity which the least-squares procedure minimizes. At this point the program prints YO, YC, WDY and other quantities associated with the observation.

The program next calculates DV(I), the partial derivative ∂YC/∂P(I), for each parameter P(I) which is to be varied. This is done numerically by adding an increment DP(I) to P(I) and calling CALC again to compute the angle now called YD. Then the derivative is DV(I) = (YD − YC)/DP(I).

The NV derivatives for each observation are used to compute the contributions of the observation to the NV × NV matrix A and the vector VEC, which make up the normal equations. For simplicity the flow chart of Fig. 6 is written as though the entire symmetric matrix A is computed, but in the PDP-5/8 program only the upper triangular part is stored.

After the contributions of all the observations have been accumulated, the matrix is inverted by subroutine SMI (Busing & Levy, 1962*b*; Busing *et al.*, 1968, pp. 186–190). The parameter changes, revised parameters, and standard errors are computed and printed.

Constraints on the lattice parameters

For triclinic crystals all six lattice parameters are varied independently. For higher symmetries, however, the appropriate constraints should be applied.

Angles which are known to be 90° or 120° are simply fixed at these values. Setting the corresponding variable selection integer KI(I) to zero with procedure /RK causes the parameter to be omitted from the normal equations.

Two or more parameters which are constrained to be equal are treated in the following way: One of them is chosen as the independent variable, and its selection integer KI(I) is set to one. The KI(I)'s corresponding to the dependent variables are set to zero because they must not be adjusted directly.

A subroutine CNSTRN is provided which sets the dependent parameters equal to the independent one. This subroutine is called each time CALC is entered (see Fig. 5), so that when the independent parameter is incremented for the derivative calculation, the dependent ones are also modified by the CNSTRN routine.

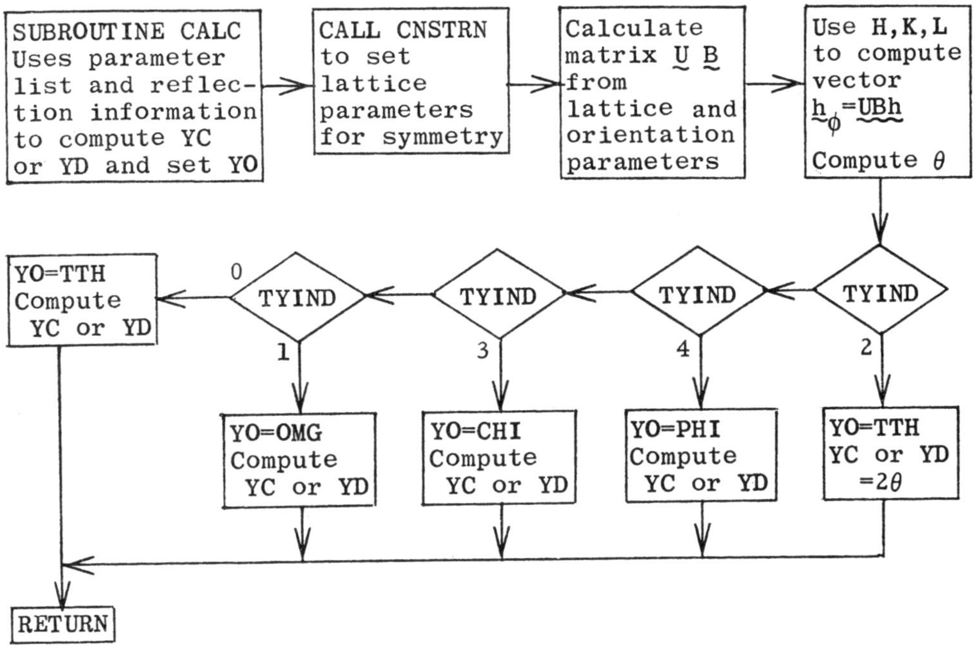

Fig. 5. Subroutine CALC *which is called by the least-squares program to calculate a quantity corresponding to an observed angle.*

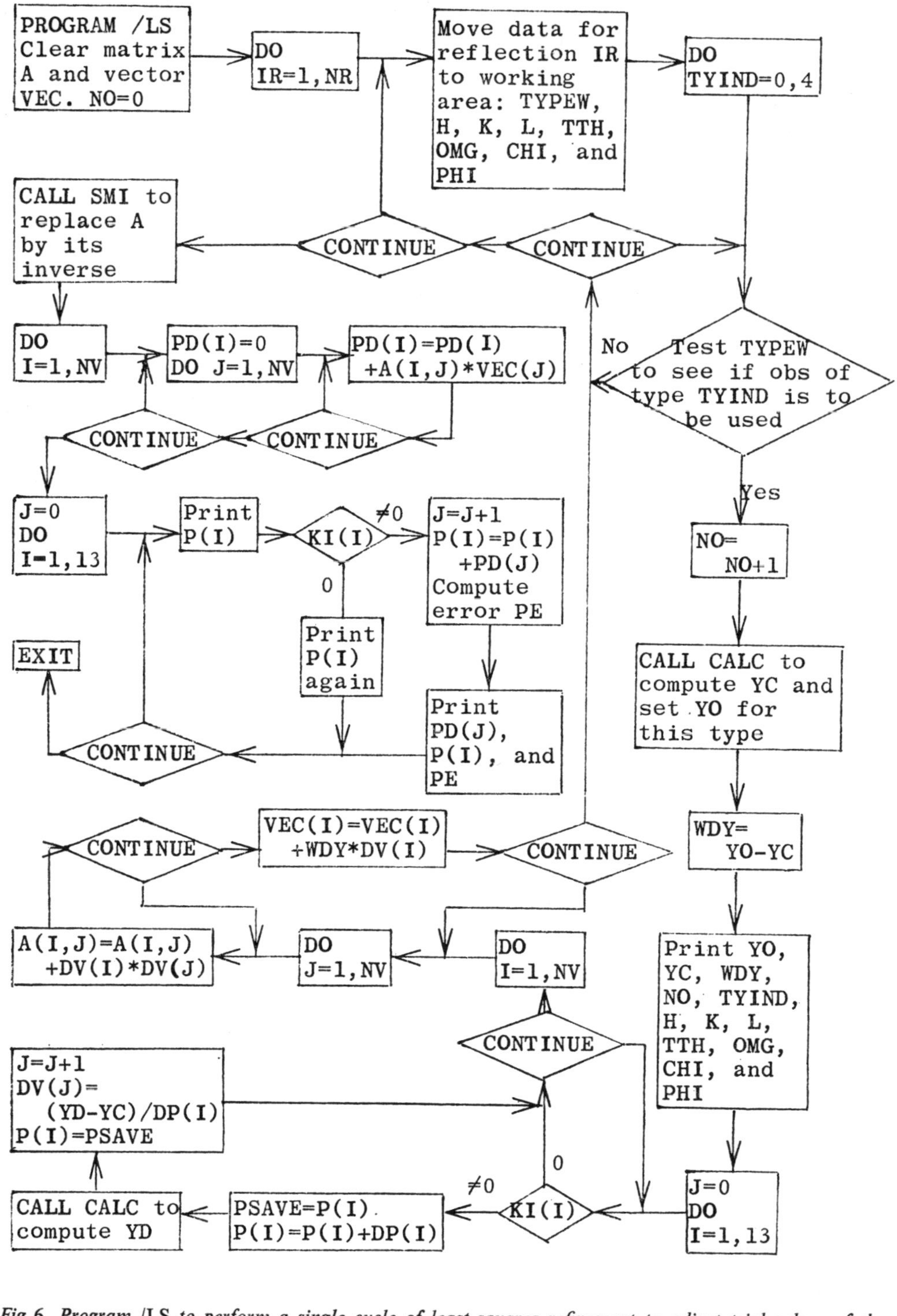

Fig.6. Program /LS to perform a single cycle of least-squares refinement to adjust trial values of the lattice and orientation parameters. The program calls subroutine CALC *shown in Fig.5.*

In this way the derivative is correctly computed to include the required constraint, and the parameter change computed for the independent variable is that which applies to the dependent ones as well.

Subroutines CNSTRN for the various unit cell symmetries are included in the programs for the Oak Ridge computer-controlled X-ray diffractometer (Busing *et al.*, 1968, p. 190).

Difficulties due to non-ideal conditions

It should be emphasized that the procedures which have been described were derived on the basis of certain ideal assumptions: a perfectly aligned diffractometer, a point source of monochromatic X-rays, and a centered, point crystal with no mosaic spread. The methods will produce satisfactory results to the extent

Table III. Glossary of symbols used in the flow charts, Figs. 5 and 6, for the least-squares program. Arrays are shown with their dimensions in parentheses

A(9,9)	Matrix of the least-squares normal equations
CHI	Angle χ for an observed reflection
DP(13)	Constant increments to be added to each parameter to obtain numerical derivatives
DV(9)	Calculated derivatives with respect to those parameters which are to be varied
H	Miller index of an observed reflection
IR	Integer which designates a particular reflection
K	Miller index of an observed reflection
KI(13)	Variable selection indicators. KI(I) = 0 to hold P(I) constant; KI(I) = 1 to vary it
L	Miller index of an observed reflection
NO	Counter for number of observations used
NR	Number of reflections measured
NV	Number of parameters varied. $NV \leq 9$
OMG	Angle ω for an observed reflection
P(13)	List of parameters: λ, a, b, c, cos α, cos β, cos γ, ω_1, χ_1, φ_1, ω_2, χ_2, φ_2
PD(9)	Calculated parameter changes
PE	Calculated standard error of a variable
PHI	Angle φ for an observed reflection
PSAVE	Value of P(I) saved while P(I) is incremented
TTH	Angle 2θ for an observed reflection
TYIND	Indicator of type of observation to be computed by subroutine CALC. This is T' (Busing *et al.*, 1968, p. 21)
TYPEW	Word which defines the values of TYIND to be used with a given observation. This is $\Sigma T''$ (Busing *et al.*, 1968, p. 21)
VEC(9)	Vector of the least-squares normal equations
WDY	YO–YC
YC	Angle calculated by subroutine CALC using trial parameters
YD	Angle calculated by subroutine CALC with one parameter incremented
YO	Value of instrument angle used as the observation

that the actual experimental conditions approximate these ideals.

In practice we find that the instrument can be aligned and the sample can be centered with sufficient accuracy to eliminate these factors as sources of error. Beam divergence due to the finite sizes of the X-ray focal spot and the sample crystal is usually not a problem. Two situations remain which occasionally cause difficulties.

First, a sample crystal may have an apparent temperature factor such that the intensity falls off rapidly with increasing counter angle, and reflections with the α-doublet resolved are too weak to be measured. Under these conditions it may be necessary to make observations of the unresolved reflections assuming a mean wavelength. Increasing the take-off angle to increase the beam divergence and broaden the peaks may produce more consistent results in this case. Alternatively, the more nearly monochromatic β component may be used if the reflections are strong enough.

Second, the sample may have a mosaic spread which is large or irregular, so that the ω value which establishes the Bragg condition is poorly defined. In this case it may be possible to improve the measurement by using a modified ω-scan routine to obtain the center of an irregular peak by averaging the angles at which the intensity drops to some specified fraction of the maximum. This modification to the /CR routine has been programmed by H. A. Levy and used in this laboratory with some success.

References

Busing, W. R. & Levy, H. A. (1962*a*). *ORGLS, a General Fortran Least-Squares Program.* ORNL-TM-271. Oak Ridge National Laboratory, Oak Ridge, Tennessee.

Busing, W. R. & Levy, H. A. (1962*b*). *Comm. Ass. Comp. Mach.* **5**, 445.

Busing, W. R. & Levy, H. A. (1967*a*). *Angle Calculations for 3- and 4-Circle X-ray and Neutron Diffractometers.* ORNL-4054. Oak Ridge National Laboratory, Oak Ridge, Tennessee.

Busing, W. R. & Levy, H. A. (1967*b*). *Acta Cryst.* **22**, 457.

Busing, W. R., Ellison, R. D., Levy, H. A., King, S. P. & Roseberry, R. T. (1968). *The Oak Ridge Computer-Controlled X-ray Diffractometer.* ORNL-4143. Oak Ridge National Laboratory, Oak Ridge, Tennessee.

Discussion

SEEMAN: Is it possible to use the centre-reflection procedure during the data collection in order to correct for crystal movement?

BUSING: We do not find that is necessary. If we make a careful alignment at the beginning, then the peaks remain centered.

COPPENS: Is the least-squares process linear?

BUSING: No, it is not linear but requires only two or three cycles for convergence.

AHMED: What sort of accuracy in cell dimensions do you obtain from your least-squares orientation matrix? Do you use only high Bragg angle reflexions?

BUSING: The cell parameters are determined using the highest angle reflections in order to ensure good α_1–α_2 splitting. We consistently obtain standard deviations of about $\pm 0{\cdot}0004$ Å.

COPPENS: Did you imply in your talk that a left-right shutter device is preferable, for measuring reflexions, to using half-peak heights?

BUSING: Not really; we did not have a left-right shutter device so we had to use something else. In any case, both of these methods ultimately depend on the accuracy of the diffractometer alignment.

Some Desirable Features for the Control of On-Line Diffractometers

E. J. GABE

Department of Energy, Mines and Resources, Ottawa, Canada.

The development of a system of on-line control for a diffractometer will obviously differ in its details from all other systems, but the overall schemes will undoubtedly show many similarities. The average user is probably not greatly concerned about precisely how the computer performs its control functions, but will be concerned with the input to and output from the program. As in all other things there is a price on convenience, and in this case the price can be rather high. This paper will deal with some ways we have developed for increasing the ease of use and reliability without raising the price too much.

The relevant equipment used is a PDP-8 computer with 8K of core storage, a Hewlett Packard model 120B oscilloscope, and a Picker 4-circle diffractometer on which the motor drives have been extensively modified.

Oscilloscope display

During the initial set-up of a crystal on a diffractometer it is very convenient to be able to read the instrument angles at all times; and when the machine is running automatically it is at least reassuring to see them. Presumably because of the cost, there has been a tendency to minimize this type of display. However, in an on-line system there is no reason why the computer should not run an oscilloscope display as the main part of a background program, with the instrumental control program as foreground. For this purpose an ordinary oscilloscope is quite adequate, although it may be necessary to modify the intensity control circuit to obtain more intensity upon request from the computer. The computer itself acts as the binary to decimal converter.

The precise details of the hardware and software will vary from machine to machine, but an XY display of 64×64 points is more than sufficient to display all the information likely to be of interest to the diffractometer user. Using 5×7 grids for each character and leaving 2 blank rows between characters and one blank row between lines, *i.e.* each character is 7×8 points, a display of eight 9-character lines is obtainable and the result is completely legible. With a machine similar to the PDP-8 which has a $1{\cdot}5 \mu s$ cycle time, a repetition rate of between 100 and 200 times per second may be achieved and the display is absolutely flicker free. A PDP-8 program to generate the characters from the internal binary form, via a table look-up type of routine, requires approximately 360 12-bit locations, of which about half is storage of the character tables. This display uses only 6 lines of characters, the first is h, k, l, the second is the number of reflections measured, and the remaining 4 lines are the labelled angles, recorded to 3 decimal places. The interface is basically two 6-bit D–A converters, which take the 12-bit word in the accumulator and decode it to position the beam on the receipt of the first I/O command and then intensify the beam via a pulse amplifier on the second command.

Indexing scheme

In order to avoid duplication of symmetry

related reflections, the method of generating the successive sets of indices, or indexing scheme, must be fairly flexible. This is perhaps not so important with most organic crystals, because they have relatively low symmetry, but for inorganic crystals and minerals it becomes essential to be able to control the volume of reciprocal space which is to be examined.

One method of doing this is illustrated in Fig. 1. The principle is that the reflections in any segment of reciprocal space may be generated successively by incrementing in directions which form the edges of the segment. Thus for example in the orthorhombic case the increment directions would be 1,0,0; 0,1,0; 0,0,1, the first step (1,0,0) is then taken as that from reflection to reflection, the second step (0,1,0) that from one line of reflections to the next, and the third step (0,0,1) that from one sheet of reflecions to the next. It is necessary to know the current reflection $\mathbf{h}_1$, the reflection at the beginning of the current line $\mathbf{h}_2$, and that at the beginning of the current sheet $\mathbf{h}_3$. Tests against a vector $\mathbf{h}_{\max}$, a vector whose components are the maximum values of h, k, and l, and against $\sin\theta_{\max}$, serve as the cut-offs in reciprocal space. By including a vector $\mathbf{h}^\circ$ which may be called an origin-defining reflection, the segment to be generated may have its origin anywhere in reciprocal space. This concept is useful where the required portion of reciprocal space must be generated in 2 segments. In the Laue group $m3$, for example, the unique portion of reciprocal space is bounded by the planes $hk0$, hkk, hkh, but the reflections on the hkk plane are identical to those on the hkh plane, and to avoid duplication the reflections must be generated in 2 segments. The first segment has

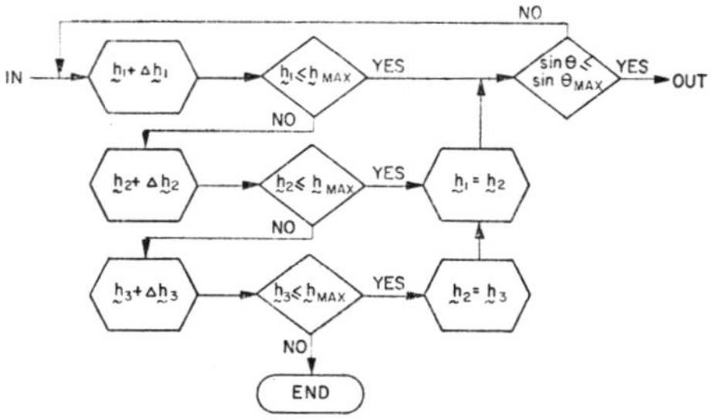

Fig. 1. *Indexing scheme.*

$\triangle \boldsymbol{h}$, the matrix whose columns are the increment directions, 1,1,1; 1,1,0; 1,0,0, and $\mathbf{h}^\circ$ 0,0,0, the second segment 1,1,1; 1,1,0; 0,1,0, and $\mathbf{h}^\circ$ 1,2,0.

The input to the scheme consists of the matrix $\triangle \boldsymbol{h}$, the origin vector $\mathbf{h}^\circ$, and the first current reflection vector $\mathbf{h}_1$. The equation

$$\mathbf{h}_1 = \mathbf{h}^\circ + \triangle \boldsymbol{h} \,.\, \mathbf{n} \tag{1}$$

will then yield the vector $\mathbf{n}$, whose components are the number of steps in each of the increment directions required to reach $\mathbf{h}_1$ from $\mathbf{h}^\circ$. When the components of $\mathbf{n}$ are known, it is possible to work out $\mathbf{h}_3$ and $\mathbf{h}_2$, the vectors to the start of the current sheet and line of reflections, respectively.

A further example will, perhaps, clarify the procedure. Suppose we wish to generate all the remaining reflections with $\mathbf{h}$ negative for the Laue group $2/m$, starting at the reflection $-3,2,4$ and indexing so that k changes fastest and l least. Then

$$\triangle \boldsymbol{h} = \begin{bmatrix} 0 & -1 & 0 \\ 1 & 0 & 0 \\ 0 & 0 & 1 \end{bmatrix}. \tag{2}$$

In order to avoid duplicating reflections with $h=0$ or $l=0$, $\mathbf{h}^\circ$ should be $-1,0,1$, and so

$$\mathbf{n} = \triangle \boldsymbol{h}^{-1} \,.\, (\mathbf{h}_1 - \mathbf{h}^\circ) = \begin{bmatrix} 0 & 1 & 0 \\ -1 & 0 & 0 \\ 0 & 0 & 1 \end{bmatrix} \cdot \begin{pmatrix} -3--1 \\ 2-\ 0 \\ 4-\ 1 \end{pmatrix} = \begin{pmatrix} 2 \\ 2 \\ 3 \end{pmatrix} \tag{3}$$

and from this

$$\mathbf{h}_3 = \mathbf{h}^\circ + \triangle\mathbf{h}_{i3} \,.\, n_3 = \begin{pmatrix} -1+0 \\ 0+0 \\ 1+3 \end{pmatrix} = \begin{pmatrix} -1 \\ 0 \\ 4 \end{pmatrix} \tag{4}$$

and

$$\mathbf{h}_2 = \mathbf{h}_3 + \triangle\mathbf{h}_{i2} \,.\, n_2 = \begin{pmatrix} -1+-2 \\ 0+\ \ 0 \\ 4+\ \ 0 \end{pmatrix} = \begin{pmatrix} -3 \\ 0 \\ 4 \end{pmatrix} \tag{5}$$

Similarly $\mathbf{h}_1$ can be shown to be $-3,2,4$.

In some cases it is necessary to impose a restriction on the data in the form of a systematic absence condition. The Laue group $4/m$ provides a good example of this. The unique region is bounded by the 3 axes, but the $h0l$ reflections are duplicated in the $0kl$ zone. To eliminate the $0kl$ zone by making $\mathbf{h}^\circ$ 1,0,0, would eliminate the $00l$ reflections however, but a systematic absence of the form $k = XN$, where X is some prime number greater than k_{max}, for the $0kl$ reflections, will eliminate all except those with $k = 0$.

A list of possible origin vectors and increment matrices for all Laue groups is given in Table I. The routines to implement this scheme require about 150 locations to set up, *i.e.* to find $\mathbf{n}$, $\mathbf{h}_2$, and $\mathbf{h}_3$, and about 60 locations to actually generate the reflections as shown in Fig. 1.

Input and output

The basic means of input and output to most diffractometers which are on line to small computers is paper tape. Usually the input is typed, and this is perhaps the most convenient method. The input section of the control program can simplify the typing by the use of a key letter print-out to label all information, and it is not difficult to arrange that the information may be typed in any order. Apart from being convenient, this also provides a built-in error correction routine. Table II shows a selection of the commands presently available with this arrangement. Key

Table I

System	Laue group	$\mathbf{h}^\circ$	Δh	Absence
Triclinic	$\bar{1}$	0, 0, 0	1, 0, 0/0, 1, 0/0, 0, 1	
		−1, 0, 1	−1, 0, 0/0, 1, 0/0, 0, 1	
		−1, 1, 0	−1, 0, 0/0, 1, 0/0, 0, −1	
		0, 1, −1	1, 0, 0/0, 1, 0/0, 0, 1	
Monoclinic	$2/m$	0, 0, 0	1, 0, 0/0, 1, 0/0, 0, 1	
		−1, 0, 1	−1, 0, 0/0, 1, 0/0, 0, 1	
Orthorhombic	mmm	0, 0, 0	1, 0, 0/0, 1, 0/0, 0, 1	
Tetragonal	$4/m$	0, 0, 0	1, 0, 0/0, 1, 0/0, 0, 1	$0kl$: $k = XN$
	$4/mmm$	0, 0, 0	1, 0, 0/1, 1, 0/0, 0, 1	
Hexagonal	$6/m$	0, 0, 0	1, 0, 0/0, 1, 0/0, 0, 1	$0kl$: $k = XN$
	$6/mmm$	0, 0, 0	1, 0, 0/1, 1, 0/0, 0, 1	
Cubic	$m3$	0, 0, 0	1, 0, 0/1, 1, 0/1, 1, 1	
		1, 2, 0	0, 1, 0/1, 1, 0/1, 1, 1	
	$m3m$	0, 0, 0	1, 0, 0/1, 1, 0/1, 1, 1	
Trigonal	$\bar{3}$	0, 0, 0	1, 0, 0/0, 1, 0/0, 0, 1	$0k0$: $k = XN$
		−1, 2, 1	0, 1, 0/−1, 1, 0/0, 0, 1	
	$\bar{3}m$	0, 0, 0	1, 0, 0/0, 1, 0/0, 0, 1	

Table II

Key letter	Message	Explanation
A	ALIGN	Align n given reflections hkl
E	EXTCND	Type sets of systematic absence codes (A to E) in the form $Ah+Bk+Cl=D+En$, for the appropriate reflection groups
G	GO	Start intensity measurement
I	INDEX	Measure particular hkl intensity
K		Stop intensity measurement
M	MTX TR	Type orientation matrix
N	NEXHKL	Set angles for particular hkl value
R	REFRFL	Type number, frequency of taking, and values of reference reflections
X	DPY ON	Turn C.R.T. display on
Y	DPY OFF	Turn C.R.T. display off

letters can also be used to cause specific operations, such as, set a particular reflection or turn to specified angles. The cheapest form of output device is paper tape, but, at least in North America, not many of the large computers have paper tape readers. There does not seem to be an inexpensive way around this problem, but an attractive alternative is an incremental magnetic tape recorder. This has several advantages:

1. It can be made compatible with any requirements of the larger computer.
2. There is virtually no limit to the amount of data which may be output for subsequent analysis, *e.g.* the intensities of the individual steps of a scan.
3. Reflection tapes may be prepared, very quickly, before the measurement run, thereby allowing more core to be used for on-line checking and may be data reduction.
4. The time taken for output is eliminated.

In our experience, however, such a device is not reliable, and it will be some time before we will be willing to trust the recorder alone.

Reliability

From a practical point of view the prime requirement of any automatic system is that it run continuously. Although the control program obviously cannot get around hardware failures, there are many other problems associated with the malfunctioning of equipment and movement of the circles which may be taken care of by the program.

One such problem which we have encountered has led to two easily programmed solutions which may have wider applicability. The problem itself was caused by electrical noise from the high voltage generator of the X-ray unit. All efforts to eliminate this noise by conventional grounding and shielding failed, and the effect of it was to reverse one or more of the flip-flops which hold the various control functions. Thus, for example, during a 2θ scan, the 2θ run flip-flop would suddenly be switched to STOP, causing a halt at that point. The initial solution was to put in a timing routine which is initialized at the start of each reflection and counts 1-second clock pulses until a preset time runs out, usually 4-5 minutes. If this happens the assumption is that 'something' has gone wrong, it does not matter what that something may be, and the computer re-initiates the program and flip-flop states and starts the measurement of the current reflection as if it were a new reflection. If the 'something' is actual failure of a component, the machine will continually try to remeasure the same reflection, in which case one could call a halt after some number of unsuccessful attempts. We now regard this routine as a last line of defence, because although it solved many of the problems it could not guarantee that the flip-flops associated with counting were set correctly during measurement. The solution to this problem which comes to mind immediately is to test all counts for significance in some way. However, this involves a fair amount of programming and it is not easy to devise tests of this sort.

A second and simpler solution is to arrange that the program continuously gives the commands necessary to keep the function flip-flops in the required state. This is most easily arranged if the control program is written so that there is one re-entrant monitor routine which supervises the flow of the program. Subroutine links may be added to or deleted from this monitor as needed and in this way all the necessary functions are monitored continuously. In many cases a single instruction will serve to control a particular function and in these cases the appropriate instructions may be added directly to the routine. This technique has proved to be very successful and because it is easy to add new control routines to such a system, it is well adapted to the situation where curing one fault will in general allow the next weakest link in the chain to become apparent. We have, for example, recently eliminated a fault which only occurred about once every two weeks.

Discussion

BUSING: The only troubles we had similar to those you discussed was occasional pickup in the computer if it was running when the X-ray generator was switched on.

BEDNOWITZ: We had a similar experience but this was due to electrical interference from a nearby work-shop.

GAFNER: We also had this problem but eliminated it by putting in our own alternator system. What sort of protection do you have against circle collision?

GABE: We have software protection as well as the usual hardware contact switches. In the control mode the program must correctly reactivate the circle drive gates every 10 milliseconds in order for the circle movement to continue.

Programming for Automatic Diffraction Data Acquisition in a Time-Shared Environment

ALLAN L. BEDNOWITZ,

IBM Watson Research Center, Yorktown Heights, New York, 10598, *U.S.A.*

Many crystallographic laboratories operate X-ray and neutron diffraction experiments under varying degrees of automation. Their instruments are controlled by plug board or hardwired programs, paper tape, or punched card controllers and dedicated small binary computers. The output of these controller-recorders may be on paper tape, punched cards, or even magnetic tape. In any case, the data output from these instruments is transmitted in some way to a larger general-purpose computer for processing and analysis. This procedure usually produces a gap in time between the logging of data and return of information to the operator. This has proved acceptable in many cases. However, there is a growing trend toward obtaining more interactive and flexible facilities in the automation of X-ray and neutron diffraction experiments. (See for example the systems described by Jarvis, 1966; Beaucage *et al.*, 1966; Corfield, 1969; Prince & Alperin, 1969; Johansson, 1969.)

The feasibility of time-sharing diffraction experiments

It is fortuitous that diffraction data acquisition is particularly well suited for time-shared computer control. The data to be acquired are 'non-volatile'. An experiment can halt and wait for reasonably long periods of time while the computer is servicing other users. The accumulated counts held in an external scaler will not be lost by a delay in computer response. If necessary, measurements can be repeated whenever required. This type of input differs considerably from the volatile data obtained with instruments such as the gas chromatograph or mass spectrometer. As a result, the disadvantages of time-sharing are minimal for the case of X-ray data acquisition.

In order to support complex control decisions and interactive data processing with additional facilities such as disk file storage, plotter graphic output, CRT displays, and large high-speed core storage, two approaches can be used. One approach uses the emerging technology of 'terminal' time-sharing operation in order to take direct advantage of a large computing center (Konnerth, 1969). A second approach which has been implemented in several laboratories is to share an intermediate-size computer (such as the IBM 1800) among several experiments (Cole & Ryniker, 1968). By sharing the cost of these larger computing systems among an appropriate number of users, it is hoped that each user will be able to satisfy most of his data-processing needs as well as his control requirements within an economical framework.

Although both approaches are currently in use in our laboratory, this discussion will be confined to programming for time sharing on an IBM 1800. One of the desirable features in a system of this kind is the ability to perform batch data-processing (complication and execution of FORTRAN programs and modification of experiment control programs) without interfering in the operation of other users. Flexibility in the addition and deletion of experiments and hardware control programs is usually desired in an experimental laboratory.

Programming costs are almost always underestimated. The use of a high-level language such as FORTRAN aids in simplifying the programming problem for most applications. Even though a high-level language may require excessive core space and execution time compared to assembly languages, the flexibility gained is well worth the cost. Also, the preparation and modification of programs is considerably facilitated by the ability of a shared system to economically support card I/O and a list printer.

Allocation of facilities

The batch processing facility has the lowest priority in a shared system. The control programs for each user rank higher in priority, exceeded by the interrupt servicing subroutines and the data channel operations.

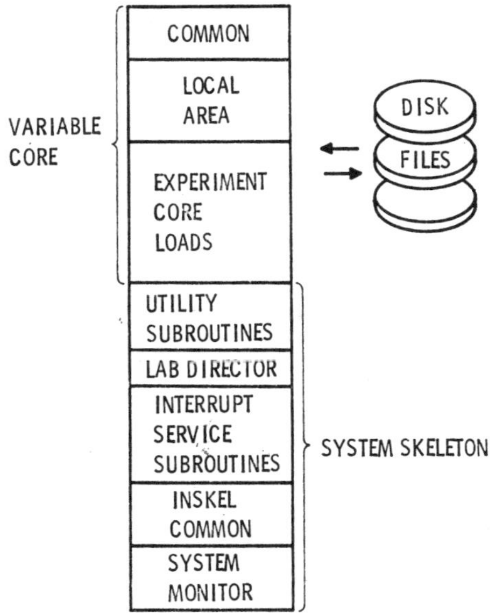

Fig. 1. *Allocation of core resources among permanent and temporary core resident programs and data arrays.*

The available core storage must be apportioned between users and the monitor in some appropriate manner. As a concrete example, consider how the core resources of an 1800 are apportioned among several user experiments by our Lab Director scheme. Each experiment receives a block of space within the core resident system SKELETON (see Fig. 1). The space allocated to each user varies according to the needs of a specific experiment. The Lab Director scheme was left open-ended as it was not initially known how many experiments would ultimately be attached to the system. The first experiment to be attached was the CCXD (Cole, Okaya & Chambers, 1963, 1965), a four-circle X-ray diffractometer which originally operated on a dedicated IBM 1620. Additional experiments added later include a pole-figure goniometer, powder diffractometer, fluorescence spectrometer, and an $x-y$ densitometer.

Data areas and space for interrupt service subroutines were assigned to each user as required. Each addition to the SKELETON reduces the size of variable core available for executing experiment or batch programs. Eventually additional blocks of core must be added as the number of users increases or other sharing schemes devised.

The data storage area in the SKELETON is referenced by a user program with a simple FORTRAN labeled COMMON statement:

COMMON/INSKEL/IDUM(550),
IDATA(300), IARRY(250) . (1)

For example, the array IDUM might represent 550 locations used by the system or other users, while the 300 locations in IDATA are used to store control data for the CCXD experiment. These data include control constants, lattice constants, orientation parameters, and hardware status indicators. IARRY is a data buffer used for storing scan data. Each subroutine requiring data stored in the permanent core area contains this COMMON statement. The data are referenced in FORTRAN by the EQUIVALENCE statement. For example, a program containing the statement:

EQUIVALENCE (IDATA(124), NW), (2)

can refer to NW (the number of steps in a step scan) in all subsequent statements, with the proper core address being supplied by the compiler.

Transfer of control between users

In a dedicated computer, a control program can go into a loop or WAIT state while a real-time operation is occurring. The computer might even be closely involved with the external operation, for example, by counting pulses from an incremental encoder or scaling circuits. If, however, these simple functions are performed by external hardware devices, considerable unused computer time is available for sharing purposes. An example of this is the time it takes an external scaler to accumulate a statistically significant intensity count. The efficiency of a time-sharing environment is heavily dependent on the system's ability to make full use of these time periods. In order to recognize the occurrence of such an interval and to avoid undue complexity and rigidity, each user can be left the responsibility of informing the monitor of his experiment's status.

In the Lab Director scheme, the status of experiments is altered by a call to the LABDR subroutine. If an experiment is about to enter a dormant state, awaiting the completion of a real-time operation, it will signal the Lab Director that it is releasing control. This is done by the statement:

CALL LABDR(0) . (3)

The zero parameter indicates an *unconditional* transfer to the Director. After this statement is executed, the only way to reactivate the experiment is upon receipt of an interrupt after the completion of some real time operation.

A user can perform a *conditional* yield to the Director and restore his active status by using an experiment number as the parameter in Statement 3; LABDR will then set the experiment 'ready' status in the Directory. It is the responsibility of each programmer to relinquish control in either fashion periodically. A maximum time slice, however, is provided. In order to maintain a simplified Director, no provision is made for saving the current condition of the user's program if his time slice should expire, or even if he should transfer control voluntarily.

Thus (3) becomes the terminal statement in each experiment control routine. Soon after initiating a real time operation, an experient program must transfer control to the Lab Director and logically vanish from variable core. Physically, the experiment control routine will be overlayed in core by the next user with 'ready' status in the Directory. If no user is in a ready condition, the Lab Director will return control to the System Director for time-sharing batch jobs. The System Director (SYSDR) is an IBM supplied monitor which contains the complex programming necessary to support time-sharing between process and non-process programs as well as the Master Interrupt Control program for sequencing priority interrupts. The System Director will interrupt a batch job, saving the current condition of all registers and filing the entire variable core area when requested to end time sharing by an interrupt subroutine. It will restore the batch job and continue execution of that job when requested to begin time-sharing by the Lab Director.

The simplicity of the Lab Director imposes additional programming requirements on each user. Each user must set his own pointers indicating current status before releasing control to the Lab Director. As the Lab Director does not keep track of the location in a program where control was released, it is up to the user to coordinate the sequence of routines to be executed upon return of control. To satisfy these requirements, a user can write a Coordinator routine complete with a status word and table of entry points. The status word acts as a pointer to the next routine required in the program sequence. An experiment program is loaded from disk storage into the variable core area by the System Director when the Lab Director executes the statement

CALL CHAIN (EXPN) . (4)

The entry point EXPN is the experiment Coordinator routine. Control must then pass from the Coordinator to the appropriate experiment control routine as indicated by the Coordinator Status Word (COSW).

Adapting a control program to the Lab Director

A time-shared diffractometer control program (AXDA – Automatic X-ray Data Acquisition) has been successfully used for several years (Bednowitz, Hamilton & Coppens, 1966). It was originally designed to operate under the Multiple Spectrometer Control System at the Brookhaven National Laboratory (Beaucage *et al.*, 1966). Subsequently AXDA was adapted for use with an IBM 1800 process controller. The latest version (III) has been modified to a modular form appropriate for operating under the Lab Director (Bednowitz, Ryniker & Cole, 1968; Bednowitz, 1968). A single module can initiate, control, and complete a major activity task of the experiment.

A module may be a subroutine of the Coordinator itself, an independent link to be loaded upon execution of a CHAIN statement, or a LOCAL (Load on Call) routine. The ultimate structure is dependent on the available core. There are significant differences between CHAIN and LOCAL transfers. A LOCAL subroutine is loaded into a reserved area by a main program CALL and must terminate in a transfer to the Lab Director or use a RETURN statement to return to the main routine. The CHAIN statement can be used to cascade main program overlays, thereby constructing control logic unlimited by the computer core size.

Communication between LOCAL subroutines and CHAIN main programs is supplied by the regular FORTRAN COMMON area in upper core. Communication between modules overlayed by other user control programs is supplied by the labeled COMMON/INSKEL/area in lower core. The INSKEL area contains control parameters and data buffers which are shared by Interrupt Service Subroutines as well as main program modules.

Each operational Module relinquishes control of the variable core area as soon as it initiates a real time operation. Upon receipt of a priority interrupt indicating the completion of that real-time operation, the associated Interrupt Service Subroutine will route control through the hierarchy of command programs until it returns to the original module. Control will continue to be routed to the same module until the basic operation is complete. Before the interrupt occurs, other experiment programs or even batch jobs may be executed by overlaying the module in variable core.

Consider the module RDMV as an example. It checks the current encoder readings of all shafts, decides on the motions necessary to reach the target angles, initiates the operation of the drive motors, and then relinquishes control to the Lab Director. On receipt of an appropriate interrupt from the last encoder to reach its target angle, the experiment ready status will be set in the Directory by a call to the entry point LBDR in the Lab Director,

$$\text{CALL LBDR(NEXP)}\,, \qquad (5)$$

where NEXP is the appropriate experiment number. This call will not relinquish control to the Lab Director, but will cause time-sharing to terminate. All interrupt subroutines terminate with a call to the Master Interrupt Control routine,

$$\text{CALL INTEX}\,. \qquad (6)$$

As soon as time-sharing is terminated and the System Director transfers control to the Lab Director, the Directory will be scanned for the next ready experiment. When a time-slice is available, the Coordinator for this experiment will be loaded from disk and executed. As indicated by the Coordinator Status Word (COSW), control will be routed to RDMV. This procedure will be repeated until RDMV is satisfied that the target angles have been achieved, whereupon it will alter the COSW to point to the sequencing module PRIMX for the next step in the experiment.

Each module performs a set of basic control, computational, or I/O operations, each of which may be repeated several times in a data run. Most modules set the COSW pointer to PRIMX before exiting. The PRIMX routines are FORTRAN-coded programs, each containing the framework necessary to sequence the steps in a specific experiment. If a change in the sequence of steps is desired, only the appropriate PRIMX subroutine needs be altered. There are currently three versions of this module. Two are for crystal structure data acquisition (PRIME, PRIMS). PRIME performs step scans over diffraction peaks, and PRIMS performs integrating scans. The third version (PRIMA) is arranged to collect TDS data along straight lines in reciprocal space. It is conceivable that by suitable construction of this module one might do powder analysis, crystal alignment, precision lattice constants, or even pole figure work on the same instrument.

Output data as well as program modules are stored in a fixed area on the disk file. The only permanent core requirements for additional experiments are for interrupt service subroutines and a SKELETON COMMON area. An attempt has been made in AXDA III to keep the complexity of interrupt service subroutines to a minimum. This necessitates additional control logic in the operational modules and more frequent CHAIN operations with an increased number of excursions to the disk. Disk file operations can use up considerable time in unproductive mechanical movements. A balance must be achieved between the size and complexity of the core resident Interrupt Service Subroutines and the time required for service of frequent requests to load control modules.

The main routines in all the modules are written in FORTRAN, making the initial programming and subsequent revision much easier than if they had been in assembly language. This has been especially useful in allowing rapid alternation of the control modules, as hardware requirements were modified. For example, the CRT display was easily added to the functions of the ICOUT module as soon as the scope was available. Also, RDMV was rapidly altered when the CCXD encoders were replaced with new and distinctly different types.

Subroutines callable by FORTRAN have been written to handle bit manipulation, shifting, masking, and digital I/O. These routines are incorporated into all the experiments on the system. As a result, the full power and flexibility of FORTRAN is available in the control routines even when they perform non-FORTRAN types of operations. Assembly language is used only where space and time are at a premium, such as for the core resident Interrupt Service Subroutines. Besides the micro-functions mentioned above, several short subroutines are shared by all users of the system. These subroutines perform data manipulation such as curve smoothing, peak searching, line shape analysis, and least-squares parameter fitting.

Using the disk storage

Output data as well as program modules can be stored in fixed areas on the disk files. As a safety measure, the data stored in the SKELETON COMMON area are periodically stored on disk. This area is restored from disk storage when a major control program error is discovered by the system. The entire system is also reloaded from disk when major errors such as illegal operation codes or storage protect violations are encountered. Each user may have a special subroutine for handling reinitialization or termination of his operation in case of a Reload. Similar errors occurring in a batch job have no effect on the control programs, which are not active during batch operations. An offending batch job is simply ejected with the appropriate error message and the next job in the hopper is read into core.

The space available for storage of programs on the disk is more than adequate. However, it seems that no matter how much space is available, it eventually becomes full of stored data. One solution has been to periodically dump the data onto binary cards for permanent storage. As an alternative, the data can be transmitted to a large system through a direct link (in our case to a 360/67) and stored on large private disk files or magnetic tapes. Continuous scanning experiments such as powder diffraction and pole figures tend to produce large files of raw data very rapidly. At some point after data reduction, analysis, and presentation of the results, someone has to make a decision as to what to do with the raw data.

Limitations

The main difficulty encountered in the Lab Director scheme has been achieving the proper balance of operations in order to avoid unnecessarily frequent disk transfers. This is especially true with regard to the extensive transfers needed when starting or terminating time sharing for non-process batch jobs. The system described here currently operates with the slowest disk drives available on the IBM 1800. This is usually not a serious problem for the batch jobs, since they are not operating in real time. The time wasted, however, results in unnecessary delays for all experiment programs. AXDA III handles the problem by releasing control *unconditionally* only when a sufficiently long dormant time period is expected (>5 sec) in order to justify the disk operation overhead. Frequent *conditional* releases allow other experiment programs to execute whenever they are ready. The problem may also be alleviated by using faster disk drives.

Certain experiments do not seem to be compatible with the simple time-sharing system described here. These experiments have *all three* of the following characteristics: short intervals between data inputs, (<10 sec); extremely long runs (hours); and the requirement for extremely large and complex interrupt handling routines (>2000 words). This type of experiment requires a multiprogramming environment or a dedicated computer. A multiprogramming system would allow several reasonably large programs to remain in core simultaneously while execution time is shared, either by time-slicing, receipt of priority interrupts, or both.

References

Beaucage, D. R., Kelley, M. A., Ophir, D., Rankowitz, S., Spinrad, R. J. & Van Norton, R. (1966). *Nucl. Instr. Methods* **40**, 26.

Bednowitz, A. L. (1968). Pittsburgh Diffraction Conference Abstracts.

Bednowitz, A. L., Hamilton, W. C. & Coppens, P. (1966). Pittsburgh Diffraction Conference Abstracts.

Bednowitz, A., Ryniker, R. & Cole, H. (1968). *IBM Research Report* RC2313.

Cole, H. (1969). *IBM J. Res. Develop.* **13**, 5–14.

Cole, H., Okaya, Y. & Chambers, F. W. (1963). *Rev. Sci. Inst.* **34**, 872–876.

Cole, H., Okaya, Y. & Chambers, F. W. (1965). *Laboratory Management* **3** (11), 21–27.

Cole, H. & Ryniker, R. (1968). Pittsburgh Diffraction Conference Abstracts.

Corfield, P. W. R. (1969). *Acta Cryst.* A**25**, Suppl. S71.

Jarvis, R. A. (1966). *J. Sci. Inst.* **43**, 899–907.

Johansson, Karl-Erik (1969). *Acta Cryst.* A**25**, Suppl. S79.

Konnerth, K. L. (1969). *IBM J. Res. Develop.* **13**, 132–138.

Prince, E. & Alperin, H. A. (1969). *Acta Cryst* A**25**, Suppl. S74.

Discussion

SCHILLING: Is there much idle time on the IBM 1800 during operation?

BEDNOWITZ: The wait light of the 1800 is almost always on since the control programs are large but fast. As the number of time-sharing users increases, the idle time will no doubt be decreased.

MASLEN: We have a diffractometer on-line to a large computer, but our experience in getting it into operation is such that I would never attempt it again. A prime difficulty is that one must often wait for other users in order to check out a fault. How do you overcome this problem?

BEDNOWITZ: We have lights which indicate which of the modular routines is being executed at a given time and which is in core at the time of a fault.

TOPIC J

PROTEIN PROGRAMS

CHAIRMAN: G. KARTHA

Some Programs used in Protein Structure Studies

GOPINATH KARTHA

Roswell Park Memorial Institute, Buffalo, N.Y. 14203, U.S.A.

During the last decade, we have been involved in writing and developing many programs for crystallographic computations on the second and third generation of electronic digital computers. Though most of these programs were written for application to the general case, their operational efficiency and flexibility in the context of protein structure studies were always kept in mind. Some of the programs used were, of course, only applicable to the specific case of the protein studies.

During the period of development of these programs, the machines on which these computations were done have undergone vast changes in speed as well as in their hardware and software capabilities, requiring changes in the philosophy involved in planning, writing, testing, and running these programs. Hence, during the last few years, the programing has been done almost entirely in non-machine-oriented symbolic language, like Fortran, to enable ease of conversion from machine to machine of different hardware configurations. However, in planning for computational economy, flexibility, and speed, at least in those programs that are extensively used, the hard- and software capabilities have to be kept in mind.

The programs that are now being used most often have been designed for a 32K, 36-bit word, tape-oriented system like the IBM 7040. However, retailoring the programs to suit another machine of equivalent, though not identical, capabilities, operating in time-shared modes, or essentially disc-oriented like the CDC 6400 has not been difficult*.

The programs used in these studies can be divided into major categories as follows:

1. Data collection
2. Data reduction
3. Fourier synthesis
4. Least-squares refinement
5. Protein phase evaluation
6. Phase refinement
7. Special purpose

In planning the programs for personal use, rigidity of input-output format requirements were sacrificed for computing speed for the specific case. Input-output formats were kept flexible but consistent enough to permit smooth flow of data from one type of computation to the next. Most manual intervention and card handling was avoided by using the output tape of one program as input to the next. Wherever reflection data card handling was inevitable, an attempt was made to prepare summary cards containing as much data from the different derivative crystals as possible for that reflection. Variable format control cards read in at object time were used to choose the required information from these summary cards or tapes. No elaborate attempts were made to rid the programs of all bugs that bedevil the unwary user ignorant of the basic aims and limitations of the program.

These programs have been successfully modified to run on RCA 70/45 spectra computer which is a 31-bit word tape disc oriented machine.

Short descriptions of some of these programs are given below:

Data collection

The reflection data from the protein and derivative crystals are collected in shells of increasing resolution. For a given crystal mount, the input to the automatic diffractometer consists of prepunched cards containing the three angles $\varphi, \chi, 2\theta$ (Furnas & Harker, 1955) of the reflection. These cards also contain the indices of the reflection and information to direct the movement of the arcs from their positions at the end of the last measurement. The input deck is punched by the program in an order that minimizes the angular movement of the orienting devices between reflections. The orientation of the crystal is flexible, so that any real or reciprocal axis may be parallel to the diffractometer axis. The program also has the option of listing reflections in sorting orders more suitable for manual methods of data collection.

Data reduction

These programs, like all similar programs, convert the reflection intensities to structure amplitudes after applying geometrical, absorption, and splitting corrections. For each reflection, an estimate of standard deviation for the amplitude is given and overall intensity statistics are listed at the end in any required scattering angle intervals.

For protein derivatives, this step also includes the scaling of the data to that of the parent protein; this scaling, if necessary, can be done within specified resolution ranges. This also gives information of amplitude variations due to heavy atom substitutions in the different derivatives and their suitability for protein phase evaluation. For any reflection from the derivative crystal, the corresponding free protein amplitude is obtained by a table look-up operation by initially loading all the free protein amplitudes into the core of the machine in a three-dimensional table, arranged on the basis of the reflection indices.

Fourier program

This is one program that is repeatedly used and much effort was taken in making it efficient and convenient. It will be discussed in some detail. The nucleus of the present program was made available to me by Prof. Y. Okaya and to him my thanks are due. Important considerations in writing all Fourier programs are the magnitude of the input-output data, computational efficiency, and extent of storage requirements. The overall operational speed of the program depends on these competing requirements and usually it is necessary to optimize these for the machine on which the program is run. These considerations occur in a very acute form for Fourier programs designed for large structures.

In protein studies, in addition to computing efficiency, the ease with which the output can be interpreted and handled becomes very important. In the study of smaller molecules where atomic resolution exists, Fourier maps are mostly used for obtaining a set of co-ordinates representing the atomic centers, either visually from the output sheets or by a peak search routine in the computer. Once a complete set of co-ordinates is obtained, focus usually shifts away from the Fourier maps to least-squares techniques. In protein studies, unfortunately, we rarely come to this ideal goal, and hence have to deal with electron density contours, for a large part. Hence it is essential to have the Fourier output in the actual linear scale in a form that is easy to contour and inspect. Furthermore the need for computing maps with different types of coefficients requires flexibility in the stage at which Fourier coefficients are prepared.

The Fourier program widely used in our work and written in Fortran IV contained the following parts.

FOUR	main program; sections SETUP, BITPRD, TRDSUM. SETUP – subroutine to prepare binary reflection tape. BITPRD reads binary reflection tape, accumulating first and second summation results – TRDSUM computes third summation
INDEX SPGRP	subroutines to prepare the coefficient tape. Called by SETUP.
CONTUR PRINT	subroutines for output in suitable form. Called by TRDSUM.

The program handles the different space groups through a short SPGRP routine which is space-group and sometimes even summation-order specific and may have to be rewritten for different kinds of jobs. The index order, range, and summation intervals are flexible so that a satisfactory and convenient linear scale can be obtained for the printed output which may be called in through the subroutine PRINT or the subroutine CONTUR as desired. In the form called by PRINT, we get the usual output of five figure numbers on a 16/16 net. However, the output that has proved adequate and convenient for most purposes of small-molecule as well as protein studies is called in through the subroutine CONTUR. This lists the density values in a compressed, coded output at points of a net up to 61/53 per printed page. Parameters, read in at object time, control the nature and range of these output codes.

In the present version, the Fourier series may be computed at cell edges up to 1/120 along the third summation direction and 1/200 along the other directions with complete flexibility. Index maximum limits at present are 40 for the third sum and 50 for the other two. The reflection input is from a binary tape, is not stored in core, and is unlimited in number. The program can perform, in addition to Fourier and difference Fourier sections, other types of summations such as section projections, bounded sections, etc. The different types of summations can be requested through input specifications. Furthermore, flexibility of multiple job operations is obtained by use of input parameters for coefficient modifications and variable format read in at object time. Almost exact scale for output was obtained by the complete freedom of choice in the summation intervals along the three directions. The trigonometric tables for the fractions of the three directions are computed at the beginning of a job and stored. At later stages of the computation, these are used in building up tables (Beevers & Lipson strip type) as need arises and computational speed demands.

For the purpose of discussion, we shall assume that the summation order is *LKH*, *i.e.* sections of map in the *YX* plane are computed with *X* running across the page and *Y* down. In actual summation, any index order can be chosen for computing efficiency or convenience of output format.

With a 32K word core machine, the approximate core area used by the program is given in Table I with more details about use of work storage in Table II. It is seen that most of the storage is used for accumulating second summation results. Area for third summation results is shared by trigonometric tables, intermediate first summation results, etc., that are needed during the accumulation of the second

Table I. Approximate storage allocation for the Fourier program using a 32K word core machine

	No. of words in storage
Program and subroutines	5K
Storage for results	22K
Input-output buffers	2K
Systems monitor (IBSYS)	4K

Table II. Storages for intermediate results and their status during program run

Storage designation	Range of parameters	Storage in words	General remarks
Cos Y, cos Z, sin Y, sin Z	All in range 0 to 1	800	Computed once per job
Cos HX, sin HX	For all H indices $X=0$ to $\frac{1}{4}$	1,000	Computed once per job
Second summation sums $C(Z,Y,H)$; $S(Z,Y,H)$	All H indices $Z=Z_L$ to Z_H; $Y=Y_L$ to Y_H	16,000	Cleared before every pass of binary reflection tape
Remaining storages are shared by BL tables and third summation results			
Cos X, sin X	$X=0$ to 1	400	Computed once per job
Cos LZ, sin LZ	All L indices; $Z=Z_L$ to Z_H	1,000	Computed once per tape pass
Cos KY, sin KY	Present K only; $Y=Y_L$ to Y_H	120	Table reprepared every time K changes
First sum results. Four types. $P(Z,H)$, $Q(Z,H)$, $R(Z,H)$, $S(Z,H)$	All indices H; $Z=Z_L$ to Z_H	2,000	Present K only. Cleared after second summation results are augmented for current K
Third summation results RHO(X,Y)	$X=0$ to 1, $Y=Y_L$ to Y_H	5,000	For present Z section. Cleared for each section and reused. Same area used as storage for BL tables and intermediate first sums during tape read

sum results and can be cleared before the start of the third summation. Some third summation area is cleared for use between calculation of Z sections after the results are put on tape for listing. This is repeated till all Z sections corresponding to the block computed during accumulation of the second sum table are finished. In this way the number of Z sections computed per single reading of binary reflection tape is maximized.

The main program in its section SETUP reads the reflection tape and writes binary tapes containing indices and coefficients prepared by calling subroutines INDEX and SPGRP. If the reflections are sorted on indices irrespective of signs, then only one entry is made for a given positive index triplet. The SPGRP routine generates the coefficients corresponding to the symmetry requirements. In the binary reflection tape, each index triplet is followed by eight coefficients needed for the most general case of Fourier series factorization (Stout & Jensen, 1965) into products of trigonometric terms. In addition, a parity number is added as the 12th entry per index to signal the non-zero coefficients of the set. This procedure, while increasing the size of the stored programs, preserves the flexibility of the program and at the same time offers faster branch points for dealing with reflections having many zero coefficients, as is the case with Patterson coefficients and centrosymmetric reflections. Intervals along the three directions can be chosen as desired, within limits, and trigonometric tables are prepared for each interval corresponding to index frequency unity. From this set, the complete Beevers Lipson (BL) Tables are prepared at the beginning of the job corresponding to all H indices and X intervals to a maximum of $\frac{1}{4}$ the cell edge. These tables are used for fast third summation by making available the necessary trigonometric terms by a fast table-look-up operation. For each block of Z sections, the program requires one pass of the binary tape. At the beginning of this pass, BL Tables are prepared for the cos LZ and sin LZ for all L indices and Z sections corresponding to Z_H to Z_L where H and L correspond to the highest and lowest Z values computed during the present pass of the binary tape. The number of sections that can be computed with a single pass of the binary tape depends on the size of the Fourier section XY computed. For the present arrangement of the program, XY section of ribonuclease maps of size 61×45 are computed at the rate of 10 sections per pass, while the increase of the size to 121×53 on a different orientation (XZ) reduced it to 6 sections per pass.

In performing the synthesis, the main program section BITPRD reads the binary reflection tape. As each term is read, four sets of one-dimensional partial sums corresponding to each third sum index H and Z sections Z_L to Z_H are augmented using the current first summation BL Tables. When the second summation index K changes (this should be made as infrequent as possible by proper sorting), input and first summation is suspended and BL strips for cos KY and sin KY are prepared for all Y corresponding to the last second summation index K. The second summation partial sums are then augmented for the next block of sections Z_L to Z_H, for all Y intervals and for all third summation indices H (this is the largest storage block needed for the program). The storage for the first partial sums are then cleared and the input operations of reading reflections and accumulating first sums continue till the next K index change is signalled. This sequence of operations proceeds till the end of the reflection tape is reached. The reflection tape is rewound. In main program section TRDSUM the third summation storage area is cleared and the third summation section for Z_L is computed using the internally stored second summation sums and third summation BL Tables. For the third summation $X=0$ to $\frac{1}{4}$, odd and even indices H, of reflections are separately summed and the complete section $X=0$ to 1 calculated by sign permutations. After one section is completely calculated, the results are output in one section by calling subroutine CONTUR. The third summation area is cleared and the next section Z_L+1 is computed using the second summation sums corresponding to that value of Z. These operations are continued till the section corresponding to the highest value Z_H in the present block is completed. If additional sections are needed, the new BL Tables for the first index are prepared for the next block starting with $Z_L=Z_H+1$ and another pass of the binary reflection tape is initiated by transferring control to the section BITPRD. The cycles of reading binary tapes and writing output sections continue till all required sections are obtained. Another job, or the same job corresponding to a new range of Y intervals, can be performed by reading additional input cards giving job specifications and re-reading the reflection tape.

Tape operation times are reduced by using two high density tape units (800BPI) and dividing the reflections into two binary tapes so that, while reading one, the other can be rewound and be ready for the next pass. Tightly blocked formats for writing the binary tapes, adequate buffering of input-output operations, and provision for ample computation between I/O instructions can appreciably reduce the running time on the computer.

Sample running time of a protein Fourier

Machine – IBM 7040, 32K word storage, 4 tape units used (high density 800 BPI)
Problem Ribonuclease 2 Å map. c-axis sections 60/88 net, 100 sections with about 9000

reflections. $HMAX=16$, $KMAX=20$, $LMAX=27$. Number of 61×45 sections per tape read = 10. Total running time for 100 sections = 52 min.

This running time is reduced by a factor of 3 on the similar but faster core machine 7044 and perhaps by another factor of 10 on the CDC 6400.

Of the above running time, further break-down can be given as below. About 9 minutes were needed for reading the BCD reflection tape under variable format, selecting reflections as specified by input parameters, and writing binary tapes through appropriate calls to subroutines INDEX and SPGRP. For a single pass of binary tape, computing the first summation and accumulating second summation sums for 10 sections required about 150 seconds, whereas 95 seconds were needed for performing the third summation for the 10 sections and outputting it in compressed, coded form through call to the subroutine CONTUR. Hence, in all, one pass of binary tape giving 10 sections needed slightly over 4 minutes.

Least-squares program

Standard programs used in small-molecule studies were also used by proper arrangement of core allocations. Storage of a large number of reflections along with their indices and standard deviations was made possible by packing the indices, amplitudes, and sigmas into one 36-bit storage location and unpacking them when needed. The packing and unpacking routines were written specifically for the present machine but replacing them by Fortran subroutines is also easy.

Protein phase angle program

This program is specific for the protein study and a block diagram (Kartha, 1964) is shown in Fig. 1. The input parameters

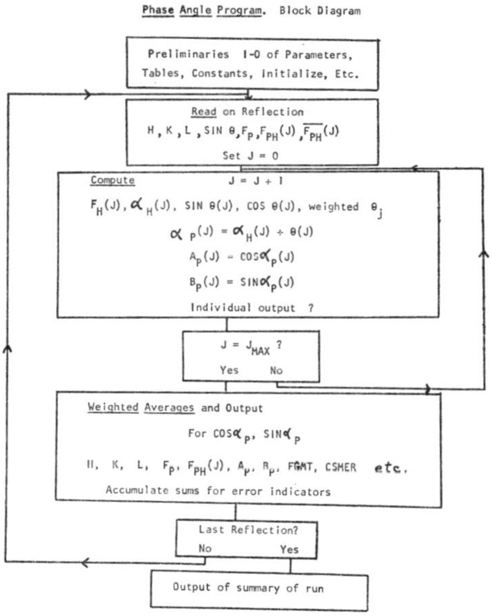

Fig. 1. A block diagram of the protein phase angle program. From each heavy atom derivative, protein phase angles are estimated using isomorphous series and anomalous scattering information. Phase estimates for all derivatives are then combined to give the final best phase estimate.

to the program consist of the protein and heavy atom derivative amplitudes which may include anomalous scattering results. At the start of the computation, heavy atom positions, occupancies, and thermal parameters as well as reliability factors for the various ranges of measurement are supplied. The program estimates the protein phase angle for every reflection read in and also assigns a figure of merit for the estimate. The program also writes an output tape in a form suitable for immediately computing Fourier or difference-Fourier maps with or without weights assigned to the reflection amplitudes. The program also gives a summary of figures of merit and other relevant results in terms of data resolution.

Phase refinement program

The appropriate co-ordinates and the corresponding modified density values from Fourier maps are written on a binary tape and the tape is used in this program in computing calculated phases for an electron density model. This program is planned to have the flexibility of a structure factor program with speed comparable to that of a Fourier program. Phase angles calculated in this program can be used for modifying phase angles of reflections of low figures of merit obtained from the phase angle program (Kartha, 1968). This phase modification is flexible and depends on the estimated figure of merit as well as the amplitude calculated on the basis of the electron density model. Figs. 2(*a*) and 2(*b*) show the density sections of ribonuclease before and after such refinement.

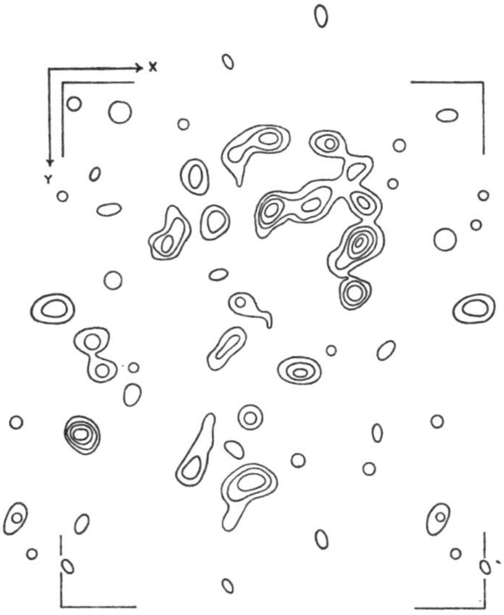

Fig. 2(*b*). *Same section of ribonuclease* 2 Å *electron density map as Fig.* 2(*a*) *but after phase angle refinement of reflections whose phases are poorly estimated by the phase angle program. The map has, in general, a cleaner background and better continuity than in Fig.* 2(*a*).

Miscellaneous programs

Many other programs that are only of interest in connection with protein studies have also been written. One such program loads the electron density points of the model with designated boundary in the machine so that a series of replotted sections in any new direction and linear scale can be obtained and printed by use of appropriate transformation matrices. Another program plots the distances between the α-carbons of the main chain of the protein in a two-dimensional display, which easily illustrates some of the features of the tertiary structure in a simple way. Programs have also been written for computing the radius of gyration and small-angle scattering properties from the protein electron density model of known molecular boundary.

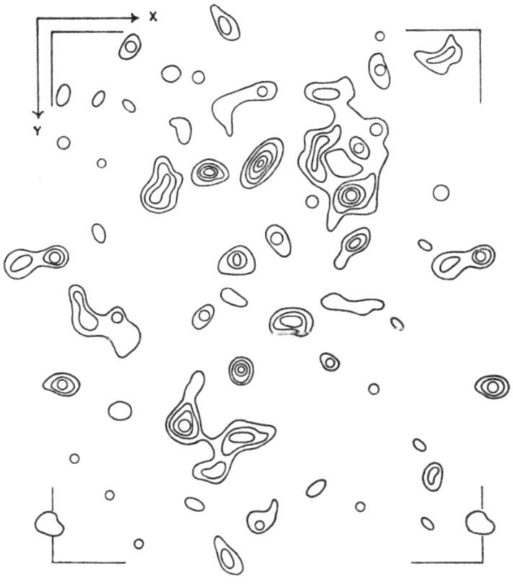

Fig. 2(*a*). *A section of ribonuclease* 2 Å *electron density map at section* $Z = 0{\cdot}170$ *before phase refinement. Contours at equal but arbitrary intervals.*

Acknowledgement

I have great pleasure in recording my thanks to the National Science Foundation and the National Institute of Health, as well as to the New York State Department of Health for supporting these studies. I also would like to thank the staff of the Roswell Park Memorial Institute computing center for their co-operation. I thank Prof. David Harker for his interest and encouragement.

References

Furnas, T. C. & Harker, D. (1955). *Rev. Sci. Instr.* **26**, 449.

Kartha, G. (1964). *Abstracts of ACA Meeting*, Bozeman.

Kartha, G. (1968). *Abstracts of ACA Meeting.* Buffalo.

Stout, G. H. & Jensen, L. H. (1965). *X-ray Structure Determination*, p. 445. New York: Macmillan.

TOPIC K

FOURIER TECHNIQUES

CHAIRMAN: F. R. AHMED

Fourier Techniques

F. R. AHMED

Biochemistry Laboratory, National Research Council of Canada, Ottawa 7, Ontario, Canada.

Generalized Fourier expression

The Fourier summations which are evaluated for three-dimensional Patterson, electron-density, and difference maps require the minimum number of arithmetical operations if computed according to the general expression

$$\varrho(XYZ) = \frac{1}{V}\Big\{F(000) + \sum_h \sum_k \sum_l \Big\{ \sum_{i=1 \to N_j}^{\text{group } j} [R(hkl) \operatorname{trg} 2\pi|h|X \times \operatorname{trg} 2\pi|k|Y \operatorname{trg} 2\pi|l|Z]_{ij} + \sum_{i=1 \to M_j}^{\text{group } j} [S(hkl) \operatorname{trg} 2\pi|h|X \times \operatorname{trg} 2\pi|k|Y \operatorname{trg} 2\pi|l|Z]_{ij} + \text{similar terms of other groups of reflexions}]\Big\} \quad (1)$$

where

V	unit-cell volume
X, Y, Z	fractional coordinates of the grid points
N_j	number of R terms in group $j(=0,1,2, \text{ or } 4)$
M_j	number of S terms in group $j(=0,1,2, \text{ or } 4)$
R	$=m[\pm A(hkl) \pm A(\bar{h}kl) \pm A(h\bar{k}l) \pm A(hk\bar{l})]$ triclinic $=m[\pm A(hkl) \pm A(\bar{h}kl)]$ or $m[\pm A(hkl) \pm A(hk\bar{l})]$ monoclinic $=m[\pm A(hkl)]$ orthorhombic
S	similar expressions to R for $B(hkl)$
trg	cos or sin
m	multiplicity $(=2,4, \text{ or } 8)$

It is important to note that *for systems higher than orthorhombic*, the electron-density expressions listed in the *International Tables for X-ray Crystallography* (1952) do not include the equivalence relationships between reflexions within the same octant of reciprocal space. As an example, in the tetragonal system $F(hkl)$ and $F(khl)$ are related but both must be substituted in the electron-density expressions. The number of terms N_j (and M_j), in general, is not necessarily the same for all groups of reflexions. The multiplicity m is meant to describe the number of equivalent reflexions in the full reciprocal space for the types $h00$, $0k0$, $00l$, $0kl$, $h0l$, $hk0$, hkl, and should have the value 2, 4, or 8.

In a generalized program, the different groups of reflexions can be described fully with a maximum of *two* expressions of the form

$$c_1h + c_2k + c_3l = c_4n + c_5 \quad (2)$$

where c_1, c_2, c_3 have the values 0, 1, or 2; $c_4 = 2$ or 4; and $c_5 = 0$, 1, 2, or 3. If the groups are to be based on the parities of the three indices, then in order to allow for all possibilities these parities must be worked out to modulus 4, since modulus 2 would suffice only for the triclinic, monoclinic, and orthorhombic systems.

Computation procedure

Fourier calculations usually involve the following main steps:

1. sorting of the data on $|h|$, $|k|$, $|l|$;
2. calculation and storage of the R_{ij}, S_{ij} amplitudes;
3. evaluation of the Fourier summations in three separate stages of single Fourier summations, and output of the results.

The sorting in step (1) is done on the *absolute values* of the three indices, regardless of the signs. The sequence in which the three indices are to be sorted is usually dependent on the preferred layout of the Fourier results. This preliminary step has a large effect on the execution time of steps (2) and (3), and does not usually have to be done more than once, at the data reduction stage. For generality of discussion, let $|h_1|$ denote the index that varies least frequently, $|h_2|$ the intermediate varying index, and $|h_3|$ that which varies most frequently in the sorted data.

The Fourier amplitudes R_{ij}, and S_{ij} if present, once calculated in step (2) can be stored and employed repeatedly for computing the different sections of the three-dimensional Fourier synthesis. In effect, they replace the original $A(hkl)$, $B(hkl)$ data.

The three separate stages of summation performed in step (3) are as follows:

I. $C_{ij} = \sum_{|h_3|} R_{ij} \operatorname{trg} 2\pi|h_3|X_3$ for constant values of $|h_1|$ and $|h_2|$

II. $D_{ij} = \sum_{|h_2|} C_{ij} \operatorname{trg} 2\pi|h_2|X_2$ for constant values of $|h_1|$

III. $G(XYZ) = \sum_{i,j} \sum_{|h_1|} D_{ij} \operatorname{trg} 2\pi|h_1|X_1$

and similarly for the S_{ij} terms, if present. Stage I is usually computed for one value of X_3 at a time (or more if there is sufficient memory); stage II for all values of X_2 (if possible) at every change in $|h_2|$; stage III for all values of X_1 (if possible) at every change in $|h_1|$. The locations holding the partial sums of stage I can be reassigned after execution of the succeeding stage II, and similarly the stage II sums do not have to be saved once the succeeding stage III is executed.

Some programs limit the summations to the range 0–0·25 for X, Y, Z as for the Beevers-Lipson strips, keeping the sums separate, then combining the results in different ways for the complete range of coordinates 0–1·0, or as needed for the asymmetric part of the unit cell. Details of this procedure have been described by Cruickshank, Pilling, Bujosa, Lovell & Truter (1961), Sly & Shoemaker (1961), and Rollett (1965).

In Fourier programs, the evaluation of trigonometric functions by power series is usually very wasteful; table look-up is much faster and should be recommended as long as a sensible compromise in the length of table can be reached. One such compromise is to store both 10000(cos and sin $(2\pi x)$] in one word as integer numbers for values of x between 0 and 0·1250, at intervals of 0·0001. In this range, both functions are always positive and hence can be combined and separated easily. The cos and sin values for $x > 0{\cdot}125$ may be obtained by simple manipulation of the stored quantities. The same trigonometric table and look-up routine can also be employed in other crystallographic programs.

Distorted and undistorted maps

Scheme 1

Computation is fastest if the grid lines are chosen parallel to the unit-cell axes, and also the space group symmetry relationships are best observed if each axial length is divided into an even number of equal parts. However, mechanical restrictions on the printer output make it impossible to print these Fourier results on a properly scaled mesh which corresponds to the unit cell. The printed grid in this case is distorted and cannot be employed directly for measurements of interatomic distances.

Scheme 2

At some sacrifice of computer time, which is not too critical with present-day fast computers, it is advantageous to have the Fourier results printed in an undistorted and properly scaled net. This can be achieved by computing the Fourier map at the points of an orthogonal grid superimposed on the sections of the asymmetric part of the unit cell as shown in Fig. 1(*a,b,c*) and by having the results printed in this orthogonal net. Thus, by stating the unit-cell parameters, the starting position (X_0, Y_0, Z_0), the upper limits (X_m, Y_m, Z_m), and the number of subdivisions per Å, the program can be arranged to make all the necessary transformations to the orthogonal net and to calculate the intervals and the lower and upper limits to cover at least all the required space. If the decimal points are printed on the result sheets, then they can also serve to identify the exact location in the unit cell where the Fourier expressions are evaluated.

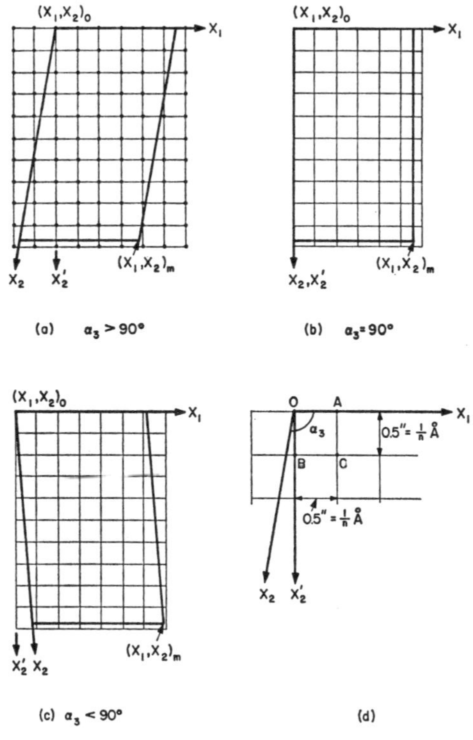

Fig. 1. *Geometry of the undistorted Fourier maps.*

These two grid options have been included in one single Fourier program written by Ahmed (1966) in FORTRAN IV for use on the IBM/360 system. In scheme 1, the sections are for X_3 constant; in each printed section the rows are for constant X_2, and the columns are for constant X_1. In scheme 2, X_3 is constant for each section, X_2 is constant for each row, but X_1 is different for points in the same column except when α_3 is 90°. Thus, in the program no distinction is made between the two schemes during the first two stages of summation, but different procedures are followed during the third stage.

For the IBM 1403 on-line printer, a $\frac{1}{2}''$ spacing between the printed mesh points is obtained in the horizontal direction by allowing 5 point positions for each column, and in the vertical direction by giving three single line feeds between the printed lines. Thus, in scheme 2, if n is the number of subdivisions required per Å (usually $n=3$ or 4), the intervals in fractional coordinates are calculated from Fig. 1(*d*) as

$$\Delta X_1 = 1/(na_1)$$
$$\Delta X_2 = 1/(na_2 \sin \alpha_3)$$
$$\Delta X_3 = 1/(na_3)$$

The fractional coordinates of the points O, A, B, C in Fig. 1(*d*) are as follows:

O:	$(X_1)_0$	$(X_2)_0$	X_3
A:	$(X_1)_0 + \Delta X_1$	$(X_2)_0$	X_3
B:	$(X_1)_0 - (\cos \alpha_3 / na_1 \sin \alpha_3)$	$(X_2)_0 + \Delta X_2$	X_3

C: $(X_1)_0-(\cos\alpha_3/na_1\sin\alpha_3)+\Delta X_1$
$(X_2)_0+\Delta X_2\quad X_3$

The map scale in this case is $(2/n)$ Å per inch.

Program options

In order to enhance the flexibility and performance of a Fourier program, it is desirable and practical to include the following options:

1. to compute for any space group without having to do the computation in the triclinic system, remembering the restriction stated earlier on systems higher than orthorhombic.
2. to allow computation of both the grid schemes described in the preceding section.
3. to select the amplitudes corresponding to a Patterson, electron-density, or difference maps as appropriate.
4. to input a sharpening function for Patterson syntheses.
5. to modify or exclude certain reflexions, either by specifying their indices, by comparing $|F_o|$ and $|F_c|$, or by stating an upper limit for $\sin^2\theta$.
6. to scale the structure amplitudes and Fourier results.
7. to select the grid intervals and limits, the number of line feeds, the field length, and the number of digits to be printed on the right of the decimal point.

Differential syntheses

Refinement of atomic positions by differential syntheses is performed usually by calculating the positions of maximum electron density for the observed and for the calculated structure factors, then combining these results to correct the observed positions for finite summation. One main assumption in the mathematical derivation of the peak positions is that the shifts $(\varepsilon_x,\varepsilon_y,\varepsilon_z)$ Å from the assumed positions (x,y,z) Å are small enough to allow omission of the second and higher powers of the shifts. In order to satisfy this condition, especially in the final stages of refinement and when the finite summation errors are large, it is necessary to assume slightly different starting coordinates for the obs and calc syntheses, which in general are slightly different from those used in calculating the structure factors and phases. Thus, in the most general case three different sets of atomic coordinates have to be read at input time, if the refinement cycle is to be performed in one run on the computer.

The shifts for each peak position are determined from three linear equations (Booth, 1946) of the form

$$\varepsilon_x\frac{\partial^2\varrho}{\partial x^2}+\varepsilon_y\frac{\partial^2\varrho}{\partial x\partial y}+\varepsilon_z\frac{\partial^2\varrho}{\partial x\partial z}=-\frac{\partial\varrho}{\partial x},\ \ldots\ \text{etc.,}\qquad(3)$$

where the electron-density partial derivatives are to be evaluated at the assumed atomic positions (x,y,z). A differential synthesis program written in FORTRAN IV by Pippy & Ahmed (1969) for use on the IBM/360 system, evaluates the electron density and its first and second order partial derivatives for a maximum of 30 atoms using both the obs and calc data simultaneously. The program then computes the shifts, applies the back-shift corrections to the observed peak positions, and outputs the results, all in the same run. With this arrangement it is not necessary to store the R_{ij} and S_{ij} amplitudes as was the case for the Fourier program. These amplitudes are calculated for one combination of $|h|$, $|k|$, $|l|$ at a time, employed in the calculations, and then discarded. The Fourier summations for ϱ, $\partial\varrho/\partial x$, ..., $\partial^2\varrho/\partial x^2$, ..., etc., are computed in three stages similar to those described for the Fourier program. Also, the same trigonometric table and look-up routine are employed in this program.

The corrected atomic positions are computed from the general expression

$$x_{\text{correct}} = (x)_{\text{S.F.}} + (x+\varepsilon_x)_{\text{obs}} - (x+\varepsilon_x)_{\text{calc}} \quad (4)$$

where $(x)_{\text{S.F.}}$ is the coordinate used for calculating the structure factors and phases. The estimated standard deviations are calculated by the general expression

$$\sigma(x) = \frac{2\pi(\Sigma h^2 m|F_o - F_c|^2)^{1/2}}{a \,.\, V\,|\partial^2\varrho/\partial x^2|}\ \text{Å} \quad (5)$$

The equations in this section are for the atomic coordinates expressed in Å, but the program works throughout in terms of the fractional atomic coordinates so that the input and output parameters are consistent with the other crystallographic programs in the system.

References

Ahmed, F. R. (1966). *Fourier Program for Distorted and Undistorted Nets. World List of Crystallographic Computer Programs.* 2nd ed. Appendix, p. 52, program no. 539.

Booth, A. D. (1946). *Trans. Faraday Soc.* **42**, 444.

Cruickshank, D. W. J., Pilling, D. E., Bujosa, A., Lovell, F. M. & Truter, M. R. (1961). *Computing Methods and the Phase Problem in X-ray Crystal Analysis.* Oxford: Pergamon Press.

Pippy, M. E. & Ahmed, F. R. (1969). Differential Syntheses Program. Unpublished.

Rollett, J. S. (1965). *Computing Methods in Crystallography.* Oxford: Pergamon Press.

Sly, G. & Shoemaker, D. P. (1961). *Computing Methods and the Phase Problem in X-ray Crystal Analysis.* Oxford: Pergamon Press.

Discussion

KARTHA: In your Fourier program, what are the limits on the number of reflexions, and do you have to pack the indices?

AHMED: Allowing 24K words for the program and data arrays, the limits on the reflexions would be 3000 reflexions for orthorhombic, 5000 for monoclinic, and 9000 for triclinic. The *absolute* values of the three indices, the group number, and an order indicator are packed into one word.

HARDGROVE: What is the most efficient method of calculating the derivatives of electron densities in differential syntheses, and would it be a good idea to compute them from the Fourier map results?

AHMED: The derivatives are best evaluated from their mathematical expressions as described by Ahmed & Cruickshank (*Acta Cryst.* (1953) **6**, 765). Using the Fourier results would be less accurate unless the mesh intervals are small.

MATTHEWS: Is there any advantage in calculating $\partial(\varrho_o - \varrho_c)/\partial x$ as a means of eliminating the series termination errors?

AHMED: This procedure saves computer time but is less accurate.

DOMENICANO: Have you considered adding an automatic peak search routine to the Fourier program?

AHMED: It is a useful routine, but low in my priority list.

KARTHA: Is there much difference in computing time of distorted and undistorted maps, and what is the practical interval?

AHMED: The time difference is small, since the first and second stages of summation are identical in both methods. Intervals of 3 points per Å are usually sufficient.

TOPIC L

DATA FILES

CHAIRMAN: J. M. STEWART

»Novel« Computing Techniques

JAMES M. STEWART

Computer Science Center, University of Maryland, College Park, Maryland 20742, U.S.A.

The topic 'novel' computing techniques is perhaps a misnomer. What we are really after here are those programming techniques which lead to codes that make efficient and economical use of computers. This should be no 'novelty'. Two areas I will stress are those dealing with 'IO', that is, input-output, and with the use of compiler languages.

Input-output considerations

In other sessions there have been hints about the necessity for the careful control of the format of data storage for convenient and efficient use of mass storage devices (tape, drum, disc, bulk core). In general, it is impossible even for the largest computers to do all operations 'in core', that is, in the immediate access storage. For all machines now in common use, the internal computing speed is much greater than the data transfer from mass storage devices. The range of 'IO' times to computing times is generally 10^2 to 10^6 for computers. This means that IO deserves very careful attention.

This attention is to two variables. The first is the amount of data stored. The second, the form in which it is stored. In the classification of data, one should think carefully about the ratio of recovery time to recalculation time. Consider the following simple example: should $\sin\theta/\lambda$ and $\sin^2\theta/\lambda^2$ be stored in addition to h, k, and l for each reflection. On the average, moving a word from mass storage device to core requires $\sim$10 microseconds. The squaring of $\sin\theta/\lambda$ requires $\sim$4 microseconds. Therefore, the strategy is to calculate and store $\sin\theta/\lambda$ and to generate $\sin^2\theta/\lambda^2$ at the time it is required. Each crystallographic quantity needs to be evaluated on a similar basis and the appropriate values saved in the data file. Once the 'list' (in the Rollett sense) of quantities to be saved on mass storage has been established in this way, its organization on the mass storage device must be considered. Dr. Rollett has discussed this problem for ALGOL users.

I will now try to describe what we have done for FORTRAN programming. The terms that are used in describing the storage are: buffer, physical record, logical record, pointer, file, and track. A buffer is a set of immediate-access storage locations of a limited size used for the temporary storage of data. A physical record is a collection of data items which will be less than or equal to the size of the buffer. A logical record is one or more physical records which constitute a 'logically' grouped set of data. A pointer is an index which is set to indicate the beginning of a given data group within the buffer. A file is a complete set of physical and logical records for a given problem. A track (or sector) is a physically distinct group of words on a mass storage device. Usually each track is contiguous in such a way that the transfer of the words it contains to immediate access storage is done as a single operation. That is, no switching or additional mechanical movements are required after the act of reading is initiated. It is interesting to note that for many crystallographic calculations, a magnetic tape is faster than a drum or disc since reflections are utilized sequentially. With the tape, the next record is ready to be read, while with the spinning disc or drum, a synchronization problem arises which is data specific. The buffer principle has another important mechanical effect. When used properly, it makes the number of mechanical data transfers less than they might otherwise be. There is an optimum ratio between core used for buffers and frequency of mechanical transfer, and this ratio should be observed in so far as is practical. Our strategy in FORTRAN has been to define a data file for each crystal under study. This file consists of logical records, which is in direct mimicry of a usual FORTRAN operating system. These logical records consist of data grouped by function (*i.e.* like Rollett's lists – cell dimensions, scattering factors, reflection information, etc.). Finally the logical records are written into and out of a buffer region in the immediate access store as physical records of controlled length.

Each physical record contains four pieces of information: the count of all the words in the record, the count of the physical record, the logical record identifier, and a packing indicator telling how many words/item (item might be a reflection or an atom).

We have then prepared a set of FORTRAN subroutines to deal with files formatted in this structure.

Compiler considerations

The second 'novel' aspect of our work has been to try to generate codes which are written in FORTRAN but retained some of the speed and generality of machine language codes. We also hope to be able to work with any machine which has FORTRAN and, therefore, try to use a simplified version of FORTRAN which may be easily used on various machines We call our dialect 'PIDGIN' FORTRAN. We carefully surveyed the various FORTRAN compilers and then defined the following as the vocabulary of 'PIDGIN' FORTRAN:

GO TO	unconditional, computed, and assigned
ASSIGN	
IF	arithmetic only
DO	
CONTINUE	
CALL	
RETURN	
STOP	
END	
FORMAT	I, X, E, F, A, and H only
DIMENSION	
COMMON	block only
EQUIVALENCE	
DATA	
FUNCTION	
SUBROUTINE	

The use of DOUBLE PRECISION is restricted to machines or calculations which demand it. All input-output is handled through READ (N, M) LIST and
WRITE (N, M) LIST
Within the body of programs only three different logical units are referred to. These correspond to (A) the input buffer containing the image of the last read card from the input file, (B) the output line-printer file, and (C) the punch file. All other input operations are placed in special subroutines as:

WRITE (N)
READ (N)

```
REWIND N
ENDFILE N
```

The statements: BACKSPACE, ENTRY, IF (logical), BLOCK DATA, INTEGER, REAL, COMPLEX, LOGICAL, EXTERNAL, PARAMETER, etc. are avoided altogether for various reasons as a result of the study.

The chosen vocabulary was derived not only to be universally acceptable but moreover to be transparent to the compiler. For programs to be used for scientific calculations which are to be executed many times, it is desirable to attempt to produce FORTRAN code which matches the efficiency of machine code. If one uses very general or elegant FORTRAN statements, the compiler writer is constrained to produce very general machine code. Therefore, 'PIDGIN' FORTRAN requires that the statements be such that the code compiled will be similar to that which one would write in machine language. The gain in using 'PIDGIN' FORTRAN is then the shortened checkout time, the efficiency of coding, and the many compiler diagnostics and cross checks. The liability is the failure to take advantage of the full power of the FORTRAN language.

We have attempted throughout to utilize the FORTRAN in such a way that efficiency and ease of adaptation will not be compromised. An example will serve best to illustrate this application. We restrict all major arrays to 'single' dimensioned variables. This results in FORTRAN coding that is far less elegant than might otherwise be the case. It also results in programs which execute faster and which may be easily adjusted for differing availability of immediate access storage. Typically a three dimensional array in XRAY67 is set up in the following way:

```
    DIMENSION ATOM(300),
      SYM(756)
    DATA IATMX, INDXC,
      ISYM/3000,3,12/
    NATOM=1-INDXC
    NSYM  =1-ISYM
```

The atomic coordinates are then loaded as, say, x, y, z and at each encountered atom, the following steps are used:

```
    IF(NATOM-IATMX) 11,...,...
 11 NATOM=NATOM+INDXC
    ATOM(NATOM)=X
    ATOM(NATOM+1)=Y
    ATOM(NATOM+2)=Z
```

The rotation matrices are loaded in a similar fashion. Later on, when it is desired to generate all the symmetrically related atoms, then the statements below may be used:

```
    IMAX=NATOM
    DO 201 J=1, NSYM, ISYM
    DO 101 N=1, IMAX, INDXC
    NATOM=NATOM+INDXC
    ATOM(NATOM)=ATOM(N)
      *SYM(J)+ATOM(N+1)
      *SYM(J+1)+ATOM(N+Z)
      *SYM(J+2)+SYM(J+3)
    ATOM(NATOM+1)=ATOM(N)
      *SYM(J+4)+ATOM(N+1)
      *SYM(J+5)+ATOM(N+2)
      *SYM(J+6)+SYM(J+7)
    ATOM(NATOM+2)=ATOM(N)
      *SYM(J+8)+ATOM(N+1)
      *SYM(J+9)+ATOM(N+2)
      *SYM(J+10)+SYM(J+11)
101 CONTINUE
201 CONTINUE
```

The above set of statements should be compared to the following:

```
      DIMENSION ATOM(3,1000),
        SYM(3,4,48)
      DATA IATMX/1000/
      NATOM=0
      NSYM=0
      IF(NATOM-IATMX) 11,...,...
   11 NATOM=NATOM+1
      ATOM (1, NATOM)=X
      ATOM (2, NATOM)=Y
      ATOM (3, NATOM)=Z
      IMAX=NATOM
      DO 401 J=1, NSYM
      DO 301 J1=1, IMAX
      NATOM=NATOM+1
      DO 201 J2=1,3
      ATOM(J2, NATOM)=SYM(J2,4,J)
      DO 101 J3=1,3
      ATOM(J2, NATOM)=
        SYM(J2,J3,J)*ATOM(J3,J1)+
        ATOM(J2, NATOM)
  101 CONTINUE
  201 CONTINUE
  301 CONTINUE
  401 CONTINUE
```

If one now examines the machine language code produced in these two cases, the following statistics can be given:

	Single Dimensioned Variables	Multiple Dimensioned Variables
No. of instructions used	127	120
Execution time in seconds	4·777	5·210

Another example of careless programming is the casual use of calling sequences. In this case, the work of establishing addresses is done at execution time. It is, in general, much safer to use COMMON storage for communication among subroutines. In this case, addresses are established at compilation and loading time.

The burden of these exercises is to show that for large-scale calculations one must approach the use of the computer in 'novel' ways. The full power of the compiler may not be utilized, but the power of the machine to carry out crystallographic calculations will be enhanced.

In actual application, these factors in XRAY67 have, we feel, 'paid off'. Typical charge times for the 1108 under EXEC VIII are shown for a complete structure solution. Allowing for no false runs, a structure analysis on a 64 atom structure in $P2_1/n$ with 5700 reflections could be carried out in the following charge time (EXEC VIII is a multiprogramming system so that IO times are 'concealed'; it has a memory access time of 375 nanoseconds and a floating-point add time of 1·75 microseconds):

Summary of the discussion

This session was characterized by a long, sometimes emotional, discussion of the problems and successes of the application of 'third' generation computers. It was clear that crystallographers from around the world have had to struggle very hard to keep up with the rapidly-evolving software and hardware now becoming available. Communications among hard-pressed systems programmers, computer-operating staffs and working crystallographers have at times been low. On the other hand, as the crystallographers have studied and learned the power and generality available to them under the new systems, and as these new systems have reached a good (even if not perfect) state of check-out, there has been a tremendous increase in computing power. The consensus was that a remarkable increase in computing power has recently taken place and is currently beginning to be exploited by practising crystallographers.

	1108 EXEC VIII Time in min.
generation of diffractometer settings cards	1·062
preliminary data treatment of diffractometer output	5·7976
data reduction, establish symmetry codes	2·2114
estimation of scale and temperature factors	4·3155
sigma-two	6·2634
phase solution	2·5774
E-map Fourier	2·1778
load atomic parameters	0·6582
structure factor	1·8270
F_o Fourier	2·0825
bond lengths and angles	0·9906
load atomic parameters	0·6444
full matrix least-squares isotropic refinement (5 cycles)	61·2925
difference Fourier	1·6044
bond lengths and angles	0·9610
load atomic parameters	2·5749
block diagonal least-squares, mixed temperature factors (H isotropic – others anisotropic) (4 cycles)	37·2584
difference Fourier	1·8410
bond lengths and angles	0·9614
least-squares plane and line	0·0286
TOTAL	137·1300

A Method for Handling Crystallographic Data Files on a Mass Storage

A. DOMENICANO

Laboratorio di Strutturistica Chimica del CNR, Città Universitaria, 00100 *Roma, Italy.*

A system of crystallographic programs for the electronic computer UNIVAC 1108 is being prepared at our laboratory, and is now at the check-out stage (Domenicano, Spagna & Vaciago, 1969). It consists of programs for routine crystal-structure analysis calculations, including a Σ_2 program. The minimum machine requirements are 65K core storage, 3FH432, and 1 Fastrand (this is a medium-speed mass storage, partially available to users on a rental basis). The system is now operating in Rome under the control of a local, modified version of EXEC II.

All the programs within the system have been written in Fortran V, particular attention being paid to the hardware and software peculiarities of the computer. The system is, therefore, particularly efficient; on the other hand, it cannot be easily adapted to, say, IBM or CDC machines.

A fuller description of the system will be given at the 3rd Congress of the Italian Crystallographic Association, to be held in Torino, 24–26 September 1969. I wish here to give only a brief account of the methods we have used for handling the I/O files and of our experience in this field.

The I/O files may be located either on magnetic tape or on Fastrand for each program in the system. We have found that the use of Fastrand as a data storage is very convenient for crystallographic work and is preferable to the use of magnetic tapes.

By using fixed-point variables and saving only the figures which are effectively significant, a reflection $(h,k,l,F_o,\sigma(F_o),A_c,B_c)$ can be stored in two or three 36-bit words. If the data files are written in such a compact way, a Fastrand area of 100,000 words can accommodate a number of files, and can be used, therefore, for storing original data as well as intermediate results for several structures.

We have associated a six-character alphameric code with each file in a Fastrand area. The code table is reported in a second Fastrand area, whose length may be kept to a minimum (a few hundred words are usually enough). The content of this table is read by all our programs before searching for the input file, and is in turn modified and rewritten when a new file is output. This makes it possible to use Fastrand as a data storage without knowing the exact positions of the requested I/O files, which are identified only by the associated alphameric code. This clearly enables several independent crystallographers to store their own files in the same Fastrand area.

The system includes a house-keeping program, by means of which the data files can be easily deleted or moved from Fastrand to tape and vice versa, thus closing any gap which may remain after deletion. Owing to random access, the execution time of each program in the system does not depend on the positions of the I/O files in the Fastrand area.

The task of programming all the I/O routines has been greatly reduced by the existence of the NTRAN subroutine, which is part of the EXEC II operating system.

Our experience, though limited, suggests that this method of handling the I/O files is particularly suited for crystallographic work. Since many files are often created or destroyed in a short period of time, it would be impractical to identify a particular file through its sequential position in a Fastrand area shared by several users. We think that corresponding methods could be devised for any other machine disk or drum mass storage.

Reference

Domenicano, A. Spagna, R. & Vaciago, A. (1969). *Rend. Accad. Lincei.* VIII, **47,** 331.

TOPIC M

ALGOL

CHAIRMAN: G. C. BASSI

The Use of ALGOL in Programming and Publication

GÉRARD C. BASSI

Laboratoire de Diffraction Neutronique, Centre d'Etudes Nucléaires de Grenoble et Laboratoire d'Electrostatique et de physique du Métal, C.N.R.S., CEDEX 85 *et* 166, *Grenoble, France.*

Since the publication of the preliminary report on the language in 1958, the use of Algol has become common in the field of computational mathematics. When the Algol committee started its definition work, the hope was to design a universal and powerful language, built in a logical way, so that no misinterpretation, either in the description of numerical processes or in the translation into the corresponding program, could remain possible. The result of this work is the 'Report on the Algorithmic Language Algol 60' (Naur, 1960), representing what was said to be 'the union of the committee's concepts and the intersection of its agreements'.

Logical structure of ALGOL 60

The language can be presented as a set of grammar rules applied to a certain number of elements, the statements. Three different levels of language must be distinguished, the Reference Language, the Publication Language, and the Hardware Representations. The last level is directly connected to the type of available computer. An important fact is the distinction between the other two levels: the Reference Language is used as the 'defining language' and is 'the basic reference and guide for compiler builders'; the Publication Language is supposed to be as close as possible to the Reference Language and a 'univocal correspondence with reference representation must be secured'. This means that the content and logical structure of any algorithmic description are independent of their representation. This twinned nature of the language leads to a very important aspect of its utility: it can be used to describe a numerical process in such a way that the process remains easily readable and understandable, and the same text may be used directly to translate this algorithm into machine language (compiling) for execution.

The Algol language is composed of statements, normally preceded by declarations. Both are composed of symbols. These symbols (of which there are 116), and the ways of combining them, are given in the basic report, in the normal form of BACKUS. It is noteworthy that this work of definition has been responsible for the progress in the formal language theory.

For instance, the definition of the logical values is given in the following way, called a metalinguistic formula:

⟨logical value⟩ : : = TRUE | FALSE

The concept written inside the bracket ⟨ ⟩ is called a metalinguistic variable. The mark : : = indicates the definition and | means simply 'or'. Both are metalinguistic connectives. The above example means that a logical value can be either TRUE or FALSE. The 'words' TRUE and FALSE are themselves described as being a part of the 116 basic symbols, and must be taken as a whole. They are different from the result obtained with a string formed with the t, r, u, e characters.

In the above example, the right side of definition includes only basic symbols, but this does not exclude the possibility of finding another metalinguistic variable in brackets:

⟨identifier⟩: : = ⟨letter⟩ |
⟨identifier⟩ ⟨letter⟩ |
⟨identifier⟩ ⟨digit⟩

Here, the succession of two brackets ⟨A⟩ ⟨B⟩ means that a possible value of the metalinguistic variable B is following a possible value of the metalinguistic variable A. Such a definition has meaning if, and only if, the metalinguistic variables appearing in the right-hand part of the definition have been already defined. In the above example, the definition is obvious.

⟨digit⟩: : = 0|1|2|3|4|5|6|7|8|9

Similarly a letter is defined as any possible letter from *a* to *z* and from *A* to *Z*. The definition of an identifier allows us to say that SINUS, OTTAWA, IUCR, B345, A1B3 are identifiers and that 3B45 and I.U.Cr are not. As a consequence, one is free to build any identifier which can be used anywhere else for identification of variables and others, provided that the metalinguistic variable ⟨identifier⟩ appears in the right-hand side of the definition of another metalinguistic variable:

⟨variable identifier⟩: : = ⟨identifier⟩
⟨procedure identifier⟩: : = ⟨identifier⟩

The whole syntax defining the language is given fully in the basic report; therefore it is unnecessary to describe it further here. This example was given to emphasize one extremely positive aspect of Algol 60, namely, the precision of its foundations.

Use of the language

This precision has had an important consequence: it was inevitable that such a complicated description would still contain some contradictions or perhaps ambiguities. In 1962, the international Algol Committee met again to solve these problems, and the result was the publication of the Revised Algol 60 Report (Naur, 1963), which contained a slightly changed description of some particular points. Despite improvements and corrections, some ambiguities still remain. They are now published (Knuth, 1967) and, as the author pointed out, the 'list is actually more remarkable for its shortness than its length. A complete critique which goes to the same level of detail would be almost out of the question for other languages comparable to Algol 60, since the list would probably be a full order of magnitude longer'.

The use of Algol is not widespread in all countries. Although supported from its early stages by the ACM Committee on Programming Language, Algol appeared, at least initially, to be used mainly in Europe. At that time there were few computers available in these countries, and the work done in symbolic languages (such as FORTRAN) was only a small fraction of what had been done in the U.S.A., where few people felt the necessity for changing their own programming language. Also, Algol is not easy to describe in a readily understandable way, and early descriptions often did not resemble the original report. A large number of books and publications are currently available which explain clearly the internal and logical structure responsible for Algol's efficiency (e.g. Dijkstra, 1962; Lecht, 1967).

In the particular case of teaching in the universities, this language is now generally taught and is often considered as the language of the scientific culture.

Up to now, a large number of algorithms have been published, using Algol as publication language. Each issue of the Communications of the ACM contains a certain number of algorithms, and more than 95% of them are written in Algol. Such Algol 'programs' are published in *Numerische Mathematik* (Germany), *Computer Journal* (U.K.), *Algorytmy* (U.S.S.R.) etc.

Books containing numerical methods in terms of Algol 60 have been published also (e.g. Hawgood, 1965). All these algorithms are normally well tested and are given in the form of procedures. The second edition of the *World List of Crystallographic Computer Programs* was issued in 1966. At that time Algol compilers were not normally available everywhere. In spite of this, we find no less than 70 Algol programs, as compared with 86 programs written in the different available Fortrans (Shoemaker, 1966).

ALGOL 68

The IFIP working group is now devising a new language called Algol 68. This new language can be considered as the logical consequence of the work done, in the field of formal languages theory, to create Algol 60. However, this latter is not a part of Algol 68 where some logical improvements have been introduced, such as general structure handling, parallel computation, notion of reference thus introducing the notation of indirect addressing to allow any formalism (lists and so on). The syntax is described in two steps: the metalanguage itself, and then the rules, are both defined in the BACKUS form. Compilers will certainly not be rapidly available but one is already announced in Munich.

To conclude, we can say that if Algol programs cannot be as efficient in time as programs written in a language close to the machine structure (Fortran, PL 1 . . .), the logical structure of the Algol 60 language makes it a really valuable tool for algorithms description, publication, communication, and teaching.

References

Dijkstra, E. W. (1962). *A Primer of ALGOL 60 Programming*. London: Academic Press.

Hawgood, J. (1965). *Numerical Methods in ALGOL*. New York: McGraw-Hill.

Knuth, D. E. (1967). *Comm. Ass. Comp. Mach.* **10**, 611.

Lecht, C. P. (1967). *The Programmer's ALGOL: A Complete Reference*. New York: McGraw-Hill.

Naur, P. (ed.) (1960*a*). Report on the Algorithmic Language ALGOL 60. *Numerische Mathematik*, **2**, 106.

Naur, P. (ed.) (1960*b*). Report on the Algorithmic Language ALGOL 60. *Comm. Ass. Comp. Mach.* **3**, 299.

Naur. P. (ed.) (1963). Revised Report on the Algorithmic Language ALGOL 60. *Comm. Ass. Comp. Mach.* **6**, 1.

Preliminary Report on an International Algebraic Language (1958). *Comm. Ass. Comp. Mach.* **1** (12), 8.

Shoemaker, D. P. (ed.) (1966). *World List of Crystallographic Computer Programs*. 2nd ed.

Discussion

STEWART: Sometimes it would be useful to be able to make an ALGOL program from a FORTRAN program. Would you comment on this?

BASSI: A program in FORTRAN can always be written more easily in ALGOL, but the opposite is not true since ALGOL has many more possibilities than FORTRAN. However, PL1 can do as much as ALGOL although it is not defined in the same logical way.

TOPIC N

STANDARD TESTS

CHAIRMAN: J. M. STEWART

Standard Tests for Crystallographic Programs

JAMES M. STEWART

Computer Science Center, University of Maryland, College Park, Maryland, 20742 *U.S.A.*

We must be assured that the various computer codes which have been written or are being written for crystallographic calculations are correct. To this end, Professor Cruickshank suggested some time ago that a set of test decks be generated which would serve as test cases and which could then be used by crystallographers to verify their programs. These tests have been set up to cover space groups $P1$, $P\bar{1}$, $P2_1/c$, $Pmn2_1$, and $I\bar{4}3d$ as general examples. In each case, sin θ/λ, scattering factors, structure factors, least squares, bond lengths and angles, and a block of a Fourier were calculated. These calculations have been carried out at a number of different locations using different computers and programs. The results have then been compared and each laboratory informed of the variation in results. The features tested so far are, of course, not exhaustive, but do represent a beginning.

Each of the test cases which now exist were made by taking an arbitrary cell and placing a small 'molecule' in it, either in general or special position, depending upon the case. Structure factors were then calculated and a set of 'observed' data were generated from the calculated structure factors. This was done by generating a quasi-Gaussian set of random errors and applying them to the calculated structure factors. The original parameters of the structure were then displaced and defined to be the trial structure. This set of $|F|$ values and the trial structure constitute the test cases. This material has been distributed to the University of Maryland, the National Research Council of Canada, Dr. W. Hamilton, Dr. A. Larson, Dr. G. Bassi, and to Professor D. W. J. Cruickshank for calculation.

It is the intention of the Commission on Crystallographic Computing to publish this first set of test results as soon as they are proven reliable. A second set of tests on the less common types of calculations will then be undertaken. One of the expected difficulties with the second set will be the verification of results, since the necessary programs may exist only in very few laboratories.

List of Lecturers

	Topics	Pages
Ahmed, F. R.	A 6.1	55
	G 3	279
	H 3	309
	K	355
Alcock, N. W.	G 2	271
Bassi, G.	M	373
Bednowitz, A. L.	A 6.2	58
	I 3	336
Busing, W. R.	I 1	319
Coppens, P.	G 1	255
Cruickshank, D. W. J.	E 1	187
Dewar, R. B. K.	A 6.3	63
Doedens, R. J.	E 2	198
Domenicano, A.	L 2	368
Gabe, E. J.	I 2	331
Gassmann, J.	A 3	26
Hall, S. R.	A 6.4	66
Hauptman, H.	A 5	41
Hornstra, J.	B 5	103
Hoppe, W.	A 3	26
Huber,R.	B 4	96
Jacobson, R. A.	B 1	81
Johnson, C. K.	F 1	207
	F 2	220
	F 3	227
Karle, I. L.	A 2	19
Karle, J.	A 1	13
	A 4	37
	A 7	75
Kartha, G.	C 2	132
	J	345
Larson, A. C.	G 5	291
Maslen, E. N.	F 4	231
Matthews, B. W.	C 3	146
Nordman, C. E.	B 6	110
Okaya, Y.	C 1	127
Pawley, G. S.	F 5	243
Rollett, J. S.	D 1	167
	H 2	302
Santoro, A.	G 4	283
Schilling, J. W.	B 6	110
	B 7	115
	E 3	201
Seeman, N. C.	B 2	87
Shiono, R.	H 4	312
Sparks, R. A.	D 2	182
Stewart, J. M.	A 6.5	71
	H 1	297
	L 1	363
	N	379
Tollin, P.	B 3	90
de Vries, A.	C 4	160
Zechmeister, K	A 3	26

List of Participants

Ahmed, F. R., Canada
Alcock, N. W., U.K.
Anderson, D. N., U.S.A.
Baglio, J. A., U.S.A.
Baird, H. W., U.S.A.
Bart, J. C. J., Italy
Bassi, G., France
Bayliss, P., Canada
Beauchamp, A., Canada
Becker, P., France
Bednowitz, A. L., U.S.A.
Berman, H., U.S.A.
Beurskens, P. T., Netherlands
Birnbaum, G. I., Canada
Birnbaum, K. B., Canada
Blount, J. F., U.S.A.
Bonnet, J. J., France
Borene, J., France
Boucher, M. L., U.S.A.
Brenner, S. A., U.S.A.
Brisse, F. R., Canada
Bruce, J., U.S.A.
Burnett, R. M., U.S.A.
Bushnell, G. W., Canada
Busing, W. R., U.S.A.
Calvert, L. D., Canada
Chandrasekaran, R., India
Chao, G. Y., Canada
Chapuis, G., Switzerland
Christoph, G. G., U.S.A.
Cody, V., U.S.A.
Coppens, P., U.S.A.
Coyle, R. A., Australia
Cruickshank, D. W. J., U.K.
Cruse, W. B. T., Canada
Debaerdemaeker, T., Belgium
Dewar, R. B. K., U.S.A.
Dexter, D. D., U.S.A.
Dickens, B., U.S.A.
Doedens, R. J., U.S.A.
Domenicano, A., Italy
Dyke, M., U.S.A.
Eiss, R., U.S.A.
Fang, J. H., U.S.A.
Fischer, J., France
Fisher, J., U.S.A.

Fredericks, R. J., U.S.A.
Gabe, E. J., Canada
Gafner, G., South Africa
Gainsford, G., U.S.A.
Garvey, R. G., U.S.A.
Gassmann, J., Germany
George, J. W., U.S.A.
Germain, G., Belgium
Giese, R. F., U.S.A.
de Gil, E. R., Venezuela
Glusker, J., U.S.A.
Golightly, J. P., Canada
Gonschorek, W., Germany
Greene, P. T., U.S.A.
Grundy, H. D., Canada
Guillermo, T. R., U.S.A.
Hall, S. R., Canada
Hanson, J. C., U.S.A.
Hardgrove, G., U.S.A.
Harkema, S., Netherlands
Hauptman, H., U.S.A.
Hayden, T. D., U.S.A.
Hazell, A. C., Denmark
Hazell, R. G., Denmark
Heck, H. G., Netherlands
Hjortaas, J. A., Norway
Hoard, L. G., U.S.A.
Hoge, R., Canada
Hoppe, W., Germany
Hornstra, J., Netherlands
Hounslow, A. W., Canada
Huber, C. P., Canada
Huber, R., Germany
Huml, K., Czechoslovakia
Hunter, F. D., U.S.A.
Jacobs, G. D., U.S.A.
Jacobson, R. A., U.S.A.
Jansonius, J. N., Netherlands
Jeannin, Y., France
Johnson, C. K., U.S.A.
Kadlec, R. J., U.S.A.
Karle, I. L., U.S.A.
Karle, J., U.S.A.
Kartha, G., U.S.A.
Kerr, K. A., Canada
Koch, M., Belgium
Krukonis, A. P., U.S.A.
Larson, A. C., U.S.A.
Lawrence, J. L., U.K.
Lenstra, A. T. H., Netherlands
Le Page, Y., Canada
Lesk, A. M., U.S.A.
Leung, K. Y., Canada
Liminga, R., Sweden
Loehlin, J. H., U.S.A.
Lundgren, J. O., Sweden
Macaulay, E. W., Canada
Machacek, H., Canada
Mackay, I. R., U.K.
Mailer, C., Canada
Maslen, E. N., Australia
Matthews, B. W., U.S.A.
McCauley, J. W., U.S.A.
Michael, L. W., U.S.A.
Migchelsen, A., Netherlands
Milliken, K. S., Canada
Mitchell, C. M., Canada
Mootz, D., Germany
Motherwell, S., U.K.
Nichols, T. D., U.S.A.
Nicol, A. W., U.K.
Nordman, C. E., U.S.A.
Novak, C., Czechoslovakia
Okaya, Y., U.S.A.
Olthof-Hazekamp, R., Netherlands
Paradies, H. H., Germany
Pawley, G. S., U.K.
Penfold, B. R., New Zealand
Peterse, W. J. A. M., Netherlands
Pierrot, M., France
Poland, V. L., U.S.A.
Pollard, D. R., Canada
de Rango, C., France
Ricard, L., Canada
Richard, P., Canada
Rietveld, H. M., Netherlands
Robinson, W. T., New Zealand
Rollett, J. S., U.K.
Rosenstein, R., U.S.A.
Santoro, A., U.S.A.
Scheidt, W. R., U.S.A.
Schenk, H., Netherlands
Schilling, J. W., U.S.A.
Scott, J. D., Canada
Seabaugh, P. W., U.S.A.
Seeman, N. C., U.S.A.
Shelton, C., U.S.A.
Shiono, R., U.S.A.
Silverman, J., U.S.A.
Smith, A. E., U.S.A.
van Soest, T. C., Netherlands
Soltzberg, L. J., U.S.A.
Sørensen, A. M., Denmark
Sparks, R. A., U.S.A.
Stephens, J. S., Canada
Stewart, J. M., U.S.A.
Sullenger, D. B., U.S.A.
Thozet, A., France
Thudium, R. N., U.S.A.
Tollin, P., U.K.
Tran Qui Duc, France
Tsoucaris, G., France
Verbist, J. J., Belgium
Vickery, B. L., U.K.
de Vries, A., U.S.A.
Watson, K. J., Denmark

Weber, H. P., Switzerland
Weeks, C. M., U.S.A.
Whitlow, S. R., Canada
Williams, G. J. B., Canada
Winter, W. T., U.S.A.
Yannoni, N. F., U.S.A.
Yerkess, J., U.K.
Zelwer, C., France